Lecture Notes in Computer Science 2904

Edited by G. Goos, J. Hartmanis, and J. van Leeuwen

Springer
Berlin
Heidelberg
New York
Hong Kong
London
Milan
Paris
Tokyo

Thomas Johansson   Subhamoy Maitra (Eds.)

Progress in Cryptology – INDOCRYPT 2003

4th International Conference on Cryptology in India
New Delhi, India, December 8-10, 2003
Proceedings

 Springer

Series Editors

Gerhard Goos, Karlsruhe University, Germany
Juris Hartmanis, Cornell University, NY, USA
Jan van Leeuwen, Utrecht University, The Netherlands

Volume Editors

Thomas Johansson
Lund University
Dept. of Information Technology
P.O. Box 118, Ole Romers vag 3
221 00 Lund, Sweden
E-mail: thomas@it.lth.se

Subhamoy Maitra
Indian Statistical Institute
Applied Statistics Unit
203, B T Road
Kolkata 700 108, INDIA
E-mail: subho@isical.ac.in

Cataloging-in-Publication Data applied for

A catalog record for this book is available from the Library of Congress.

Bibliographic information published by Die Deutsche Bibliothek
Die Deutsche Bibliothek lists this publication in the Deutsche Nationalbibliografie;
detailed bibliographic data is available in the Internet at <http://dnb.ddb.de>.

CR Subject Classification (1998): E.3, G.2.1, D.4.6, K.6.5, F.2.1-2, C.2

ISSN 0302-9743
ISBN 3-540-20609-4 Springer-Verlag Berlin Heidelberg New York

Springer-Verlag is a part of Springer Science+Business Media

springeronline.com

© Springer-Verlag Berlin Heidelberg 2003
Printed in Germany

Typesetting: Camera-ready by author, data conversion by Boller Mediendesign
Printed on acid-free paper   SPIN: 10973035   06/3142   5 4 3 2 1 0

Preface

The INDOCRYPT conference series started in 2000, and INDOCRYPT 2003 was the fourth one in this series. This series has been accepted by the international research community as a forum for presenting high-quality crypto research, as is evident from the 101 submissions this year, spread over 21 countries and all five continents. The accepted papers were written by authors from 16 countries, covering four continents.

A total of 101 papers were submitted for consideration to the program committee, and after a careful reviewing process 30 were accepted for presentation. One of the conditionally accepted papers was withdrawn by the authors as they found an error in the paper that could not be repaired in the short time between the notification of the review and the final version submission. Thus the final list contains 29 accepted papers. We would like to thank the authors of all submitted papers, including both those that were accepted and those which, unfortunately, could not be accommodated.

The reviewing process for INDOCRYPT was very stringent and the schedule was extremely tight. The program committee members did an excellent job in reviewing and selecting the papers for presentation. During the review process, the program committee members communicated using a review software package developed by Bart Preneel, Wim Moreau and Joris Claessens. We acknowledge them for providing this software. These proceedings include the revised versions of the 29 selected papers. Revisions were not checked by the program committee and the authors bear the full responsibility for the contents of the respective papers. Our thanks go to all the program committee members and the external reviewers (a list of them is included in the proceedings) who put in their valuable time and effort in providing important feedback to the authors.

This year the invited talks were presented by Prof. Harald Niederreiter and Prof. Jennifer Seberry. They do not need any introduction. Prof. Niederreiter presented a talk on "Linear Complexity and Related Complexity Measures for Sequences" and the talk of Prof. Seberry was on "Forensic Computing." Both talks have been included in these proceedings.

The organization of the conference involved many individuals. We would like to thank the general co-chairs Prof. Rajeeva L. Karandikar and Dr. P.K. Saxena for taking care of the actual hosting of the conference. They were ably assisted by the organizing committee, whose names are included in the proceedings. Additionally, we would like to thank Tanmoy Kanti Das for handling all the submissions, and Avishek Adhikari and Madhusudan Karan for putting together this proceedings in its final form. Finally we would like to acknowledge Springer-Verlag for active cooperation and timely production of the proceedings.

December 2003   Thomas Johansson
                                                                Subhamoy Maitra

Organization

Indocrypt 2003 was organized by the Indian Statistical Institute, Delhi and the Scientific Analysis Group, Delhi.

General Co-chairs

Rajeeva L. Karandikar   Indian Statistical Institute, Delhi, India
Dr. P.K. Saxena   Scientific Analysis Group, New Delhi, India

Program Co-chairs

Thomas Johansson   Department of Information Technology, Lund University, Sweden
Subhamoy Maitra   Indian Statistical Institute, Kolkata, India

Program Committee

R. Balasubramanian   Institute of Mathematical Sciences, India
Rana Barua   Indian Statistical Institute, Kolkata, India
Simon R. Blackburn   University of London, UK
Anne Canteaut   INRIA-Rocquencourt, France
John Clark   University of York, UK
Cunsheng Ding   Hong Kong University of Science and Technology, China
Yvo Desmedt   Florida State University, USA
Tor Helleseth   University of Bergen, Norway
Charanjit Singh Jutla   IBM Watson, USA
Thomas Johansson   Lund University, Sweden
Kaoru Kurosawa   Ibaraki University, Japan
Andrew Klapper   University of Kentucky, USA
Arjen K. Lenstra   Citibank, USA
Helger Lipmaa   Helsinki University of Technology, Finland
C.E. Veni Madhavan   Indian Institute of Science, Bangalore, India
Subhamoy Maitra   Indian Statistical Institute, Kolkata, India
Alfred John Menezes   University of Waterloo, Canada
Phong Nguyen   École Normale Supérieure, France
C. Pandu Rangan   Indian Institute of Technology, Chennai, India
Bart Preneel   Katholieke Universiteit Leuven, Belgium
Bimal Roy   Indian Statistical Institute, Kolkata, India
Vincent Rijmen   Cryptomathic, Belgium
P.K. Saxena   Scientific Analysis Group, Delhi, India
Jennifer Seberry   University of Wollongong, Australia

Pantelimon Stanica   Auburn University, Montgomery, USA
Kapaleeswaran Viswanathan   Queensland University of Technology, Australia
Moti Yung   Columbia University, USA

Organizing Committee

Dr. K.K. Bajaj   Office of the Controller of CA, MIT, Delhi
Prof. B.K. Dass   University of Delhi
Dr. B.K. Gairola   NIC, Delhi
Dr. Arup Pal   ISI, Delhi
Dr. Anish Sarkar   ISI, Delhi
Dr. Ashutosh Saxena   IDRBT, Hyderabad
Dr. Meena Kumari   SAG, Delhi
Mr. Jagdish Prasad   SAG, Delhi
Dr. Shri Kant   SAG, Delhi
Mr. G.S. Gupta   SAG, Delhi
Mr. Parimal Kumar   SAG, Delhi
Mr. Rajeev Thaman   SAG, Delhi
Mr. N. Rajesh Pillai   SAG, Delhi
Ms. Sarvjeet Kaur   SAG, Delhi
Ms. Roopika Chaudhary   SAG, Delhi
Ms. Noopur Shrotriya   SAG, Delhi

External Referees

Daniel Augot
S.S. Bedi
Siddika Berna Örs
Sanjay Burman
Sucheta Chakrabarti
Suresh Chari
Thomas W. Cusick
Amites Dasgupta
Alex Dent
Vu Dong To
Ratna Dutta
Caroline Fontaine
Steven Galbraith
Gagan Garg
Indivar Gupta
Helena Handschuh
Darrel Hankerson
Florian Hess
Kouichi Itoh
Tetsu Iwata

Goce Jakimoski
Pascal Junod
Tanmoy Kanti Das
Meena Kumari
Pradeep Kumar Mishra
Françoise Levy-dit-Vehel
Keith Martin
Shin'ichiro Matsuo
Yi Mu
Partha Mukhopadhyay
Hirofumi Muratani
Sean Murphy
Jorge Nakahara, Jr.
Mridul Nandi
Laxmi Narain
Wakaha Ogata
Yasuhiro Ohtaki
Markku-Juhani O. Saari-
nen
Elisabeth Oswald

Saibal K. Pal
Matthew Parker
Kenny Paterson
N. Rajesh Pillai
David Pointcheval
Michaël Quisquater
Malapati Raja Sekhar
Matt Robshaw
Pankaj Rohatgi
Atri Rudra
Kouichi Sakurai
Palash Sarkar
Hervé Sibert
Nigel Smart
Martijn Stam
Michael Steiner
H.V. Kumar Swamy
Jacques Traoré
Frederik Vercauteren
Johan Wallén

Yejing Wang   Tianbing Xia   Xianmo Zhang
Peter Wild   Pratibha Yadav   Weiliang Zhao

Sponsoring Institutions

DRDO, Government of India
Shoghi Communications Ltd., New Delhi, India
Technocab India Ltd., Bangalore, India

Table of Contents

Secrect Sharing

Bilinear Pairing

Public Key

Signature Scheme

Protocol

Elliptic Curve & Algebraic Geometry

Implementation & Digital Watermarking

Authentication

Author Index

Linear Complexity and Related Complexity Measures for Sequences

Harald Niederreiter

Department of Mathematics, National University of Singapore,
2 Science Drive 2, Singapore 117543, Republic of Singapore
nied@math.nus.edu.sg

Abstract. We present a survey of recent work on the linear complexity, the linear complexity profile, and the k-error linear complexity of sequences and on the joint linear complexity of multisequences. We also establish a new enumeration theorem on multisequences and state several open problems. The material is of relevance for the assessment of keystreams in stream ciphers.

1 Introduction

We recall that a *stream cipher* is a symmetric cryptosystem in which plaintexts and ciphertexts are strings (or, in other words, finite sequences) of elements of a finite field $\mathbb{F}_q$ (with q elements), and encryption and decryption proceed by termwise addition, respectively subtraction, of the same string of elements of $\mathbb{F}_q$. The latter string serves as the (encryption and decryption) key; it is called the *keystream* in the context of stream ciphers. Ideally, the keystream should be a "truly random" string of elements of $\mathbb{F}_q$. In practice, keystreams are obtained from certain secret seed data by a (perhaps even publicly available) deterministic algorithm. We refer to Rueppel [52] for a survey of algorithms for keystream generation and also for general background on stream ciphers.

A central issue in the security analysis of stream ciphers is the quality assessment of keystreams. In other words, we need to know how close the keystream is to true randomness. Keystreams guaranteeing an adequate security level must meet various requirements such as possessing good statistical randomness properties and a high complexity (in a suitable sense), so that the keystream cannot be inferred from a small portion of its terms.

Most practical keystream generators are implemented in hardware and use linear feedback shift registers (LFSRs) as basic components. Algorithmically, an LFSR is described by a linear recurrence relation (see Definition 1). In the system-theoretic approach to the assessment of keystreams one seeks to find out to what extent the keystream can be simulated by LFSRs. Clearly, if a keystream can be simulated by a short LFSR, then the keystream has to be discarded as insecure. This approach leads naturally to the notion of linear complexity of a string or of an ultimately periodic sequence of elements of $\mathbb{F}_q$ (see Definition 2).

T. Johansson and S. Maitra (Eds.): INDOCRYPT 2003, LNCS 2904, pp. 1–17, 2003.

In this paper we survey and discuss the linear complexity and related complexity measures for sequences from the viewpoint of stream ciphers. The emphasis is on recent work. Earlier surveys of this topic were given by Niederreiter [42], [43] and Rueppel [52]. We will not cover complexity measures that are too far removed from the linear complexity, such as higher-order complexity measures, 2-adic complexity measures, and complexity measures based on pattern counting. For these complexity measures not treated here we refer the reader to the survey in Niederreiter [43].

2   Linear Complexity

From now on, as a matter of convenience we speak of strings, respectively sequences, over $\mathbb{F}_q$ when we mean strings, respectively sequences, of elements of $\mathbb{F}_q$.

As mentioned in Section 1, LFSRs are the basic hardware components of keystream generators. The output sequence of an LFSR is ultimately periodic and can be described by a linear recurrence relation.

Definition 1. *Let k be a positive integer. Then the sequence $s_1, s_2, \ldots$ over $\mathbb{F}_q$ is said to satisfy a* linear recurrence relation *over $\mathbb{F}_q$ (of order k) if there exist $a_0, a_1, \ldots, a_{k-1} \in \mathbb{F}_q$ such that*

$$s_{i+k} = \sum_{h=0}^{k-1} a_h s_{i+h} \qquad for \ i = 1, 2, \ldots.$$

The sequence $s_1, s_2, \ldots$ in Definition 1 is uniquely determined by the linear recurrence relation and by the initial values $s_1, \ldots, s_k$. Conversely, the linear recurrence relation and the initial values can be recovered from the first $2k$ terms of the sequence by the Berlekamp-Massey algorithm. By convention, the zero sequence $0, 0, \ldots$ over $\mathbb{F}_q$ is meant to satisfy also a linear recurrence relation over $\mathbb{F}_q$ of order 0. The following notions of complexity are fundamental for the system-theoretic approach to the assessment of keystreams.

Definition 2. *Let n be a positive integer and let S be a finite or infinite sequence over $\mathbb{F}_q$ containing at least n terms. Then the nth* linear complexity $L_n(S)$ *of S is the least k such that the first n terms of S can be generated by a linear recurrence relation over $\mathbb{F}_q$ of order k. If S is ultimately periodic, then its* linear complexity $L(S)$ *is defined by*

$$L(S) = \sup_{n \geq 1} L_n(S).$$

It is clear that we always have $0 \leq L_n(S) \leq n$ and $L_n(S) \leq L_{n+1}(S)$. The extreme values of $L_n(S)$ correspond to highly nonrandom behavior, for if $s_1, \ldots, s_n$ are the first n terms of S, then $L_n(S) = 0$ if and only if $s_i = 0$ for $1 \leq i \leq n$, whereas $L_n(S) = n$ if and only if $s_i = 0$ for $1 \leq i \leq n-1$ and $s_n \neq 0$. Note also that if S is ultimately periodic, then we have $L(S) < \infty$.

An important tool for the analysis of the linear complexity of periodic sequences is the discrete Fourier transform. Let N be a given positive integer. Note that a finite field $\mathbb{F}_r$ contains a primitive Nth root of unity if and only if $r \equiv 1$ mod N. Now let $\mathbb{F}_q$ be such that there exists a primitive Nth root of unity α in some finite extension of $\mathbb{F}_q$. This condition is equivalent to $\gcd(N, q) = 1$. Then the *discrete Fourier transform* of a time-domain N-tuple $(t_0, \ldots, t_{N-1}) \in \mathbb{F}_q^N$ is the frequency-domain N-tuple $(f_0, \ldots, f_{N-1})$ with

$$f_j = \sum_{i=0}^{N-1} t_i \alpha^{ij} \qquad \text{for } j = 0, \ldots, N-1.$$

We have $f_j \in \mathbb{F}_q(\alpha)$ for all j.

The discrete Fourier transform is a linear transformation that can be represented by the Vandermonde matrix V for $1, \alpha, \ldots, \alpha^{N-1}$. Since α is a primitive Nth root of unity, the matrix V is nonsingular, and so the discrete Fourier transform can be inverted. This means that enumeration procedures may be carried out in the time domain or in the frequency domain, wherever they are more transparent.

For a positive integer N, let us say that a sequence $s_1, s_2, \ldots$ over $\mathbb{F}_q$ is N-*periodic* if $s_{i+N} = s_i$ for all $i \geq 1$. Note that N is then a period of the sequence, but not necessarily the least period. It is an important fact that the linear complexity of an N-periodic sequence can be determined via the discrete Fourier transform. In the case $\gcd(N, q) = 1$ this is made possible by a classical theorem according to which the linear complexity of the N-periodic sequence $s_1, s_2, \ldots$ is given by the Hamming weight of the discrete Fourier transform of the N-tuple $(s_1, \ldots, s_N)$; see e.g. [23, Theorem 6.8.1]. An analogous relationship holds when $\gcd(N, q) > 1$, but in this case one has to work with the generalized discrete Fourier transform as explained in Massey and Serconek [33].

A fundamental question for stream ciphers is the following: if we pick an N-periodic sequence S over $\mathbb{F}_q$ at random, what is the expected linear complexity $L(S)$? For fixed q and $N \geq 1$, the number of N-periodic sequences over $\mathbb{F}_q$ is q^N. In the natural stochastic model where all these q^N sequences are considered equally likely, the expected value $E_N^{(q)}$ of the linear complexity is given by

$$E_N^{(q)} = \frac{1}{q^N} \sum_S L(S),$$

where S runs through all N-periodic sequences over $\mathbb{F}_q$. Rueppel [51] considered $E_N^{(q)}$ for $q = 2$ and obtained the following partial results: if $N = 2^n$, then

$$E_N^{(2)} = N - 1 + 2^{-N};$$

if $N = 2^n - 1$ with a prime n, then

$$E_N^{(2)} \gtrapprox e^{-1/n}\left(N - \frac{1}{2}\right).$$

He conjectured that $E_N^{(2)}$ is always close to N.

Rueppel's conjecture was proved 16 years later by Meidl and Niederreiter [36] for arbitrary q. To describe this result, we need to recall the notion of a cyclotomic coset. If $w \geq 1$ and $j \geq 0$ are integers with $\gcd(w, q) = 1$, then a *cyclotomic coset* C_j mod w (relative to powers of q) is defined by

$$C_j = \{0 \leq k \leq w - 1 : k \equiv jq^r \bmod w \text{ for some } r \geq 0\}.$$

Theorem 1. *Let* $N = p^v w$, *where* p *is the characteristic of* $\mathbb{F}_q$, $v \geq 0$, *and* $\gcd(p, w) = 1$. *Let* $B_1, \ldots, B_h$ *be the different cyclotomic cosets mod* w *and put* $b_i = |B_i|$ *for* $1 \leq i \leq h$. *Then*

$$E_N^{(q)} = N - \sum_{i=1}^{h} \frac{b_i(1 - q^{-b_i p^v})}{q^{b_i} - 1}.$$

Corollary 1. *For any* N *we have*

$$E_N^{(q)} > N - \frac{w}{q - 1},$$

and if $\gcd(N, q) = 1$, *then we get the improved bound*

$$E_N^{(q)} \geq \left(1 - \frac{1}{q}\right) N.$$

Corollary 1 can be refined for small q (see [36]). The proof of Theorem 1 proceeds by transforming the problem of evaluating $E_N^{(q)}$ from the time domain to the frequency domain via the generalized discrete Fourier transform. Some special cases of Theorem 1 were already proved a bit earlier in Meidl and Niederreiter [34].

Various authors have recently studied the linear complexity of special sequences. We mention some papers that are of cryptologic interest. Caballero-Gil [3], [4], Garcia-Villalba and Fúster-Sabater [14], and Tan [59] considered the linear complexity of typical hardware keystream generators. Helleseth *et al.* [19] studied the linear complexity of Lempel-Cohn-Eastman sequences. Konyagin *et al.* [29] and Meidl and Winterhof [39] treated the linear complexity of the discrete logarithm function. Shparlinski [54] and Shparlinski and Silverman [57] considered the linear complexity of the Naor-Reingold function and Shparlinski [55] the linear complexity of the power generator. Further information on the linear complexity of special sequences can be found in the books of Shparlinski [53], [56].

3   Linear Complexity Profile

The following concept is meaningful not only for ultimately periodic sequences, but for arbitrary infinite sequences over $\mathbb{F}_q$.

Definition 3. *For an infinite sequence S over $\mathbb{F}_q$, let $L_n(S)$ denote again the nth linear complexity of S. Then the sequence $L_1(S), L_2(S), \ldots$ is called the* linear complexity profile *of S.*

The linear complexity profile of S is a nondecreasing sequence of nonnegative integers. Thus, the linear complexity profile is fully determined if we know where and how large its jumps are. These data can be conveniently given in terms of the continued fraction expansion of the generating function of S. Recall that if $s_1, s_2, \ldots$ are the terms of S, then the *generating function* of S is given by

$$G := \sum_{i=1}^{\infty} s_i x^{-i} \in \mathbb{F}_q((x^{-1})),$$

where $\mathbb{F}_q((x^{-1}))$ is the field of formal Laurent series over $\mathbb{F}_q$ in the variable x^{-1}. The continued fraction expansion of G has the form

$$G = 1/(A_1 + 1/(A_2 + \cdots)),$$

where the partial quotients $A_1, A_2, \ldots$ are polynomials over $\mathbb{F}_q$ of positive degree. Then the important fact is that the jumps in the linear complexity profile of S are exactly the degrees $\deg(A_1), \deg(A_2), \ldots$ of the partial quotients, and the locations of the jumps are also uniquely determined by the degrees of the partial quotients (see [50, Section 7.1] for an elementary proof of this fact).

In a natural probabilistic framework, the jumps in the linear complexity profile are i.i.d. random variables. This leads to the result that for a random sequence S over $\mathbb{F}_q$ we have

$$L_n(S) = \frac{n}{2} + O(\log n) \qquad \text{for all } n \geq 2,$$

and furthermore deviations from $n/2$ of the order of magnitude $\log n$ must appear for infinitely many n (see [41]). This suggests to study the deviations of $L_n(S)$ from $n/2$. There has been a strong interest in sequences S for which these deviations are bounded.

Definition 4. *If d is a positive integer, then a sequence S over $\mathbb{F}_q$ is called* d-perfect *if*

$$|2L_n(S) - n| \leq d \qquad \text{for all } n \geq 1.$$

A 1-perfect sequence is also called perfect. *A sequence is called* almost perfect *if it is d-perfect for some d.*

An alternative characterization of d-perfect sequences can be based on continued fractions (see [50, Theorem 7.2.2]).

Theorem 2. *A sequence S over $\mathbb{F}_q$ is d-perfect if and only if the generating function G of S is irrational and the partial quotients A_j in the continued fraction expansion of G satisfy $\deg(A_j) \leq d$ for all $j \geq 1$.*

It is an interesting problem to construct almost perfect sequences explicitly. Sophisticated methods of constructing such sequences are based on the use of algebraic curves over finite fields. This approach was initiated by Xing and Lam [64] and Xing et al. [67]. A systematic investigation of examples of almost perfect sequences that can be obtained from these methods was carried out by Kohel et al. [26]. Further comments on, and surveys of, this approach can be found in Niederreiter and Xing [50, Section 7.2], Xing [62], [63], and Xing and Niederreiter [66].

Niederreiter and Vielhaber [47] obtained enumeration theorems for the number $A_n^{(q)}(d)$ of strings S over $\mathbb{F}_q$ of length n with the property that $|2L_t(S)-t| \leq d$ for $1 \leq t \leq n$. This leads to a formula for the Hausdorff dimension of the set of d-perfect sequences over $\mathbb{F}_q$. Equivalently, in view of Theorem 2, this provides a formula for the Hausdorff dimension of the set of generating functions over $\mathbb{F}_q$ for which the degrees of all partial quotients are at most d. A detailed survey of further probabilistic properties of continued fraction expansions of formal Laurent series over $\mathbb{F}_q$ is given in Berthé and Nakada [2]. In the binary case, Carter [5] determined the number of bit strings of given length n with a prescribed linear complexity profile. Houston [20], [21] studied the density of 1's in perfect binary sequences.

Griffin and Shparlinski [15] studied the linear complexity profile of the power generator and Gutierrez et al. [17] and Meidl and Winterhof [40] considered the linear complexity profile of other nonlinear pseudorandom number generators.

It is an important requirement that a good keystream S should have an acceptable linear complexity profile (i.e., one close to that of a random sequence) for every starting point in the sequence. In other words, if S has the terms $s_1, s_2, \ldots$ and S_r is the shifted sequence $s_{r+1}, s_{r+2}, \ldots$, then S_r should have an acceptable linear complexity profile for every integer $r \geq 0$. The determination of the linear complexity profile of shifted sequences leads to the following algorithmic problem: given the continued fraction expansion of a generating function $G(x) \in \mathbb{F}_q((x^{-1}))$, compute the continued fraction expansion of $(G(x) + a)/x$ with $a \in \mathbb{F}_q$ in an efficient manner. This problem was solved by Niederreiter and Vielhaber [48] who designed an algorithm of arithmetic complexity $O(n^2)$ for computing the continued fraction expansions of $G_n(x), \ldots, G_1(x)$, where $G_r(x)$ is the generating function of the string $(s_r, \ldots, s_n)$ for $1 \leq r \leq n$. It is easily seen that this complexity bound is in general best possible. We mention here the following related open problem.

Open Problem 1. *Construct a sequence S over $\mathbb{F}_q$ such that, for some $d \geq 1$, all shifted sequences of S (including S itself) are d-perfect.*

Recent investigations have led to the discovery of interesting connections between the linear complexity profile and the lattice structure of sequences. Such connections were first obtained by Niederreiter and Winterhof [49] for periodic sequences and later by Dorfer and Winterhof [11], [12] in the general case. The relevant notion in this work is that of the lattice profile which is defined as follows. First of all, given a sequence S with terms $s_1, s_2, \ldots$ in $\mathbb{F}_q$ and integers

$t \geq 1$ and $N \geq 2$, we say that S passes the t-dimensional N-lattice test if the vectors $\mathbf{s}_n - \mathbf{s}_0$, $n = 1, \ldots, N - t$, span $\mathbb{F}_q^t$, where

$$\mathbf{s}_n = (s_n, s_{n+1}, \ldots, s_{n+t-1}) \qquad \text{for } 0 \leq n \leq N - t.$$

The greatest value of t such that S passes the t-dimensional N-lattice test is called the *lattice profile* at N. With this definition, close relationships between the linear complexity profile and the lattice profile arise. The related concept of interval linear complexity profile was introduced and analyzed by Balakirsky [1].

4   The k-Error Linear Complexity

A cryptographically strong keystream must not only have a large linear complexity, but also altering a few terms should not cause a significant decrease of the linear complexity (otherwise the keystream becomes vulnerable to cryptanalytic attacks based on the Berlekamp-Massey algorithm). This requirement leads to the concept of k-error linear complexity (see [10, Chapter 4], [13]). It is convenient here to use a notation from coding theory, namely $d(S, T)$ for the Hamming distance of two strings S and T over $\mathbb{F}_q$ of the same length.

Definition 5. *Let S be a string over $\mathbb{F}_q$ of length $n \geq 1$ and let k be an integer with $0 \leq k \leq n$. Then the k-error linear complexity $L_{n,k}(S)$ of S is defined by*

$$L_{n,k}(S) = \min_{T : d(S,T) \leq k} L_n(T),$$

where the minimum is over all strings T over $\mathbb{F}_q$ of length n with $d(S, T) \leq k$. If S is a finite or infinite sequence over $\mathbb{F}_q$ containing at least n terms, then we put $L_{n,k}(S)$ for the k-error linear complexity of the first n terms of S.

Thus, if S has at least n terms, then $L_{n,k}(S)$ is the least nth linear complexity that can be obtained by altering at most k terms among the first n terms of S. It is clear that $L_{n,0}(S) = L_n(S)$. The notion of k-error linear complexity is basic in the stability theory of stream ciphers developed by Ding *et al.* [10]. In this theory one studies the behavior of the (nth) linear complexity under changes of terms in a sequence. Just to indicate briefly that stability is a valid concern, here is an extremely bad instance of instability: the string $S = (0, \ldots, 0, 1) \in \mathbb{F}_q^n$ has $L_n(S) = n$, but if we change the last term to 0, then the nth linear complexity drops down to 0. Obviously, this string S is also unsuitable as a keystream.

Investigations of the distribution of values of the k-error linear complexity over strings of fixed length were initiated by Niederreiter and Paschinger [45] for $q = 2$ and continued for arbitrary q by Meidl and Niederreiter [35]. These results also lead to information on the expected value $E_{n,k}^{(q)}$ of $L_{n,k}(S)$ for random $S \in \mathbb{F}_q^n$. In fact, in [35] one can find bounds which are approximately of the form

$$\frac{n}{2} - \frac{k}{2} \log_q n \overset{\leq}{\approx} E_{n,k}^{(q)} \overset{\leq}{\approx} \frac{n}{2} - \frac{k}{2},$$

where $\log_q$ denotes the logarithm to the base q. There is still a considerable gap between the lower and upper bound.

Since we now have two parameters n and k in $L_{n,k}(S)$, various notions of k-error linear complexity profiles can be considered, but little work has been done in this direction.

Open Problem 2. *Determine the behavior of $L_{n,k}(S)$ for random sequences S over $\mathbb{F}_q$ and for meaningful ways in which n and k are related, e.g. for fixed k and $n \to \infty$ or for k and n tending to ∞ at a fixed ratio.*

There is also a version of the k-error linear complexity that is tailored to periodic sequences and which was first considered by Stamp and Martin [58].

Definition 6. *Let S be an N-periodic sequence over $\mathbb{F}_q$ and let k be an integer with $0 \leq k \leq N$. Then the k-error linear complexity $L_{N,k}^{\mathrm{per}}(S)$ of S is defined by*

$$L_{N,k}^{\mathrm{per}}(S) = \min_T L(T),$$

where the minimum is over all N-periodic sequences T over $\mathbb{F}_q$ for which the first N terms differ in at most k positions from the corresponding terms of S.

Efficient algorithms for the computation of $L_{N,k}^{\mathrm{per}}(S)$ for various special values of N were designed by Kaida *et al.* [24], Lauder and Paterson [31], Stamp and Martin [58], and Xiao and Wei [60]. The case $q = 2$ and $k = 1$ was studied in detail by Kolokotronis *et al.* [27], [28].

Open Problem 3. *Design fast algorithms for computing the k-error linear complexities $L_{n,k}(S)$ and $L_{N,k}^{\mathrm{per}}(S)$ in the general case.*

Lower bounds on the expected value of the k-error linear complexity $L_{N,k}^{\mathrm{per}}(S)$ of random N-periodic sequences S over $\mathbb{F}_q$ were established by Meidl and Niederreiter [36]. The special case where N is a prime different from the characteristic of $\mathbb{F}_q$ was considered earlier by the same authors in [34].

An important issue for practical keystream generation is the relationship between the linear complexity $L(S)$ and the k-error linear complexity $L_{N,k}^{\mathrm{per}}(S)$ for N-periodic sequences S over $\mathbb{F}_q$, in particular the question whether these complexities can be made simultaneously large. We note the trivial bound $L(S) \leq N$ and the bound $L_{N,k}^{\mathrm{per}}(S) \leq N - 1$ for $1 \leq k \leq N$ which follows easily from a more refined upper bound in [44]. The latter paper contains also results which show that the above complexities can indeed be made large at the same time and can even reach their maxima simultaneously. The following is a sample result from [44].

Theorem 3. *Let $N \geq 2$ with $\gcd(N, q) = 1$ and let ℓ be the smallest cardinality of a nonzero cyclotomic coset mod N. Suppose that k is an integer with $1 \leq k \leq N$ satisfying*

$$\sum_{j=0}^{k} \binom{N}{j} (q-1)^j < q^\ell.$$

Then there exists an N-periodic sequence S over $\mathbb{F}_q$ with

$$L(S) = N \quad and \quad L_{N,k}^{\mathrm{per}}(S) = N - 1.$$

For related results and discussions we refer to Cusick *et al.* [6, Chapter 3], Ding [9], and Niederreiter [44]. More can be proved in the case where N is prime by using techniques of analytic number theory. The following theorem of Niederreiter and Shparlinski [46] was shown by such methods.

Theorem 4. *For any fixed prime power q, there are infinitely many primes r such that for $N = r$ there is an N-periodic sequence S over $\mathbb{F}_q$ with*

$$L(S) = N \quad and \quad L_{N,k}^{\mathrm{per}}(S) = N - 1 \quad for \ 1 \le k \le \frac{N^{0.677}}{2 \log N}.$$

Theorems 3 and 4 are existence theorems only, i.e., they guarantee that at least one N-periodic sequence over $\mathbb{F}_q$ with the desired properties exists. However, if only a small number of N-periodic sequences over $\mathbb{F}_q$ had these properties, it would not be wise to choose any of these sequences as a keystream since an attacker could perform an exhaustive search for these sequences and thus break the stream cipher. Thus, it is important to also establish results which show that actually a large number of N-periodic sequences S over $\mathbb{F}_q$ satisfy $L(S) = N$ and have a large k-error linear complexity $L_{N,k}^{\mathrm{per}}(S)$. Results of this type were obtained by Meidl and Niederreiter [38].

Further relationships between the linear complexity and the k-error linear complexity of periodic sequences were studied by Kurosawa *et al.* [30]. For instance, they determined the least value of k such that $L_{N,k}^{\mathrm{per}}(S) < L(S)$ when S is an N-periodic sequence over $\mathbb{F}_q$ with N a power of the characteristic of $\mathbb{F}_q$. For the same types of N-periodic sequences, Kaida *et al.* [25] proved related results on the values of $k \ge 1$ for which $L_{N,k}^{\mathrm{per}}(S) < L_{N,k-1}^{\mathrm{per}}(S)$.

Dai and Imamura [7] and Jiang *et al.* [22] considered the change in the linear complexity of periodic sequences that results from the substitution, insertion, or deletion of terms.

5 Joint Linear Complexity of Multisequences

Recent developments in stream ciphers point towards an interest in word-based or vectorized stream ciphers; see e.g. Dawson and Simpson [8] and Hawkes and Rose [18]. The theory of such stream ciphers requires the study of the complexity of multisequences, i.e., of parallel streams of finitely many sequences. We will denote a multisequence consisting of m parallel streams of sequences $S_1, \ldots, S_m$ over $\mathbb{F}_q$ by $\mathbf{S} = (S_1, \ldots, S_m)$.

In the framework of linear complexity theory, the appropriate complexity measure for multisequences is obtained by looking at the linear recurrence relations that $S_1, \ldots, S_m$ satisfy simultaneously. Thus, one speaks of the joint linear complexity in this context.

Definition 7. *Let n be a positive integer and let $\mathbf{S} = (S_1, \ldots, S_m)$ be an m-fold multisequence over $\mathbb{F}_q$ such that each sequence S_j, $1 \le j \le m$, contains at least n terms. Then the nth joint linear complexity $L_n(\mathbf{S}) = L_n(S_1, \ldots, S_m)$ is the least order of a linear recurrence relation over $\mathbb{F}_q$ that simultaneously generates the first n terms of each sequence S_j, $1 \le j \le m$. If $S_1, \ldots, S_m$ are ultimately periodic, then the joint linear complexity $L(\mathbf{S}) = L(S_1, \ldots, S_m)$ is defined by*

$$L(\mathbf{S}) = \sup_{n \ge 1} L_n(\mathbf{S}).$$

We again have $0 \le L_n(\mathbf{S}) \le n$ and $L_n(\mathbf{S}) \le L_{n+1}(\mathbf{S})$, and for ultimately periodic $\mathbf{S}$ we have $L(\mathbf{S}) < \infty$. Since the $\mathbb{F}_q$-linear spaces $\mathbb{F}_q^m$ and $\mathbb{F}_{q^m}$ are isomorphic, the given m-fold multisequence $\mathbf{S}$ over $\mathbb{F}_q$ can also be identified with a single sequence $\mathcal{S}$ having its terms in the extension field $\mathbb{F}_{q^m}$. The (nth) joint linear complexity of $\mathbf{S}$ can then also be interpreted as the (nth) $\mathbb{F}_q$-*linear complexity* of $\mathcal{S}$, which is the least order of a linear recurrence relation over $\mathbb{F}_q$ that the terms of $\mathcal{S}$ satisfy (see [10, pp. 83–85]). This viewpoint is often convenient in proofs.

If $\mathbf{S}$ is an infinite multisequence, then in analogy with Definition 3 we may call the sequence $L_1(\mathbf{S}), L_2(\mathbf{S}), \ldots$ the *joint linear complexity profile* of $\mathbf{S}$. As in the case of single sequences, one can consider multisequences with perfect or almost perfect joint linear complexity profile. Some work in this direction was done by Xing [61] and Xing *et al.* [65].

As for single sequences, it is of great interest to determine the expected value of the joint linear complexity of periodic multisequences. If $\mathbf{S} = (S_1, \ldots, S_m)$ with periodic sequences $S_1, \ldots, S_m$ over $\mathbb{F}_q$, then we can assume w.l.o.g. that they have the common period N. The appropriate expected value is then

$$E_N^{(q,m)} = \frac{1}{q^{mN}} \sum_{\mathbf{S}} L(\mathbf{S}),$$

where the sum is over all q^{mN} N-periodic m-fold multisequences $\mathbf{S}$ over $\mathbb{F}_q$. Meidl and Niederreiter [37] obtained a formula for $E_N^{(q,m)}$ which is quite analogous to that in Theorem 1 for the case $m = 1$. In fact, with the notation in Theorem 1 we have

$$E_N^{(q,m)} = N - \sum_{i=1}^{h} \frac{b_i(1 - q^{-mb_i p^v})}{q^{mb_i} - 1}.$$

This leads to similar lower bounds as in Corollary 1, namely

$$E_N^{(q,m)} > N - \frac{w}{q^m - 1},$$

and for $\gcd(N, q) = 1$ to the improved lower bound

$$E_N^{(q,m)} \ge \left(1 - \frac{1}{q^m}\right) N.$$

For given integers $m \geq 1$, $n \geq 1$, and $0 \leq L \leq n$, let $N_n^{(m)}(L)$ denote the number of m-fold multisequences $(S_1, \ldots, S_m)$ over $\mathbb{F}_q$ of length n with nth joint linear complexity L. It is trivial that $N_n^{(m)}(0) = 1$. For $m = 1$ there is the classical formula of Gustavson [16] (see also [50, Theorem 7.1.6]) according to which we have

$$N_n^{(1)}(L) = (q-1)q^{\min(2L-1, 2n-2L)} \qquad \text{for } 1 \leq L \leq n. \tag{1}$$

The determination of $N_n^{(m)}(L)$ for $m \geq 2$ seems to be a difficult problem. However, we are able to generalize the formula (1) in the range $1 \leq L \leq n/2$. This result (to be proved in Theorem 5 below) is an original contribution of the present paper.

First we need some notation and auxiliary considerations. Let $\mathcal{M}_q$ denote the set of all monic polynomials over $\mathbb{F}_q$ and let μ_q be the Moebius function on $\mathcal{M}_q$ (see [32, Exercise 3.75]) which is defined in analogy with the Moebius function in elementary number theory.

Lemma 1. *For $b = 0, 1, \ldots$ put*

$$\sigma_b = \sum_{\substack{d \in \mathcal{M}_q \\ \deg(d) = b}} \mu_q(d).$$

Then we have $\sigma_0 = 1$, $\sigma_1 = -q$, and $\sigma_b = 0$ for $b \geq 2$.

Proof. Consider the power series

$$H(z) = \sum_{b=0}^{\infty} \sigma_b z^b \tag{2}$$

over the complex numbers. Since $|\mu_q(d)| \leq 1$ for all $d \in \mathcal{M}_q$, we have $|\sigma_b| \leq q^b$ for all $b \geq 0$. Therefore $H(z)$ is absolutely convergent for $|z| < \frac{1}{q}$, and so the following manipulations of power series are justified. With $\mathcal{P}_q$ denoting the set of all monic irreducible polynomials over $\mathbb{F}_q$, we obtain

$$H(z) = \sum_{b=0}^{\infty} \left(\sum_{\substack{d \in \mathcal{M}_q \\ \deg(d) = b}} \mu_q(d) \right) z^b = \sum_{d \in \mathcal{M}_q} \mu_q(d) z^{\deg(d)}$$

$$= \prod_{f \in \mathcal{P}_q} \left(1 + \mu_q(f) z^{\deg(f)} + \mu_q(f^2) z^{\deg(f^2)} + \cdots \right)$$

$$= \prod_{f \in \mathcal{P}_q} \left(1 - z^{\deg(f)} \right).$$

Thus,

$$\frac{1}{H(z)} = \prod_{f \in \mathcal{P}_q} \left(1 - z^{\deg(f)} \right)^{-1} = \prod_{f \in \mathcal{P}_q} \left(1 + z^{\deg(f)} + z^{\deg(f^2)} + \cdots \right)$$

$$= \sum_{d \in \mathcal{M}_q} z^{\deg(d)} = \sum_{b=0}^{\infty} \left(\sum_{\substack{d \in \mathcal{M}_q \\ \deg(d)=b}} 1 \right) z^b = \sum_{b=0}^{\infty} q^b z^b = \frac{1}{1-qz}.$$

It follows that $H(z) = 1 - qz$, and so a comparison with (2) proves the lemma.
$\square$

Now we consider an arbitrary nonzero polynomial $h \in \mathbb{F}_{q^m}[x]$. Let

$$h = a g_1^{e_1} \cdots g_r^{e_r} \tag{3}$$

with $a \in \mathbb{F}_{q^m}^*$ be the canonical factorization of h in $\mathbb{F}_{q^m}[x]$. Each monic irreducible factor $g_i \in \mathbb{F}_{q^m}[x]$, $1 \le i \le r$, divides a uniquely determined polynomial in $\mathcal{P}_q$, where $\mathcal{P}_q$ is as in the proof of Lemma 1. Let $f_1, \ldots, f_t$ be the complete list of polynomials in $\mathcal{P}_q$ that arise by considering all $g_1, \ldots, g_r$. For each $k = 1, \ldots, t$ let

$$f_k = g_{k,1} \cdots g_{k,u_k}$$

be the canonical factorization of f_k in $\mathbb{F}_{q^m}[x]$. Then the factorization (3) can be written in the form

$$h = a \prod_{k=1}^{t} \prod_{j=1}^{u_k} g_{k,j}^{e_{k,j}}$$

with all $e_{k,j} \ge 0$. For $1 \le k \le t$ put

$$m_k = \min_{1 \le j \le u_k} e_{k,j}.$$

Now define

$$K_q(h) = \prod_{k=1}^{t} f_k^{m_k} \in \mathbb{F}_q[x]$$

and call $K_q(h)$ the $\mathbb{F}_q$-*core* of h. The important fact is that a polynomial in $\mathbb{F}_q[x]$ divides $h \in \mathbb{F}_{q^m}[x]$ if and only if it divides $K_q(h)$.

We are now ready to prove the generalization of (1) for the range $1 \le L \le n/2$.

Theorem 5. *For $m \ge 1$ and $n \ge 2$ we have*

$$N_n^{(m)}(L) = (q^m - 1)q^{(m+1)L-m} \qquad \text{for } 1 \le L \le \frac{n}{2}.$$

Proof. It is convenient to work equivalently with the $\mathbb{F}_q$-linear complexity of strings over $\mathbb{F}_{q^m}$ of length n (compare with the discussion following Definition 7). With a string $\mathcal{S} = (s_1, \ldots, s_n)$ over $\mathbb{F}_{q^m}$ of length n we associate the formal Laurent series

$$\mathcal{G} = \sum_{i=1}^{n} s_i x^{-i} \in \mathbb{F}_{q^m}((x^{-1})).$$

Let deg denote the canonical extension of the degree function from $\mathbb{F}_{q^m}[x]$ to $\mathbb{F}_{q^m}((x^{-1}))$. Then $\mathcal{S}$ has $\mathbb{F}_q$-linear complexity $\leq L$ if and only if there exists a rational function g/f with $f \in \mathbb{F}_q[x]$, $g \in \mathbb{F}_{q^m}[x]$, and $\deg(f) \leq L$ such that

$$\deg\left(\mathcal{G} - \frac{g}{f}\right) < -n. \qquad (4)$$

Now we consider the case $1 \leq L \leq n/2$. Then we claim that g/f in (4) is uniquely determined. For if also $g_1/f_1 \in \mathbb{F}_{q^m}(x)$ with $\deg(f_1) \leq L$ satisfies

$$\deg\left(\mathcal{G} - \frac{g_1}{f_1}\right) < -n,$$

then

$$\deg\left(\frac{g}{f} - \frac{g_1}{f_1}\right) < -n.$$

This implies

$$\deg(gf_1 - g_1 f) < -n + \deg(f) + \deg(f_1) \leq 0,$$

hence $gf_1 - g_1 f = 0$, and so $g_1/f_1 = g/f$. The claim is thus established.

Consequently, the strings $\mathcal{S} \in \mathbb{F}_{q^m}^n$ with $\mathbb{F}_q$-linear complexity L can be identified with the rational functions g/f with $f \in \mathbb{F}_q[x]$, f monic, $\deg(f) = L$, $g \in \mathbb{F}_{q^m}[x]$, $\deg(g) < L$, and f and g having no common factor in $\mathbb{F}_q[x]$ of positive degree. Thus, $N_n^{(m)}(L)$ is equal to the number of ordered pairs (f, g) of polynomials with the above properties. Note that the condition that f and g have no common factor in $\mathbb{F}_q[x]$ of positive degree is equivalent to $\gcd(f,g) \in \mathbb{F}_{q^m}[x]$ having no factor in $\mathbb{F}_q[x]$ of positive degree. By the considerations following Lemma 1, this is in turn equivalent to the $\mathbb{F}_q$-core of $\gcd(f,g)$ being equal to 1. Then, using a property of the Moebius function μ_q on $\mathcal{M}_q$ (see [32, Exercise 3.75(a)]), we can write

$$N_n^{(m)}(L) = \sum_{\substack{f \in \mathcal{M}_q \\ \deg(f)=L}} \sum_{\substack{g \in \mathbb{F}_{q^m}[x] \\ \deg(g)<L, K_q(\gcd(f,g))=1}} 1 = \sum_{\substack{f \in \mathcal{M}_q \\ \deg(f)=L}} \sum_{\substack{g \in \mathbb{F}_{q^m}[x] \\ \deg(g)<L}} \sum_{\substack{d \in \mathcal{M}_q \\ d \mid K_q(\gcd(f,g))}} \mu_q(d)$$

$$= \sum_{\substack{f \in \mathcal{M}_q \\ \deg(f)=L}} \sum_{\substack{g \in \mathbb{F}_{q^m}[x] \\ \deg(g)<L}} \sum_{\substack{d \in \mathcal{M}_q \\ d \mid f, d \mid g}} \mu_q(d).$$

By rearranging the summation, we get

$$N_n^{(m)}(L) = \sum_{\substack{d \in \mathcal{M}_q \\ \deg(d) \leq L}} \mu_q(d) \left(\sum_{\substack{f \in \mathcal{M}_q \\ \deg(f)=L, d \mid f}} 1 \right) \left(\sum_{\substack{g \in \mathbb{F}_{q^m}[x] \\ \deg(g)<L, d \mid g}} 1 \right)$$

$$= \sum_{\substack{d \in \mathcal{M}_q \\ \deg(d) \leq L}} \mu_q(d) q^{L-\deg(d)} q^{m(L-\deg(d))}$$

$$= q^{(m+1)L} \sum_{b=0}^{L} q^{-(m+1)b} \sum_{\substack{d \in \mathcal{M}_q \\ \deg(d)=b}} \mu_q(d).$$

With the notation in Lemma 1 we can then write

$$N_n^{(m)}(L) = q^{(m+1)L} \sum_{b=0}^{L} \sigma_b q^{-(m+1)b},$$

and the result of Lemma 1 yields now the desired formula for $N_n^{(m)}(L)$.   □

Open Problem 4. *Determine $N_n^{(m)}(L)$ for $n/2 < L \le n$.*

Acknowledgment

The research of the author was partially supported by the grant R-394-000-011-422 with Temasek Laboratories in Singapore.

References

1. V.B. Balakirsky, Description of binary sequences based on the interval linear complexity profile, *Sequences and Their Applications – SETA '01* (T. Helleseth, P.V. Kumar, and K. Yang, eds.), pp. 101–115, Springer, London, 2002.
2. V. Berthé and H. Nakada, On continued fraction expansions in positive characteristic: equivalence relations and some metric properties, *Expositiones Math.* **18**, 257–284 (2000).
3. P. Caballero-Gil, Regular cosets and upper bounds on the linear complexity of certain sequences, *Sequences and Their Applications* (C. Ding, T. Helleseth, and H. Niederreiter, eds.), pp. 161–170, Springer, London, 1999.
4. P. Caballero-Gil, New upper bounds on the linear complexity, *Comput. Math. Appl.* **39**, no. 3–4, 31–38 (2000).
5. G. Carter, Enumeration results on linear complexity profiles, *Cryptography and Coding II* (C. Mitchell, ed.), pp. 23–34, Oxford University Press, Oxford, 1992.
6. T.W. Cusick, C. Ding, and A. Renvall, *Stream Ciphers and Number Theory*, Elsevier, Amsterdam, 1998.
7. Z.D. Dai and K. Imamura, Linear complexity of one-symbol substitution of a periodic sequence over GF(q), *IEEE Trans. Inform. Theory* **44**, 1328–1331 (1998).
8. E. Dawson and L. Simpson, Analysis and design issues for synchronous stream ciphers, *Coding Theory and Cryptology* (H. Niederreiter, ed.), pp. 49–90, World Scientific, Singapore, 2002.
9. C. Ding, Binary cyclotomic generators, *Fast Software Encryption* (Leuven, 1994), Lecture Notes in Computer Science, Vol. **1008**, pp. 29–60, Springer, Berlin, 1995.
10. C. Ding, G. Xiao, and W. Shan, *The Stability Theory of Stream Ciphers*, Lecture Notes in Computer Science, Vol. **561**, Springer, Berlin, 1991.
11. G. Dorfer and A. Winterhof, Lattice structure and linear complexity profile of nonlinear pseudorandom number generators, *Applicable Algebra Engrg. Comm. Comput.* **13**, 499–508 (2003).
12. G. Dorfer and A. Winterhof, Lattice structure of nonlinear pseudorandom number generators in parts of the period, *Monte Carlo and Quasi-Monte Carlo Methods 2002* (H. Niederreiter, ed.), Springer, Berlin, to appear.

13. H.J. Fell, Linear complexity of transformed sequences, *Eurocode '90* (Udine, 1990), Lecture Notes in Computer Science, Vol. **514**, pp. 205–214, Springer, Berlin, 1991.
14. L.J. Garcia-Villalba and A. Fúster-Sabater, On the linear complexity of the sequences generated by nonlinear filterings, *Inform. Process. Lett.* **76**, no. 1–2, 67–73 (2000).
15. F. Griffin and I.E. Shparlinski, On the linear complexity profile of the power generator, *IEEE Trans. Inform. Theory* **46**, 2159–2162 (2000).
16. F.G. Gustavson, Analysis of the Berlekamp-Massey linear feedback shift-register synthesis algorithm, *IBM J. Res. Develop.* **20**, 204–212 (1976).
17. J. Gutierrez, I.E. Shparlinski, and A. Winterhof, On the linear and nonlinear complexity profile of nonlinear pseudorandom number generators, *IEEE Trans. Inform. Theory* **49**, 60–64 (2003).
18. P. Hawkes and G.G. Rose, Exploiting multiples of the connection polynomial in word-oriented stream ciphers, *Advances in Cryptology – ASIACRYPT 2000* (T. Okamoto, ed.), Lecture Notes in Computer Science, Vol. **1976**, pp. 303–316, Springer, Berlin, 2000.
19. T. Helleseth, S.-H. Kim, and J.-S. No, Linear complexity over F_p and trace representation of Lempel-Cohn-Eastman sequences, *IEEE Trans. Inform. Theory* **49**, 1548–1552 (2003).
20. A.E.D. Houston, Densities of perfect linear complexity profile binary sequences, *Applications of Combinatorial Mathematics* (Oxford, 1994), IMA Conference Series, Vol. **60**, pp. 119–133, Oxford University Press, Oxford, 1997.
21. A.E.D. Houston, On the limit of maximal density of sequences with a perfect linear complexity profile, *Designs Codes Cryptogr.* **10**, 351–359 (1997).
22. S. Jiang, Z.D. Dai, and K. Imamura, Linear complexity of a sequence obtained from a periodic sequence by either substituting, inserting, or deleting k symbols within one period, *IEEE Trans. Inform. Theory* **46**, 1174–1177 (2000).
23. D. Jungnickel, *Finite Fields: Structure and Arithmetics*, Bibliographisches Institut, Mannheim, 1993.
24. T. Kaida, S. Uehara, and K. Imamura, An algorithm for the k-error linear complexity of sequences over $GF(p^m)$ with period p^n, p a prime, *Inform. and Comput.* **151**, 134–147 (1999).
25. T. Kaida, S. Uehara, and K. Imamura, On the profile of the k-error linear complexity and the zero sum property for sequences over $GF(p^m)$ with period p^n, *Sequences and Their Applications – SETA '01* (T. Helleseth, P.V. Kumar, and K. Yang, eds.), pp. 218–227, Springer, London, 2002.
26. D. Kohel, S. Ling, and C.P. Xing, Explicit sequence expansions, *Sequences and Their Applications* (C. Ding, T. Helleseth, and H. Niederreiter, eds.), pp. 308–317, Springer, London, 1999.
27. N. Kolokotronis, P. Rizomiliotis, and N. Kalouptsidis, First-order optimal approximation of binary sequences, *Sequences and Their Applications – SETA '01* (T. Helleseth, P.V. Kumar, and K. Yang, eds.), pp. 242–256, Springer, London, 2002.
28. N. Kolokotronis, P. Rizomiliotis, and N. Kalouptsidis, Mimimum linear span approximation of binary sequences, *IEEE Trans. Inform. Theory* **48**, 2758–2764 (2002).
29. S. Konyagin, T. Lange, and I.E. Shparlinski, Linear complexity of the discrete logarithm, *Designs Codes Cryptogr.* **28**, 135–146 (2003).
30. K. Kurosawa, F. Sato, T. Sakata, and W. Kishimoto, A relationship between linear complexity and k-error linear complexity, *IEEE Trans. Inform. Theory* **46**, 694–698 (2000).

31. A.G.B. Lauder and K.G. Paterson, Computing the error linear complexity spectrum of a binary sequence of period 2^n, *IEEE Trans. Inform. Theory* **49**, 273–280 (2003).
32. R. Lidl and H. Niederreiter, *Finite Fields*, Cambridge University Press, Cambridge, 1997.
33. J.L. Massey and S. Serconek, Linear complexity of periodic sequences: a general theory, *Advances in Cryptology – CRYPTO '96* (N. Koblitz, ed.), Lecture Notes in Computer Science, Vol. **1109**, pp. 358–371, Springer, Berlin, 1996.
34. W. Meidl and H. Niederreiter, Linear complexity, k-error linear complexity, and the discrete Fourier transform, *J. Complexity* **18**, 87–103 (2002).
35. W. Meidl and H. Niederreiter, Counting functions and expected values for the k-error linear complexity, *Finite Fields Appl.* **8**, 142–154 (2002).
36. W. Meidl and H. Niederreiter, On the expected value of the linear complexity and the k-error linear complexity of periodic sequences, *IEEE Trans. Inform. Theory* **48**, 2817–2825 (2002).
37. W. Meidl and H. Niederreiter, The expected value of the joint linear complexity of periodic multisequences, *J. Complexity* **19**, 61–72 (2003).
38. W. Meidl and H. Niederreiter, Periodic sequences with maximal linear complexity and large k-error linear complexity, *Applicable Algebra Engrg. Comm. Comput.*, to appear.
39. W. Meidl and A. Winterhof, Lower bounds on the linear complexity of the discrete logarithm in finite fields, *IEEE Trans. Inform. Theory* **47**, 2807–2811 (2001).
40. W. Meidl and A. Winterhof, On the linear complexity profile of explicit nonlinear pseudorandom numbers, *Inform. Process. Lett.* **85**, 13–18 (2003).
41. H. Niederreiter, The probabilistic theory of linear complexity, *Advances in Cryptology – EUROCRYPT '88* (C.G. Günther, ed.), Lecture Notes in Computer Science, Vol. **330**, pp. 191–209, Springer, Berlin, 1988.
42. H. Niederreiter, Finite fields and cryptology, *Finite Fields, Coding Theory, and Advances in Communications and Computing* (G.L. Mullen and P.J.-S. Shiue, eds.), pp. 359–373, Dekker, New York, 1993.
43. H. Niederreiter, Some computable complexity measures for binary sequences, *Sequences and Their Applications* (C. Ding, T. Helleseth, and H. Niederreiter, eds.), pp. 67–78, Springer, London, 1999.
44. H. Niederreiter, Periodic sequences with large k-error linear complexity, *IEEE Trans. Inform. Theory* **49**, 501–505 (2003).
45. H. Niederreiter and H. Paschinger, Counting functions and expected values in the stability theory of stream ciphers, *Sequences and Their Applications* (C. Ding, T. Helleseth, and H. Niederreiter, eds.), pp. 318–329, Springer, London, 1999.
46. H. Niederreiter and I.E. Shparlinski, Periodic sequences with maximal linear complexity and almost maximal k-error linear complexity, *Cryptography and Coding* (Cirencester, 2003), Lecture Notes in Computer Science, Springer, Berlin, to appear.
47. H. Niederreiter and M. Vielhaber, Linear complexity profiles: Hausdorff dimensions for almost perfect profiles and measures for general profiles, *J. Complexity* **13**, 353–383 (1997).
48. H. Niederreiter and M. Vielhaber, An algorithm for shifted continued fraction expansions in parallel linear time, *Theoretical Computer Science* **226**, 93–104 (1999).
49. H. Niederreiter and A. Winterhof, Lattice structure and linear complexity of nonlinear pseudorandom numbers, *Applicable Algebra Engrg. Comm. Comput.* **13**, 319–326 (2002).

50. H. Niederreiter and C.P. Xing, *Rational Points on Curves over Finite Fields: Theory and Applications*, London Math. Soc. Lecture Note Series, Vol. **285**, Cambridge University Press, Cambridge, 2001.
51. R.A. Rueppel, *Analysis and Design of Stream Ciphers*, Springer, Berlin, 1986.
52. R.A. Rueppel, Stream ciphers, *Contemporary Cryptology: The Science of Information Integrity* (G.J. Simmons, ed.), pp. 65–134, IEEE Press, New York, 1992.
53. I.E. Shparlinski, *Number Theoretic Methods in Cryptography: Complexity Lower Bounds*, Birkhäuser, Basel, 1999.
54. I.E. Shparlinski, Linear complexity of the Naor-Reingold pseudo-random function, *Inform. Process. Lett.* **76**, no. 3, 95–99 (2000).
55. I.E. Shparlinski, On the linear complexity of the power generator, *Designs Codes Cryptogr.* **23**, 5–10 (2001).
56. I.E. Shparlinski, *Cryptographic Applications of Analytic Number Theory: Complexity Lower Bounds and Pseudorandomness*, Birkhäuser, Basel, 2003.
57. I.E. Shparlinski and J.H. Silverman, On the linear complexity of the Naor-Reingold pseudo-random function from elliptic curves, *Designs Codes Cryptogr.* **24**, 279–289 (2001).
58. M. Stamp and C.F. Martin, An algorithm for the k-error linear complexity of binary sequences with period 2^n, *IEEE Trans. Inform. Theory* **39**, 1398–1401 (1993).
59. C.H. Tan, Period and linear complexity of cascaded clock-controlled generators, *Sequences and Their Applications* (C. Ding, T. Helleseth, and H. Niederreiter, eds.), pp. 371–378, Springer, London, 1999.
60. G. Xiao and S. Wei, Fast algorithms for determining the linear complexity of period sequences, *Progress in Cryptology – INDOCRYPT 2002* (A. Menezes and P. Sarkar, eds.), Lecture Notes in Computer Science, Vol. **2551**, pp. 12–21, Springer, Berlin, 2002.
61. C.P. Xing, Multi-sequences with almost perfect linear complexity profile and function fields over finite fields, *J. Complexity* **16**, 661–675 (2000).
62. C.P. Xing, Applications of algebraic curves to constructions of sequences, *Cryptography and Computational Number Theory* (K.Y. Lam et al., eds.), pp. 137–146, Birkhäuser, Basel, 2001.
63. C.P. Xing, Constructions of sequences from algebraic curves over finite fields, *Sequences and Their Applications – SETA '01* (T. Helleseth, P.V. Kumar, and K. Yang, eds.), pp. 88–100, Springer, London, 2002.
64. C.P. Xing and K.Y. Lam, Sequences with almost perfect linear complexity profiles and curves over finite fields, *IEEE Trans. Inform. Theory* **45**, 1267–1270 (1999).
65. C.P. Xing, K.Y. Lam, and Z.H. Wei, A class of explicit perfect multi-sequences, *Advances in Cryptology – ASIACRYPT '99* (K.Y. Lam, E. Okamoto, and C.P. Xing, eds.), Lecture Notes in Computer Science, Vol. **1716**, pp. 299–305, Springer, Berlin, 1999.
66. C.P. Xing and H. Niederreiter, Applications of algebraic curves to constructions of codes and almost perfect sequences, *Finite Fields and Applications* (D. Jungnickel and H. Niederreiter, eds.), pp. 475–489, Springer, Berlin, 2001.
67. C.P. Xing, H. Niederreiter, K.Y. Lam, and C. Ding, Constructions of sequences with almost perfect linear complexity profile from curves over finite fields, *Finite Fields Appl.* **5**, 301–313 (1999).

Forensic Computing

Xiang Li and Jennifer Seberry

Centre for Computer Security Research, SITACS, University of Wollongong,
Wollongong, NSW, 2522, Australia.
jennie@uow.edu.au

Abstract. Technology is rapidly changing the speed and manner in which people interact with each other and with the world. As technology helps criminals to operate more easily and quickly across borders, so law enforcement capability must continuously improve to keep one step ahead. Computer forensics has become a specialized and accepted investigative technique with its own tools and legal precedents that validate the discipline. Specially designed forensic software is also widely used during the whole process of computer forensic investigation. This article introduces computer forensic and computer evidence, introduces and compares some forensic software, and summarizes its likely future development.

1 Background

1.1 Computer Forensics Defined

Computer forensics or forensic computing has become a popular topic in computer security circles and in the legal community. So what is computer forensics?

Dorothy A. Lunn's (Dorothy) definition of computer forensics is, "The employment of a set of predefined procedures to thoroughly examine a computer system using software and tools to extract and preserve evidence of criminal activity." Judd Robbins (Judd), a computer forensics investigator, defines computer forensics as "Simply the application of computer investigation and analysis techniques in the interest of determining potential legal evidence." James Borck (James), in his article "Leave the cybersleuthing to the experts". Defines Computer forensics as "the equivalent of surveying a crime scene or performing an autopsy on a victim".

From these descriptions, we can see computer forensics can be defined as the application of computer investigation and analysis techniques in the interests of determining potential evidence. It deals with the application of law to a science and it involves the use of sophisticated technology tools and procedures that must be followed to guarantee the accuracy of the preservation of evidence and the accuracy of results concerning computer evidence processing.

1.2 Requirements for Computer Forensics

Nowadays more and more criminals have been shifting their attention from armed robbery to computer crime. Criminals may use computers in one of two

T. Johansson and S. Maitra (Eds.): INDOCRYPT 2003, LNCS 2904, pp. 18–35, 2003.

ways in support of their actions. Either as the repository for information relating to their criminal activity, which is called computer related crime, or as a tool in actually committing a crime, this is so called computer crime. For them, a computer is much more powerful than a knife or a gun. More seriously, the probability of being caught on their actions is very low. For example, a hacker can break into a bank's online transaction system and steal $250,000 without being caught or charged for lack of enough evidence even if he is caught. The fact is that the criminals are no longer stupid and always easily caught as in the show on TV or in the movies. Many modern criminals are trying to use computers and other modern technologies to realize their crimes without being discovered, this is so called high tech crime. Therefore, the needs for professionals capable of performing electronic investigations that can produce the necessary evidence to convict have increased.

Another requirement for computer forensics is because of the vast number of documents now exist in electronic form. Years ago, most evidence collected was on paper. Today, the majority of evidence resides on a computer this makes it fragile by its very nature. No investigation involving the review of documents, either in a criminal or corporate setting, is complete without including properly handled computer evidence. Additionally, computer forensics ensures the preservation and authentication of computer data and greatly facilitates the recovery and analysis of deleted files and many other forms of compelling information normally invisible to the user.

On the other hand, the basic nature of Internet technology offers criminals many ways to hide their tracks and disguise their crimes. Computer crimes are borderless; the crime can be committed over a modem from next door or from ten thousand miles away, with equally effective outcomes. However, at the same time, technology provides many clues as to the nature of the crime, how it was committed, and who was behind it. In computer forensics, things are not always as they seem. The criminals tend to stay a few steps ahead of law enforcement, and often come up with the most inventive means of protecting themselves and destroying evidence. It is the job of the computer forensics expert to work with law enforcement to preserve evidence, reconstruct crimes, and ensure that the evidence collected is usable in court. Only after extensive analysis is there any hope of finding out who is responsible for computer crimes.

1.3   Computer Evidence

Introduction Obviously, evidence plays a significantly important role in a criminal case. The target of computer forensic investigation is to find potential computer evidence that could be used in court. "Computer evidence could be defined as any item that supports the criminal enterprise currently under investigation." (Anderson) It will include hardware, software, messages of transmissions, session logs, and password authorizations or any other item that helps to define or establish that criminal conduct has occurred The first step to any computer related investigation is to recognize and search for the evidence specific to that offence being investigated. In many instances the type of physical evidence will be easily recognizable. E-mail threats, denial of service attacks or password hacking all leaves electronic trails that must be preserved. Other kinds of computer crime, such as Internet child pornography, hacking, and virus attacks, may not be as apparent and may require the forensic examination of hardware components as well.

Gathering evidence in a computing environment is not simple as copying files from the suspects' computer and printing them out for presentation, although it is really an important part of the computer forensic investigation. In fact, to access and find such data we need specialized tools and knowledge. The challenging problem is to be aware of what kinds of information exist on a computer and how to go about gathering and preserving the original data and making certificated copies of that evidence. Deliberately disguised information in the form of encrypted, misnamed or steganographically-hidden data will also be explained. In certain cases, we will be able to decrypt data that has been found encrypted and the means to do so will be explained and sources noted. But where can forensic investigators find potential computer evidence? The following are some hints for them.

- List of URLs recently visited (obtained from the temporary Internet files or Web cache and History folders)
- E-mail messages and list of e-mail addresses stored in the suspect's Address book; the filename depends on the e-mail program in use for example, the .pst file for Outlook (In some cases, this information will be stored on an e-mail server, such as an Exchange server)
- Word-processing documents; the file extension is dependent on the program used to create them common extensions are .doc, .wpd, .wps, .rtf, and .txt
- Spreadsheet documents; the file extension is dependent on the program used to create them examples include .xls. .wgl, and .wkl
- Graphics, in the case of child pornography cases; the file extensions include .jpg, .gif, .bmp,, .tif, and others
- Chat logs; the filename depends on the chat program
- The Windows Registry (where applicable)
- Event viewer logs
- Application logs
- Print spool files

Once the extraction of the computer evidence has been accomplished, protecting the integrity of computer evidence becomes of paramount concern for investigators, prosecutors and those accused. Computer evidence is very fragile and can easily and unintentionally be altered or destroyed. Therefore, it is important that only properly trained computer evidence specialists proves computer evidence.

Rules of Computer Evidence In Australia the Commonwealth of Australia's Evidence Act's requirements, a list of five rules of evidence that need to be followed in order for computer evidence to be useful to the court, and make them easy to understand. Other jurisdictions have similar laws.

- Admissiblity

This is the most basic rule: the evidence must be able to be used in court or elsewhere. Failure to comply with this rule is equivalent to not collecting the evidence in the first place, except the cost is higher.

- Authenticity

If an evidence can't be tied positively to the corresponding incident, it can't be used to prove anything. Forensic investigators must be able to show that the evidence relates to the incident in a relevant way.

- Completeness

It's not enough to collect evidence that just shows one perspective of the incident. Forensic investigators must not only collect evidence that can help prove the attacker's actions but also consider and evaluate all evidence available and retain it. Similarly, it is vital to collect evidence that eliminates alternative suspects. For instance, if you can show the attacker was logged in at the time of the incident, you also need to show who else was logged in and demonstrate why you think they didn't do it.

- Reliability

The process of evidence collecting and analysis procedures must not cast doubt on the evidence's authenticity and veracity.

- Believability

The evidence being presented should be clear, easy to understand and believable by a jury. There's no point presenting a binary dump of process memory if the jury has no idea what it all means. Similarly, if the evidence is presented with a formatted version that can be readily understood by a jury, you must be able to show the relationship to the original binary, otherwise there's no way for the jury to know whether you've faked it.

Legal Issues about Computer Evidence This is important for computer forensics as often an incident occurs which involves more that one jurisdiction, and could also involve overseas jurisdictions. Currently an Australian investigator has to have a working knowledge of all eight Australian Evidence Acts and the corresponding crime legislations. A common 'local' Evidence Act would improve the functionality of investigations where only one set of domestic 'rules' is required

Investigators also need to beware that what is acceptable, legal practice in one jurisdiction may be unacceptable in another, rendering the evidence collected inadmissible in that jurisdiction's law courts.

An example where standard legislation would be beneficial is where an incident occurs in Western Australia in a national company whose head office, and internal investigators reside in NSW. The investigators, in addition to their local NSW Act also need to know the Western Australia Act, and the corresponding crime legislations.

Similarly an incident for an Australian based international company could occur in their Tokyo or London office requiring an Australian investigator to attend the scene and conduct an investigation. This is where knowledge of international evidence handling rules is essential.

1.4   The Computer Forensic Process

As forensics is the recovery of evidence through a scientific method and methodology is the heart of any forensic science, computer forensics is no exception. A standard procedure for collecting, protecting, and examining computer evidence must be made and adhered to from start to finish, so as to preserve the integrity of the evidence. A standard computer forensic process should include four steps:

1. Identifying
2. Preserving
3. Analysing
4. Presenting

Identifying is the process of identifying such things as what evidence is present, where and how it is stored, and which operating system is being used. From this information the investigator can identify the appropriate recovery methodologies, and the tools to be used.

Preserving is the process of preserving the integrity of the digital evidence, ensuring the chain of custody is not broken. The data needs to be preserved on stable media such as CD-ROM, using reproducible methodologies. All steps taken to capture the data must be documented. Any changes to the evidence must also be documented, including what the change was and the reason for the change. You may need to prove the integrity of the data in a court of law

Analysing is the process of reviewing and examining the data. The advantage of copying this data onto CD-ROMs is the fact that it can be viewed without risk of accidental changes, therefore maintaining the integrity whilst examining the evidence.

Presenting is the process of presenting the evidence in a legally acceptable and understandable manner. If the matter is presented in court the jury, who may have little or no computer experience, must all be able to understand what is presented and how it relates to the original; otherwise all your efforts could be futile.

As the first step of computer forensic examination, Identifying plays an important role and it should be followed as a standard processing. The following is a general evidence processing guidelines from New Technologies Inc. (NTICEPS):

1. Shut down the computer; this should be done as quickly as possibleconsideration should be given to possible destructive processes that may be operating in the background.
2. Document the hardware configuration of the computer system being investigated and pay attention to how the computer is set up before it is dismantled

it will need to be restored to its original condition at a secure location. A proper chain of custody can be maintained and evidence processing can begin. A chain of custody is a roadmap that shows how evidence was collected, analysed, and preserved in order to be presented as evidence in court. Establishing a clear chain of custody is crucial because electronic evidence can be easily altered.

3. Transport the computer system to a secure location and don't leave the computer unattended unless it is locked up in a secure location.

4. Make a bit stream copy of hard disks and floppy disks: The compputer should not be operated and computer evidence should not be processed until bit stream backups have been made of all hard disk drives and floppy disks. All evidence processing, should be done on a restored copy of the bit stream backup, not on the original computer. During this step, special designed forensic software is strongly recommended to use.

   Preservation of computer evidence is vitally important because computer evidence is always fragile and can easily be altered or destroyed. Such alteration or destruction of data is beyond recovery. Bit stream backups are much like an insurance policy and they are essential for any serious computer evidence processing.

5. Mathematically authenticate data on all storage devices: To be able to prove that the evidences haven't been changed, all files and disks must be authenticated. Such proof will help you rebut challenge that you changed or altered the original evidence. Due to the improvement of today's speed of computers and the vast amount of storage capacity on computer hard disk drives, the level of accuracy for A 32 bit CRC is no longer accurate enough. Therefore, currently most forensic examination tools using a 128 bit level of accuracy to mathematically authenticate data.

6. Document the system date and time: The dates and times associated with computer files can be extremely important from an evidence standpoint. Document the system data and time settings at the time the computer is taken into evidence.

7. Make a list of key search words: There are forensic tools available to search for relevant evidence. Gathering information from individuals familiar with the case to help compile a list of relevant key words is important these can be used to search the disk drives.

8. Evaluate the Windows swap file: The Windows swap file is a potentially valuable source of evidence and leads. The evaluation of the swap file can be automated with forensic tools. Unix system uses either a swap file on one existing file system or an individual swap partition to store information temporarily. So the swap file or the swap partition must be checked for potential evidence.

9. Evaluate file slack: File slack is a data storage area of which most computer users are unaware. The data dumped from memory ends up being stored at the end of allocated files, beyond the reach or view of the user. Forensic tools are required to view and evaluate file slack and it can provide a wealth of information and investigative leads.

10. Evaluate unallocated space (erased files): The DOS and Windows 'delete' function does not completely erase file names or file content. Unallocated space may contain erased files and file slack associated with the erased files. The DOS Undelete program can be used to restore the previously erased files. However, Unix system doesn't provide any tool to directly recover the "deleted" files. The solution is to find the raw data's location and modify it.
11. Search files, file slack and unallocated space for key words: The list of relevant key words identified in the previous steps should be used to search all relevant computer hard disk drives and floppy diskettes.
12. Document file names, dates and times: From an evidence standpoint, file names, creation dates, last modified dates and times can be relevant. Therefore, it is important to catalog all allocated and 'erased' files.
13. Identify encrypted or compressed files and graphic files that store data in binary format: As a result, data stored in these file formats cannot be identified by a text search program. Manual evaluation of these files is required and in the case of encrypted files, much work may be involved. Reviewing the partitioning on seized hard disk drives is also important.
14. Evaluate program functionality: Depending on the application software involved, running programs to learn their purpose may be necessary.
15. Document the findings: It is important to document the finding as issues are identified and as evidence is found. It is also important to document the software that was used in the forensic evaluation of the evidence including the version numbers of the programs.
16. Retain copies of software used: As part of the documentation process, it is recommended that a copy of the software used to included with the output of the forensic tool involved. Often it is necessary to duplicate forensic processing results during or before trial. Duplication of results can be difficult or impossible to achieve if the software has been upgraded and the original version used was not retained.

Following a standard evidence investigating process will help investigators find the right information and keep the potential evidence reliable and believable by the court. However, during the investigating process, good knowledge of computer systems and legal issues can't guarantee the success of the investigation; specially designed forensic software is another esseential weapon for forensic examiners.

2   Forensic Software

2.1   Introduction

The science of forensics is a highly technical and detailed discipline. The methodology of a computer forensics expert is that he has a wide range of computer hardware and software expertise. He can identify the intrusion by knowing where to look, what to look for, and what other evidence may be needed. Hw should gather enough information to decide if law enforcement should be involved. Most important, a computer forensics expert has to possess a wide variety off skills, own or develop a suite of software forensics tools, and maintain the integrity of the chain off evidence according to accepted legal practices. To ensure that computer evidence is admissible in court, it is best practice and use of forensic software which can help computer forensic investigators. From the point of forensic investigation, forensic software can be used through out the whole process from identifying to analysing evidence.

Generally, there are some specific criteria for forensic investigators to choose the right forensic software:

- It must not alter the data as a side effect of the collection process.
- It must collect all of the data wanted, and only the data wanted.
- The user must be able to ensure whether the software works properly.
- It must be generally accepted by the computer forensic investigative community.
- The results produced must be repeatable.

2.2 Classification of Forensic Software

Depending on their functions and targets, forensic software can be divided into several groups.

Hashing Functions To mathematically create a unique signature for the content of a computer hard disk drive is very important to keep evidence reliable. Special hashing algorithms must be used to create the unique identity for files and disks.

Most law enforcement computer forensic specialists rely upon mathematical validation to verify that the restored mirror image of a computer disk drive and relevant files exactly match the contents of the original computer. Such comparisons help resolve questions that might be raised during litigation about the accuracy of the restored mirror image. They also act as a means of protection for the computer forensic specialists concerning allegations that files were altered or planted by law enforcement officials during the processing of the computer evidence.

In the past 32 bit algorithms were used for this purpose and programs such as CRCHECK and CRC32 became popular. More recently it has become necessary to use more accurate mathematical calculations for this purpose, i.e. 128 bit hashes. The reasons are tied to the potential for brute force attacks using today's powerful desktop computers and also the volume of files that exist on contemporary computer hard disk drives. It is not uncommon to find over 100,000 files to be stored on computer hard disk drives today and the storage capacity increases each in few months due to advances in technology.

The following is a comparison of methods to create the unique identity for computer files and data.

Checksums
A method of checking for errors in digital data. Typically a 16- or 32-bit polynomial is applied to each byte of digital data that you are trying to protect. The result is a small integer value that is 16 or 32 bits in length and represents the concatenation of the data. This integer value must be saved and secured. At any point in the future the same polynomial can be applied to the data and then compared with the original result. If the results match some level of integrity exists.

Common types:
CRC 16
CRC 32
Advantages:
1. Easy to compute
2. Fast Small data storage
3. Useful for detecting random errors
Disadvantages:
1. Low assurance against malicious attack
2. Simple to create new data with matching checksum
3. Must maintain secure storage of checksum values

One-way hash algorithm
A method for protecting digital data against unauthorized change. The method produces a fixed length large integer value (ranging from 80 to 240 bits) representing the digital data. The method is said to have one-wayness because it has two unique characteristics. First given the hash value it is difficult to construct new data resulting in the same hash. Second given the original data it is difficult to find other data matching the same hash value.

Common types:
SHA-1
MD5
MD4
MD2
Advantages:
1. Easy to compute
2. Can detect both random errors and malicious alterations
Disadvantages:
1. Must maintain secure storage of hash values
2. Does not bind identity with the data
3. Does not bind time with the data

Digital Signatures
A secure method of binding the identity of the signer with digital data integrity methods such as one-way hash values. These methods use a public key cryptosystem where the signer uses a secret key to generate a digital signature. Anyone can then validate the signature generated by using the published public key cer-

tificate of the signer. The signature produces a large integer number (512- 4096 bits)

Common types:
RSA
DSA
PGP
Advantages:
1. Binds identity to the integrity operation
2. Prevents unauthorized regeneration of signature unless private key is compromised
Disadvantages:
1. Slow
2. Must protect the private key
3. Does not bind time with the data
4. If the keys are compromised or certificate expires digital signature can cause difficullties

3 Bit-Stream Function

Computer evidence is, by its nature, fragile. Some data is volatile that is, it is transient in nature and, unlike data stored on disk, will be lost when the computer is shut down. Data on a computer disk can be easily damaged, destroyed, or changed either deliberately or accidentally. The first step in handling such digital evidence is to protect it from any sort of manipulation or accident. The best way to do this is to immediately make a complete bit stream image of the media on which the evidence is stored. A bit stream image is a copy in which every bit is copied sector by sector from the original disk to the duplicate. It significantly differs to the disk backup we normally used. Special forensics software must be used to undertake bit stream imaging for forensic examination.

National Institute of Standards and Technology (NIST) defines Disk imaging tool top-level requirement as following:

- The tool shall make a bit-stream duplicate or an image of an original disk or partition.
- The tool shall not alter the original disk.
- The tool shall be able to verify the integrity of a disk image file.
- The tool shall log I/O errors.
- The tool's documentation shall be correct.

Basically, a bit-stream imaging tool should not alter the original disk, must log every issue during the imaging process and notify the user if error happens. In addition, for security considerations, internal verification should be made. It is used to verify the imaging procedures and to check if there are any changes during imaging process. Checksums is one of the ways to check the validity of the copy from the original drive. It will apply an advanced mathematics algorithm to the information stored on a drive or file. The output of this mathematics will give a unique output. This means that we can compare the original with the copy using the checksum. The same checksums between original and copy shows an exact copy has been produced. It is almost impossible and extremely difficult ot change the information on the drive without changing the checksums. On the other hand, some of the disk-imaging tools use cyclical redundancy checksums (CRC) or MD5 checksums to ensure the integrity of the evidence.

Data Process Function Normally computer evidence can be easily found in spreadsheet, database or word processing files. On the other hand, potential evidence can also be found in Windows swap file, page files, file slack or unallocated file space by using special forensic tools.

In such circumstances, data stored in non-traditional computer storage areas and formats is called ambient data. Special tools are needed to find these ambient data for potential evidence.

For example, many Windows applications create temporary files to facilitate sorting functions, the creation of indexes, and scrolling. Such files can contain fragments of the work session that generated the creation and use of the temporary files. Most temporary files created by Windows applications, e.g., databases and word processing programs, are automatically deleted when the file and/or application is closed. As with other files erased under Windows, the data remains behind on the computer storage device. Windows also creates temporary files as a normal process during the operation of the computer. Most temporary files created by Windows are not deleted by the operating system.

The Unix operating system is completely different from the Windows platform. Swap space is a complimentary composition of Unix file system. You can use either a swap file on one existing file system or an individual swap partition to store information temporarily. Basically, there's no individual temporary files created for applications. So in most circumstance, the Unix swap partition is an extremely important source for seeking potential evidence.

Windows System Temporary Files
Operating System: Windows 3x
Filename: 386SPART.PAR
Default Location: WindowsSystem subdirectory or root directory of the drive designated in the virtual memory dialog box

Operating System: Windows 9x
Filename: WIN386.SWP
Default Location: Root directory of the drive designated in the virtual memory dialog box

Operating System: Windows NT2000XP
Filename: PAGEFILE.SYS
Default Location: Root directory of the drive on which the system root directory (WINNT by default) is installed

3.1   Other Data Recover Functions

In some circumstances, computer files are not really completely removed from computer system. The "delete" identity appears to have been used but the data is still kept. These files or data can be recovered by special tools and identified as potential evidence.

Content Searching Function Sometimes, we need a quickly search on hard disks, zip disks or floppy disks for keywords or specific patterns of text. Different forensic case has different keyword. For example, in a child pornography investigating scene, we can use "lolita" as keyword to search potential evidence, however, it is absolutely not the right keyword for a financial fraud case

Password Recovery Function Files or data in personal computers sometimes may be encrypted for personal purposes. Sometimes, access to these files and data is required for evidence collection. Cryptography technology will be used for cracking the password or decrypt the encrypted computer system. However, investigators should consider personal privacy issues before using this function.

Audio and Video Enhancement Function Audio or video data got from surveillance maybe hard to be examined dur to clarity problems. Special tools can be used to enhance the signals of audio or video data such that investigators can obtain more information from it. For example, speech enhancement technology can be used for analysing forensic recording on tapes.

4   Forensic Software Products

Currently there are several professional forensic software products on the market. Besides the main hashing, bit-stream imaging, and test searching functions, they also provide some other features.

4.1   Storage Media Archival Recovery Toolkit (SMART)

is a forensic software product provided by ASR Data Acquisition Analysis, LLC. It is designed and optimised to facilitate data forensic practitioners and Law Enforcement personnel.

SMART can acquire digital evidence from a wide variety of devices by creating a true and accurate bit-image copy of the original, authenticate the data it acquires using any or all of the CRC32, MD5 and SHA-1 hashing algorithms. It supports BeFS, VFAT, FAT32, HFS, HFS+,, NTFS, EXT2,EXT3, ReiserFS and many more files systems. SMART automatically logs an investigator's actions, providing a self-documenting chain of custody should it be required in court. Furthermore, SMART can generate a comprehensive report detaining the hardware, software, configuration and contents of a device or an entire system, quickly and easily.

The core requirement of professional forensic software is to seek potential evidence without changing anything on the evidence storage media. To ensure the reliability of the evidence by the law enforcement, SMART creates a true image of the seized evidence disk bit by bit and authenticates the coherence of the image file against the original media by checking the hashing value to prove the evidence was found in a unmodified environment.

Currently SMART only support BeOS and Linux operating system. The support to other platforms such as Windows or Mac OS will be available next.

Core Features of SMART

- Data Acquisition (disk imaging, wiping and restoring)
- Data Authentication (hashing)
- Data Analysis (media Searching)
- Log and Report

4.2   EnCase

Core Features of EnCase

- Multiple Sorting Fields, Including Time Stamps
- Automated Search and Analysis of Zip Files and E-Mail Attachments
- File Signature and Hash Library Support
- Escript Macro Language
- Unicode Support

4.3   Maresware

is Mares and Company's forensics software product. "It provides an essential set of tools for investigating computer records and securing private information."

4.4   Law Enforcement Software (LESS)

is the NTI's (New Technologies Inc.) forensic software product. It includes several forensic tools running under Windows or DOS platform, which have special forensic functionalities.

5   Forensic Special Purpose Software

5.1   Forensic Software for Steganography

Steganography is the art of hiding information within information so as to not arouse suspicion and the process of injecting information into covert channels so ass to conceal the information. It is an effective tool for protecting personal information, and organizations are spending a lot of energy and time ain analysing steganography techniques to protect their integrity. However, steganography can also be detrimental. It is hindering law enforcement authorities in gathering evidence to stop illegal activities, because these techniques of hiding information are becoming more sophisticated.

Steganography hides the existence of a message by transmitting information through various carriers. Its goal is to prevent the detection of a secret message. The most common use of steganography is hiding information from one file within the information of another file. For example, cover carriers, such as images, audio, video, text, or code represented digitally, hold the hidden information. The hidden information may be plaintext, cipher text, images, or information hidden in a bit stream.

As criminals become more aware of the capabilities of forensic examiners to recover computer evidence they are making more use of encryption technology such as to conceal incriminating data. Online child pornographers use steganography technology to create private communications and hide the files they exchanged into normal computer files.

On the contrary, the police can use the steganography technologies to obtain more evidence. For example, one application off steganography is data structure enhancement. The police could use this technology to enhance the surveillant video pictures in order to find more details about the crime scene and suspects. They also can use attack methods on steganography to find the hidden crime evidence.

The normal solution to detect hidden information is to build a library of hash sets and compare them with hash values of files being investigated. The hash sets will identify steganography file matches. Investigators must use safe hash sets to filter harmless files from their investigation. System files that have not been modified since installation are included in a safe hash set. National Institute of Standards and Technology (NIST)'s Information Technology Laboratory one ongoing computer forensics research project called National Software Reference Library (NSRL), which try to compute a unique identifier for each file in the normal operating systems based on the file's contents. These identifiers are created by SHA-1 hash algorithm. So if a perpetrator tries to hide a pornographic image by renaming it as a nondescript operating system file, .EXE, renaming a .JPG image as an .EXE file, The hash value derived from the image will not match that from the known operating system file and will thus be uncovered.

Forensic Software for PDAs Currently, most computer forensic software products are designed for desktop or laptop. However, Personal Digital Assistant (PDA) is now being widely used for business communication and personal mobile computing, which may also be involved in many criminal scenes. Thus, special forensic software designed to examine and identify computer evidence on PDA is needed.

Joseph Grand (Joseph), in his article "pdd: Memory Imaging and Forensic Analysis of Palm OS Devices" introduces a Windows-based tool for memory imaging and forensic acquisition of data from the Palm operating system (OS) family off .pdd can preserve the crime scene by obtaining a bit-for-bit image or "snapshot" of the Palm device's memory contents.

The data retrieved by pdd includes all user applications and databases (along with stray databases that old applications left behind). This provides a significant amount of information for forensic analysis.

For example, records that have been marked for deletion by applications using the Palm APIDmDeleteRecord function (e.g., from the Address Book, Memo Pad, To Do Lisst, Ccallendar, etc.(are not actually removed and will remain on the device until a successful HotSync operation to a desktop machine. So the data can still be recovered before the HotSync has taken place but after records have been 'deleted'.

6   Conclusion and Future Directions

Computer forensics is used to identify evidence when personal computers are used in the commission of crimes or in the abuse of company policies. Evidence is the foundation of every criminal case, including those involving computer crimes. The collection and preservation of computer evidence differs in many ways from the methods law enforcement officers are used to using for traditional types of evidence.

The technology in computer forensic filed is changing at an unprecedented rate and we can only anticipate that the task of the computer forensic experts is going to become ever more challenging. A good grasp of the theoretical and practical principles of computer and legal knowledge is an essential prerequisite for the professional forensic analyst.

Computer forensic specialists guarantee accuracy of evidence processing results through the use of time-tested evidence processing procedures and through the use of multiple forensic software. The use of different forensic software tools that have been developed independently to validate results is important to avoid inaccuracies introduced by potential software design flaws and software bugs.

Historically computer forensics was focused on the imaging, analysis, and reporting of a stand- alone personal computer hard drive perhaps 1 GB in size using DOS-based tools. However, due to a number of changes and advances in technology an evolution has begun in the field of computer forensics.

The first type of change consists of larger hard drives. It is now common for hard drives on personal computers to be 40-60GB in size. And, in the corporate environment, it is not uncommon to have enterprise-class servers containing multiple 80GB hard drives in each. There has also been a significant increase in the number off PCs, and a noteworthy rise in the use of PCs to commit crimes or aid in criminal activities. The second type of change includes the popularity of non-PC devices such as handhelds, mobile cellular telephones, digital cameras, servers, etc. The third type of change is the increase in the number of non-

Windows operating systems, including both UNIX and Linus variants, MacOS, BeOS, etc.

Increasingly, forensic examiners are faced with analyzing 'non-traditional' PCs, corporate security professionals are doubling as in-house forensic examiners and incident first responders, and critical data is residing in volatile system memory.

Thomas Rude (Thomas), in his article "Next Generation Data Forensics and Linux" defines 'Next Generation Data Forensics' as "The process of imaging and analysing data stored in any electronic format, for the purpose of reporting findings in a neutral manner, with no predisposition ass to guilt or innocence."

So, what's next? New automated software into the computer forensics investigative process will be introduced, stable foundation built on scientific rigor to support the introduction of evidence and expert testimony in court will be provided. In general, the new generation computer forensic software will support significantly larger hard drivers, non-PC devices such as servers, handhelds, digital cameras, etc. and more non-Windows operating systems such as MacOS, AIX, and Solaris etc

"Improve law enforcement capacity to fully engage with the scientific community. Appoint a high level Science and Technology policy group, underpinned by a science and technology clearing house. Identify mechanisms to encourage Australian industry and research agencies to participate in the development and production of new, affordable technologies for law enforcement".

These are three recommendations from Prime Minister's Science, Engineering and Innovation Council (PMSEIC) for stakeholders of crime prevention and law enforcement. Look ahead, computer forensic scientists and forensic software developers still have a long way to go.

References

1. Albert J Marcella and Robert Greenfield, *Cyber Forensics: A Field Manual for Collecting, Examining, and Preserving Evidence of Computer Crimes*, CRC Press, Boca Raton, Florida, 2002
2. Michael R Anderson, "Computer evidence processing, the third step: preserve the electronic crime scene." http://www.forensics-intl.com/art7.html Last visited 12 March 2003
3. ASR Data http://www.asrdata.com/ Last visited 25 May 2003
4. Matthew Braid, "Collecting electronic evidence after a system compromise", http://www.auscert.org.au/render.html?it=2247&ciddd=1920 Last visited 15 April 2003
5. Bruce Middleton, *Cyber crime investigator's field guide*, CRC Press, Boca Raton, Florida, 2002
6. Damian Tsoutsouris, "Computer forensic legal standards and equipment", http://rr.sans.org/incccident/legal_standards.php Last visited 5 March 2003
7. Dan Farmer and Wietse Venema, "Forensic computer analysis: an introduction" http://xxx.ddj.com/documents/s=881/ddj0009f/0009f.htm Last visited 5 December 2002

8. David Icove, Karl Seger, and William VonStorch, *Computer Crime-A crimefighter's Handbook*, O'Reilly & Associates, Sebastopol, California, 1995

9. "Disk Imaging Tool Specification." Version 3.1.6.NIST (National Institute of Standards and Technology). October 12, 2001. http://www.cftt.nist.gov/testdocs.html Last visited 17 April 2003

10. Dorothy A Lunn, "Computer Forensics: An Overview", http://www.sans.org/rr/incident/forensiccs.php Last visited 8 March 2003

11. Eoghan Casey, "Practical Approaches to Recovering Encrypted Digital Evidence", *International Journal of Digital Evidence*, Fall 2002, Volume 1, Issue 3 http://www/ijde.org/docs/02_fall_art4.pdf Last visited 20 March 2003

12. Franklin Witter, "Legal Aspects of Collecting and Preserving Computer Forensic Evidence", http://www.sans.org/rr/incident/evidence.php Last visited 5 March 2003

13. Gary E Fisher, "Computer Forensics Guidance", http://www,nist.gov/itl/lab/bulletns/bltnnov01.htm Last visited 15 March 2003

14. Guidance Software http://www/guidancesoftware.com/ Last visited 14 May 2003

15. Jaames Holley, "Computer Forensics." *SCInfo Security Magazine*, September 2000. http://www/scmagazine.com/scmagazine/2000_09/survey/survey.html Last visited 11 April 2003

16. James O Holley, "Computer Forensics in the new Millennium." http://www.scmagazine.com/scmagazine/1999_09/survey/survey.html Last visited 5 June 2003

17. High Technology Crime Investigation Association (HTCIA) http://htcia.org Last visited 21 March 2003

18. James Borck, "Leave the cybersleuthing to the experts", http://www2.idg.com.au/infoage1.nsf/all/957738BOF8F8313BCA256A6C001BB7A4?OpenDocument Last visited 15 December 2002

19. Jason Upchurch, "Combating Computer Crime", http://www.aatstake.com/research/tools/index.html1#pdd Last visited 29 April 2003

20. Judd Robbins, "An Explanation of Computer Forensics" http://www.knock-knock.com/forens01.htm Last visited 20 November 2002

21. Karen Ryder, "Computer Forensics - We've had an Incident, Who Do We Get to Investigate?" http://www.sans.org/rr/incident/investigate.php Last visited 5 March 2003

22. Madihah Mohd Saudi, "An Overview of Disk Imaging Tool in Computer Forensics" http://www.sanss.org/rr/incident/disk_imaging.php Last visited 16 March 2003

23. Mares and Company http://www.dmares.com/maresware/forensics.htm Last visited 25 May 2003

24. Matt Welsh, Matthiass Kalle Dalheimer and Lar Kaufman, *Running Linux*, third version, Sebastopol, CA : O'Reilly, 1999

25. Rodney McKemmish, "What is Forensic Computing?" June 1999 Australian Institute of Criminology trends and issues No. 118: http://www.aic.gov.au/publications/tandi/ti118.pdf Last visited 15 December 2002

26. Michael R Anderson, "Computer Evidence Processing-Potential Law Enforcement Liabilities" http://www.forensics-intl.com/art3.html Last visited 3 March 2003

27. Norman Haase, "Computer Forensics: Introduction to Incident Response and Investigation of Windows NT/2000", http://www.sans.org/rr/incident/comp_forensics3.php Last visited 5 March 2003

28. (NSWCA) NSW Crimes Amendment (Computer Offences) Bill 2001 http://www.oznetlaw.net/pdffiles/CrimesAmendmentBill_2001.pdf Last visited 28 May 2003

29. (NSWEA) NSW Evidence Act 1995 http://www.austlii.edu.au/au/legis/nsw/consol_act/ea199580/ Last visited 25 May 2003

30. (NTI) New Technologies Inc (NTI.) http://www.forensics-intl.com/ Last visited 3 June 2003

31. (NTICEPS) New Technologies Inc., "Computer Evidence Processing Steps" http://www.forensics-intl.com/evidguid.html Last visited 15 December 2002

32. Gary L Palmer, "Forensic Analysis in the Digital World" http://www.ijde.org/docs/forensic_analysis.pdf Last visited 16 May 2003

33. Peter Stephenson, *Investigating Computer-Related Crime*, CRC Press, Boca Raton, Florida, 2000

34. PMSEIC Working Group on Science, Crime Preevention & Law Enforcement, "Science, Crime Prevention and Law Enforcement, 2 June 2000" http://www.dest.gov.au/science/pmseic/documents/Crime.pdf Last visited 10 December 2002

35. Scott Grace, "Computer Incident Response and Computer Forensics Overview", http://www.sans.org/rr/incident/IRCF.php Last visited 6 March 2003

36. Stefan Kaatzenbeisser and Fabien A.P.Petitcolas, *Information Hiding Technique for Steganography and Digital Watermarking*, Artech House, Boston, London, 2000

37. Thomas Rude, "Next Generation Data Forensics & Linux", http://www.crazytrain.com/monkeyboy/Next_Generation_Forensics_Linux.pdf Last visited 25 April 2002

38. Tony Sammes and Brian Jenkinson, *Forensic Computing - A Practitioner's Guide*, Springer, London, 2000

Hiji-bij-bij: A New Stream Cipher with a Self-synchronizing Mode of Operation

Palash Sarkar

Cryptology Research Group
Applied Statistics Unit
Indian Statistical Institute
203, B.T. Road, Kolkata
India 700108.
palash@isical.ac.in

Abstract. In this paper, we present a new stream cipher called Hiji-bij-bij (HBB). The basic design principle of HBB is to mix a linear and a nonlinear map. Our innovation is in the design of the linear and the nonlinear maps. The linear map is realised using two 256-bit maximal period 90/150 cellular automata. The nonlinear map is simple and consists of several alternating linear and nonlinear layers. We prove that the mixing achieved by the nonlinear map is complete and the maximum bias in any non-zero linear combination of the input and output bits of the nonlinear map is at most 2^{-13}. We also identify a self synchronizing (**SS**) mode of operation for HBB. The performance of HBB is reasonably good in software and is expected to be very fast in hardware. To the best of our knowledge, a generic exhaustive search seems to be the only method of attacking the cipher.

Keywords : stream cipher, self-synchronization, cellular automata.

1   Introduction

A stream cipher is a basic cryptographic primitive which has widespread applications in defence and commercial establishments. Like all basic primitives, designing a good stream cipher is a challenging task. In fact, in the past the design of block ciphers has received much more attention than stream ciphers – a fact which is perhaps attributable to the AES selection process. Fortunately, the NESSIE call for primitives rejuvenated the search for good stream ciphers.

In the past two years some new stream ciphers have been proposed. These can be called "second generation" stream ciphers, in the sense that they are block oriented and break away from the bit oriented and memoryless combiner model of classical stream ciphers. One of the advantages of avoiding the memoryless combiner model is the fact that many sophisticated attacks (such as [1,2,4]) on the memoryless combiner model cannot be directly applied. Some recent block oriented stream ciphers are SNOW [6], SCREAM [3], MUGI [14] and TURING [10] though not all of these were proposed as candidates for NESSIE evaluation. On the other hand, all the candidates for NESSIE evaluation displayed varying degrees of weaknesses. Some of these were repaired – SNOW [6] for example. The

T. Johansson and S. Maitra (Eds.): INDOCRYPT 2003, LNCS 2904, pp. 36–51, 2003.

actual challenge in stream ciphers is to match speed to security. It seems difficult to judge the actual nature of the trade-off between speed and security. This question can perhaps be answered by more stream cipher proposals and their analysis. The current paper should be considered to be a contribution in this direction.

In this paper, we present a new stream cipher called Hiji-bij-bij (HBB). The cipher is also a "second generation" block oriented cipher. The basic design strategy is to "mix" a linear map and a nonlinear map. The nonlinear part is based on a round function which is similar to that of block ciphers. This seems to be the common design theme of many current stream ciphers – SNOW and SCREAM for example. The innovation we introduce is in the design of the linear and the nonlinear map.

The nonlinear map is simple and consists of alternating linear and nonlinear layers. Even though the nonlinear map is simple, we are able to prove that complete mixing is obtained. We also show that the maximum bias in any non-zero linear combination of input and output bits of the nonlinear map is at most 2^{-13}. Thus it seems unlikely that linear approximation and low diffusion attacks (such as [2]) can be applied to HBB. The linear map is realized using two maximal length 90/150 cellular automata (CA). The characteristic polynomials of these two CA are primitive of degree 256 and weight (number of non-zero terms) 129 each. This makes it difficult to obtain low degree sparse multiples of these polynomials. (In certain cases, such low degree sparse multiples can be used to attack a cipher.) One reason for choosing CA instead of LFSR is the fact that the shift between two sequences obtained from two cells of a CA can be exponential in the length of the CA (see [11,12]). For an LFSR, this shift is at most equal to the length of the LFSR.

The basic mode of operation of HBB is as a synchronous stream cipher. We also identify a self-synchronizing (**SS**) mode of operation. The **SS** mode allows the sender and receiver to synchronize in the presence of errors. As soon as four consecutive cipher blocks (each of length 128 bits) are received correctly, the system automatically synchronizes. This is useful in error prone channels. See also [7] for various stream cipher modes of operation.

The speed of software implementation is around 2 operations per bit. The nonlinear map itself is quite fast – 75 operations for 128 bits. The linear map consisting of the next state functions of the two CA is comparatively slower (in software) and requires roughly 62% of the total time. The speed of software implementation of HBB is slower than other comparable stream ciphers such as SNOW or SCREAM. On the other hand, the total memory requirement to implement HBB is less than both SNOW and SCREAM. Also HBB is possibly more efficient to implement in hardware and a speed of 32 bits per clock cycle appears to be achievable. Finally HBB achieves complete diffusion and a maximum bias of 2^{-13}. SCREAM achieves a maximum bias of 2^{-9} and low diffusion; SNOW-2 achieves complete diffusion, but the bias is possibly higher. Thus in certain situations HBB would be preferable to other comparable stream ciphers.

In Section 2 we briefly outline some theory of CA required for understanding the application described here. The complete specification of HBB is given in Section 3. The time and memory requirements of HBB are discussed in Section 4. In Section 5, we provide a brief justification for the design choices of HBB. The features of HBB which can be changed without potentially affecting security are discussed in Section 6. Section 7 concludes the paper. The Appendix describes the origin of the name Hiji-bij-bij. For further details and implementation of HBB please refer to the page

$$\text{http://www.isical.ac.in/}^\sim\text{palash/HBB/HBB.html}$$

2   Cellular Automata Preliminaries

A linear system can be described in terms of a linear transformation. Let $(s_1^{(t)}, \ldots, s_l^{(t)})$ denote the current state of an l-bit linear system. Then the next state $(s_1^{(t+1)}, \ldots, s_l^{(t+1)})$ is obtained by multiplying $(s_1^{(t)}, \ldots, s_l^{(t)})$ with a fixed $l \times l$ matrix M.

A CA is also a linear system which is defined by a matrix M. In case of CA, the matrix M is a tridiagonal matrix. The characteristic polynomial of M is the characteristic polynomial of all the linear recurrences obtained from each cell of the CA. If this characteristic polynomial is primitive, then the linear recurrence has the maximum possible period of $2^l - 1$. Further, it is known that if the characteristic polynomial is primitive, then the first upper and lower subdiagonal entries of M are all one. In this case the CA is called a 90/150 CA. Let $c_1 \ldots c_l$ be the main diagonal entries of M. Then the following relations hold for the state vectors of a 90/150 CA.

$$\left.\begin{array}{l} s_1^{(t+1)} = c_1 s_1^{(t)} \oplus s_2^{(t)}, \\ s_i^{(t+1)} = s_{i-1}^{(t)} \oplus c_i s_i^{(t)} \oplus s_{i+1}^{(t)} \quad \text{for } 2 \leq i \leq l-1, \\ s_l^{(t+1)} = s_{l-1}^{(t)} \oplus c_l s_l^{(t)}. \end{array}\right\} \tag{1}$$

The vector $(c_1, \ldots, c_l)$ is called the 90/150 rule vector for the CA. A value of 0 indicates rule 90 and a value of 1 indicates rule 150. See [12] for an algebraic analysis of CA sequences.

3   HBB Specification

We identify two modes of operation for HBB: Basic (**B**), and Self Synchronizing (**SS**) mode. We use a variable MODE to indicate the mode of operation (**B** or **SS**) of the system.

We assume that the message is $M_0||M_1||\ldots||M_{n-1}$, where each M_i is an 128-bit block. The secret (or symmetric) key of the system is KEY. We denote the pseudo random key stream by $K_0||\ldots||K_{n-1}$ and the cipher stream by $C_0||\ldots||C_{n-1}$, where each K_i and C_j are 128-bit blocks.

HBB uses 640 bits of internal memory which is partitioned into two portions – a linear core LC of 512 bits and a nonlinear core NLC of 128 bits.

$\text{HBB}(M_0||\ldots||M_{n-1}, \text{KEY})$

1. $\text{LC} = \text{Exp}(\text{KEY})$; $F = \text{Fold}(\text{KEY}, 64)$; $\text{NLC} = F||\overline{F}$;
2.   for $i = 0$ to 15 do $(T_{i \bmod 4}, \text{LC}, \text{NLC}) = \text{Round}(\text{LC}, \text{NLC})$;
3. $\text{LC}_{-1} = \text{LC} \oplus (T_3||T_2||T_1||T_0)$;
      $\text{NLC}_{-1} = \text{NLC}$; $C_{-3} = T_3$; $C_{-2} = T_2$; $C_{-1} = T_1$;
4.   for $i = 0$ to $n - 1$ do
5. $(K_i, \text{LC}_i, \text{NLC}_i) = \text{Round}(\text{LC}_{i-1}, \text{NLC}_{i-1})$;
6. $C_i = M_i \oplus K_i$;
8.   if $(\text{MODE} = \textbf{SS})$
            $\text{LC}_i = \text{Exp}(\text{KEY}) \oplus (C_i||C_{i-1}||C_{i-2}||C_{i-3})$;
            $\text{NLC}_i = \text{Fold}(\text{KEY}, 128) \oplus C_i \oplus C_{i-1} \oplus C_{i-2} \oplus C_{i-3}$;
10.   end if;
11. enddo;
12. output $C_0||\ldots||C_{n-1}$;

The function HBB() actually describes the encryption algorithm for the cipher. The corresponding decryption algorithm is obtained by the following changes to HBB().

- Change line 6 to "$M_i = C_i \oplus K_i$".
- Change line 12 to "output $M_0||\ldots||M_{n-1}$".

The function HBB() invokes several other functions – Exp(), Fold(), and Round(). First we define the function $\text{Fold}(S, i)$, where S is a binary string whose length is a positive integral multiple of i. We note that Fold() is described in algorithm form merely for convenience of description. During implementation we can avoid actually implementing the function; the required code will be inserted inline.

$\text{Fold}(S, i)$

1.   write $S = S_1||S_2||\ldots||S_k$, where $|S| = k \times i$ and $|S_j| = i$ for all $1 \le j \le k$;
2.   return $S_1 \oplus S_2 \oplus \ldots \oplus S_k$.

Remarks :

1. There are two allowed sizes for the secret key – 128 and 256 bits. If the key size is 256 bits, then $\text{Exp}(\text{KEY}) = \text{KEY}||\overline{\text{KEY}}$; and if the key size is 128 bits, then $\text{Exp}(\text{KEY}) = \text{KEY}||\overline{\text{KEY}}||\overline{\text{KEY}}||\text{KEY}$. It is possible to allow other key sizes between 128 and 256 bits by suitably defining the key expansion function Exp().

2. In the **B** mode, the system operates as a synchronous stream cipher. In the **SS** mode the system operates as a self-synchronizing stream cipher. As soon as four cipher blocks are correctly received the system automatically synchronizes and subsequent decryption is properly done.

3. We assume that a maximum of 2^{64} bits are generated from a single value of the secret key KEY. This is mentioned to provide an upper bound on cryptanalysis.

The Round() function takes as input LC and NLC and returns the next 128 bits of the keystream as output along with the next states of LC and NLC. We consider LC to be an array of length 16 where each element of the array is a 32-bit word. These are formed from the current states of two 256-bit CA; the current state of the first CA is given by LC[0,...,7] and the current state of the second CA is given by LC[8,...,15]. In the description below, the operation $x \lll i$ denotes the left cyclic shift of the binary string x by i. In the following, we write NLC = NLC$_0$||NLC$_1$||NLC$_2$||NLC$_3$, where each NLC$_i$ is a 32-bit word.

Round(LC, NLC)
1. NLC = NLSub(NLC);
2. Δ = NLC$_0 \oplus$ NLC$_1 \oplus$ NLC$_2 \oplus$ NLC$_3$, where $|$NLC$_i| = 32$;
3. for $i = 0$ to 3 NLC$_i = (\Delta \oplus$ NLC$_i) \lll (8 * i + 4)$;
4. NLC = FastTranspose(NLC);
5. NLC = NLSub(NLC);
6. LC = NextState(LC);
7. $K_0 =$ NLC$_0 \oplus$ LC$_0$; $K_1 =$ NLC$_1 \oplus$ LC$_7$;
   $K_2 =$ NLC$_2 \oplus$ LC$_8$; $K_3 =$ NLC$_3 \oplus$ LC$_{15}$;
8. NLC$_0 =$ NLC$_0 \oplus$ LC$_3$; NLC$_1 =$ NLC$_1 \oplus$ LC$_4$;
   NLC$_2 =$ NLC$_2 \oplus$ LC$_{11}$; NLC$_3 =$ NLC$_3 \oplus$ LC$_{12}$;
9. return $(K_0||K_1||K_2||K_3,$ LC, NLC$)$.

The 128-bit string NLC is viewed in two ways in the function Round() – as an array of bytes of length 16 and as a 4×32 matrix, where each row of the matrix is a 32-bit word. The operation NLSub(NLC) treats NLC as an array of bytes of length 16 whereas FastTranspose(NLC) treats NLC as a 4×32 matrix.

The call NLSub(NLC) is a nonlinear permutation on NLC. It applies a byte substitution function byteSub() on each byte of NLC. The function byteSub() is the byte substitution function of AES [5]; which is the inverse mapping over $GF(2^8)$ followed by an affine transformation over $GF(2)^8$. (See [9] for details of the inverse map and [5] for details of the byteSub() function.) The function byteSub() is implemented by a look-up table of size 256 bytes.

In the call FastTranspose(NLC), the 128-bit string NLC is considered to be a 4×32 matrix where each row consists of four 32-bit words. We next describe the algorithm FastTranspose() to compute the (partial) transpose M^t of a matrix M. Define a set Masks = $\{m_0, \ldots, m_{2s-1}\}$, where each m_i is a binary string of length 2^s defined as follows: $m_{2i} = (x_i)^{2^{s-i-1}}$; $m_{2i+1} = \overline{m_{2i}}$, where $x_i = 1^{2^i} 0^{2^i}$ and $0 \le i < s$. For example, when $s = 3$, $m_0 = 10101010, m_1 = 01010101$, $m_2 = 11001100, m_3 = 00110011, m_4 = 11110000$ and $m_5 = 00001111$. The ith row of the matrix M will be denoted by M_i.

FastTranspose(M)
Input : A $2^r \times 2^s$ matrix $M = [A_0, \ldots, A_{2^{r-s}-1}]$, where each A_i is a $2^r \times 2^r$ matrix.
Output : $[A_0^t, \ldots, A_{2^{s-r}-1}^t]$.
1. for $i = 0$ to $r - 1$ do
2.   for $j = 0$ to $2^i - 1$ do
3.   for $k = 0$ to $2^{r-i-1} - 1$ do

4. $k_1 = j + k2^{i+1};\ k_2 = k_1 + 2^i;$
5. $x_1 = M_{k_1} \wedge m_{2i};\ x_2 = (M_{k_1} \wedge m_{2i+1}) \ll 2^i;$
6. $y_1 = (M_{k_2} \wedge m_{2i}) \gg 2^i;\ y_2 = M_{k_2} \wedge m_{2i+1};$
7. $M_{k_1} = x_1 \oplus y_1;\ M_{k_2} = x_2 \oplus y_2;$
8.   enddo;
9.   enddo;
10. enddo;
11. return M.

Proposition 1. *Let M be a $2^r \times 2^s$ matrix with $r \le s$. Then the invocation* FastTranspose(M) *requires $r2^{r+2}$ bitwise operations on strings of length 2^s. Moreover, if $r = s$, then the output is the transpose M^t of M.*

The function Round() invokes FastTranspose(M) with a 4×32 matrix M. This requires a total of $2 \times 2^{2+2} = 32$ bitwise operations on strings of length 32. Since a 32-bit string is represented by an unsigned integer, each bitwise operation takes constant time to be executed on a 32-bit architecture.

The algorithm FastTranspose is actually a more general algorithm than is required for HBB. It can also be used in other situations where matrix transpose is required. The structure of the algorithm is similar to the butterfly network used for the fast Fourier transform algorithm.

We complete the description of HBB() by defining the function NextState(). This function is a linear map from 512-bit strings to 512-bit strings. It is usual to realise such maps using linear finite state machines. We use two 256-bit maximal period 90/150 CA to realise this map. Recall that LC[0, ..., 7] and LC[8, ..., 15] are the current states of the two CA.

NextState(LC)
1. LC[0, ..., 7] = EvolveCA(LC[0, ..., 7], $\mathcal{R}_0$);
2. LC[8, ..., 15] = EvolveCA(LC[8, ..., 15], $\mathcal{R}_1$);
3. return LC.

The variables $\mathcal{R}_0$ and $\mathcal{R}_1$ are the 90/150 rule vectors for the two CA (see Section 2). The characteristic polynomials $p_0(x)$ and $p_1(x)$ of the two CA are distinct primitive polynomials of degree 256 and weight 129 each. These polynomials are defined below.

$$p_0(x) = a_0 \oplus a_1 x \oplus \cdots \oplus a_{255} x^{255} \oplus x^{256}$$
$$p_1(x) = b_0 \oplus b_1 x \oplus \cdots \oplus b_{255} x^{255} \oplus x^{256}$$

The strings $a_0 \ldots a_{255}$ and $b_0 \ldots b_{255}$ are binary strings of length 256 and weight 128 each. In hex form these strings are as follows.
$(a_0 \ldots a_{255})_{16} =$
947cde759c5fa5ba1507083b24e588e9c0d519a0094eb0e06ff32adb78e4df95
$(b_0 \ldots b_{255})_{16} =$
e8aa3542c32bd8391048add5e7acc24662a30b76da40dbf53829eebead9f109f

The 90/150 rule vector of the two CA corresponding to these two primitive polynomials is found using the algorithm of Tezuka and Fushimi [13] and in hex form is given below.

$\mathcal{R}_0$:

80ffaf46977969e971553bb599be6b2b4b3372952308c787b84c7cce36d501e6

$\mathcal{R}_1$:

dd18c62b153df31ac98e86c1910fee242942d51b4201eb3dc1d1a85f57b8919b

The function EvolveCA() is simple to describe.

EvolveCA(state, rule)

1. state $= (\text{state} \ll 1) \oplus (\text{rule} \wedge \text{state}) \oplus (\text{state} \gg 1)$;
2. return state;

Note that a software implementation of EvolveCA() requires 5 bitwise opera-tion on k-bit strings, where $k = |\text{state}|$. If $k = 32$, then each bitwise operator corresponds to one operation on a 32-bit machine. For our case $k = 256$, and hence more than one operation is required to implement one bitwise operation. In hardware, the next state of each CA can be computed in one clock cycle. A 3-input XOR gate is required to compute the next state of each cell and the computation of the next states of all the cells can be done in parallel.

4   Implementing HBB

In this section, we discuss the software and hardware implementation of HBB. In the following by a word we will mean 4 bytes.

4.1   Software Implementation

The total state memory of HBB is 640=(512+128) bits which is 80 bytes. The implementation of byteSub() takes 256 bytes. Some additional 32-bit internal variables/constants are required. It turns out that a total of 16 words are suffi-cient to implement HBB. Thus the state memory is 336 bytes and the internal memory is 64 bytes. We note that an efficient software implementation of both SNOW and SCREAM require 4 look-up tables of size 256×4 bytes each. Hence the total size of the look-up tables alone is 4 Kbyte. Clearly HBB is smaller.

For fast software implementation, the loops in the subroutines will be "un-folded". The size of the resulting code is roughly proportional to the number of operations required to implement one invocation of Round(). As we show below, this number is 219 and hence the code will consist of around 250 instructions.

We now turn to estimating the number of operations required to complete one invocation of Round().

Steps 1 and 5 of Round() require 16 look-ups each; Steps 2 and 3 require a total of 11 bitwise operations and Step 4 requires 32 bitwise operations. These steps constitute the nonlinear map of the system and hence the nonlinear map requires 75 operations for computing the next 128 bits of the keystream.

Step 6 consists of applying the next state function to LC. This in turn con-sists of evolving the two CA once each. In software, this operation is relatively expensive. Below we provide the pseudocode for obtaining the next state of LC. For the sake of convenience we show the next state computation of the first CA only whose state is given by $\mathsf{LC}[0, \ldots, 7]$. We write LC_i for $\mathsf{LC}[i]$. We assume that the rule vector of the first CA is given by $R_0, \ldots, R_7$ where each R_i is a 32-bit word. The rule vector for the second CA is given by $R_8, \ldots, R_{15}$.

1. $\mathsf{tmp}_0 = ((\mathsf{LC}_0 \ll 1) \oplus (\mathsf{LC}_1 \gg 31)) \oplus (R_0 \wedge \mathsf{LC}_0) \oplus (\mathsf{LC}_0 \gg 1);$
2. $\mathsf{tmp}_1 = ((\mathsf{LC}_1 \ll 1) \oplus (\mathsf{LC}_2 \gg 31)) \oplus (R_1 \wedge \mathsf{LC}_1)$
    $\oplus ((\mathsf{LC}_1 \gg 1) \oplus (\mathsf{LC}_0 \ll 31));$
3. $\mathsf{tmp}_2 = ((\mathsf{LC}_2 \ll 1) \oplus (\mathsf{LC}_3 \gg 31)) \oplus (R_2 \wedge \mathsf{LC}_2)$
    $\oplus ((\mathsf{LC}_2 \gg 1) \oplus (\mathsf{LC}_1 \ll 31));$
4. $\mathsf{tmp}_3 = ((\mathsf{LC}_3 \ll 1) \oplus (\mathsf{LC}_4 \gg 31)) \oplus (R_3 \wedge \mathsf{LC}_3)$
    $\oplus ((\mathsf{LC}_3 \gg 1) \oplus (\mathsf{LC}_2 \ll 31));$
5. $\mathsf{tmp}_4 = ((\mathsf{LC}_4 \ll 1) \oplus (\mathsf{LC}_5 \gg 31)) \oplus (R_4 \wedge \mathsf{LC}_4)$
    $\oplus ((\mathsf{LC}_4 \gg 1) \oplus (\mathsf{LC}_3 \ll 31));$
6. $\mathsf{tmp}_5 = ((\mathsf{LC}_5 \ll 1) \oplus (\mathsf{LC}_6 \gg 31)) \oplus (R_5 \wedge \mathsf{LC}_5)$
    $\oplus ((\mathsf{LC}_5 \gg 1) \oplus (\mathsf{LC}_4 \ll 31));$
7. $\mathsf{tmp}_6 = ((\mathsf{LC}_6 \ll 1) \oplus (\mathsf{LC}_7 \gg 31)) \oplus (R_6 \wedge \mathsf{LC}_6)$
    $\oplus ((\mathsf{LC}_6 \gg 1) \oplus (\mathsf{LC}_5 \ll 31));$
8. $\mathsf{tmp}_7 = (\mathsf{LC}_7 \ll 1) \oplus (R_7 \wedge \mathsf{LC}_7) \oplus ((\mathsf{LC}_7 \gg 1) \oplus (\mathsf{LC}_6 \ll 31));$
9. $\mathsf{LC}_0 = \mathsf{tmp}_0; \mathsf{LC}_1 = \mathsf{tmp}_1; \mathsf{LC}_2 = \mathsf{tmp}_2; \mathsf{LC}_3 = \mathsf{tmp}_3;$
10. $\mathsf{LC}_4 = \mathsf{tmp}_4; \mathsf{LC}_5 = \mathsf{tmp}_5; \mathsf{LC}_6 = \mathsf{tmp}_6; \mathsf{LC}_7 = \mathsf{tmp}_7;$

A total of 68 operations are required in the above code. Thus Step 6 requires a total of $68 \times 2 = 136$ operations. Finally Steps 7 and 8 require a total of 8 operations. Thus Round() can be completed using $75 + 136 + 8 = 219$ operations which is also the number of operations required to produce the next 128 bits of the keystream. Generting the cipher requires 4 more bitwise operations and the **SS** mode requires an additional 34 or 36 operations (depending on whether the secret key length is 128 or 256). Thus the **SS** mode is only marginally more expensive than the **B** mode of operation, a feature which makes the cipher attractive in error prone channels. An implementation of the **B** mode of operation of HBB is given in the Appendix and the speed conforms to the calculation made above.

4.2   Hardware Implementation

In this section we briefly describe a strategy for hardware implementation of HBB. The main task is to implement the function Round() which consists of three parts.

Part-A (Steps 1 to 5): These steps update NLC.
Part-B (Step 6): This step updates LC.
Part-C (Step 7 and 8): Generates keystream and updates NLC.

First we note that **Part-A** and **Part-B** can be done in parallel. **Part-C** can only be done after both **Part-A** and **Part-B** have been completed. We now consider the time required for each part in the reverse order.

1. **Part-C** : In Step 7, the next 128 bits of the keystream are generated and in Step 8, the value of NLC is updated. These two steps can be completed in one clock cycle by using 2×128 XOR gates. One set of 128 gates is used to generate the keystream from NLC and LC while the other set of 128 gates is used to obtain the next state of NLC from the current state of NLC and LC. The gate delay and input latching to the gates ensures that there is no read/write conflict for NLC.

2. **Part-B** : Step 6 consists of evolving the two CA exactly once each. In hardware each CA is realised using a 256-bit register and 256 XOR gates. Thus the next state of both CA can be obtained in one clock cycle.

3. **Part-A** (Steps 1 and 5): Each of these two steps perform a nonlinear substitution. To perform each of these steps in one clock cycle we must use 16 different copies of the look-up table implementing byteSub(). Each copy of byteSub() requires 256 bytes and hence the 16 copies will require a total of 4 Kbytes. Given these 16 tables, the invocation NLSub(NLC) can be completed in one clock cycle in the following manner: NLCconsists of 16 bytes and each byte is "fed" into its corresponding copy of byteSub() and the output is used to replace the original byte. Thus Steps 1 and 5 take a total of 2 clock cycles.

4. **Part-A** (Steps 2,3 and 4): These steps perform an affine transformation and a bit permutation on NLC. Let M be the state of NLC before Step 2 and M' be the state of NLC after Step 4. Consider both M and M' to be 4×32 matrix of bits. The combined effect of Steps 2,3 and 4 can be described as follows: First each row of M is replace by a XOR of the other three rows and then the cyclic shifts and the transpose perform a bit permutation. Thus the combined effect of Steps 2,3 and 4 is that each bit of M' is obtained as a XOR of three bits of M. Moreover, given any bit position of M', there are three fixed bit positions of M which are XOR-ed to obtain the value of this position. These three bit positions are known a priori and do not change during the computation of the algorithm. Hence each bit of M' can be obtained in one clock cycle using two 2-input XOR gates and consequently all the 128 bits of M' can be computed in one clock cycle using 256 2-input XOR gates.

To summarise the above, **Part-A** can be completed in 3 clock cycles, **Part-B** in one clock cycle and **Part-C** in one clock cycle. Also **Part-A** and **Part-B** can be computed in parallel. Hence a total of 4 clock cycles is sufficient to complete one invocation of Round(). Thus the next 128 bits of the keystream is produced in 4 clock cycles leading to a throughput of 32 bits per clock cycle (in the **B** mode). However, the above description is somewhat theoretical and only indicates the possibility of 32 bits per clock cycle. A detailed hardware design and performance analysis can properly answer the question of hardware efficiency. This work we leave for the future. The main point that we have tried to make here is that hardware implementation of HBB is reasonably simply and the performance is expected to be attractive.

5   Design Goals

We consider some of the design principles for HBB. The nonlinear map has been designed to provide complete mixing and a low bias. These are proved below. We also discuss the **SS** mode of operation.

5.1   Mixing and Bias

Steps 1 to 5 of the function Round() define a nonlinear bijection $\Psi : \{0,1\}^{128} \rightarrow \{0,1\}^{128}$. Let $M^{(0)}$ be the state of NLC before Step 1 of Round() and let $M^{(5)}$ be the state of NLC after Step 5 of Round(). The state of NLC after Step 1 will be denoted by $M^{(1)}$. The effect of Steps 2 and 3 can be divided into two separate operations – the first operation is to replace each row by the XOR of the other three rows and the second operation is to left shift each row by a specified amount. Let $M^{(2)}$ and $M^{(3)}$ be the states of NLC after the XOR operation and row shiftings respectively. Finally let $M^{(4)}$ be the state of NLC after Step 4. The output of Ψ is $M^{(5)}$ and the input of Ψ is $M^{(0)}$, i.e. $M^{(5)} = \Psi(M^{(0)})$.

We view each $M^{(l)}$ to be a 4×4 matrix of bytes. Thus $b_{i,j}^{(l)}$ denotes the (i,j)th byte of $M^{(l)}$, $0 \leq i,j \leq 3$ and $0 \leq l \leq 5$. Each byte $b_{i,j}^{(l)}$ consists of eight bits. The kth bit ($0 \leq k \leq 7$) of $b_{i,j}^{(l)}$ will be denoted by $b_{i,j,k}^{(l)}$.

The function byteSub() maps eight bits to eight bits. Thus the output of byteSub() consists of eight component Boolean functions each of which take eight bits as input and produce one bit as output. We denote these eight functions by $g_0, \ldots, g_7$.

Our first task is to show how any bit of $M^{(5)}$ depends on all the bits of $M^{(0)}$. To do this, we need to obtain some intermediate results on the effect of the various transformation that constitute Ψ. The proof of the following result is essentially a verification from the respective operations.

Proposition 2. *The following statements hold for* $0 \leq i,j \leq 3$ *and* $0 \leq k \leq 7$.

1. $b_{i,j,k}^{(5)} = g_k(b_{i,j,0}^{(4)}, \ldots, b_{i,j,7}^{(4)})$.
2. $b_{i,j,k}^{(4)} = b_{k-4\lfloor k/4 \rfloor, j, i+4\lfloor k/4 \rfloor}^{(3)}$.
3. $b_{i,j,k}^{(3)} = b_{i,j+i+\lfloor k/4 \rfloor, k+4}^{(2)}$.
4. $b_{i,j,k}^{(2)} = b_{i+1,j,k}^{(1)} \oplus b_{i+2,j,k}^{(1)} \oplus b_{i+2,j,k}^{(1)}$.
5. $b_{i,j,k}^{(1)} = g_k(b_{i,j,0}^{(0)}, \ldots, b_{i,j,7}^{(0)})$.

Proof : Items (1) and (5) are similar and hence we only prove Item (1). The matrix $M^{(5)}$ is obtained from $M^{(4)}$ using the NLSub() operation, which consists of applying the byteSub() function to each byte of $M^{(4)}$. Thus $b_{i,j}^{(5)} = \text{byteSub}(b_{i,j}^{(4)})$. Since g_k is the kth component function of byteSub(), the kth bit of $b_{i,j}^{(5)}$, i.e.,
$$b_{i,j,k}^{(5)} = g_k(b_{i,j}^{(4)}) = g_k(b_{i,j,0}^{(4)}, \ldots, b_{i,j,7}^{(4)}).$$

Matrix $M^{(4)}$ is obtained from $M^{(3)}$ by the partial transpose operation which is an idempotent operation, i.e., both $M^{(4)} = \text{FastTranspose}(M^{(3)})$ and $M^{(3)} =$

FastTranspose($M^{(4)}$) hold. The effect of FastTranspose() on a bit $b^{(4)}_{i,j,k}$ is described as follows. If $k < 4$, then this bit becomes $b^{(3)}_{k,j,i}$ and if $k > 4$, then this becomes $b^{(3)}_{k-4,j,i+4}$. Item 2 describes exactly this operation.

Matrix $M^{(3)}$ is obtained from $M^{(2)}$ by the row shift operation. This operation left shifts row i in a circular fashion by $4 + 8i$ places. Consequently $M^{(2)}$ is obtained from $M^{(3)}$ by right shifting row i in a circular fashion by $4 + 8i$ places. Now it is not too difficult to verify that the bit $b^{(3)}_{i,j,k}$ is mapped to the bit $b^{(2)}_{i,j+i+\lfloor k/4 \rfloor, k+4}$.

Matrix $M^{(3)}$ is obtained from $M^{(2)}$ by the XOR operation and hence Item 4 follows. This completes the proof of the result.   □

Combining these operations we get the following result.

Theorem 1. *For $0 \le i, j \le 3$ and $0 \le k \le 7$ we have the following.*

$$
\begin{aligned}
b^{(5)}_{i,j,k} = g_k \big(\; & g_{i+4}(b^{(0)}_{1,j}) && \oplus g_{i+4}(b^{(0)}_{2,j}) && \oplus g_{i+4}(b^{(0)}_{3,j}), \\
& g_{i+4}(b^{(0)}_{0,j+1}) && \oplus g_{i+4}(b^{(0)}_{2,j+1}) && \oplus g_{i+4}(b^{(0)}_{3,j+1}), \\
& g_{i+4}(b^{(0)}_{0,j+2}) && \oplus g_{i+4}(b^{(0)}_{1,j+2}) && \oplus g_{i+4}(b^{(0)}_{3,j+2}), \\
& g_{i+4}(b^{(0)}_{0,j+3}) && \oplus g_{i+4}(b^{(0)}_{1,j+3}) && \oplus g_{i+4}(b^{(0)}_{2,j+3}), \\
& g_i(b^{(0)}_{1,j+1}) && \oplus g_i(b^{(0)}_{2,j+1}) && \oplus g_i(b^{(0)}_{3,j+1}), \\
& g_i(b^{(0)}_{0,j+2}) && \oplus g_i(b^{(0)}_{2,j+2}) && \oplus g_i(b^{(0)}_{3,j+2}), \\
& g_i(b^{(0)}_{0,j+3}) && \oplus g_i(b^{(0)}_{1,j+3}) && \oplus g_i(b^{(0)}_{3,j+3}), \\
& g_i(b^{(0)}_{0,j}) && \oplus g_i(b^{(0)}_{1,j}) && \oplus g_i(b^{(0)}_{2,j}) \\
\big). &
\end{aligned}
$$

We would like to show that $b^{(5)}_{i,j,k}$ depends on each of the bits $b^{(0)}_{i,j,k}$. In other words, we would like to show that each bit of the output of Ψ is non-degenerate on all the input bits. (A Boolean function $f(x_1, \dots, x_n)$ is said to be degenerate on variable x_i, if $f(a_1, \dots, a_{i-1}, 0, a_{i+1}, \dots, a_n) = f(a_1, \dots, a_{i-1}, 1, a_{i+1}, \dots, a_n)$ for all choices of $(a_1, \dots, a_{i-1}, a_{i+1}, \dots, a_n) \in \{0,1\}^{n-1}$. Otherwise it is non-degenerate on the variable x_i. The function f is non-degenerate if it is non-degenerate on all the variables.)

We first note that *each of the output component functions $g_0, \dots, g_7$ of* byte-Sub*() are non-constant and non-degenerate Boolean functions.* This fact will be used to prove our required result.

Theorem 2. *Let $y_1, \dots, y_{128}$ be the output bits of the nonlinear map Ψ. Then each y_i depends on all the input 128 bits.*

Proof : First note that all the 16 bytes $b^{(0)}_{i,j}$ occur in the expression for $b^{(5)}_{i,j,k}$ given in Theorem 1. The value of $b^{(5)}_{i,j,k}$ is obtained by invoking g_k on eight bits, where each of these bits is obtained as XOR of three other bits. Let us number the eight input bits of g_k as $a_0, \dots, a_7$ corresponding to the row in which it occurs in Theorem 1. We take a closer look at how the bytes $b^{(0)}_{i,j}$ influence the a_is.

The bytes

$$b_{3,j}^{(0)}, b_{0,j+1}^{(0)}, b_{1,j+2}^{(0)}, b_{2,j+3}^{(0)}, b_{1,j+1}^{0}, b_{2,j+2}^{(0)}, b_{3,j+3}^{(0)}, b_{0,j}^{(0)}$$

occur exactly once each in the expression for $b_{i,j,k}^{(5)}$. (The other 8 bytes occur exactly twice each.) Further, no two of these bytes occur in any of the a_is and hence it is possible to set up an 1-1 correspondence between the a_is and these bytes. Using this 1-1 correspondence we rename these bytes as $B_0, \ldots, B_7$, i.e., $B_0 = b_{3,j}^{(0)}$, $B_1 = b_{0,j+1}^{(0)}$ and so on.

Now we turn to the actual proof. We show that $b_{i,j,k}^{(5)}$ is non-degenerate on $b_{i_1,j_1,k_1}^{(0)}$ for all choices of i_1, j_1, k_1 with $0 \le i_1, j_1 \le 3$ and $0 \le k_1 \le 7$. There are two cases to consider.

Case $b_{i_1,jj}^{(0)} = B_l$ for some $0 \le l \le 7$: Since g_k is non-degenerate on all variables and hence on the lth variable, there is a combination $c_0, \ldots, c_{l-1}, c_{l+1}, \ldots, c_7$ such that $g_k(c_0, \ldots, c_{l-1}, 0, c_{l+1}, \ldots, c_7) \ne g_k(c_0, \ldots, c_{l-1}, 1, c_{l+1}, \ldots, c_7)$. We first set all the bits of $b_{i,j}^{(0)}$ which are not in $B_0, \ldots, B_7$ to arbitrary values. This determines two of the bits of each a_i. Denote the XOR of these two bits by d_i. For $0 \le i \le 7$, $i \ne l$, if $d_i = c_i$, then set the bits of B_i such that the third bit of a_i evaluates to 0; and if $d_i \ne c_i$, then set the bits of B_i such that the third bit of a_i evaluates to 1. Since the functions g_i and g_{i+4} are non-constant functions, this can always be done. This ensures that the inputs to g_k other than the lth input has the desired values $c_0, \ldots, c_{l-1}, c_{l+1}, \ldots, c_7$. Now consider the lth input. The bit d_l has already been fixed. The other bit which determines the value of a_l is obtained by invoking one of g_i or g_{i+4} on the byte B_l. Let us call this function g_p. Since g_p is a non-degenerate function, it is also non-degenerate on the k_1th input. The bit $b_{i_1,j_1,k_1}^{(0)}$ is the k_1th bit of B_l. Hence there is a combination $e_1, \ldots, e_{k_1-1}, e_{k_1+1}, \ldots, e_7$ such that $g_p(e_1, \ldots, e_{k_1-1}, 0, e_{k_1+1}, \ldots, e_7) \ne g_p(e_1, \ldots, e_{k_1-1}, 1, e_{k_1+1}, \ldots, e_7)$. Thus, fixing the other bits of B_l to the values $e_1, \ldots, e_{k_1-1}, e_{k_1+1}, \ldots, e_7$ and toggling just the value of $b_{i_1,j_1,k_1}^{(0)}$ we can toggle the value of d_l and consequently toggle the value of the output of g_k. This in turn means that by toggling only the value of $b_{i_1,j_1,k_1}^{(0)}$ and keeping all other bits to some fixed values we can toggle the output of g_k. Thus $b_{i,j,k}^{(5)}$ is non-dengerate on the bit $b_{i_1,j_1,k_1}^{(0)}$ as required.

Case $b_{i_1,jj}^{(0)} \ne B_l$ for all $0 \le l \le 7$: The analysis of this case is similar to the previous case and hence we do not provide the details.

This completes the proof of the theorem. □

Theorem 2 assures us that the mixing done in HBB is total and hence it seems improbable that low diffusion attacks can be applied to HBB. In the following we will use the notion of bias of a binary random variable. Let X be a binary random variable. We define the bias of X to be $|\Pr[X = 0] - (1/2)|$.

Theorem 3. *Let $x_1, \ldots, x_{128}$ be the input 128 bits of Ψ and $y_1, \ldots, y_{128}$ be the output 128 bits of Ψ. Let $\alpha \in \{0,1\}^{128}$ and $\beta \in \{0,1\}^{128}$ be nonzero binary strings. Then*

$$\left| \Pr[\langle \alpha, (x_1, \ldots, x_{128}) \rangle = \langle \beta, (y_1, \ldots, y_{128}) \rangle] - \frac{1}{2} \right| \leq \frac{1}{2^{13}}. \qquad (2)$$

Proof : The result is obtained from a property of the function byteSub(). It is known [9,5] that the absolute value of the bias of any nontrivial linear combination of the input and output bits of byteSub() is at most 2^{-4}. This fact and the Piling-Up Lemma (PUL) is used to prove the result.

The essential structure of Ψ is the following. First one round of NLSub() is applied. This is followed by a mixing of the columns of NLC. Then a bit permutation (row shift followed by partial transpose) is applied to NLC which is again followed by an invocation of NLSub(). The mixing of the columns of NLC ensures that each bit after mixing depends upon exactly three bits before mixing. These input three bits are obtained by three separate invocations of byteSub() on three distinct 8-bit strings. Also the output "mixed" bit is subjected to one invocation of byteSub(). Hence any linear combination λ of the input and output of Ψ can be broken up into a linear combination of at least 4 random variables T_1, T_2, T_3 and T_4, where each T_i is a linear combination of the inputs and outputs of one invocation of byteSub(). Further, the invocations of byteSub() for the different T_i's are different. From the property of byteSub(), the bias of each T_i is at most 2^{-4}. Assuming that the T_i's are independent we apply the PUL to obtain the result that the bias of λ is at most $2^3 \times 2^{-16} = 2^{-13}$. □

Theorem 3 provides a bound on the correlation probability of the best affine approximation of the nonlinear map Ψ. This bound may be further improved by introducing additional layers of NLSub() and an affine transformation. Suppose that the following lines are inserted between Steps 5 and 6 of the function Round().

5a. for $i = 1$ to NMIX
5b. $\Delta = \text{NLC}_0 \oplus \text{NLC}_1 \oplus \text{NLC}_2 \oplus \text{NLC}_3$, where $|\text{NLC}_i| = 32$;
5c.   for $j = 0$ to 3 do $\text{NLC}_i = (\Delta \oplus \text{NLC}_i) \lll (8 * i + 4)$;
5d. $\text{NLC} = \text{NLSub}(\text{NLC})$;
5e. enddo;

This will introduce an additional NMIX layers of NLSub() and the affine transformation. The effect of this will be to lower the bias even further. The bias will decrease as NMIX increases before finally plateauing off. However, the downside will be more operations per 128 bits of the key stream. The new lines will require $11 * \text{NMIX}$ bitwise operations and $16 * \text{NMIX}$ look-ups. We do not suggest these lines as part of HBB since we believe that a bias of 2^{-13} is low enough to resist cryptanalytic attacks. Hence the slowdown associated with positive values of NMIX can be avoided. However, a user might choose a small value of NMIX without affecting the speed too much.

5.2   Self-synchronizing Mode

The **SS** mode is a self synchronizing mode. This mode is useful in error prone communication channels. In this mode, the next 128 bits of the key stream depend upon the secret key KEY and the previous four 128-bit cipher blocks. Whenever the receiver correctly receives four consecutive 128-bit blocks of the cipher the next 128 bits of the key stream are correctly generated. From this point onwards as long as the cipher blocks are correctly received, the receiver will be able to correctly generate the key stream. In the **SS** mode the linear part of the state memory LC is updated after generating each 128-bit key block. This frequent re-initialization can be costly, especially in hardware. It is possible to modify the description such that the re-initialization is done after longer intervals. The corresponding synchronization scheme will be different and the exact details will depend on the chosen error model. We leave this as future work.

6 Variability

We briefly discuss the features of HBB which can be varied without potentially changing the security of the system.

1. The choice of the two 256-bit primitive polynomials can be changed. These polynomials were randomly generated and the only consideration was the weight of the polynomials. Thus any two primitive polynomials of degree 256 and weight 129 should be sufficient. Also the weight need not exactly be 129; a value somewhere near 129 should suffice.
2. The choice of the irreducible polynomial to realise $GF(2^8)$ for the map byte-Sub() does not affect security. This polynomial can be changed.
3. The expansion function $E()$ has been chosen to output a balanced string. There are many other possible choices of $E()$ which will achieve the same effect.

Any one of the above change will change the generated key stream. Hence users who prefer customized systems may incorporate one or more of the above changes to obtain their "unique" system.

7 Conclusion

In this paper we have described a new stream cipher HBB. Compared to existing ciphers, the new cipher has both advantages and disadvantages. For example, it is slower in software implementation, but requires lesser memory. Hardware implementation is expected to be comparable if not more efficient. Further, HBB provides a good combination of security features. However, the actual security of HBB can be judged only after the research community has carefully analysed it and explored possible weaknesses.

Acknowledgements : Thanks to Charanjit Jutla for reading the paper and to Vincent Rijmen and Bill Millan for some comments. Thanks also to Sibsankar Haldar for going through the code of one of the implementations of HBB. Finally, thanks to Chandan Biswas for providing the figure of Hiji-bij-bij. The current proof of Theorem 2 is considerably more detailed than the proof in the submitted version. This is in response to a comment from a reviewer for Indocrypt. I thank him for the comment.

References

1. V. V. Chepyzhov, T. Johansson and B. Smeets. A Simple Algorithm for Fast Correlation Attacks on Stream Ciphers. *Proceedings of FSE 2000*, 181-195.
2. D. Coppersmith, S. Halevi and C. Jutla. Cryptanalysis of stream ciphers with linear masking, *Proceedings of Crypto 2002*.
3. D. Coppersmith, S. Halevi and C. Jutla. Scream: a software efficient stream cipher, *Proceedings of Fast Software Encryption 2002*, LNCS vol. 2365, 195-209.
4. N. Courtois and W. Meier. Algebraic attacks on stream ciphers with linear feedback, *Eurocrypt 2003*, pp 345-359.
5. J. Daemen and V. Rijmen. *The design of Rijndael.* Springer Verlag Series on Information Security and Cryptography, 2002. ISBN 3-540-42580-2.
6. P. Ekdahl and T. Johansson. SNOW - a new stream cipher. *Proceedings of SAC, 2002.*
7. J. Dj. Golic. Modes of Operation of Stream Ciphers. *Proceedings of Selected Areas in Cryptography 2000*, pp 233-247.
8. C. S. Jutla. Encryption Modes with Almost Free Message Integrity. *Proceedings of EUROCRYPT 2001*, 529-544.
9. K. Nyberg. Differentially Uniform Mappings for Cryptography. *Proceedings of EUROCRYPT 1993*, 55-64.
10. G. Rose and P. Hawkes. Turing, a high performance stream cipher. *Proceedings of Fast Software Encryption, 2003*, to appear. Also available as IACR technical report, http://eprint.iacr.org, number 2002/185.
11. P. Sarkar. The filter-combiner model for memoryless synchronous stream ciphers. In *Proceedings of Crypto 2002*, Lecture Notes in Computer Science.
12. P. Sarkar. Computing shifts in 90/150 cellular automata sequences. *Finite Fields and their Applications*, Volume 9, Issue 2, April 2003, pp 175-186.
13. S. Tezuka and M. Fushimi. A method of designing cellular automata as pseudo random number generators for built-in self-test for VLSI. In *Finite Fields: Theory, Applications and Algorithms*, Contemporary Mathematics, AMS, 363-367, 1994.
14. D. Watanabe, S. Furuya, H. Yoshida and B. Preneel. A new keystream generator MUGI. *Fast Software Encryption, 2002*, LNCS 2365, Springer, 179-194.
15. M. Zhang, C. Caroll and A. Chan. The software-oriented stream cipher SSC2. *Fast Software Encryption, 2000*, LNCS 1978, Springer 2001, 31-48.

A   Who Is Hiji-bij-bij?

Hiji-bij-bij is a character in the story Ha-ja-ba-ra-la, a brilliant creation by Sukumar Ray[1]. Ha-ja-ba-ra-la is written in Bangla[2] and falls in the class of nonsense

[1] He is the father of Satyajit Ray, a noted Indian film director.
[2] One of the top ten most spoken languages in the world

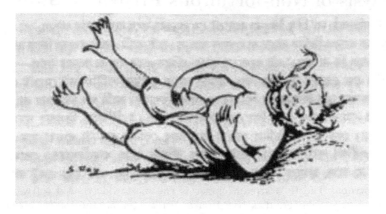

Fig. 1. Hiji-bij-bij

literature along the lines of "Alice's Adventures in Wonderland" by Lewis Carroll. The story is about a small boy's dream wanderings, where he interacts with "unreal" characters. The tale is told in an inimitable manner and is full of satire. Hiji-bij-bij (see Figure 1 for the original sketch of Hiji-bij-bij by its creator) is one of the characters that the boy meets and whose speciality is to laugh until his sides burst at absurd jokes that he himself is constantly cooking up. The word Hiji-bij-bij was probably intended as a pun on the Bangla word hiji-biji, which means nonsense scribblings, especially that of a child. This is perhaps an appropriate name for a stream cipher, which is supposed to produce random "nonsense".

Analysis of Non-fortuitous Predictive States of the RC4 Keystream Generator*

Souradyuti Paul and Bart Preneel

Katholieke Universiteit Leuven, Dept. ESAT/COSIC,
Kasteelpark Arenberg 10,
B–3001 Leuven-Heverlee, Belgium
{Souradyuti.Paul, Bart.Preneel}@esat.kuleuven.ac.be

Abstract. The RC4 stream cipher is the most widely used software based stream cipher. It is based on a secret internal state of $N = 256$ bytes and two pointers. This paper proposes an efficient algorithm to compute a special set of RC4 states named *non-fortuitous predictive states*. These special states increase the probability to guess part of the *internal state* in a known plaintext attack and present a cryptanalytic weakness of RC4. The problem of designing a practical algorithm to compute them has been open since it was posed by Mantin and Shamir in 2001. We also formally prove a slightly corrected version of the conjecture by Mantin and Shamir of 2001 that only a known elements along with the two pointers at any RC4 round cannot predict more than a outputs in the next N rounds.

1 Introduction

RC4 is the most widely used software based stream cipher. The cipher has been integrated into SSL and WEP implementations. RC4 is extremely fast and its design is simple. The cipher was designed by Ron Rivest in 1987 and kept as a trade secret until it was leaked out in 1994.

In this paper we formally prove the conjecture (due to Mantin and Shamir [1]) that only a known elements along with i and j at any RC4 round cannot predict more than a output bytes in the next N rounds. The set of *non-fortuitous predictive states* reduces the data and time complexity of the *branch and bound attack* on RC4 [1,7]. So far there was no efficient algorithm to obtain those states. The main achievement of this paper is that we design a practical two-phase recursive algorithm to determine the *non-fortuitous predictive states*. The complexity is far less than the trivial exhaustive search for small values of a.

1.1 Description of RC4

RC4 runs in two phases (description in Fig. 1). The first part is the key scheduling algorithm KSA which takes an array S to derive a permutation of $\{0, 1, 2, \ldots, N-$

---

* This work was partially supported by the Concerted Research Action GOA-MEFISTO-666 of the Flemish government.

T. Johansson and S. Maitra (Eds.): INDOCRYPT 2003, LNCS 2904, pp. 52–67, 2003.

1} using a variable size key K. The second part is the output generation part PRGA which produces pseudo-random bytes using the permutation derived from KSA. Each iteration or 'round' produces one output value. Plaintext bytes are *bit-wise* XOred with the output bytes to produce ciphertext. In most of the applications RC4 is used with word length $n = 8$ bits and $N = 256$. The symbol l denotes the *byte-length* of the secret key.

KSA (K, S)

for $i = 0$ to $N - 1$
$S[i] = i$
$j = 0$

for $i = 0$ to $N - 1$
$j = (j + S[i] + K[i \bmod l]) \bmod N$
Swap$(S[i], S[j])$

PRGA(S)

$i = 0$
$j = 0$
Output Generation loop
$i = i + 1$
$j = (j + S[i]) \bmod N$
Swap$(S[i], S[j])$
Output$= S[(S[i] + S[j]) \bmod N]$

Fig. 1. The Key Scheduling Algorithm (KSA) and the Pseudo-Random Generation Algorithm (PRGA)

1.2 Previous Attacks on RC4

RC4 came under intensive scrutiny after it has been made public in 1994. Finney showed in [3] a class of states that RC4 will never enter. The class contains all the states for which $j = i + 1$ and $S[j] = 1$. A fraction of approximately N^{-2} of all possible states fall under Finney's forbidden states. It is simple to show that these states are connected by a cycle of length $N(N - 1)$. We know that RC4 states are also connected in a cycle and the initial state, where $i = 0$ and $j = 0$, is not one of the Finney's forbidden states. Finney's forbidden states play a significant role in the analysis of *non-fortuitous predictive states*.

Jenkins detected in [5] a probabilistic correlation between the secret information (S, j) and the public information (i, output). Golić in [6] showed a positive correlation between the second binary derivative of the least significant bit output sequence and 1. Fluhrer and McGrew in [2] observed stronger correlations between consecutive bytes. Properties of the state transition graph of RC4 were analyzed by Mister and Tavares [9]. Grosul and Wallach demonstrated a related key attack that works better on very long keys. Andrew Roos also discovered in [10] classes of weak keys. Knudsen *et al.* have attacked versions of RC4 with $n < 8$ by their backtracking algorithm in which the adversary guesses the internal state and checks if an anomaly occurs in later stage [7]. In the case of contradiction the algorithm backtracks through the internal states and re-guesses.

The most serious weakness in RC4 was observed by Mantin and Shamir in [1] where they found that the probability of occurrence of zero at the second round is twice as large as expected. In broadcast applications a practical ciphertext only attack can exploit this weakness.

Fluhrer *et al.* in [11] have recently shown that if some portion of the secret key is known then RC4 can be broken completely. This is of practical importance because in the Wired Equivalence Privacy Protocol (WEP in short) a fixed secret key is concatenated with IV modifiers to encrypt different messages. In [12] it is shown that the attack is feasible.

Mironov, in [4], modelled RC4 as a Markov chain and recommended to dump the initial $12 \times N$ bytes of the output stream (at least $3 \times N$) in order to obtain uniform distribution of the initial permutation of elements.

1.3   Non-fortuitous Predictive RC4 States

Definitions of *Fortuitous state* and *Predictive State* are given in [2] and [1] respectively. We restate the definition of a *Fortuitous state*.

Definition 1. *The RC4 state (i.e., S-Box elements, i and j), in which only m known consecutive S-Box elements participate in producing the next m successive outputs, is defined to be a* fortuitous state *of length m.*

In [1] a set of special RC4 states, known as *Predictive States*, has been conceptualized. Below we give the definition of a *Predictive State* with a little modification to the one given in [1] to suit our analysis.

Definition 2. *Let A be an a state (i.e., a elements of the S-Box, i and j at some round which is defined to be round 0) which predicts b outputs (not necessarily consecutive) at rounds 1, r_2, r_3, ..., $r_b \leq N$. Then A is said to be b-predictive.*

A b-predictive a-state necessarily means that the execution of RC4 does not stop before b known outputs are produced. The RC4 execution only stops when i or j is not available at some round. In the definition, the first assertion is that an a-*state* is the snapshot of the RC4 state immediately before the first predicted output (a elements of the S-Box that do not produce any output are of no importance in the present context). Secondly, we set the upper bound on r_b to N instead of $2N$ as mentioned in [1].

It is clear from the above definitions that the set of *a-predictive a-states* is a superset of the set of *fortuitous states of length a* as any *fortuitous state of length a* is clearly an *a-predictive a-state*. Below we give the definition of a *non-fortuitous predictive state of length a*.

Definition 3. *If an a*-predictive *a*-state *is not a* fortuitous state of length a *then the state is a* non-fortuitous predictive state of length a.

We apologize that the term *"non-fortuitous"* is a misnomer. The term was coined to contrast it with the *fortuitous states* defined in [2]. In fact, *"non-fortuitous"* predictive states are fortuitous too and are far more complex to derive. As the term *non-fortuitous predictive state* is too long, in the rest of the paper we will use *non-fortuitous state* synonymously with it.

2   Importance of Predictive States

As mentioned in [1], the existence of b-predictive a-states is important for the cryptanalyst as a elements of the S-Box including j (note that the i value is always available to the cryptanalyst) can be extracted with non-trivial probability by observing b specific output bytes in the output segment.

Let the events E_A and E_B denote the occurrences of an a-$state$ and the corresponding b outputs when the i value of an a-state is known. We assume uniformity of the *internal state* and the corresponding *external state* for any fixed i value of the *internal state*. Assuming a much smaller than N and disregarding the small bias induced in E_B due to E_A, we apply Bayes' Rule to get,

$$P[E_A|E_B] = \frac{P[E_A]}{P[E_B]}P[E_B|E_A] \approx \frac{N^{-(a+1)}}{N^{-b}}.1 = N^{b-a-1} \qquad (1)$$

On the average one of the N^{a+1-b} occurrences of the event E_B is caused by the event E_A. From the above equation it is evident that a cryptanalyst will be interested in those b-$predictive$ a-states for which $P[E_A|E_B]$ is maximum. In such case the number of false hits, for which the occurrences of b specific words in the output stream (here denoted by the event E_B) are not induced by that specific internal state (here denoted by the event E_A), will be minimum. Maximizing $P[E_A|E_B]$ is equivalent to maximizing $b - a$ (see Eqn. (1)).

To determine the maximum value of b for a given a, we first formally prove the very important theorem that if a b-$predictive$ a-state exists then $a \geq b$. This was left as a conjecture in [1].[1] So, the best attack along this line is to obtain all the a-$predictive$ a-states, keep the information a priori in a database indexed by the values of the i pointer (sorted by outputs) and look up the output sequence for a possible match.

Throughout the paper $S_r[l]$ denotes the element of the S-Box indexed by l at the rth round (whether $S_r[l]$ is before or after the swapping at the rth round, should be understood in the context); the first predicted output corresponds to round 1. Similarly, i_t and j_t denote the values of i and j respectively at round t. Unless otherwise stated $S[i]$ and $S[j]$ denote the elements of the S-Box pointed to by i and j respectively at the round in question. All arithmetic operations are done modulo N.

Theorem 1. *If any b-predictive a-state exists then $a \geq b$.*

Proof. The theorem is trivially true when $a = N$. Below we consider the case when $a < N$. Let us assume that an a-state produces b outputs and $b > a$. As shown in Fig. 2, all the b outputs are generated, as the i pointer sweeps through all the positions from the index $k+1$ till the index $k+c$, in c rounds ($b \leq c \leq N$).

---

[1] In [1], the upper bound on r_b (see Def. 2 and [1]) is given to be $2N$. But within this bound the conjecture is not true, as we see that, for the trivial case of an N-state, $2N$ outputs can be predicted in $2N$ rounds.

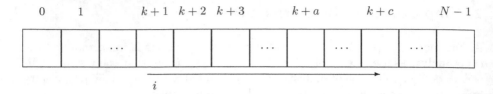

Fig. 2. Prediction of b outputs in c rounds where only a elements are known in the S-Box at round 0.

The first and the last outputs are produced at the $(k+1)^{th}$ and $(k+c)^{th}$ rounds respectively. In each of the c rounds i, j should be available otherwise RC4 halts forever (see PRGA in Fig. 1) because the swapping operation cannot be executed deterministically.

Lemma 1. *Starting from the a-state (i.e., at round 0) an adversary is unable to determine any information about the unknown elements of the S-Box in the subsequent rounds.*

Proof (Lemma). Let us assume that some information about the unknown elements becomes available to the adversary at some RC4 round using 'other' information for the first time. Then the adversary simply reverses the RC4 execution, i.e., she starts backtracking. As RC4 state is reversible (see PRGA in Fig. 1) and the only transformation the internal state goes through is the transposition of elements due to swapping, eventually she should go back to the initial state after backtracking as many steps as taken forward. But now she has some information about the unknown elements. As the adversary starts with the knowledge of only a elements of the S-Box (see Defn. 2), this is clearly not the initial state she started with. Thus we reach a contradiction, which proves the lemma.   □

Availability of j at each of these rounds implies that $S[i]$ is also available because the previous value of j is known (note, $j = j + S[i]$). From the above lemma, $S[i]$ is one of the known elements of the a-state in each of the c rounds. In each of the output producing b rounds a necessary condition is that $S[i] + S[j]$ is known. This fact along with the above lemma imply that $S[i]$ and $S[j]$ are individually known in each of those b rounds. No empty cells[2] in the S-Box can be filled with a known value in any of the output producing rounds as the swapping takes place between the known values. As i takes c different values of index in c rounds and every time $S[i]$ is known, we should have a *maximum* of $(c - b)$ swaps where $S[j]$ is unknown. In each of these $(c - b)$ swaps (where $S[i]$ is known but $S[j]$ is unknown) at most one new-cell[3] will be assigned a known value. So, a *maximum* of $(c - b)$ different new-cells between $k + 1$ and $k + c$ will be assigned known

---

[2] By an empty cell we denote an S-Box location with unknown value at the current round.

[3] A new-cell is an S-Box location with unknown value at round 0.

values at least once in the aforementioned c rounds. Consequently, a *minimum* of b different indices in the S-Box between $k+1$ and $k+c$ must be occupied with as many known values at round 0, otherwise we reach a scenario where we have at least one new-cell which is never assigned a known value. This is impossible as $S[i]$ is known in each of the c rounds as a necessary condition. As we have only a known values where $a < b$, we reach a contradiction. Therefore, our assumption (i.e., $a < b$) is wrong. We conclude that $b \leq a$, i.e., an a-*state* can not predict more than a outputs in the next N rounds. $\qquad\square$

We have already defined an a-state as the partially specified state of RC4 (i.e., a elements of S-Box, i and j) just before the 1st output is predicted. We emphasize that a elements of the S-Box that predict b outputs may change its positions as RC4 runs. So taking the snapshot of the a elements of the S-Box immediately before the round of the 1st predicted output is cryptanalytically equivalent to those a elements at some other rounds.

Corollary 1. *Only a known elements of the S-Box along with two pointers at any RC4 round cannot predict more than a outputs in the next N rounds.*

Proof. The proof is immediate from Theorem 1.

2.1   Necessary and Sufficient Condition for a Predictive State to Be Non-fortuitous

Quite understandably, the larger the number of a-*predictive a-states*, the larger will be the probability to obtain one of them under the reasonable assumption of uniformity of RC4 states.

Fortuitous states can be easily obtained using the state counting algorithm described in [2]. Our objective is to develop a method to determine *non-fortuitous states* of any length. The following theorem and the corollary divide a-*predictive a-states* into *fortuitous states* and *non-fortuitous states*.

Theorem 2. *An a-predictive a-state is a fortuitous state of length a if and only if the predicted output words are consecutive.*

Proof. The *only if* part of the theorem is direct from the definition of a *fortuitous state*. Now we prove the *if* part.

We prove it by contradiction. Let us assume that there exists at least one a-*predictive a-state* for which the outputs are consecutive but the elements of the S-Box are *not consecutive*. We assume that the generation of outputs starts when i points to $S_1[k+1]$ (see Fig. 2) and the ath output is issued when i points to $S_a[k+a]$. During the passage of i from the index $(k+1)$ to the index $(k+a)$, the number of outputs generated is also a. So, each of the a rounds produces output. Let us assume that there is at least one empty cell, say $S_0[k+t]$, between the indices $k+1$ and $k+a$ at round 0. At the tth round, $S_t[k+t]$ should contain a known value otherwise j cannot be updated at this round and consequently no output can be predicted. As $S_0[k+t]$ was initially empty, there is a round, say the rth round where $r < t$, when one known value will be swapped into the empty $S_r[k + t]$ for the first time (note there may be many such rounds). As a consequence, the rth round does not produce output because swapping takes place when j points to $S_r[k + t]$ which is unknown at that time. But we assumed that each of the a rounds produces output. Eventually we reach a contradiction. Therefore, our assumption, i.e., the S-Box elements are *not consecutive*, is wrong.

As things stand, we have a consecutive elements that are fixed in the S-Box and we get a outputs in the next a rounds. This is just the case of a fortuitous state of length a. □

Corollary 2. *An a-predictive a-state is a non-fortuitous state of length a if and only if the predicted output words are not consecutive.*

Proof. The proof is immediate from Theorem 2. □

3 Structure of Non-fortuitous States

The most interesting question that remains is whether such states really exist. Is it possible to get *non-fortuitous states* of any length a?

The two examples given by Mantin and Shamir in [1] to establish their claim to the existence of *non-fortuitous states* need some scrutiny. The first example where $S_0[2] = 0$, $i_0 = 0$ and $j_0 = 0$ which gives zero as the second output is claimed to be a *1-predictive 1-state*. The claim is not true in the strict sense of the definition (see Def. 2 and [1]) of *1-predictive 1-state* because we impose one more constraint, i.e., $S_0[1] \neq 2$, in addition to $S_0[2] = 0$. However, this inaccuracy is unrelated to the statistical bias in the second output word detected by the authors.

The second example provided in [1] is the RC4 state compatible with $S_0[-2] = 1$, $S_0[-1] = 2$, $S_0[1] = -1$, $i_0 = -3$ and $j_0 = -1$.[4] As the outputs are consecutive, then by Theorem 2 this is a *fortuitous state*. In Fig. 3 we show that from the 3rd round when $i_3 = 0$, $j_3 = 3$, $S_3[1] = -1$, $S_3[2] = 2$, $S_3[3] = 1$ the *3-predictive 3-state* has become a fortuitous state of length 3.

3.1 Determination of Non-fortuitous States: A Step Forward

Theorem 3. *Any 1-predictive 1-state is a fortuitous state of length 1.*

Proof. Any 1-predictive 1-state is of the form $i_0 = 2x - 1$, $j_0 = x$ and $S_0[2x] = x$. This is clearly a fortuitous state of length 1 (x is an integer chosen from 0 to $N - 1$). □

---

[4] The symbols -1, -2, -3 are used as shorthand for $N - 1$, $N - 2$, $N - 3$ respectively throughout this paper.

i	j	S						$S[i]$	$S[j]$	$S[i]+S[j]$	Output
		-2	-1	0	1	2	3				
-3	-1	1	2	*	-1	*	*	/	/	/	/
-2	0	*	2	1	-1	*	*	*	1	*	*
-1	2	*	*	1	-1	2	*	*	2	*	*
Fortuitous State											
0	3	*	*	*	-1	2	1	*	1	*	*
1	2	*	*	*	2	-1	1	2	-1	1	2
2	1	*	*	*	-1	2	1	2	-1	1	-1
3	2	*	*	*	-1	1	2	2	1	3	2

Fig. 3. The 3-predictive 3-state becomes a fortuitous state from the 3rd round.

The above theorem implies that the number of *1-predictive 1-states* is N and there is no *non-fortuitous state of length 1*.

Determination of *non-fortuitous states* of length a, where $a > 1$, has inherent difficulties on two counts. Firstly, the relative positions of the S-Box elements at the beginning are not known. Secondly, the set of indices containing known elements of the S-Box changes as RC4 runs. The most straightforward and naive method would be to select all possible a indices, assign to the elements pointed to by those indices and j all possible values and finally select those states which generate a non-consecutive outputs (by Corollary 2, non-consecutiveness of outputs is a necessary and sufficient condition for an *a-predictive a-state* to be *non-fortuitous*). But the cost of computation, which is $O(N^{2a+1})$ for $a \ll N$, makes such method too costly when $N = 256$ and $a = 2$ and completely impractical when $a > 2$. At this point Finney's forbidden state (see [3]) comes to our rescue.[5]

Proposition 1. *The first two elements of the S-Box of any a-predictive a-state always occupy consecutive places if $a \geq 2$, i.e., $S_0[k+1]$ and $S_0[k+2]$ (see Fig. 2) always contain known values if $a \geq 2$.*

Proof. According to Def. 2, the 1st round always produces output. So, $S_0[k + 1]$ is occupied with a known element. Assume $S_0[k + 2]$ is empty. Then $S_1[k + 2]$ is also empty because any output producing round does not put a known value in any new-cell. In such case $S[i]$ is empty in the 2nd round. Therefore, j cannot be updated and the execution of RC4 stops. Therefore, $S_0[k + 2]$ is not empty.   □

Theorem 4. *Any 2-predictive 2-state is a fortuitous state of length 2.*

Proof. Let us assume that there is at least one *2-predictive 2-state* which is *non-fortuitous*. Now we try to predict two elements (that must not be consecutive) from this state. From Proposition 1, $S_0[p_1]$ and $S_0[p_2]$ should contain known

---

[5] The cost of computation is measured in terms of the number of value assignments.

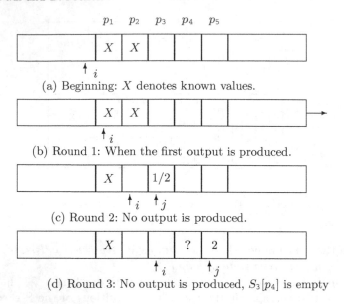

p_1 p_2 p_3 p_4 p_5

(a) Beginning: X denotes known values.

(b) Round 1: When the first output is produced.

(c) Round 2: No output is produced.

(d) Round 3: No output is produced, $S_3[p_4]$ is empty

Fig. 4. Impossibility of getting a *non-fortuitous state* of length 2. $S_2[p_3] = 1$ leads to Finney's forbidden state [3]. The j value will be lost from the 4th round. The symbol $1/2$ denotes "either 1 or 2".

elements (see Fig. 4). At round 1 we will have the same positions occupied with known values as round 0 because no new-cell will get any value as output is produced at round 1. Output cannot be produced at round 2 otherwise we will have two consecutive outputs and clearly that will be a case of a fortuitous state by Theorem 2. Therefore, at round 2, i points to $S_2[p_2]$ and j should point to $S_2[p_3]$ as shown in Fig. 4(c). For the j value to be available at round 3, $S_1[p_2]$ should be either 1 (if $j_1 = p_2$) or 2 (if $j_1 = p_1$). But $S_1[p_2] = 1$ gives rise to Finney's forbidden state [3], hence it is not possible. The S-Box arrangement at round 2 is shown in Fig. 4(c). So, at round 3, we can not get any output because swapping takes place when j points to an empty cell $S_3[p_5]$ (see Fig. 4(d)). After that, i points to $S_4[p_4]$ which is still empty. As a consequence, j_4 cannot be determined and the execution of RC4 halts. Therefore, we can not get any *non-fortuitous state* of length 2. Thus the theorem is proved.   □

Although we are unable to discover any *non-fortuitous states* so far but with the above results we are confident enough that no such state exists of length 1 or 2.

3.2   Determination of Non-fortuitous States: A General Approach

As mentioned before, two important factors make the determination of *non-fortuitous states* all the more difficult. Firstly, the relative positions of the a elements at round 0 are not known and secondly, in the subsequent rounds the indices containing the known elements change. Our algorithm is a two-phase one. The first part determines the possible relative positions of the a elements at round 0. The second part is a state counting algorithm that determines the individual *non-fortuitous* states. Note, that for a *fortuitous* state the elements are always consecutive and the set of their indices does not change in the later rounds.

Let d_t denote the *inter-element gap* between the tth element and the $(t+1)^{th}$ element. We measure $d_t = p_{t+1} - p_t - 1$ where the tth element is indexed by p_t.[6] Let T_a denote one such sequence $(d_1, d_2, d_3, ..., d_{a-1})$. Note, that the total number of such sequences is $\sum_{x=0}^{N-a} \binom{x+a-2}{a-2}$. So our first step is to sort out the sequences from the exhaustive set; more precisely, we try to reduce the search space. The problem of "relative positions" hinges on an important combinatorial problem: what is the maximum value of d_t (we will call this d_t^{max} henceforth) such that, given $(d_1, d_2, d_3, ..., d_{t-1})$, there exists a sequence $(S_0[p_1], S_0[p_2], ..., S_0[p_t], j_0)$ such that i always points to a known value till it reaches the index p_{t+1}? Proposition 3 of the Appendix A.1 implies that the i pointer has to reach at least p_{t+1} to predict the $(t+1)^{th}$ output.

There is no known method to determine the exact value of d_t^{max} other than exhaustive search on t elements and j_0. However, using recursion a loose upper bound, say $\tilde{d}_t^{max}$, can be easily made such that $d_t^{max} \leq \tilde{d}_t^{max}$ and $\tilde{d}_t^{max}$ is much smaller than the trivial maximum value for small values of a.[7] We will address the problem with respect to *non-fortuitous states*. The upper bound $\tilde{d}_t^{max}$ will be referred to as d_t^{max} henceforth.

Let $L_{a=t}$ denote the set of all sequences $(d_1, d_2, d_3, ..., d_{t-1})$ that represent the possible *inter-element gaps* of the *non-fortuitous states* of length t. Let d_t^{max} correspond to the sequence $(d_1, d_2, d_3, ..., d_{t-1}) \in L_{a=t}$. Then clearly $\{(d_1, d_2, d_3, ..., d_{t-1}, x) \mid 0 \leq x \leq d_t^{max}\} \subseteq L_{a=t+1}$. Each sequence in $L_{a=t}$ generates a subset and the union of them results in $L_{a=t+1}$. Using Propositions 2 and 3 in Appendix A.1, it can be shown that no sequence of length $(t-1)$ outside $L_{a=t}$ can be a prefix of a sequence in $L_{a=t+1}$.

Now, we outline how d_t^{max} can be evaluated. In fact, we will calculate the maximum value of the index of the $(t+1)^{th}$ element, say p_{t+1}^{max}. As d_t^{max} does not depend on the values of the indices of the S-Box elements, we fix $p_1 = 0$ to ease the computation. We know that $d_t^{max} = p_{t+1}^{max} - p_t - 1$. The algorithm is a function **Mainfunc** which takes a sequence $(0, d_2, d_3, ..., d_{t-1})$ as input and calls a recursive function **Round** in a loop to compute the corresponding p_{t+1}^{max}. Both the functions work 'almost' similarly (the small difference between the functions should not be overlooked). The functions **Mainfunc** and **Round** are shown in Appendix A.3 and Fig. 5. $L_2, L_3, .., L_t$ are all known. Now we describe the recursive function **Round**. We simulate RC4 without assigning any values to the

---

[6] The 1st element is $S_0[i_0 + 1]$. All known elements in the direction of the movement of i are numbered accordingly.

[7] The trivial maximum value of $d_t = N - (\sum_1^{t-1} d_i + t + 1)$.

elements. When $S[i]$ is known which is not assigned a value, the function **Maxj** computes the maximum value of j (denoted by j_{max}) in the current round by matching the sequence of the *inter-element gaps* of the elements between $i + 1$ and $N - 1$ with the suitable member from the global lists $L_2, L_3,.., L_t$. If j_{max} goes beyond $N - 1$ then $j_{max} = N - 1$. The function **Range** determines the range of j (denoted by J) such that i may reach up to the location j without pointing to any empty cell in the rounds in between, thereby we reduce the search space. Propositions 2, 3 and 4 in Appendix A.1 establish why it is necessary that i should be able to reach j in the later rounds. The set J includes all the indices containing unknown elements between $i + 1$ and j_{max} excluding those which violate Finney's criterion and uniqueness of permutation elements. $S[i]$ is assigned values such that j takes all possible values within the range. For every value of $S[i]$, the function **Round** calls itself. In essence, we form a search tree. The function **Round** returns one of the three types of values. If several branches are originated from the **Round** function (the case when $S[i]$ is known but not assigned a value before) then the maximum of all the returned values to the current function is stored. It is then compared with j_{max} at the current round. The greater of the two is returned to the predecessor.[8] If, at any round function, $S[i]$ is known and already assigned a value then it simply passes on the value from its successor to the predecessor. If, at any round function, $S[i]$ is unknown then the branch is terminated and i is returned. The branch is also terminated if i sweeps through N indices (in this case d_t^{max} is trivial). The cost of the algorithm is substantially less than the exhaustive search on t elements.

int **Round**(S, i, j)

1. $i = i + 1$
2. if $i = N - 1$ then return(i)
3. if $S[i]$ empty return (i)
4. if $S[i]$ known and already assigned a value
       1. $j = j + S[i]$
       2. Swap$(S[i], S[j])$
       3. return(**Round**(S, i, j))
5. if $S[i]$ known but not assigned any value
       1. $j_{max} =$**Maxj**(S, i)
       2. $J =$**Range**(S, i, j_{max})
       3. $M = 0$
       4. For each $x \in J$
              1. $S[i] = x - j$
              2. Swap$(S[i], S[x])$
              3. $M =$max$(M, $**Round**$(S, i, x))$
       5. return(max(M, j_{max}))

Fig. 5. The round function that computes p_{t+1}^{max}.

---

[8] Note, if j takes any values outside the range J at some round then i cannot cross j_{max} in the subsequent rounds and therefore, j_{max} is returned.

We have already observed that, for $a = 2$, $d_1 = 0$ (see Proposition 1). Therefore, $L_{a=2} = \{(0)\}$. We cannot determine d_2^{max} from $L_{a=2}$ as it is constructed following the constraint that the elements are forcefully made consecutive. Applying Finney's forbidden state (described in Sect. 1.2) we determine $d_2^{max} = 1$. Therefore, $L_{a=3} = \{(0,0),(0,1)\}$. In the Appendix A.2 we show that $d_2^{max} = 1$. Applying the recursion described above we derive $d_3^{max_1} = 3$ and $d_3^{max_2} = 3$ that correspond to two different members of $L_{a=3}$. Therefore, $L_{a=4} = \{(0,0,0),(0,0,1),(0,0,2),(0,0,3),$
$(0,1,0),(0,1,1),(0,1,2),(0,1,3)\}$.

To determine the individual *non-fortuitous states* of length $a = t + 1$ by a state counting recursive algorithm (the algorithm works in a similar manner as the one described before) we take members from $L_{a=t+1}$ one by one, vary p_1 from 0 to $N - 1$ and simulate RC4 without directly assigning values to the elements. In this case, at the first round, $S[j]$ should be one of the known values. The value of j_{max} at each of the other rounds is also determined from the lists L_2, L_3,.., L_t. The range of the values of j (other than the first round) is all the indices between $i + 1$ and j_{max} plus the indices containing known values between p_1 and i. That is how the search space is reduced. In this state counting algorithm, using Propositions 3 and 4 in Appendix A.1 it can be shown that if $S[j]$ is known at some round then output has to be produced in that round, i.e., $S[S[i] + S[j]]$ should be a known element as well. This condition effectively reduces the search space again. The algorithm stops whenever a outputs are produced or $S[i]$ is unknown. Among all the states, so obtained, we take only those with *non-consecutive* outputs.

We computed that the number of *non-fortuitous states* of length 3 and 4 are 7 and 1727 for $N = 256$. In Appendix A.4 we list all the *non-fortuitous states* of length 3. It is possible that many members in L_a may not eventually produce any *non-fortuitous states*. But the relative positions of the elements of any *non-fortuitous state* of length a must correspond to an entry in L_a. Our attempt is directed to develop a technique to eliminate all the trivial *a-states* which are impossible to predict a elements.

4   Cryptanalytic Significance of Non-fortuitous States

Although we mostly dealt with *non-fortuitous predictive states*, one can see that the algorithm is more robust, that is, it can as well be used to determine the set of b-predictive a-states for any a and b.

The average number of outputs, needed for any *a-predictive a-state* to occur, is reduced by knowing the number of *non-fortuitous states* of length a, in addition to that of the *fortuitous states* of length a [1]. Denote the number of *fortuitous states* and *non-fortuitous states* of length a by the symbol A and B. Assuming uniformity of the *external states* and the *internal states*, a specific elements occur as outputs at specific rounds with probability $N^{-(a+1)}$. The knowledge of *non-fortuitous states* reduces the length of the output segment required for any *a-predictive a-states* to happen by $N.N^{a+1}(\frac{1}{A} - \frac{1}{A+B})$. Here, the additional factor N comes because of $N-1$ false hits on the average (see Eqn. (1)).

If we use the *branch and bound* attack [7] on $RC4_{n=8}$ with a priori information about 3 elements in the S-Box the attack is not much improved (an **improvement of 2.36%** on the number of required output bytes) as we have only 7 *non-fortuitous states* compared to 290 *fortuitous states* of the length 3. If we use a priori information about 4 elements, then with the complete information about *non-fortuitous states*, we require approximately $2^{34.98}$ output bytes (which is around **21% less** than the earlier estimate of $2^{35.2}$ bytes based on only *fortuitous states*) for any *4-predictive 4-state* to happen. Note, that the number of *fortuitous states* and *non-fortuitous states* of length 4 are 6540 and 1727 respectively.

5   Directions for Future Work and Conclusions

Our current work leaves room for more research. We proved Corollary 1 with a bound on the number of rounds. For small values of a, we observe that the j value is lost much earlier than the Nth round. So, in such cases, Corollary 1 is true even without any bound on the number of rounds. A more challenging combinatorial problem, therefore, is what is the maximum value of a for which Corollary 1 is true without any bound on the number of rounds. From the point of view of cryptanalysis, although we are interested in small values of a (because a small increase in a drastically increases the required outputs for any attack), the problem seems alluring.

Another way to improve the present work is to suggest some elegant algebraic means to determine d_t^{max} when the values of d_1, d_2, ..., d_{t-1} are given: more precisely, one should try to build an algebraic structure for the function f which determines d_t^{max}. We see that the set L_a contains redundant members which increase the time and space complexity. We are convinced that the algorithm can be further improved.

Our work in this paper is a purely combinatorial analysis of RC4. We developed a practical scheme to derive a special set of RC4 states known as *non-fortuitous predictive states*. Apart from that many interesting properties of this cipher (e.g. known a elements cannot predict more than a elements in the next N rounds) are established. We hope these observations will lead to better understanding of the cipher.

Acknowledgements

We are grateful to Scott Fluhrer for helping us understand the state counting algorithm for *fortuitous states*. We are thankful to Christophe De Cannière for kindly going through different technical details of the paper and making valuable comments. We also thank Ilya Mironov of Stanford University and Sankardas Roy of George Mason University for useful discussions.

References

1. I. Mantin, A. Shamir, "A Practical Attack on Broadcast RC4," *Fast Software Encryption 2001* (M. Matsui, ed.), vol. 2355 of *LNCS*, pp. 152-164, Springer-Verlag, 2001.
2. S. Fluhrer and D. McGrew, "Statiscal Analysis of the Alleged RC4 Keystream Generator," *Fast Software Encryption 2000* (B. Schneier, ed.), vol. 1978 of *LNCS*, pp. 19-30, Springer-Verlag, 2000.
3. H. Finney, "An RC4 cycle that can't happen," Post in `sci.crypt`, September 1994.
4. I. Mironov, "Not (So) Random Shuffle of RC4," *Crypto 2002* (M. Yung, ed.), vol. 2442 of *LNCS*, pp. 304-319, Springer-Verlag, 2002.
5. R. Jenkins, "Isaac and RC4," Published on the Internet at `http://burtleburtle.net/bob/rand/isaac.html`.
6. J. Golić, "Linear Statistical Weakness of Alleged RC4 Keystream Generator," *Eurocrypt '97* (W. Fumy, ed.), vol. 1233 of *LNCS*, pp. 226-238, Springer-Verlag, 1997.
7. L. Knudsen, W. Meier, B. Preneel, V. Rijmen and S. Verdoolaege, "Analysis Methods for (Alleged) RC4," *Asiacrypt '98* (K. Ohta, D. Pei, ed.), vol. 1514 of *LNCS*, pp. 327-341, Springer-Verlag, 1998.
8. A. Grosul and D. Wallach, "A related key cryptanalysis of RC4," *Department of Computer Science, Rice University, Technical Report TR-00-358*, June 2000.
9. S. Mister and S. Tavares, "Cryptanalysis of RC4-like Ciphers," *SAC '98* (S. Tavares, H. Meijer, ed.), vol. 1556 of *LNCS*, pp. 131-143, Springer-Verlag, 1999.
10. A. Roos, "Class of weak keys in the RC4 stream cipher," Post in `sci.crypt`, September 1995.
11. S. Fluhrer, I. Mantin, A. Shamir, "Weaknesses in the Key Scheduling Algorithm of RC4," *SAC 2001* (S. Vaudenay, A. Youssef, ed.), vol. 2259 of *LNCS*, pp. 1-24, Springer-Verlag, 2001.
12. A. Stubblefield, J. Ionnidis and A. Rubin, "Using the Fluhrer, Mantin and Shamir attack to break WEP," *NDSS 2002*.

A   Appendix

A.1   Criteria for i to Reach an Index to Produce an Output

The fact that the i pointer can move from the index x to the index y implies that the value of j is always available in each of the intermediate $(y - x + 1)$ rounds. The S-Box region between the indices x and y is all the $(y - x + 1)$ indices from x in the direction of the movement of i.

Proposition 2. *If, at a particular round r (when $i = i_r$), j_r and some elements of the S-Box are known, then the fact that i can reach the index $i_r + k$ (where $0 < k \leq N$) from the round r depends only on j_r and the known S-Box elements between the indices $i_r + 1$ and $i_r + k$ at round r.*

Proposition 3. *Let the number of known elements of the S-Box at the rth round between the indices $i_r + 1$ and $i_r + k$ (where $0 < k \leq N$) be m and at the rth round the tth element to the right of i be indexed by p_t ($0 < t \leq m$). Then, starting from the rth round, the pointer i must reach at least p_t to predict the tth output.*

Proposition 4. *If the number of rounds, at which $S[j]$ is known during the passage of i from $i = i_r$ to $i = i_r + k$ (where $0 < k \leq N$), is m, then the number of known elements, between the indices $i_r + 1$ and i_{r+k} at round $r + k$, is also m.*

A.2 Evaluation of the Maximum Value of d_2

Theorem 5. *For any non-fortuitous state of length 3, $p_3 - p_2 < 3$ where the tth element is indexed by p_t.*

Proof. Let us assume $p_3 - p_2 = 3$. Now we try to generate 3 *non-consecutive* outputs, in a similar manner as Theorem 4. The execution of the first three rounds are shown in Figure 6. At the 2nd and the 3rd rounds j must point to $S_2[g_1]$ and $S_3[g_2]$ respectively, in order for j to be available at the third and the fourth rounds. But such conditions lead to Finney's forbidden state at round three. So our assumption is wrong. Hence, $p_3 - p_2 \neq 3$. It is easy to see that the same situation arises for $p_3 - p_2 > 3$. Therefore, $p_3 - p_2 < 3$. We know that $d_2 = p_3 - p_2 - 1$. Hence, $d_2^{max} = 1$.   □

One can see that if we relax the condition of the 1st round producing output always, then the maximum *inter-element gap* between the first two elements of the S-Box is also 1. This basic fact will be used in the determination of d_t^{max} when $t > 2$.

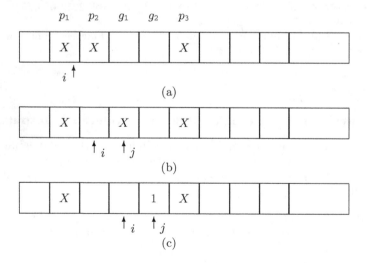

Fig. 6. *A non-fortuitous state* of length 3 with $d_2 = 2$. (a) Round 1: After production of the 1st output; X indicates known value. (b) Round 2: No output. (c) Round 3: We reach Finney's forbidden state as $j_3 = i_3 + 1$ and $S_3[j_3] = 1$.

A.3   The Function That Calls the Round Function

Mainfunc$((0, d_2, \ldots, d_{t-1}))$

1. Set $p_1 = 0$
2. $i = 1$
3. Mark the indices of the S-Box that contain known elements from the sequence $(0, d_2, \ldots, d_{t-1})$.
4. $j_{max} =$**Maxj**(S, i)
5. $J =$**Range**(S, i, j_{max})
6. $M = 0$
7. For each $x \in J$
    1. Swap$(S[i], S[x])$
    2. $M =$max$(M,$ **Round**$(S, i, x))$
8. $p_{t+1}^{max} =$max(M, j_{max})

A.4   List of Non-fortuitous States of Length 3

We list in the form of 5-tuple $(S_0[i + 1], S_0[i + 2], S_0[i + 3], i_0, j_0)$.

1.(0, 1, -2, -3, 0), 2.(0, 1, -1, -3, 0),
3.(1, 3, 0, -1, -1), 4.(128, 3, 0, -1, -128)
5.(129, 3, 0, -1, -129), 6.(2,-2,0,-1,-1), 7.(2,-1,0,-1,-1)

Nonlinearity Properties of the Mixing Operations of the Block Cipher IDEA[*]

Hamdi Murat Yıldırım[**]

Department of Mathematics, Middle East Technical University
06531 Ankara, Turkey
murat@math.metu.edu.tr

Abstract. In this paper we study the nonlinearity properties of the mixing operations $\odot$, $\boxplus$ and $\oplus$ used in IDEA. We prove that the nonlinearity of the vector function corresponding to the multiplication operation $\odot$ is zero for some key points. The Multiplication-Addition (MA) structure of IDEA is slightly changed to avoid the linearities due to these points and we suggest a new structure called RMA. The nonlinearity of MA, RMA and their composition are compared.

1 Introduction

In 1991 Lai et al. [3,4,5] introduced the block cipher IDEA (International Data Encryption Algorithm) as a slightly modified version of PES (Proposed Encryption Standard) to increase immunity against differential cryptanalysis. IDEA and PES are based on the design concept of "mixing (arithmetic) operations from different algebraic groups." IDEA encrypts 64-bit plaintext blocks to 64-bit ciphertext blocks with 128-bit key blocks.

The designers of IDEA expressed the mixing operations $\boxplus$ and $\odot$ as the functions over the $\mathbb{Z}_{2^n}$ and $\mathbb{Z}^*_{2^n+1}$, respectively and considered their nonlinearity in terms of these expressions. In [6] a measure for the nonlinearity that is based on the Hamming distance from the set of affine functions and involves all nontrivial linear combinations of the coordinate functions of a vector function is provided. In this paper, we study the nonlinearity properties of these operations by using this nonlinearity measure. For this purpose, we view them as functions on the corresponding vector spaces $\mathbb{Z}^n_2 (\cong \mathbb{Z}_{2^n})$ and $\mathbb{Z}^n_2 (\cong \mathbb{Z}^*_{2^n+1})$ respectively. We prove that the nonlinearity of the vector function corresponding to the multiplication operation $\odot$ is zero for some key points. Moreover we observe the effects of these points on the linearity of MA (Multiplication-Addition) structure of IDEA. By changing MA structure slightly we introduce so-called RMA (Reverse MA) structure. In this note the nonlinearities of RMA and MA and several compositions of

---

[*] This work is a part of my M.Sc. thesis which is carried out under the guidance of Professor Ersan Akyıldız.
[**] Partially supported by the grant TBAG ÇG/2 from The Scientific and Technical Research Council of Turkey.

T. Johansson and S. Maitra (Eds.): INDOCRYPT 2003, LNCS 2904, pp. 68–81, 2003.

MA and RMA structures are calculated and compared. These calculations suggest that RMA structure has a better nonlinearity property than MA structure used in IDEA.

The paper is organized as follows: Section 2 gives the details of the block cipher IDEA and its operations and necessary definitions. In Section 3 we discuss the nonlinearity of mixing operations of IDEA, MA and RMA structure and propose the block cipher RIDEA, a variant of IDEA.

2   Preliminaries

2.1   The Block Cipher IDEA

IDEA operates on 64-bit plaintext/ciphertext blocks, and 128-bit key and consists of 8 iterated rounds and a final transformation. The rounds and final transformation are arranged to achieve the desired confusion and diffusion by successive usage of three incompatible group operation and its chosen special structure. The design of IDEA is completely based on the concept of "mixing operations from different algebraic groups such as multiplication modulo $2^{16} + 1$ ($\odot$), addition modulo 2^{16} ($\boxplus$) and bitwise XOR ($\oplus$, bitwise addition on modulo 2) on 16 bit blocks".

Description of the Block Cipher IDEA. For the description of the encryption algorithm of IDEA, let n be a positive integer such that $2^n + 1$ is prime. $\mathbb{Z}_{2^n} = \{0, 1, \ldots, 2^n - 1\}$ denotes the ring of integers modulo 2^n, $\mathbb{Z}_{2^n+1}^*$ denotes the multiplicative group of the non-zero elements of the field $\mathbb{Z}_{2^n+1}$ modulo $2^n + 1$. We will view $\mathbb{Z}_{2^n}$ as an additive group and $\mathbb{Z}_{2^n+1}$ as a multiplicative group for the rest of the discussion. The operations discussed above can be viewed as three vector-valued functions $f, g, h : \mathbb{Z}_2^n \times \mathbb{Z}_2^n \to \mathbb{Z}_2^n$. Let $v : \mathbb{Z}_{2^n} \to \mathbb{Z}_2^n = \mathbb{Z}_2 \times \ldots \times \mathbb{Z}_2$ (n-times $\mathbb{Z}_2$) and (the direct mapping) $d : \mathbb{Z}_{2^n+1}^* \to \mathbb{Z}_{2^n}$ be the maps given by $v(x) = (x_{n-1}, \ldots, x_1, x_0)$ such that $x = \sum_{i=0}^{n-1} x_i \cdot 2^i$ and

$$d(x) = \begin{cases} 0 & if \quad x = 2^n \\ x & if \quad x \neq 2^n \end{cases} \tag{1}$$

are bijective with inverses d^{-1}, v^{-1}. Let us denote $v(x) = \boldsymbol{x} = (x_{n-1}, \ldots, x_1, x_0)$ for any $x \in \mathbb{Z}_{2^n}$.

With this convention for any $\boldsymbol{x} = v(x)$, $\boldsymbol{y} = v(y)$, $x, y \in \mathbb{Z}_{2^n}$,

(i) the multiplication operation $\odot$;

$$g(\boldsymbol{x}, \boldsymbol{y}) = \boldsymbol{x} \odot \boldsymbol{y} = v(d(d^{-1}(x) \cdot d^{-1}(y) \bmod (\, 2^n + 1)))$$

(ii) the addition operation $\boxplus$;

$$f(x, y) = x \boxplus y = v(x + y \bmod (2^n))$$

(iii) the exclusive-OR operation $\oplus$

$$h(x, y) = x \oplus y = v(\sum_{i=0}^{n-1}((x_i + y_i) \bmod 2) \cdot 2^i) = (x_{n-1} + y_{n-1} \pmod 2), \ldots, x_0 + y_0 \pmod 2)$$

For the encryption, this cipher divides 64-bit plaintext block X into four 16 bit subblocks, X_1, X_2, X_3 and X_4. They are transformed into four 16-bit ciphertext subblocks Y_1, Y_2, Y_3 and Y_4 by 8 computationally identical rounds and a final output transformation using 52 16-bit key subblocks (see Fig. 1).

For the construction of these subblocks, the first step is to partition the given 128-bit key into 8 pieces and assign them as the first 8 subblock keys of the 52 subblocks. Then, the given 128-bit key is cyclically shifted to the left by 25 positions and partitioned into eight subblocks that are assigned to the next eight subblock keys. This process is repeated until all 52 subblock keys are derived.

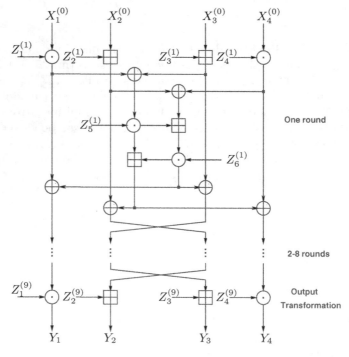

Fig. 1. Computational graph for the encryption process of the IDEA cipher

Low-High Algorithm for Multiplication. We shall use the following lemma that is given for the implementation issues of IDEA and PES in [3] and [4] :

Lemma 1 (Low-High algorithm for $\odot$).
Let a, b be two n-bit non-zero integers in $\mathbb{Z}_{2^n+1}$, then

$$ab \bmod (2^n+1) = \begin{cases} (ab \bmod 2^n) - (ab \text{ div } 2^n) & \text{if } (ab \bmod 2^n) \geq (ab \text{ div } 2^n) \\ (ab \bmod 2^n) - (ab \text{ div } 2^n) + 2^n + 1 & \text{if } (ab \bmod 2^n) < (ab \text{ div } 2^n) \end{cases}$$

where $(ab \text{ div } 2^n)$ denotes the quotient when ab is divided by 2^n.

2.2 Nonlinearity Criteria

Given $X = (x_{n-1}, x_{n-2}, \ldots, x_0) \in \mathbb{Z}_2^n$ and $A = (a_{n-1}, a_{n-2}, \ldots, a_0) \in \mathbb{Z}_2^n$, their binary inner product is defined by

$$A \cdot X = a_{n-1}x_{n-1} + a_{n-2}x_{n-2} + \ldots + a_0x_0$$

where the usual addition and multiplication operation are carried out over $\mathbb{Z}_2$. Let us define the set of all affine functions from $\mathbb{Z}_2^n$ to $\mathbb{Z}_2$:

$$\mathcal{A} = \{g_A \mid g_{A,c}(X) = A \cdot X + c, \text{ and } X \in \mathbb{Z}_2^n, c \in \mathbb{Z}_2\}$$

Note that $g_{A,c}$ is a linear function when $c = 0$.

Here is the measurement for the nonlinearity of a function $f : \mathbb{Z}_2^n \to \mathbb{Z}_2$ defined in [6]:

Definition 1. *The nonnegative integer*

$$H(f) = min_{g_{A,c} \in \mathcal{A}} \#\{X \in \mathbb{Z}_2^n \mid f(X) \neq g_{A,c}(X)\}$$

is the Hamming distance of f from the affine functions $\mathcal{A}$.

Here $\#Y$ denotes the cardinality of the set Y. The concept of nonlinearity of arbitrarily vector function $F : \mathbb{Z}_2^n \to \mathbb{Z}_2^m$ ($F = (f_1, \ldots, f_m)$ and $f_i : \mathbb{Z}_2^n \to \mathbb{Z}_2$) can be defined as follows:

Definition 2.

$$N(F) = min_{C \in \mathbb{Z}_2^m \setminus \{0\}} \{H(C \cdot F) \mid C \cdot F = \sum_{i=1}^{m} c_i f_i \}$$

All nonlinearity calculations in the following section are completely based on these definitions and in order to speed up these calculations, the algorithm based on Fast Walsh Transform [7] is used.

3 The Nonlinearity Properties of MA and RMA Structures

We observe that one point on the security of IDEA [4] depends on the fact that for $n = 2, 4, 8, 16$:

(i) for each $a \in \mathbb{Z}_{2^n+1} \backslash \{0, 2^n\}$, $l(x, a)$ is a polynomial in x over the ring $\mathbb{Z}_{2^n+1}$ with degree $2^n - 1$, where

$$l(x, y) = \begin{cases} d^{-1}[d(x) + d(y) \mod (2^n)] & \text{for all } x \text{ and } y \in \mathbb{Z}_{2^n+1} \\ 0 & \text{otherwise} \end{cases}$$

(ii) for each $a \in \mathbb{Z}_{2^n} \backslash \{0, 2^n\}$, $g(x, a)$ can not be written as a polynomial in x over the ring $\mathbb{Z}_{2^n}$, where

$$g(x, y) = d[d^{-1}(x).d^{-1}(y)) \mod (2^n + 1)] \text{ for all } x \text{ and } y \text{ in } \mathbb{Z}_{2^n}.$$

The inventors of IDEA, due to the properties above, viewed these operations highly "nonlinear" [4]. Our aim in this paper is to view these operations on the corresponding vector spaces $\mathbb{Z}_2^n (\cong \mathbb{Z}_{2^n+1}^*)$ and $\mathbb{Z}_2^n (\cong \mathbb{Z}_{2^n})$ respectively, and discuss the nonlinearity of them in the sense of the previous section.

Let $\boldsymbol{f}_a = \boldsymbol{f}(a, \boldsymbol{x})$ for every $a \in \mathbb{Z}_{2^n}$. Then the nonlinearity $N(\boldsymbol{f}_a)$ is zero. In fact for every $n \geq 1$, the right-end component $\boldsymbol{f}_0$ of the vector function $\boldsymbol{f}_a(\boldsymbol{x}) = \boldsymbol{f}(a, \boldsymbol{x})$, is $a_0 + x_0 \mod (2)$ where $\boldsymbol{a} = v(a) = (a_{n-1}, a_{n-2}, \dots, a_0)$, $\boldsymbol{x} = (x_{n-1}, x_{n-2}, \dots, x_0)$ and $\boldsymbol{f}_a = (\boldsymbol{f}_{n-1}, \boldsymbol{f}_{n-2}, \dots, \boldsymbol{f}_0)$.

In addition, for every $a \in \mathbb{Z}_{2^n}$ and $i = 0, 1, \dots, n - 1$ each component $h_i = a_i + x_i \mod (2)$ of the function $\boldsymbol{h}_a(\boldsymbol{x}) = \boldsymbol{h}(a, \boldsymbol{x})$, is an affine function. Therefore we have

Lemma 2. *For $n \geq 1$, the nonlinearity $N(\boldsymbol{f}_a)$ and $N(\boldsymbol{h}_a)$ of $\boldsymbol{f}_a$ and $\boldsymbol{h}_a$ equal to zero for every $a \in \mathbb{Z}_{2^n}$.*

Theorem 1. *For $n \geq 2$, the nonlinearity $N(\boldsymbol{g}_a)$ of the vector function $\boldsymbol{g}_a(\boldsymbol{x}) = \boldsymbol{g}(a, \boldsymbol{x})$ is zero for $a = 0, 1, 2, 2^{n-1}, 2^{n-1} + 1, 2^n - 1$.*

Proof. The case $a = 1$ is trivial. We shall prove the theorem for $a = 2$ and $a = 2^n - 1$. The remaining cases can be proved similarly.

For any $a, x \in \mathbb{Z}_{2^n}$, let us denote $g(a, \boldsymbol{x})$ by $\boldsymbol{g}_a(\boldsymbol{x}) = \boldsymbol{a} \odot \boldsymbol{x} = v(d(d^{-1}(a) \cdot d^{-1}(x) \mod (2^n + 1)))$, and $d^{-1}(x)$ by $\tilde{x}$.

For a $= 2$, Note that $d^{-1}(2) = \tilde{2} = 2$ and therefore $g_2(\boldsymbol{x}) = v(d(2 \cdot \boldsymbol{x} \mod (2^n + 1)))$ for any $x \in \mathbb{Z}_{2^n}$.

Since $\tilde{x} = \sum_{i=0}^n x_i \cdot 2^i$, $2 \cdot \tilde{x} \mod (2^n) = \sum_{i=0}^n x_i \cdot 2^{i+1} \mod (2^n) = \sum_{i=0}^{n-2} x_i \cdot 2^{i+1}$ and $2 \cdot \tilde{x} \text{ div } (2^n) = \sum_{i=0}^n x_i \cdot 2^{i+1} \text{ div } (2^n) = x_{n-1} + 2 \cdot x_n$.

Case 1: For $x = 0$ (i.e. $\tilde{x} = 2^n$), By the inequality $2 \cdot \tilde{x}$ div $(2^n) > 2 \cdot \tilde{x}$ mod (2^n) and Lemma 1, we get
$g(2, \boldsymbol{x}) = v(d(2 \cdot \tilde{x} \bmod (2^n) - 2 \cdot \tilde{x} \text{ div } (2^n) + 2^n + 1)) = v(d(2^n + 1 - 2)) = v(2^n - 1)$
and $\boldsymbol{g}_2(\boldsymbol{x}) = (1, 1, \ldots, 1)$.

Case 2: Let us consider the case $x \neq 0, x_i = 0$ for $i \in \{0, 1, \ldots, n-2\}$ and $x_{n-1} = 1$. Then one can observe $2 \cdot \tilde{x}$ div $(2^n) > 2 \cdot \tilde{x}$ mod (2^n) and conclude that $\boldsymbol{g}(2, \boldsymbol{x}) = d(2^n) = 0$ by Lemma 1. Hence $\boldsymbol{g}_2(\boldsymbol{x}) = (0, 0, \ldots, 0)$.

Case 3: For $x_{n-1} = 0$ and there exists $i \in \{0, 1, \ldots, n-2\}$ such that $x_i \neq 0$. In this case one gets $\boldsymbol{g}_2(\boldsymbol{x}) = (x_{n-2}, x_{n-3}, \ldots, x_0, 0)$.

Case 4: For $x_{n-1} = 1$ and $x_0 = 0$ and $x_i \neq 0$ for at least one $i \in \{1, \ldots, n-2\}$, we consider an index t such that $x_0 = x_1 = \ldots = x_{t-1} = 0$ and $x_t = 1$, we have $2 \cdot \tilde{x} \bmod (2^n) = \sum_{i=0}^{n-2} x_i \cdot 2^{i+1}$ and $2 \cdot \tilde{x}$ div $(2^n) = x_{n-1} = 1$ by using Lemma 1, we obtain $2 \cdot x \bmod (2^n + 1) = \sum_{i=0}^{n-2} x_i \cdot 2^{i+1} - 1$. Thus one can check that $\boldsymbol{g}_2(\boldsymbol{x}) = (x_{n-2}, x_{n-3}, \ldots, x_{t+1}, 0, 1, 1, \ldots, 1)$.

The case $x_{n-1} = 1$ and $x_0 = 1$ is similar to *Case 4*, and one gets $\boldsymbol{g}_2(\boldsymbol{x}) = (x_{n-2}, x_{n-3}, \ldots, x_1, 0, 1)$.

We can summarize all these cases as follows:

$$\boldsymbol{g}_2(\boldsymbol{x}) = \begin{cases} (1, 1, \ldots, 1) & \text{, if } \boldsymbol{x} = (0, 0, \ldots, 0) \\ (0, 0, \ldots, 0) & \text{, if } \boldsymbol{x} = (1, 0, \ldots, 0) \\ (x_{n-2}, x_{n-3}, \ldots, x_0, 0) & \text{, if } \boldsymbol{x} = (0, x_{n-2}, \ldots, x_0) \\ (x_{n-2}, \ldots, x_{t+1}, 0, 1, 1, \ldots, 1) & \text{, if } \boldsymbol{x} = (1, x_{n-2}, \ldots, x_{t+1}, 1, 0, \ldots, 0) \\ (x_{n-2}, \ldots, x_1, 0, 1) & \text{, if } \boldsymbol{x} = (1, x_{n-2}, \ldots, x_1, 1) \end{cases}$$

If $\boldsymbol{g}_2 = (p_{n-1}, \ldots, p_0)$, namely $p_i, i = \{0, 1, \ldots, n-1\}$, are the components of $\boldsymbol{g}_2 : \mathbb{Z}_2^n \to \mathbb{Z}_2^n$, then it is not difficult to see from above that $p_0(x_{n-1}, x_{n-2}, \ldots, x_0) + p_1(x_{n-1}, x_{n-2}, \ldots, x_0) = x_0$ is affine, and thus $N(\boldsymbol{g}_2) = 0$.

For a $= 2^{n-1}$,
$g(2^{n-1}, \boldsymbol{x}) = v(d(2^{n-1} \cdot \tilde{x} \bmod (2^n + 1)))$ for any $x \in Z_{2^n}$.
$2^{n-1} \cdot \tilde{x} \bmod (2^n) = \sum_{i=0}^{n} x_i \cdot 2^{n+i-1} \bmod (2^n) = x_0 \cdot 2^{n-1}$
and $2^{n-1} \cdot \tilde{x}$ div $(2^n) = \sum_{i=0}^{n} x_i \cdot 2^{n+i-1}$ div $(2^n) = \sum_{i=1}^{n} x_i \cdot 2^{i-1}$.
To compute the component functions $h_i : \mathbb{Z}_2^n \to \mathbb{Z}_2$ of $\boldsymbol{g}_{2^{n-1}} = (h_{n-1}, \ldots, h_0)$, we need to look at several cases:

Case 1: For $x \neq 0$ (i.e. $\tilde{x} = x$) and $x_0 \neq 0$,
$2^{n-1} = x_0 \cdot 2^{n-1} = 2^{n-1} \cdot x \bmod (2^n) \geq 2^{n-1} \cdot x$ div $(2^n) = \sum_{i=1}^{n-1} x_i \cdot 2^{i-1}$.
Then by Lemma 1, $2^{n-1} \cdot x \bmod (2^n + 1) = \sum_{i=0}^{n-2} 2^i + 1 - \sum_{i=1}^{n-1} x_i \cdot 2^{i-1} = 1 + \sum_{i=1}^{n-1} (1 - x_i) \cdot 2^{i-1}$ and from that equation the right-end component of the $\boldsymbol{g}_{2^{n-1}}(v(x))$, $h_0(v(x)) = (2 - x_1) \bmod (2) = x_1$.

Case 2: For $x \neq 0$ and $x_0 = 0$, $\sum_{i=1}^{n-1} x_i \cdot 2^{i-1} = 2^{n-1} \cdot x$ div $(2^n) > 2^{n-1} \cdot x \bmod (2^n) = 2^{n-1} \cdot x_0 = 0$. From Lemma 1, $2^{n-1} \cdot x \bmod (2^n + 1) =$

$2^n + 1 - (\sum_{i=1}^{n-1} x_i \cdot 2^{i-1})$. Hence, the right-end component of the $\boldsymbol{g}_{2^{n-1}}(\boldsymbol{x})$ is nothing but $h_0(\boldsymbol{x}) = 1 - x_1 \mod (2) = 1 + x_1$.

Case 3: For $x = 0$, (i.e. $\tilde{x} = 2^n$), By Lemma 1, $2^{n-1} \cdot \tilde{x} \mod (2^n + 1) = 2^n + 1 - 2^{n-1} = 2^{n-1} + 1$ since $2^{n-1} \cdot \tilde{x} \mod (2^n) = 0$ and $2^{n-1} \cdot \tilde{x} \operatorname{div} (2^n) = 2^{n-1}$ and we have $\boldsymbol{g}_{2^{n-1}}(\boldsymbol{0}) = (1, 0, \ldots, 0, 1)$. With this we computed explicitly all the values $\boldsymbol{g}_{2^{n-1}}(\boldsymbol{x})$ and leave the reader to check that $h_0(\boldsymbol{x}) = x_0 + x_1 + 1$. This gives immediately $N(\boldsymbol{g}_{2^{n-1}}) = 0$. $\square$

Remark 1. It is important to note that the large classes of weak keys for IDEA was attained by Daeman et al. in [2] using the fact that $N(\boldsymbol{g}_a) = 0$ when $a = 0$ and $a = 1$.

Now we are going to look at the nonlinearity of the combinations of the operation $\boxplus$ and $\odot$. We start with the nonlinearity of the functions $\boldsymbol{k}_{a,b}(x) = (\boldsymbol{a} \odot \boldsymbol{x}) \boxplus \boldsymbol{b}$ and $\boldsymbol{l}_{a,b}(x) = (\boldsymbol{a} \boxplus \boldsymbol{x}) \odot \boldsymbol{b}$, respectively. In the case of $n = 2, 4$ and 8 we observe that $N(\boldsymbol{k}_{a,b}) = 0$ when a $= 0, 1, 2, 2^{n-1}, 2^{n-1} + 1, 2^n - 1$, and $N(\boldsymbol{l}_{a,b}) = 0$ when b$= 0, 1, 2, 2^{n-1}, 2^{n-1} + 1, 2^n - 1$. These calculations show that the linearity of the operation $\odot$ is carried over the operation $\boxplus$.

The combinations of the operations $\boxplus$ and $\odot$ are successively used in each round of IDEA (see Fig. 1,2). The operation used there are called "Multiplication-Addition" (MA) structure and it can be viewed as a transformation $MA : \mathbb{Z}_2^n \times \mathbb{Z}_2^n \times \mathbb{Z}_2^n \times \mathbb{Z}_2^n \to \mathbb{Z}_2^n \times \mathbb{Z}_2^n$ given by

$$MA(P_1, P_2, K_1, K_2) = (C_1, C_2).$$

where (P_1, P_2), (C_1, C_2) and (K_1, K_2) are two n-bit input, output and key sub-blocks, respectively.

In order to discuss the nonlinearity properties of MA structure in the sense of Definition 2, for each fixed $K = (K_1, K_2) \in \mathbb{Z}_2^{2n}$ one can consider the transformation $M : \mathbb{Z}_2^{2n} \to \mathbb{Z}_2^{2n}$ given by

$$M(P_1, P_2) = MA(P_1, P_2, K_1, K_2) = (C_1 C_2).$$

where $C_2 = ((K_1 \odot P_1) \boxplus P_2) \odot K_2$ and $C_1 = (K_1 \odot P_1) \boxplus C_2$.

For each $K = (K_1, K_2)$ this gives us to look at this transformation as a $2n \times 2n$ S-Box. Now in addition to that transformation, we are going to discuss a slightly different one which we call Reverse MA, and denoted it by $RMA : \mathbb{Z}_2^{2n} \times \mathbb{Z}_2^{2n} \to \mathbb{Z}_2^{2n}$. In fact,

$$RMA(P_1, P_2, K_1, K_2) = MA(P_1, K_2, K_1, P_2) = (C_1', C_2').$$

where $C_2' = ((K_1 \odot P_1) \boxplus K_2) \odot P_2$ and $C_1' = (K_1 \odot P_1) \boxplus C_2'$. As above for each fixed $K = (K_1, K_2)$ the resulting transformation is denoted by $RM : \mathbb{Z}_2^{2n} \to \mathbb{Z}_2^{2n}$, namely

$$RM(P_1, P_2) = RMA(P_1, P_2, K_1, K_2).$$

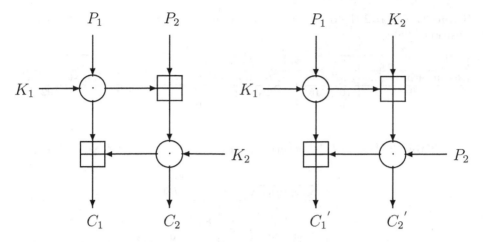

Fig. 2. Computational graph of the MA and RMA structure

MA structure actually consists of the function $k_{a,b}$ and $l_{a,b}$, and we have observed the effects of the linearity of the multiplication operation $\odot$ over these combinations.

Each round of IDEA contains one MA structure but one should consider the iteration of this structure due to the existence of 8 iterated rounds. Hence MA structure also iterates 8 times. For this reason, we shall take some compositions MA's and RMA's. Note that similar to M and RM transformations, each composition is viewed as a transformation from $\mathbb{Z}_2^{2n} \to \mathbb{Z}_2^{2n}$. Table 1 illustrates the occurrence of the nonlinearity values of M, RM and some of their compositions for the case $n = 2$ and all possible keys values. First column of the following tables shows all possible nonlinear values attained, other columns shows the number of transformations achieving the corresponding nonlinear values. For example in Table 1 for $n=2$, the possible nonlinearity values are 0, 2 and 4. Among 256 transformations in the form $M \circ M$ [res. $RM \circ RM$], 224 [res. 0] of them have nonlinearity 0, 0 [res. 192] of them have nonlinearity 2 and 32 [res. 64] of them have nonlinearity 4.

For each composition given above, when $n=4$ the nonlinearity of 2^{16} transformations can be computed for all 2^{16} key values. Table 3 shows all the possible calculated nonlinearity values and their occurrences. Among all these compositions, the number of $M \circ M$ transformations whose nonlinearity value zero is the largest, there are no zero values for $RM \circ RM$ and this composition attains the highest nonlinearity values. On the other hand, $RM \circ M$ and $M \circ RM$ give better nonlinearity values than $M \circ M$. Moreover, when $n=4$ the nonlinearities for the compositions $M \circ M \circ M$, $RM \circ M \circ RM$, $M \circ RM \circ M$ and $RM \circ RM \circ RM$ are also calculated. In this case, for each composition there are 2^{24} possible transformations for all key values (see Tables 4 and 5). It is easily observed that the number of $M \circ M \circ M$ transformations whose nonlinearity value zero is more than other transformations and there is no $RM \circ RM \circ RM$ transformation having nonlinear value 0.

Table 1. For n=2 the occurrence of the nonlinearity of M, RM and their compositions' values

Nonlinearity Values	Occurrence					
	$\#M$	$\#RM$	$\#(M \circ M)$	$\#(RM \circ RM)$	$\#(RM \circ M)$	$\#(M \circ RM)$
0	16	16	224	0	192	192
2	0	0	0	192	0	64
4	0	0	32	64	64	0

Table 2. For n=4 the occurrence of the nonlinearity of M and RM

Nonlinearity Values	Occurrence	
	$\#M$	$\#RM$
0	112	64
32	80	64
40	8	0
48	16	0
56	8	0
64	32	128

Table 3. For n=4 the occurrence of the nonlinearity of $M \circ M$, $M \circ RM$, $RM \circ M$ and $RM \circ RM$

N(M ∘ M)'s values	#(M ∘ M)	N(M ∘ RM)'s values	#(M ∘ RM)	N(RM ∘ M)'s values	#(RM ∘ M)	N(RM ∘ RM)'s values	#(RM ∘ RM)
0	4200	0	1664	0	1920	72	1
16	166	32	1152	16	80	74	1
20	144	52	256	20	80	76	8
24	56	56	128	24	16	80	30
26	108	62	264	28	48	82	121
28	188	64	7544	30	16	84	549
30	16	66	365	32	4496	86	1270
32	6652	68	2518	34	16	88	4733
34	98	70	598	36	112	90	10820
36	380	72	5979	38	80	92	22383
38	170	74	1026	40	432	94	20801
40	708	76	3466	42	288	96	4772
42	533	78	1993	44	512	98	47
44	1221	80	8776	46	48		
46	229	82	3271	48	1552		
48	3089	84	6936	50	368		
50	1026	86	4925	52	2384		
52	3281	88	7665	54	880		
54	1643	90	3918	56	2720		
56	4099	92	2604	58	720		
58	1800	94	467	60	2000		
60	3036	96	21	62	2688		
62	3325			64	8000		
64	10486			66	3200		
66	3069			68	4368		
68	4134			70	2612		
70	2379			72	5856		
72	3897			74	3824		
74	1957			76	4704		
76	1967			78	3038		
78	828			80	3967		
80	488			82	2033		
82	146			84	1584		
84	17			86	528		
				88	351		
				90	15		

Table 4. For n=4 the occurrence of the nonlinearity of $M \circ M \circ M$, $M \circ RM \circ M$ and $RM \circ M \circ RM$

N($M \circ M \circ M$)'s values	#($M \circ M \circ M$)	N($M \circ RM \circ M$)'s values	#($M \circ RM \circ M$)	N($RM \circ M \circ RM$)'s values	#($RM \circ M \circ RM$)
0	213870	0	52992	0	3584
8	8	16	1280	16	4608
12	424	20	1280	20	1536
16	14879	24	256	28	2048
20	6500	28	768	30	512
22	536	30	256	32	24576
24	4987	32	104320	34	512
26	5193	34	256	36	3072
28	14897	36	1792	38	2048
30	3243	38	1292	40	10240
32	460354	40	13004	42	3584
34	6235	42	4712	44	3072
36	32066	44	9525	46	2048
38	19146	46	1236	48	22124
40	86298	48	31725	50	6720
42	36227	50	7054	52	25520
44	80307	52	62233	54	15584
46	40097	54	18234	56	24914
48	304144	56	78828	58	22272
50	95721	58	16663	60	19248
52	288161	60	50731	62	40272
54	176661	62	74139	64	95834
56	491875	64	416428	66	59862
58	262221	66	100284	68	65246
60	418722	68	224917	70	70538
62	450788	70	112694	72	97117
64	1366824	72	429767	74	87605
66	611928	74	211967	76	102661
68	873628	76	454482	78	104916
70	756097	78	334759	80	149660
72	1241526	80	725598	82	155942
74	980569	82	587343	84	240900
76	1204923	84	916279	86	391034
78	1143217	86	1160094	88	977211
80	1279955	88	1806335	90	2309620
82	1124983	90	2654850	92	4862819
84	1052438	92	3394120	94	5357186
86	813146	94	2338294	96	1394305
88	527923	96	373853	98	16666
90	230671	98	2576		
92	52393				
94	3414				
96	21				

Table 5. For n=4 the occurrence of the nonlinearity of $RM \circ RM \circ RM$

N($RM \circ RM \circ RM$)'s values	# ($RM \circ RM \circ RM$)
74	8
76	236
78	613
80	2955
82	11559
84	49630
86	197030
88	713073
90	2243561
92	5291427
94	6432751
96	1809487
98	24883
100	3

On the basis of these computations, we suggest the usage of new structure Reverse MA in IDEA encryption algorithm instead of MA structure used in IDEA. It is clear that this modification is minor and it does not change any other structure of IDEA. We call this slightly modified version of IDEA having the Reverse MA structure as RIDEA (Reverse IDEA). The computational graph of RIDEA is shown in Fig. 3. In [4] it was stated that the round function of IDEA is "complete" (i.e. every output bits of the first round depends on every bit of the plaintext and on every bit key used for that round.) and this diffusion is provided by MA structure. In order to analyze the diffusion properties of our RIDEA, we use the Avalanche Weight Distribution (AWD) [1], Strict Avalanche Criteria (SAC) [8] and their test procedures (Appendix). For both criteria, the same set of plaintexts generated by MD5 Message Digest Algorithm and fixed round subkeys were used. For 1-round RIDEA we obtain nearly the same average AWD curve as it is obtained for 1-round IDEA (see Fig. 4). It is interesting to note that when the number of trials is 2^{25}, the minimum and maximum of all SAC table's entries of 1-round IDEA [res. RIDEA] are approximately 0,01064 and 0,99469 [res. 0,49963 and 0,50029], the average and variance of all SAC table's entries of 1-round IDEA [res. RIDEA] are roughly 0,48412 and 0,02113 [res. 0,5 and $6,05761.10^{-6}$]. From these observations we conclude that the required diffusion for RIDEA is provided at its first round.

4   Conclusion

We viewed the MA structure as a transformation from $\mathbb{Z}_2^{2n} \to \mathbb{Z}_2^{2n}$, and looked at its nonlinearity properties. We realized that by slightly changing the MA structure which we call RMA structure, we obtained far better nonlinearity properties. We believe that this structure not only provides the required diffusion but also increases the nonlinearity of IDEA.

References

1. E. Aras and M.D. Yücel, Performance Evaluation of Safer K-64 and S-boxes of Safer Family. Turkish Journal of Electrical Engineering & Computer Sciences, Vol.9, No.2, pp. 161-175, 2001.
2. J.Daeman, R. Govaerts and J. Vandewalle, Weak Keys for IDEA. Advances in Cryptology, Proc. EUROCRYPTO'93, LNCS 773, Springer-Verlag, pp. 224-231, 1994.
3. X. Lai and J. L. Massey, A Proposal for a New Block Encryption Standard. Advances in Cryptology - EUROCRYPTO'90, Proceedings, LNCS 473, pp. 389-404, Springer-Verlag, Berlin, 1990.
4. X. Lai, On the design and security of block cipher. ETH Series in Information Processing, V.1, Konstanz: Hartung-Gorre Verlag, 1992.
5. X. Lai, J. L. Massey and S. Murphy, Markov Cipher and Differential Cryptanalysis. Advances in Cryptology, EUROCRYPTO'91, Lecture Notes in Computer Science, Springer Verlag, Berlin-Heidelberg, 547, pp. 17-38, 1991.
6. K. Nyberg, On the construction of highly nonlinear permutations. In Extended Abstracts – EUROCRYPTO'92, pages 89-94, May 1992.

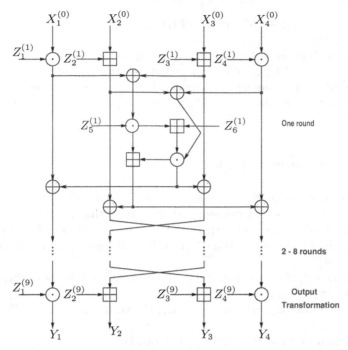

Fig. 3. Computational graph for the encryption process of the RIDEA cipher

7. Measuring Boolean Function Nonlinearity by Walsh Transform, http://www.ciphersbyritter.com/ARTS/MEASNONL.HTM.
8. A. F. Webster and S.E. Tavares, On the design of S-Boxes. Advances in Cryptology: CRYPTO'85, proceedings, Springer, 1986.

5 Appendix

5.1 Avalanche Weight Distribution

Avalanche Weight Distribution (AWD) criterion was proposed in [1] for fast rough analysis of the algorithm of the diffusion properties of block ciphers and stated as: *Even for quite plaintext pairs* (P_1, P_2), *i.e. the Hamming weight of the differences of plaintext pairs* (P_1, P_2) *is small, the distribution of the Hamming weight of the differences of corresponding ciphertext pairs* (C_1, C_2) *should be a Binomial distribution around n /2 for a good block cipher with a block length of* n.

For IDEA [res. RIDEA] cipher we have the following test procedure for this criterion [1]:

1. A plaintext P is chosen at random and the pair of that plaintext P_i is calculated so that the difference between P and P_i is e_i, i.e., $P_i = P \oplus e_i$ and P and P_i differ only in bit i, where e_i is a 64-bit unit vector with a 1 in position i, and $i \in \{1, 2, \ldots, 64\}$,

2. P and P_i are submitted to 1-round of IDEA [res. RIDEA] for encryption under a random key,

3. From the resultant ciphertexts C and C_i, the Hamming weight of the avalanche vector $wt(C_d) = wt(C \oplus C_i) = j$ is calculated, where $j \in \{0, 1, 2, \ldots, 64\}$

4. The value of the jth element of an avalanche weight distribution array with a size of 65 is incremented by 1, i.e., $AWD\_array[i,j] = AWD\_array[i,j]+1$,

5. The steps above are repeated 2^{25} times.

5.2 Strict Avalanche Criterion

Here is the test procedure for SAC (its idea was introduced in [8]):

1. A plaintext P is chosen at random and the pair of that plaintext P_i is calculated so that the difference between P and P_i is e_i, i.e., $P_i = P \oplus e_i$ and P and P_i differ only in bit i, where e_i is a 64-bit unit vector with a 1 in position i, and $i \in \{1, 2, \ldots, 64\}$,

2. P and P_i are submitted to 1-round of IDEA [res. RIDEA] for encryption under a random key,

3. From the resultant ciphertexts C and C_i, the vector $C_d = (C \oplus C_i)$ is calculated,

4. C_d is summed up to an avalanche sum array, i.e.,
avalanche_sum_array[i]=avalanche_sum_array[i]+C_d,

5. The steps above are repeated 2^{25} times.

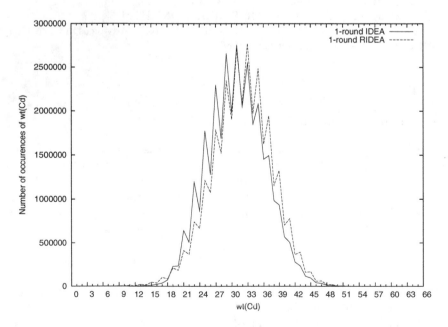

Fig. 4. Average AWD curve of 1-round IDEA and 1-round RIDEA

In order to obtain the average AWD curve for 1-round of IDEA [res. RIDEA], the average of jth column entries of the matrix $(AWD\_array)_{64 \times 64}$ is calculated and these values are sketched versus j values.

For 1-round of IDEA [res. RIDEA], the above SAC test procedure is carried out for all $i \in \{1, 2, \ldots, 64\}$ and all entries of avalanche_sum_array[i] divided by 2^{25} and this modified array taken as an ith row of the SAC table.

Impossible Differential Cryptanalysis for Block Cipher Structures

Jongsung Kim[1], Seokhie Hong[1], Jaechul Sung[2], Sangjin Lee[1], Jongin Lim[1],
and Soohak Sung[3]

[1] Center for Information Security Technologies(CIST),
Korea University, Seoul, Korea
{joshep,hsh,sangjin,jilim}@cist.korea.ac.kr
[2] Korea Information Security Agency(KISA), Seoul, KOREA,
sjames@kisa.or.kr
[3] Beajea University, Deajoan, KOREA,
sungsh@mail.paichai.ac.kr

Abstract. Impossible Differential Cryptanalysis(IDC) [4] uses impossible differential characteristics to retrieve a subkey material for the first or the last several rounds of block ciphers. Thus, the security of a block cipher against IDC can be evaluated by impossible differential characteristics. In this paper, we study impossible differential characteristics of block cipher structures whose round functions are bijective. We introduce a widely applicable method to find various impossible differential characteristics of block cipher structures. Using this method, we find various impossible differential characteristics of known block cipher structures: Nyberg's generalized Feistel network, a generalized CAST256-like structure [14], a generalized MARS-like structure [14], a generalized RC6-like structure [14], and Rijndael structure.

Keyword : Impossible Differential Cryptanalysis(IDC), impossible differential characteristic, block cipher structures

1 Introduction

The most powerful known attacks on block ciphers are differential cryptanalysis(DC) [3] and linear cryptanalysis(LC) [12]. These attacks have been applied to many known ciphers very efficiently. So, one has tried to make a block cipher secure against DC and LC. Nyberg and Knudsen first proposed the conception of a provable security against DC and gave a provable security for a Feistel structure in 1992 [16]. Since then, many block cipher structures with a provable security against DC and LC have been studied [8,9,13,16,17,19]. However, a provable security against DC and LC is not enough to give the security of block ciphers, because other cryptanalyses may be applied to them not vulnerable to DC and LC. For instance, the 3-round Feistel structure whose round functions are bijective has a provable security against DC and LC [2], but there exists a 5-round

T. Johansson and S. Maitra (Eds.): INDOCRYPT 2003, LNCS 2904, pp. 82–96, 2003.
© Springer-Verlag Berlin Heidelberg 2003

impossible differential characteristic [10]. This fact is also applied to some other structures. In this paper, we focus on IDC for block cipher structures whose round functions are bijective. We provide a general tool, called $\mathcal{U}$-method, which can find various impossible differential characteristics of block cipher structures with a certain property. We also provide an algorithm to compute the maximum length of impossible differential characteristics that can be found in the $\mathcal{U}$-method. (By modifying the algorithm, we can find the specific forms of impossible differential characteristics.) We use it to find various impossible differential characteristics of known block cipher structures. See Table 1 for a summary of our results. Furthermore, we use impossible differential characteristics of Rijndael which can be found in our algorithm to improve the previous result [6].

This paper is organized as follows. In Section 2, we describe a generalized Feistel network and Rijndael structure. In Section 3, we introduce some basic notions for IDC and the $\mathcal{U}$-method. In Section 4, we propose an algorithm to compute the maximum length of impossible differential characteristics which can be found in the $\mathcal{U}$-method. In Section 5, we find various impossible differential characteristics of known block cipher structures. In Section 6, we discuss how to apply the $\mathcal{U}$-method to an integral attack.

Table 1. Summary of our cryptanalytic results. (A: The number(r) of rounds to have the property that the maximum average of differential probability is bounded by p^{2n} where p is the maximum average of differential probability of a round function. B: The number(r) of rounds for impossible differential characteristics. See Section 2 for the details of GFN_n, Rijndael$_{128}$, Rijndael$_{192}$, and Rijndael$_{256}$ and refer to [14] for the details of a generalized CAST256-like structure, a generalized MARS-like structure, and a generalized RC6-like structure.)

Block Cipher Structure	DC (A)	comment
GFN_n	$r \geq 3n$	conjecture([17])

Block Cipher Structure	IDC (B)	comment
GFN_2	$r = 7$	This paper
GFN_n	$r = 3n + 2\ (n \geq 3)$	This paper
Rijndael$_{128}$	$r = 3$	[6]
Rijndael$_{192}$	$r = 4$	This paper
Rijndael$_{256}$	$r = 5$	This paper
Generalized CAST256	$r = n^2 - 1\ (n \geq 3)$	[19]
Generalized MARS	$r = 2n - 1\ (n \geq 3)$	This paper
Generalized RC6	$r = 4n + 1$	This paper

2   Descriptions of Block Cipher Structrues

2.1   A Generalized Feistel Network

A generalized Feistel network was introduced by Nyberg [17]. Let $(X_0, X_1, \cdots, X_{2n-1})$ be the input to one round of the network. Given n round functions

$F_0, F_1, \cdots, F_{n-1}$ and n round keys $K_0, K_1, \cdots, K_{n-1}$, the output of the round $(Y_0, Y_1, \cdots, Y_{2n-1})$ is computed by the following formulas.

$$Y_{2j} = X_{2j-2}, \quad \text{for } 1 \le j \le n-1$$
$$Y_{2j-1} = F_j(X_{2j} \oplus K_j) \oplus X_{2j+1}, \quad \text{for } 1 \le j \le n-1$$
$$Y_0 = F_0(X_0 \oplus K_0) \oplus X_1, \quad Y_{2n-1} = X_{2n-2}$$

We call X_i and Y_j as the ith subblock of input and the jth subblock of output, respectively. We denote this generalized Feistel network with n round functions by GFN_n. If F_j is regarded as a keyed-round function F and $n = 4$, a round of GFN_4 is depicted in Figure 1.

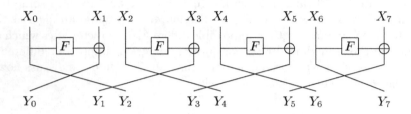

Fig. 1. A round of GFN_4

2.2   Rijndael Structure

Rijndael is a block cipher composed of SPN structure. The length of the data block can be specified to be 128, 192, or 256 bits. They are expressed as arrays of 4×4 bytes, 4×6 bytes, and 4×8 bytes, respectively. A round of Rijndael consists of 4 transformations, i.e., ByteSubstitution(BS), ShiftRow(SR), Mix-Colmn(MC), and AddroundKey(AK). A round of Rijndael with 128-bit data block is depicted in Figure 2. Here, $(f \circ g)(x)$ represents $f(g(x))$.

X_0	X_1	X_2	X_3
X_4	X_5	X_6	X_7
X_8	X_9	X_{10}	X_{11}
X_{12}	X_{13}	X_{14}	X_{15}

$AK \circ MS \circ SR \circ BS$ →

Y_0	Y_1	Y_2	Y_3
Y_4	Y_5	Y_6	Y_7
Y_8	Y_9	Y_{10}	Y_{11}
Y_{12}	Y_{13}	Y_{14}	Y_{15}

Fig. 2. A round of Rijndael with a 128-bit data block

In this paper we observe the structures of Rijndael whose nonlinear byte-wise substitutions, S-boxes, are considered as bijective black boxes. The S-boxes can be viewed as round functions F, e.g., each round includes 16 F functions. We call these structures Rijndael$_{128}$ structure, Rijndael$_{192}$ structure, and Rijndael$_{256}$ structure, respectively.

3 New Basic Notions for IDC

In this section, we will introduce some notions, which are a bit intricate but very available. We assume that a block cipher structure S has n data sub-blocks, e.g., the input and the output of one round are $(X_0, X_1, \cdots, X_{n-1})$ and $(Y_0, Y_1, \cdots, Y_{n-1})$, respectively. Throughout the paper, we consider S whose round function F is bijective, and operation to connect a subblock with another one is $\oplus$.

Definition 1. *For a block cipher structure S, the $n \times n$ Encryption Characteristic Matrix $\mathcal{E}$ and the $n \times n$ Decryption Characteristic Matrix $\mathcal{D}$ are defined as follows. If Y_j is affected by X_i, the (i, j) entry of $\mathcal{E}$ is set to 1, and if not, the (i, j) entry is set to 0. Especially, if Y_j is affected by $F(X_i)$, the (i, j) entry of $\mathcal{E}$ is set to 1_F instead of 1. Reversely, if X_j is affected by Y_i, the (i, j) entry of $\mathcal{D}$ is set to 1, and if not, the (i, j) entry is set to 0. Especially, if X_j is affected by $F(Y_i)$ or $F^{-1}(Y_i)$, the (i, j) entry of $\mathcal{D}$ is set to 1_F instead of 1. If the number of entry $1(\neq 1_F)$ in each column of the matrix is zero or one, we call it* **1-property matrix**.

For example, $\mathcal{E}$ and $\mathcal{D}$ of the Feistel structure depicted in Figure 3 are as follows. (According to Definition 1, the Feistel structure has 1-property matrices $\mathcal{E}$ and $\mathcal{D}$.)

$$\mathcal{E} = \begin{pmatrix} 1_F & 1 \\ 1 & 0 \end{pmatrix} \quad , \quad \mathcal{D} = \begin{pmatrix} 0 & 1 \\ 1 & 1_F \end{pmatrix}$$

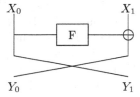

Fig. 3. A round of a Feistel structure

If S has 1-property matrices $\mathcal{E}$ and $\mathcal{D}$, we can easily find various impossible differential characteristics of S in our method. In this section, we assume that S has 1-property matrices $\mathcal{E}$ and $\mathcal{D}$.

Definition 2. *Given an input difference* $\alpha = (\alpha_0, \alpha_1, \cdots, \alpha_{n-1})$, *the input difference vector* $\boldsymbol{a} = (a_0, a_1, \cdots, a_{n-1})$ *corresponding to* α *is defined as follows.*

$$a_i = \begin{cases} 0 & \text{if } \alpha_i = 0 \\ 1^* & \text{otherwise} \end{cases}$$

We denote the output difference after r rounds for α by α^r, and denote the value of the ith subblock of α^r by α_i^r. Naturally, the difference vector which corresponds to α^r is denoted $\boldsymbol{a}^r$ (The meaning of the values of $\boldsymbol{a}^r$ will be explained below.), and the ith entry of $\boldsymbol{a}^r$ which corresponds to α_i^r is denoted a_i^r. If the same work is performed to decryption process, we use the notations β, β^r, β_i^r, $\boldsymbol{b}$, $\boldsymbol{b}^r$, and b_i^r instead of α, α^r, α_i^r, $\boldsymbol{a}$, $\boldsymbol{a}^r$, and a_i^r, respectively.

Given an input difference, the possible output differences of each subblock after r rounds can be classified by five types of differences: zero difference, a nonzero nonfixed difference, a nonzero fixed difference, exclusive-or of a nonzero fixed difference and a nonzero nonfixed difference, and a nonfixed difference. As the extended one of the notations used in Definition 2, the five types of differences stated above are denoted by the entries of difference vectors in Table 2.

Table 2. Entries of difference vectors and corresponding differences.

Entry (a_i^r or b_i^r)	Corresponding difference (α_i^r or β_i^r)
0	zero difference (denoted 0)
1	nonzero nonfixed difference (denoted δ)
1^*	nonzero fixed difference (denoted γ)
2^*	nonzero fixed difference $\oplus$ nonzero nonfixed difference ($\gamma \oplus \delta$)
$t(\geq 2)$	nonfixed difference (denoted ?)

Throughout this paper, we will use the notations 0, δ, γ, and ? as the differences stated in Table 2 (sometimes δ' (resp., γ') is used as the same kind of a difference δ (resp., γ)). According to Table 2, in the case of the entry t, we cannot predict the corresponding difference, in other words, we cannot know the difference to which the entry t does not correspond. On the other hand, in the cases of the entries $0, 1, 1^*$, and 2^*, we can predict the difference to which each entry does not correspond. For example, 2^* cannot correspond to γ, since $\gamma \oplus \delta \neq \gamma$. These facts are of use to find impossible differential characteristics of $\mathcal{S}$. In our method, we concentrate on difference vectors rather than the specific forms of differences.

In order to compute $\boldsymbol{a}^r$, we need to define an multiplication between a difference vector and an encryption characteristic matrix. (We omit the explanation for the decryption process, since it is the same work as that of the encryption process.) A difference vector $\boldsymbol{a}^r$ can be successively computed as like Equation (1).

$$\boldsymbol{a^r} = \overbrace{((((\boldsymbol{a} \cdot \mathcal{E}) \cdot \mathcal{E}) \cdots) \cdot \mathcal{E})}^{r \text{ times}} = \overbrace{((((\boldsymbol{a^1} \cdot \mathcal{E}) \cdot \mathcal{E}) \cdots) \cdot \mathcal{E})}^{r - 1 \text{ times}} = \cdots = \boldsymbol{a^{r-1}} \cdot \mathcal{E} \quad (1)$$

Without loss of generality, we define a multiplication of $\boldsymbol{a}$ and $\mathcal{E}$ (e.g., $\boldsymbol{a} \cdot \mathcal{E} = (a_i)_{1 \times n} \cdot (\mathcal{E}_{i,j})_{n \times n} = (\sum_i a_i \cdot \mathcal{E}_{i,j})_{1 \times n}$)

First, we consider a multiplication between an entry of difference vector a_i and an entry of matrix $\mathcal{E}_{i,j}$. The multiplication $a_i \cdot \mathcal{E}_{i,j}$ represents the relation between the input difference of the ith subblock and the output difference of the jth subblock. Table 3 illustrates the meaning of the multiplication.

Table 3. Multiplication between an entry of difference vector and an entry of matrix. ($k \in \{0, 1, 1^*, 2^*, t\}$)

$a_i \cdot \mathcal{E}_{i,j}$	Meaning
$k \cdot 0 = 0$	The output difference of the jth subblock is not affected by the input difference of the ith subblock.
$k \cdot 1 = k$	The output difference of the jth subblock is affected by the input difference of the ith subblock.
$k \cdot 1_F$	The output difference of the jth subblock is affected by the difference after F for the input difference of the ith subblock.
$0 \cdot 1_F = 0$	For zero difference, the output difference after F is also zero.
$1^* \cdot 1_F = 1$	For a difference γ, the output difference after F is δ.
$1 \cdot 1_F = 1$	For a difference δ, the output difference after F is δ'.
$2^* \cdot 1_F = 2$	For a difference $\gamma \oplus \delta$, the output difference after F is ?.
$t \cdot 1_F = t$	For a difference ?, the output difference after F is also ?.

Second, we define an addition of $a_i \cdot \mathcal{E}_{i,j}$ and $a_{i'} \cdot \mathcal{E}_{i',j}$ where $i \neq i'$. Since the addition of entries represents exclusive-or of corresponding differences, it can be naturally defined as follows.

1. The addition of two entries which have not $*$ is defined over the integer.
2. If one entry, denoted e, has not $*$ and the other has $*$, then the addition of these two entries is defined as follows.
   - If the entry e is 0 or 1, then $e + 1^* = (e + 1)^*$, otherwise, $e + 1^* = e + 1$.
   - If the entry e is 0, $e + 2^* = (e + 2)^*$, otherwise, $e + 2^* = e + 2$.

According to Table 3, $*$ is preserved (e.g, $x^* \cdot 1 = x^*$ where x^* represents the entry 1^* or 2^*.) only if $\mathcal{E}_{i,j} = 1$. And $\mathcal{E}$ is 1-property matrix by our assumption. Thus, there does not exist the addition of two entries which have $*$. Table 4 illustrates the relation between the addition of entries and exclusive-or of corresponding differences. We can easily verify that these operations, $\cdot$ and $+$, are well defined.

To help understand new operations, we consider the Feistel structure. If the input difference vectors $\boldsymbol{a}$ and $\boldsymbol{b}$ are $(0, 1^*)$ and $(1^*, 0)$, respectively, then $\boldsymbol{a^r}$ and $\boldsymbol{b^r}$ are computed by the Equations (2) and (3), respectively. Figure 4 describes these equations.

Table 4. Relation of addition and exclusive-or ($k \in \{0, 1, 1^*, 2^*, t\}$, $t' \geq 2$, and Δ is the corresponding difference for k.)

Addition	Exclusive-or
$0 + k = k$	$0 \oplus \Delta = \Delta$
$1 + 1 = 2$	$\delta \oplus \delta' = ?$
$1 + 1^* = 2^*$	$\delta \oplus \gamma = \delta \oplus \gamma$
$1 + 2^* = 3$	$\delta \oplus (\delta' \oplus \gamma) = ?$
$1 + t = 1 + t$	$\delta \oplus ? = ?$
$1^* + t = 1 + t$	$\gamma \oplus ? = ?$
$2^* + t = 2 + t$	$(\gamma \oplus \delta) \oplus ? = ?$
$t + t' = t + t'$	$? \oplus ? = ?$

$$
\begin{aligned}
\boldsymbol{a}^1 &= \boldsymbol{a} \cdot \mathcal{E} = (0 \cdot 1_F + 1^* \cdot 1, \ 0 \cdot 1 + 1^* \cdot 0) = (0 + 1^*, 0 + 0) = (1^*, 0) \\
\boldsymbol{a}^2 &= \boldsymbol{a}^1 \cdot \mathcal{E} = (1^* \cdot 1_F + 0 \cdot 1, \ 1^* \cdot 1 + 0 \cdot 0) = (1 + 0, 1^* + 0) = (1, 1^*) \\
\boldsymbol{a}^3 &= \boldsymbol{a}^2 \cdot \mathcal{E} = (1 \cdot 1_F + 1^* \cdot 1, \ 1 \cdot 1 + 1^* \cdot 0) = (1 + 1^*, 1 + 0) = (2^*, 1) \quad (2) \\
\boldsymbol{a}^4 &= \boldsymbol{a}^3 \cdot \mathcal{E} = (2^* \cdot 1_F + 1 \cdot 1, \ 2^* \cdot 1 + 1 \cdot 0) = (2 + 1, 2^* + 0) = (3, 2^*) \\
\boldsymbol{a}^5 &= \boldsymbol{a}^4 \cdot \mathcal{E} = (3 \cdot 1_F + 2^* \cdot 1, \ 3 \cdot 1 + 2^* \cdot 0) = (3 + 2^*, 3 + 0) = (5, 3)
\end{aligned}
$$

$$
\begin{aligned}
\boldsymbol{b}^1 &= \boldsymbol{b} \cdot \mathcal{D} = (1^* \cdot 0 + 0 \cdot 1, \ 1^* \cdot 1 + 0 \cdot 1_F) = (0 + 0, 1^* + 0) = (0, 1^*) \\
\boldsymbol{b}^2 &= \boldsymbol{b}^1 \cdot \mathcal{D} = (0 \cdot 0 + 1^* \cdot 1, \ 0 \cdot 1 + 1^* \cdot 1_F) = (0 + 1^*, 0 + 1) = (1^*, 1) \\
\boldsymbol{b}^3 &= \boldsymbol{b}^2 \cdot \mathcal{D} = (1^* \cdot 0 + 1 \cdot 1, \ 1^* \cdot 1 + 1 \cdot 1_F) = (0 + 1, 1^* + 1) = (1, 2^*) \quad (3) \\
\boldsymbol{b}^4 &= \boldsymbol{b}^3 \cdot \mathcal{D} = (1 \cdot 0 + 2^* \cdot 1, \ 1 \cdot 1 + 2^* \cdot 1_F) = (0 + 2^*, 1 + 2) = (2^*, 3) \\
\boldsymbol{b}^5 &= \boldsymbol{b}^4 \cdot \mathcal{D} = (2^* \cdot 0 + 3 \cdot 1, \ 2^* \cdot 1 + 3 \cdot 1_F) = (0 + 3, 2 + 3) = (3, 5)
\end{aligned}
$$

Now, we show how to use the entries of difference vectors for finding impossible differential characteristics of $\mathcal{S}$. We denote a r-round impossible differential characteristic with an input difference $\alpha = (\alpha_0, \alpha_1, \cdots, \alpha_{n-1})$ and an output difference $\beta = (\beta_0, \beta_1, \cdots, \beta_{n-1})$ by $\alpha \nrightarrow_r \beta$. Using the forgoing definitions and the encryption process, we can get the following four types of impossible differential characteristics. (We can also get the other four types of impossible differential characteristics by using the decryption process.)

- If $a_i^r = 0$, then there exists $\alpha \nrightarrow_r \beta$ where $\beta_i \neq 0$.
- If $a_i^r = 1$, then there exists $\alpha \nrightarrow_r \beta$ where $\beta_i = 0$.
- If $a_i^r = 1^*$, say γ, then there exists $\alpha \nrightarrow_r \beta$ where $\beta_i \neq \gamma$.
- If $a_i^r = 2^*$, say $\gamma \oplus \delta$, then there exists $\alpha \nrightarrow_r \beta$ where $\beta_i = \gamma$.

As mentioned before, if $a_i^r \geq 2$, we cannot predict the corresponding difference for a_i^r. It follows that we cannot find an impossible differential characteristic using the entry $t(\geq 2)$. However, the entries $0, 1, 1^*$, and 2^* are useful to find impossible differential characteristics of $\mathcal{S}$. We denote the set of these entries by

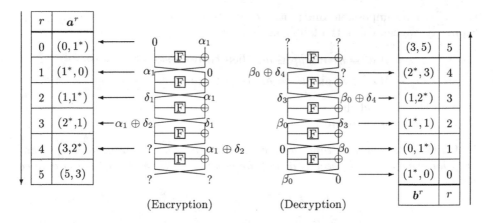

Fig. 4. Corresponding differences to a^r and b^r where $a = (0, 1^*)$ and $b = (1^*, 0)$ (α_1 is a nonzero fixed difference and δ_i are nonzero nonfixed differences.)

$\mathcal{U} = \{0, 1, 1^*, 2^*\}$. Other entries except for the elements of $\mathcal{U}$ will not be concerned.

If $a_i^r \in \mathcal{U}$, as stated above, there exist r-round impossible differential characteristics. Furthermore, $\mathcal{S}$ may have impossible differential characteristics for more than r rounds when $a_i^r \in \mathcal{U}$. To find these long characteristics, we need to define an auxiliary set $\bar{m}$ with respect to the entry $m \in \mathcal{U}$. $\bar{m}$ has following two properties. First, $\bar{m}$ is a subset of $\mathcal{U}$. Second, the elements of $\bar{m}$ correspond to the differences which can not be represented by the entry m. Consider $\bar{1}^*$. Assume that the entry 1^* corresponds to a nonzero fixed difference γ. Then, the entry 1^* can not correspond to the differences such as zero, $\gamma'(\neq \gamma)$, or $\gamma \oplus \delta$. So, we have $\bar{1}^* = \{0, 1^*, 2^*\}$. Similarly, we have $\bar{m}$ for other element m as like Table 5.

Table 5. Corresponding differences to the entries $m \in \mathcal{U}$ and the entry sets $\bar{m}$.

Entry (m)	Difference	Entry set ($\bar{m}$)	Differences
0	0	$\bar{0} = \{1, 1^*\}$	δ or γ
1	δ	$\bar{1} = \{0\}$	0
1^*	γ	$\bar{1}^* = \{0, 1^*, 2^*\}$	0 or $\gamma'(\neq \gamma)$ or $\gamma \oplus \delta$
2^*	$\gamma \oplus \delta$	$\bar{2}^* = \{1^*\}$	γ

How can we find impossible differential characteristics for more than r rounds using the notations $m \in \mathcal{U}$ and $\bar{m}$, when $a_i^r \in \mathcal{U}$? For example, assume $a_i^r = 2^*$ and $b_i^{r'} \in \bar{2}^*$. (Recall that only if $\mathcal{E}_{i,j} = 1$, $*$ is preserved, and $\mathcal{E}$ is 1-property matrix by our assumption.) $a_i^r = 2^*$ means $\alpha_i^r = \alpha_j \oplus \delta$ where $\alpha_j \neq 0$ for some j, and $b_i^{r'} \in \bar{2}^*$ means $\beta_i^{r'} = \beta_k$ where $\beta_k \neq 0$ for some k. Hence, there exists a

$(r + r')$-round impossible differential characteristic $\alpha \not\to_{r+r'} \beta$ where $\alpha_j = \beta_k$. Similarly, we can check the following properties.

- If $a_i^r = m$ and $b_i^{r'} \in \bar{m}$, then there exists $\alpha \not\to_{r+r'} \beta$.
- If $a_i^r \in \bar{m}$ and $b_i^{r'} = m$, then there exists $\alpha \not\to_{r+r'} \beta$.

We call this method that uses the elements of $\mathcal{U}$ to find impossible differential characteristics as $\mathcal{U}$-**method**.

Definition 3. *Given an input difference vector $\boldsymbol{a}$, the maximum number of encryption rounds with respect to $\boldsymbol{a}$ and the entry $m \in \mathcal{U}$ (or the set $\bar{m}$) is defined by*

$$\mathcal{ME}_i(\boldsymbol{a}, m) \triangleq max_r\{r | a_i^r = m\},$$

$$\mathcal{ME}_i(\boldsymbol{a}, \bar{m}) \triangleq max_{u \in \bar{m}}\{\mathcal{ME}_i(\boldsymbol{a}, u)\}.$$

Also, the maximum number of encryption rounds with respect to $m \in \mathcal{U}$ (or the set $\bar{m}$) is defined by

$$\mathcal{ME}_i(m) \triangleq max_{\boldsymbol{a} \neq 0}\{\mathcal{ME}_i(\boldsymbol{a}, m)\},$$

$$\mathcal{ME}_i(\bar{m}) \triangleq max_{\boldsymbol{a} \neq 0}\{\mathcal{ME}_i(\boldsymbol{a}, \bar{m})\}.$$

Similarly, each maximum number of decryption rounds is defined by

$$\mathcal{MD}_i(\boldsymbol{b}, m) \triangleq max_r\{r | b_i^r = m\}, \quad \mathcal{MD}_i(\boldsymbol{b}, \bar{m}) \triangleq max_{u \in \bar{m}}\{\mathcal{MD}_i(\boldsymbol{b}, u)\}.$$

$$\mathcal{MD}_i(m) \triangleq max_{\boldsymbol{b} \neq 0}\{\mathcal{MD}_i(\boldsymbol{b}, m)\}, \quad \mathcal{MD}_i(\bar{m}) \triangleq max_{\boldsymbol{b} \neq 0}\{\mathcal{MD}_i(\boldsymbol{b}, \bar{m})\}.$$

We denote $max_{i,m}\{\mathcal{ME}_i(\boldsymbol{a}, m) + \mathcal{MD}_i(\boldsymbol{b}, \bar{m})\}$ by $\mathcal{M}(\boldsymbol{a}, \boldsymbol{b})$. Then, clearly it holds that $\mathcal{M}(\boldsymbol{a}, \boldsymbol{b}) = max_{i,m}\{\mathcal{ME}_i(\boldsymbol{a}, \bar{m}) + \mathcal{MD}_i(\boldsymbol{b}, m)\}$. Let $max_{\boldsymbol{a} \neq 0, \boldsymbol{b} \neq 0}\{\mathcal{M}(\boldsymbol{a}, \boldsymbol{b})\}$ be denoted $\mathcal{M}$, then $\mathcal{M}$ can be computed by Equation (4). This equation will be used in the next section.

$$\mathcal{M} = max_{i,m}\{\mathcal{ME}_i(m) + \mathcal{MD}_i(\bar{m})\} = max_{i,m}\{\mathcal{ME}_i(\bar{m}) + \mathcal{MD}_i(m)\} \quad (4)$$

So, we have the following theorem.

Theorem 1. *If a round function of a block cipher structure $\mathcal{S}$ is considered as a bijective black box and $\mathcal{S}$ has 1-property matrices $\mathcal{E}$ and $\mathcal{D}$, then the maximum number of rounds for impossible differential characteristics that can be found in the $\mathcal{U}$-method is $\mathcal{M}$.*

(**Toy Example**) If a round function of the Feistel structure is bijective, then the length $\mathcal{M}$ for the cipher is 5.

Using the Equations (2) and (3), we have $\mathcal{M}((0, 1^*), (1^*, 0)) = 5$. Similarly, we can solve the equations related to other difference vectors, $\boldsymbol{a}$ and $\boldsymbol{b}$. Using the equations, we can check $\mathcal{M}(\boldsymbol{a}, \boldsymbol{b}) \leq 4$. So, we have $\mathcal{M} = max_{\boldsymbol{a} \neq 0, \boldsymbol{b} \neq 0}\{\mathcal{M}(\boldsymbol{a}, \boldsymbol{b})\} = 5$. Hence the Feistel structure has a 5-round impossible differential characteristic whose form is $(0, \alpha_1) \not\to_5 (\beta_0, 0)$ where $\alpha_1 = \beta_0 \neq 0$ (Refer to Figure 4).

4   Algorithm to Compute the Length $\mathcal{M}$

In this section, we propose an algorithm to compute the maximum number of rounds for the impossible differential characteristics which can be found in the $\mathcal{U}$-method. The algorithm is applied to a block cipher structure $\mathcal{S}$ whose round function is bijective, and encryption characteristic matrix $\mathcal{E}$ and decryption characteristic matrix $\mathcal{D}$ are 1-property matrices. [4] We assume that a round function of $\mathcal{S}$ is bijective and $\mathcal{S}$ has 1-property matrices $\mathcal{E}$ and $\mathcal{D}$.

To perform the algorithm, we need some variables. Table 6 illustrates the meaning of variables used in the algorithm. The main part of the algorithm is to distinguish between entries x and x^* where $x = 1$ or 2. Using the variable s_i in Table 7, we can distinguish between entries x and x^*, i.e., the jth entry of output difference vector for the rth round has $*$ if and only if $s_0 + s_1 + \cdots + s_{n-1} = -1$, because $\mathcal{E}$ and $\mathcal{D}$ are 1-property matrices.

Table 6. The meaning of variables used in Algorithm 1. $(y \geq 0)$

Variables	Meanings
$e_{i,j} = 0$	$\mathcal{E}_{i,j} = 0$
$e_{i,j} = 1$	$\mathcal{E}_{i,j} = 1$ or 1_F
$\widetilde{e}_{i,j} = 0$	$\mathcal{E}_{i,j} = 1$ $(x^* \cdot \mathcal{E}_{i,j} = x^*$ preserves $*$.$)$
$\widetilde{e}_{i,j} = 1$	$\mathcal{E}_{i,j} = 0$ $(x^* \cdot \mathcal{E}_{i,j} = 0)$ or $\mathcal{E}_{i,j} = 1_F$ $(x^* \cdot \mathcal{E}_{i,j} = x)$ (These equations do not preserve $*$.)
$a_i^r = y$ (resp., x)	The ith entry of difference vector $\boldsymbol{a^r}$ is y (resp., x^*).
$\hat{a}_i^r = 0$	The ith entry of difference vector $\boldsymbol{a^r}$ has not $*$.
$\hat{a}_i^r = -1$	The ith entry of difference vector $\boldsymbol{a^r}$ has $*$.

Table 7. Multiplication between an entry of difference vector and an entry of matrix in Algorithm 1.

A entry $c,(\hat{a}_i^r)$ of difference vectors	A entry $d,(\widetilde{e}_{i,j})$ of $\mathcal{E}$	$c \cdot d$	$\hat{a}_i^r + \widetilde{e}_{i,j} = s_i$ if$(s_i = 1)$ $s_i \leftarrow 0$
$x^*, (-1)$	$0, (1)$	0	0
$x^*, (-1)$	$1_F, (1)$	x	0
$x^*, (-1)$	$1, (0)$	x^*	-1
$x, (0)$	$0, (1)$	0	0
$x, (0)$	$1_F, (1)$	x	0
$x, (0)$	$1, (0)$	x	0

---

[4] In fact, we may compute the number of rounds $\mathcal{M}$ by modifying the algorithm even though $\mathcal{E}$ and $\mathcal{D}$ are not 1-property matrices.

Step 1 : Input the encryption characteristic matrix $\mathcal{E} = (\mathcal{E}_{i,j})_{n \times n}$

for $i = 0$ to $n - 1$
    for $j = 0$ to $n - 1$
        if $\mathcal{E}_{i,j} = 0$, then $e_{i,j} \leftarrow 0$ and $\widetilde{e}_{i,j} \leftarrow 1$
        if $\mathcal{E}_{i,j} = 1$, then $e_{i,j} \leftarrow 1$ and $\widetilde{e}_{i,j} \leftarrow 0$
        if $\mathcal{E}_{i,j} = 1_F$, then $e_{i,j} \leftarrow 1$ and $\widetilde{e}_{i,j} \leftarrow 1$

Step 2 : Compute the values of $\mathcal{ME}_i(m)$ *where* $0 \leq i \leq n - 1$ *and* $m \in \mathcal{U}.$

$\mathcal{ME}_i(m) \leftarrow 0$, for $0 \leq i \leq n - 1$, $0 \leq m \leq 3$
/* The $m's$ values 0,1,2, and 3 indicate the entries $0, 1, 1^*$, and 2^*, respectively. */
F or eac h input difference vector $\boldsymbol{a}$ /* $\boldsymbol{a}$ represen ts $\boldsymbol{a}^0$. */
    for $i = 0$ to $n - 1$
        if $(a_i^0 = 0)$ $\hat{a}_i \leftarrow 0$
        else if $(a_i^0 = 1)$ $\hat{a}_i \leftarrow -1$
        for $m = 0$ to 3
            $\mathcal{ME}_i(\boldsymbol{a}, m) \leftarrow 0$
    $r \leftarrow 0$
    while (there exists some index l such that $a_l^r \leq 2$.)
        for $j = 0$ to $n - 1$
            $t_j \leftarrow 0, \hat{t}_j \leftarrow 0$
            /* t_j and $\hat{t}_j$ are the temporary parameters to compute $\boldsymbol{a}^{r+1}$ and $\hat{\boldsymbol{a}}^{r+1}$. */
            for $i = 0$ to $n - 1$
                $t_j \leftarrow t_j + a_i^r \cdot e_{i,j}$
                $s_i \leftarrow \hat{a}_i^r + \widetilde{e}_{i,j}$
                if $(s_i = 0)$ $s_i \leftarrow 0$
                $\hat{t}_j \leftarrow \hat{t}_j + s_i$
        $r \leftarrow r + 1$
        $a_i^r \leftarrow t_i, \ \hat{a}_i^r \leftarrow \hat{t}_i$, for $0 \leq i \leq n - 1$
        for $i = 0$ to $n - 1$
            if $(a_i^r = 0)$ $\mathcal{ME}_i(\boldsymbol{a}, 0) \leftarrow r$
            if $(a_i^r = 1$ and $\hat{a}_i^r = 0)$ $\mathcal{ME}_i(\boldsymbol{a}, 1) \leftarrow r$
            if $(a_i^r = 1$ and $\hat{a}_i^r = -1)$ $\mathcal{ME}_i(\boldsymbol{a}, 2) \leftarrow r$
            if $(a_i^r = 2$ and $\hat{a}_i^r = -1)$ $\mathcal{ME}_i(\boldsymbol{a}, 3) \leftarrow r$
    for $i = 0$ to $n - 1$
        for $m = 0$ to 3
            if $(\mathcal{ME}_i(m) \leq \mathcal{ME}_i(\boldsymbol{a}, m))$ $\mathcal{ME}_i(m) \leftarrow \mathcal{ME}_i(\boldsymbol{a}, m)$

Step 3 : Compute the values of $\mathcal{MD}_i(\bar{m})$ *wher e* $0 \leq i \leq n - 1$ *and* $m \in \mathcal{U}.$

Compute the values of $\mathcal{MD}_i(m)$ by inserting the matrix $\mathcal{D}$ into *Steps* 1 and 2.
for $i = 0$ to $n - 1$
    $\mathcal{MD}_i(\bar{0}) \leftarrow max\{\mathcal{MD}_i(1), \mathcal{MD}_i(2)\}$
    $\mathcal{MD}_i(\bar{1}) \leftarrow \mathcal{MD}_i(0)$
    $\mathcal{MD}_i(\bar{2}) \leftarrow max\{\mathcal{MD}_i(0), \mathcal{MD}_i(2), \mathcal{MD}_i(3)\}$
    $\mathcal{MD}_i(\bar{3}) \leftarrow \mathcal{MD}_i(2)$
    /* Note $\bar{0}, \bar{1}, \bar{2}$, and $\bar{3}$ represen t $\bar{0}, \bar{1}, \bar{1}^*$, and $\bar{2}^*$ in T able 5, respectiv ely. */

Step 4 : Output the length $\mathcal{M}$. *(Equation (4))*

Output $max_{0 \leq i \leq n-1, 0 \leq m \leq 3}(\mathcal{ME}_i(m) + \mathcal{MD}_i(\bar{m}))$

Algorithm 1 to compute the length $\mathcal{M}$.

5 Results for Some Block Cipher Structures

In this section, we present the specific forms of impossible differential characteristics for some block cipher structures such as a generalized Feistel network, a generalized CAST256-like structure, a generalized MARS-like structure, and a generalized RC6-like structure, and Rijndael structures. We experimentally find the impossible differential characteristics within a finite number of subblocks. However, we can generalize our simulation results, because a generalized block cipher structures has a regular structural feature.

5.1 Nyberg's Generalized Feistel Network (GFN_n)

GFN_n has 1-property matrices $\mathcal{E}$ and $\mathcal{D}$, so we can apply the network to Algorithm 1. The running time of Algorithm 1 is dominated by *Steps* 2 and 3. However, using the fact that the encryption process of GFN_n is almost same as the decryption process, $\mathcal{MD}_i(m)$ in Step 3 can be easily computed from the values of $\mathcal{ME}_i(m)$. So, we can reduce half of the running time. For finding the length $\mathcal{M}$ for GFN_{16}, we executed a program written in visual C 6.0 and running on a set of 10 PCs under Windows. From this, we found it in one computer with about 4 hours. The following proposition is a result based on our simulation.

Proposition 1. *(1) If a round function of GFN_2 is bijective, then the length $\mathcal{M}$ for the cipher is 7. (2) If a round function of GFN_n is bijective and $n \geq 3$, then the length $\mathcal{M}$ for the cipher is $(3n + 2)$.*

By modifying Algorithm 1, we can get the specific forms of various impossible differential characteristics of GFN_n ($2 \leq n \leq 16$), and we can generalize such characteristics as like Table 8.

Table 8. Impossible differential characteristics for GFN_n ($\alpha_i \neq 0$, $\beta_i \neq 0$, $(\beta'_0, \beta'_2) \neq (0,0)$, and $\beta''_0 = \alpha_2$.)

GFN_2	GFN_n ($n \geq 3$)
$(0,0,0,\alpha_3) \nrightarrow_7 (\beta'_0, 0, \beta'_2, 0)$	$(0,0,0,\cdots,0,\alpha_{2n-1}) \nrightarrow_{3n+2} (\beta'_0, 0, \beta'_2, 0, \cdots, 0)$
$(0,0,\alpha_2,0) \nrightarrow_7 (\beta'_0, 0, \beta'_2, 0)$	$(0,0,\cdots,0,\alpha_{2n-2},0) \nrightarrow_{3n+2} (\beta_0, 0, 0, \cdots, 0, 0)$
$(0,0,\alpha_2,\alpha_3) \nrightarrow_7 (\beta_0, 0, 0, 0)$	$(0,\cdots,0,\alpha_{2n-2},\alpha_{2n-1}) \nrightarrow_{3n+2} (\beta_0, 0, 0, 0, \cdots 0)$
$(0,0,\alpha_2,\alpha_3) \nrightarrow_7 (0, 0, \beta_2, 0)$	.
$(0,\alpha_1,\alpha_2,0) \nrightarrow_7 (\beta''_0, 0, 0, 0)$	.
$(0,\alpha_1,\alpha_2,\alpha_3) \nrightarrow_7 (\beta''_0, 0, 0, 0)$	.

We also performed Algorithm 1 for other generalized Feistel networks, e.g., a generalized CAST256-like structure, a generalized MARS-like structure, and a generalized RC6-like structure described in [14]. Table 9 shows the specific forms of impossible differential characteristics on each structure. Based on our experiment, we stress that the generalized MARS-like structure among the foregoing four structures has the most strong resistance against IDC.

Table 9. Impossible differential characteristics for other generalized Feistel networks $(\alpha_{n-1} \neq 0, \beta_0 \neq 0.$ $\alpha_i = \beta_i \neq 0$ and i is an odd number.)

Structure	Impossible Differential Characteristic	Condition
Generalized CAST256	$(0,\cdots,0,\alpha_{n-1}) \not\rightarrow_{n^2-1} (\beta_0,0,\cdots,0)$	$n \geq 3$
Generalized MARS	$(0,\cdots,0,\alpha_{n-1}) \not\rightarrow_{2n-1} (\beta_0,0,\cdots,0)$	$n \geq 3$
Generalized RC6	$(0,\cdots,0,\alpha_i,0,\cdots,0) \not\rightarrow_{4n+1} (0,\cdots,0,\beta_i,0,\cdots,0)$	$\cdot$

5.2   Rijndael Structure

The output subblock Y_i of Rijndael is affected by four input subblocks due to the linear layer composed of ShiftRow transformation and MixColumn transformation. Y_i is also affected by four subblocks after ByteSubstitution transformation. Thus, Rijndael structure has 1-property matrices $\mathcal{E}$ and $\mathcal{D}$ whose column has all zeros but four 1_F. It follows that Rijndael structure can be applied to Algorithm 1. Following is our simulation result.

Proposition 2. *(1) (Rijndael$_{128}$ structure [6]) Given plaintext pair which are equal at all bytes but one, the ciphertexts after 3 rounds cannot be equal in any of a column. (2) (Rijndael$_{196}$ structure) Given plaintext pair which are equal at all bytes but one, the ciphertexts after 4 rounds cannot be equal in any of three columns. (3) (Rijndael$_{256}$ structure) Given plaintext pair which are equal at all bytes but one, the ciphertexts after 5 rounds cannot be equal in any of seven columns.*

Note : Cheon et. al. [6] proposed an attack algorithm which uses the 3-round impossible differential characteristics of $Rijndael_{128}$ structure. (Note that the 4-round impossible differential characteristics proposed in [6] do not include the MixColumn and AddRoundKey transformations of the last round. These are the same characteristics for 3 rounds stated in Proposition 2.) In [6], sixteen of them are used to attack 6-round Rijndael with a data complexity of $2^{91.5}$ chosen plaintexts (CP) and a time complexity of 2^{122} encryptions. However we found other 3-round impossible differential characteristics to allow attacking 6-round Rijndael with less complexities. If we apply the 3-round impossible differential characteristics, $\alpha \not\rightarrow \beta$ or $\alpha \not\rightarrow \beta'$ [5] to the attack algorithm used in [6], then we can attack 6-round Rijndael which uses 128-bit data as like Table 10.

6   Further Research

An interesting property of the $\mathcal{U}$-method is that they can be converted to a tool of 1st order integral attack. Consider a block cipher structure whose round

---

[5] $\alpha = (0,\cdots,0,\alpha_i,0,\cdots,0)$ where $\alpha_i \neq 0$, and i is 0,4,8, or 12.
$\beta = (\beta_0,\beta_1,0,0,0,\beta_5,\beta_6,0,0,0,\beta_{10},0,\beta_{12},0,0,0),(\beta_0,\beta_1,0,0,0,\beta_5,0,0,0,0,0,\beta_{11},\beta_{12},0,0,\beta_{15}),$
$(\beta_0,0,0,0,0,0,\beta_6,0,0,0,\beta_{10},\beta_{11},\beta_{12},0,0,\beta_{15})$ or $(0,\beta_1,0,0,0,\beta_5,\beta_6,0,0,0,\beta_{10},\beta_{11},0,0,\beta_{15}).$
$\beta' = (0,0,0,0,0,\beta_5',0,0,0,0,\beta_{10}',0,0,0,0,\beta_{15}'),(\beta_0',0,0,0,0,0,0,0,0,0,0,0,\beta_{10}',0,0,0,0,\beta_{15}'),$
$(\beta_0',0,0,0,0,\beta_5',0,0,0,0,0,0,0,0,0,\beta_{15}'),$ or $(\beta_0',0,0,0,0,\beta_5',0,0,0,0,\beta_{10}',0,0,0,0,\beta_{15}')$ where $\beta_i \neq 0$ and $\beta_i' \neq 0$.

Table 10. Complexity of Impossible Differential Cryptanalysis on 6-Round Rijndael

3-round distinguishers used in the attack	Condition [a]	Data (CP)	Time (encryptions)
$\alpha \nrightarrow \beta$	8-byte	$2^{75.5}$	$2^{116.4}$
	9-byte	$2^{83.4}$	$2^{108.4}$
	10-byte	$2^{91.3}$	$2^{100.4}$
	11-byte	$2^{99.2}$	$2^{92.4}$
	12-byte	$2^{107.1}$	$2^{84.4}$
$\alpha \nrightarrow \beta'$	12-byte	$2^{99.1}$	$2^{84.4}$
	13-byte	$2^{107.0}$	$2^{76.4}$
	14-byte	$2^{114.9}$	$2^{68.4}$

[a] The number of bytes of ciphertext to be used for filtering out wrong pairs.

functions are bijective. If a saturated input subblock is considered as the entry 1^* and a constant input subblock is considered as the entry 0, then the entry 0 in the set $\mathcal{U}$ corresponds to a constant value, and the entries 1 and 1^* corresponds to a saturated set, and the entries 2 and 2^* corresponds to a balanced set. Based on this fact, we performed simulations for the block cipher structures which were dealt with in this paper and found 1st order integrals for less rounds than impossible differential characteristics on each cipher. (For example, GFN_n ($n \geq 2$) has a $(2n + 3)$-round 1st order integral.)

Although we do not know of any other appliances of the method using the matrix, we expect that the possibility to apply the method to other attacks may be of interest.

References

1. C.M. Adams, *The CAST-256 Encryption Algorithm*, AES Proposal, 1998.
2. K. Aoki and K. Ohta, *Strict evaluation of the maximum average of differential probability and the maximem average of linear probability*, IEICE Transactions fundamentals of Electronics, Communications and Computer Sciences, No.1, 1997, pp 2-8.
3. E. Biham and A. Shamir, *Differential cryptanalysis of DES-like cryptosystems*, Advances in Cryptology - CRYPTO'90, LNCS 537, Springer-Verlag, 1991, pp 2-21.
4. E. Biham, A. Biryukov, and A. Shamir, *Cryptanalysis of skipjack reduced to 31 rounds using impossible differentials*, Advances in Cryptology - EUROCRYPT'99, LNCS 1592, Springer-Verlag, 1999, pp 12-23.
5. C. Burwick, D. Coppersmith, E. D'Avignon, R. Gennaro, S. Halevi, C. Jutla, S.M. Matyas, L. O'Connor, M. Peyravian, D. Safford, and N. Zunic, *MARS - A Candidate Cipher for AES*, AES Proposal, 1998.
6. J. Cheon, M. Kim, K. Kim, and J. Lee, *Improved Impossible Differential Cryptanalysis of Rijndael and Crypton*, ICISC'01, LNCS 2288, Springer-verlag, 2001, pp 39-49.
7. J. Daemen and V. Rijndael, *The Rijndael block cipher*, AES proposal, 1998.

8. S. Hong, S. Lee, J. Lim, J. Sung, D. Choen, and I. Cho, *Provable Security against Differential and Linear Cryptanalysis for the SPN structure*, FSE'00, LNCS 1978, Springer-Verlag, 2000, pp 273-283.
9. S. Hong, J. Sung, S. Lee, J. Lim, and J. Kim, *Provable Security for 13 round Skipjack-like Structure*, Information Processing Letters, vol 82, 2002, pp 243-246.
10. L.R. Knudsen, *DEAL - A 128-bit Block Cipher*, AES Proposal, 1998.
11. L. Knudsen and D. Wagner, *Integral cryptanalysis*, FSE'02, LNCS 2365, Springer-Verlag, 2002, pp 112-127.
12. M. Matsui, *Linear cryptanalysis method for DES cipher*, Advances in Cryptology - EUROCRYPT'93, LNCS 765, Springer-Verlag, 1994, pp 386-397.
13. M. Matsui, *New structure of block ciphers with provable security against differential and linear cryptanalysis*, FSE'96, LNCS 1039, Springer-Verlag, 1996, pp 205–218.
14. S. Moriai, S. Vaudenay , *On the Pseudorandomness of Top-Level Schemes of Block Ciphers*, Advances in Cryptology - ASIACRYPT'00, LNCS 1976, Springer-Verlag, 2000, pp 289-302.
15. National Security Agency. NSA Releases Fortezza Algorithms. Press Release, June 24, 1998. Available at http://csrc.ncsl.nist.gov/encryption/skipjack-1.pdf.
16. K. Nyberg and Lars R. Knudsen, *Provable security against differential cryptanalysis*, Advances in Cryptology - CRYPTO'92, LNCS 740, Springer-Verlag, 1992, pp 566–574.
17. K.Nyberg *Generalized Feistel Networks*, Advances in Cryptology - ASIACRYPT'96, LNCS 1163, Springer-Verlag, 1996, pp 91-104.
18. R.L. Rivest, M.J.B. Robshaw, R. Sidney, and Y.L. Yin, *The RC6 block cipher*, AES Proposal, 1998.
19. J. Sung, S. Lee, J. Lim, S. Hong, S. Park, *Provable Security for the Skipjack-like Structure against Differential Cryptanalysis and Linear Cryptanalysis*, Advances in Cryptology - ASIACRYPT'00, LNCS 1976, Springer-Verlag, 2000, pp 274-288.

Impossible Differential Attack on 30-Round SHACAL-2

Seokhie Hong[1], Jongsung Kim[1], Guil Kim[1], Jaechul Sung[2], Changhoon Lee[1], and Sangjin Lee[1]

[1] Center for Information Security Technologies(CIST),
Korea University, Seoul, Korea
{hsh,joshep,okim912,crypto77,sangjin}@cist.korea.ac.kr
[2] Korea Information Security Agency(KISA), Seoul, KOREA,
sjames@kisa.or.kr

Abstract. SHACAL-2 is a 256-bit block cipher with various key sizes based on the hash function SHA-2. Recently, it was recommended as one of the NESSIE selections. Up to now, no security flaws have been found in SHACAL-2. In this paper, we discuss the security of SHACAL-2 against an impossible differential attack. We propose two types of 14-round impossible characteristics and using them, we attack 30-round SHACAL-2 with 512-bit key. This attack requires 744 chosen plaintexts and has time complexity of $2^{495.1}$ 30-round SHACAL-2 encryptions.

Keywords : Block Cipher, Impossible Differential Attack, SHACAL-2

1 Introduction

SHACAL-2[4] is a 64-round block cipher proposed by H. Handschuch and D. Naccache as a submission of the NESSIE(New European Schemes for Signatures, Integrity, and Encryption) project. The cipher is based on the compression function of SHA-2. Recently, SHACAL-2 was recommended as one of the NESSIE selections.

There are several advantages for using SHACAL-2. We already know that the primitive SHA-2[11] can be implemented in various environments, so we can share most of codes of the SHA-2 as a block cipher with only small modification. Also, there are no attacks reported on SHA-2 in the open literature. These facts increase the trust in SHACAL-2 as a good block cipher.

In this paper, however, we show the reduced-version of SHACAL-2 (e.g. the reduced 30-round out of the full 64 rounds) is vulnerable to an impossible differential attack. We propose two types of 14-round impossible characteristics by combining two types of 11-round impossible differential characteristics and a nonlinear equation of 3 rounds. Using them, we attack 30-round SHACAL-2 with data complexity of 744 chosen plaintexts and time complexity of $2^{495.1}$ 30-round SHACAL-2 encryptions.

T. Johansson and S. Maitra (Eds.): INDOCRYPT 2003, LNCS 2904, pp. 97–106, 2003.

The paper is organized as follows: In section 2, we describe the block cipher SHACAL-2. In section 3, we show how to construct two types of 14-round impossible characteristics. In section 4, we present an impossible differential attack on 30-round SHACAL-2. Finally, section 5 summarizes the paper.

2 Description of SHACAL-2

SHACAL-2[4] is a 256-bit block cipher with various key sizes. It is based on the compression function of the hash function SHA-2[11] introduced by NIST.

In the SHACAL-2, a 256-bit plaintext is divided into eight 32-bit words - A, B, C, D, E, F, G and H. We denote by X^i the value of word X before the ith round, that is, the plaintext P is divided into $A^0, B^0, C^0, D^0, E^0, F^0, G^0$, and H^0, and the ciphertext is composed of $A^{64}, B^{64}, C^{64}, D^{64}, E^{64}, F^{64}, G^{64}$ and H^{64}. The ith round of encryption is performed as follows.

$$T_1^{i+1} = H^i + \Sigma_1(E^i) + Ch(E^i, F^i, G^i) + K^i + W^i \tag{1}$$
$$T_2^{i+1} = \Sigma_0(A^i) + Maj(A^i, B^i, C^i) \tag{2}$$
$$H^{i+1} = G^i \tag{3}$$
$$G^{i+1} = F^i \tag{4}$$
$$F^{i+1} = E^i \tag{5}$$
$$E^{i+1} = D^i + T_1^{i+1} \tag{6}$$
$$D^{i+1} = C^i \tag{7}$$
$$C^{i+1} = B^i \tag{8}$$
$$B^{i+1} = A^i \tag{9}$$
$$A^{i+1} = T_1^{i+1} + T_2^{i+1} \tag{10}$$

for $i = 0, ..., 63$ where $+$ means the addition modulo 2^{32} of 32-bit words, W^i are the 32-bit round subkeys, and K^i are the 32-bit round constants which are different in each of the 64 rounds. (See [11] for the details of round constants.) The functions used in the above encryption process are defined as follows.

$$Ch(X, Y, Z) = (X \& Y) \oplus (\neg X \& Z)$$
$$Maj(X, Y, Z) = (X \& Y) \oplus (X \& Z) \oplus (Y \& Z)$$
$$\Sigma_0(X) = S_2(X) \oplus S_{13}(X) \oplus S_{22}(X)$$
$$\Sigma_1(X) = S_6(X) \oplus S_{11}(X) \oplus S_{25}(X)$$

, where $\neg X$ means the complement of 32-bit word X, and $S_i(X)$ means the right rotation of X by i bit positions.

Using the property $X - Y = X + (2^{32} - 1 - Y) + 1 = X + (\neg Y) + 1$, we have the ith round of decryption as follows.

$$T_1^{i+1} = A^{i+1} - \Sigma_0(B^{i+1}) - Maj(B^{i+1}, C^{i+1}, D^{i+1}) \tag{11}$$
$$= A^{i+1} + (\neg \Sigma_0(B^{i+1})) + (\neg Maj(B^{i+1}, C^{i+1}, D^{i+1})) + 2$$
$$H^i = T_1^{i+1} - \Sigma_1(F^{i+1}) - Ch(F^{i+1}, G^{i+1}, H^{i+1}) - K^i - W^i \tag{12}$$
$$= T_1^{i+1} + (\neg \Sigma_1(F^{i+1})) + (\neg Ch(F^{i+1}, G^{i+1}, H^{i+1})) + (\neg K^i) + (\neg W^i) + 4$$
$$G^i = H^{i+1} \tag{13}$$

$$F^i = G^{i+1} \tag{14}$$
$$E^i = F^{i+1} \tag{15}$$
$$D^i = E^{i+1} - T_1^{i+1} = E^{i+1} + (\neg T_1^{i+1}) + 1 \tag{16}$$
$$C^i = D^{i+1} \tag{17}$$
$$B^i = C^{i+1} \tag{18}$$
$$A^i = B^{i+1} \tag{19}$$

The key scheduling of SHACAL-2 takes a maximum 512-bit key and shorter keys than 512 bits may be used by padding the key with zeros to a 512-bit string. However, SHACAL-2 is not intended to be used with a key shorter than 128 bits. Let the 512-bit key string be denoted $W = [W^0 \| W^1 \| \cdots \| W^{15}]$. The key expansion of 512 bits W to 2048 bits is defined by

$$W^i = \sigma_1(W^{i-2}) + W^{i-7} + \sigma_0(W^{i-15}) + W^{i-16}, \ 16 \le i \le 63.$$
$$\sigma_0(x) = S_7(x) \oplus S_{18}(x) \oplus R_3(x)$$
$$\sigma_1(x) = S_{17}(x) \oplus S_{19}(x) \oplus R_{10}(x)$$

where $R_i(X)$ means the right shift of 32-bit word X by i bit positions.

3 Impossible Characteristics of SHACAL-2

In this section, we propose two types of 14-round impossible characteristics of SHACAL-2. These characteristics are made by combining two types of 11-round impossible differential characteristics and a nonlinear equation of 3 rounds. We describe how to construct two types of 11-round impossible differential characteristics and then a nonlinear equation of 3 rounds. These two types of 11-round impossible differential characteristics are determined by the m.s.b. of the 0th round key W^0. That is, if the value of the key bit is 0, one of them holds with probability 1, otherwise, the other holds with probability 1.

To begin with, we introduce the notations which will be used throughout this paper.

- P : The 256-bit plaintext, i.e., $P = (A^0, \cdots, H^0)$.
- P^r : The 256-bit text before the rth round, i.e., $P^r = (A^r, \cdots, H^r)$.
- x_i^r : The ith bit of X^r where $X^r \in \{A^r, B^r, \cdots, H^r, W^r, K^r\}$.
- $t_{1,i}^r$: The ith bit of T_1^r defined in (1).
- $?$: An unknown value of difference in the 32-bit word.
- e_i : A 32-bit word that has zeros in all bit positions except for bit i.
- $\sim e_i$: A 32-bit word that has zeros in the positions of bits $0 \sim (i-1)$, one in the position of bit i, and unknown values in the positions of bits $(i+1) \sim 31$.
- $e_{i_1, \cdots, i_k}$: $e_{i_1} \oplus \cdots \oplus e_{i_k}$.

3.1   11-Round Impossible Differential Characteristics

SHACAL-2 has two types of 11-round truncated differential characteristics with probability 1 from rounds r to $r + 10$. For the sake of clarity, we show how to construct two types of 11-round impossible differential characteristics for rounds 0-10 which will be used in our attack. Firstly, we describe a 9-round impossible differential characteristic for rounds 2-10, and secondly, we explain how to construct the input differences of rounds 0 and 1.

Table 1 illustrates a 9-round truncated differential characteristic for rounds 2-10. In Table 1, each row represents a value of exclusive-or difference. We can check Table 1 by using the following properties.

1. If $X \oplus X^* = e_i$, then it holds that $\Sigma_0(X) \oplus \Sigma_0(X^*) = e_{i_0,i_1,i_2}$ where i_0, i_1 and i_2 are equal to $i - 2$, $i - 13$ and $i - 22$ (mod 32), respectively.
2. If $X \oplus X^* = e_i$, then it holds that $\Sigma_1(X) \oplus \Sigma_1(X^*) = e_{i_0,i_1,i_2}$ where i_0, i_1 and i_2 are equal to $i - 6$, $i - 11$ and $i - 25$ (mod 32), respectively.
3. If $X \oplus X^* =\sim e_i$, $Y \oplus Y^* =\sim e_j$ and $i > j$, then it holds that $(X + Y) \oplus (X^* + Y^*) =\sim e_j$.

Round (r)	ΔA^r	ΔB^r	ΔC^r	ΔD^r	ΔE^r	ΔF^r	ΔG^r	ΔH^r
Input $(r = 2)$	0	0	0	e_{31}	0	0	0	e_{31}
3	e_{31}	0	0	0	0	0	0	0
4	$\sim e_9$	e_{31}	0	0	0	0	0	0
5	?	$\sim e_9$	e_{31}	0	0	0	0	0
6	?	?	$\sim e_9$	e_{31}	0	0	0	0
7	?	?	?	$\sim e_9$	e_{31}	0	0	0
8	?	?	?	?	$\sim e_6$	e_{31}	0	0
9	?	?	?	?	?	$\sim e_6$	e_{31}	0
10	?	?	?	?	?	?	$\sim e_6$	e_{31}
Output	?	?	?	?	?	?	?	$\sim e_6$

Table 1. A 9-round truncated differential characteristic with probability 1.

Table 1 shows a 9-round truncated differential characteristic whose form is $(0, 0, 0, e_{31}, 0, 0, 0, e_{31}) \rightarrow (?, ?, ?, ?, ?, ?, ?, \sim e_6)$. This characteristic implicates the fact that for given any pair with an input difference of the form $(0, 0, 0, e_{31}, 0, 0, 0, e_{31})$, the l.s.b. of output difference in the eighth word after 9 rounds, Δh_0^{11} cannot be 1. This fact means that this characteristic includes a 9-round impossible differential characteristic whose form is

$$(0, 0, 0, e_{31}, 0, 0, 0, e_{31}) \nrightarrow (?, ?, ?, ?, ?, ?, ?, \Delta H^{11}) \qquad (20)$$

where the l.s.b. of ΔH^{11} is 1. In our attack, we use this impossible differential characteristic instead of the 9-round truncated differential characteristic shown in Table 1.

From now on we show how to construct the input differences of rounds 0 and 1. First, we consider the input difference of round 1. Denote two values of a plaintext pair by (P, P^*). To get the difference $P^2 \oplus P^{*2} = (0, 0, 0, e_{31}, 0, 0, 0, e_{31})$, it is a necessary condition that $\Delta A^1 = \Delta B^1 = \Delta E^1 = \Delta F^1 = 0$ and $\Delta C^1 = \Delta G^1 = e^{31}$. But, we can not make such a difference of P^2 and P^{*2} by giving conditions on ΔD^1 and ΔH^1 without considering the values a^1_{31}, b^1_{31}, e^1_{31} a^{*1}_{31}, b^{*1}_{31}, and e^{*1}_{31}. So, we should give conditions on a^1_{31}, b^1_{31}, e^1_{31} a^{*1}_{31}, b^{*1}_{31}, and e^{*1}_{31} as well as ΔD^1 and ΔH^1. In our observation, there are eight types of conditions as like Table 2 satisfying $P^2 \oplus P^{*2} = (0, 0, 0, e_{31}, 0, 0, 0, e_{31})$.

Type	ΔA^1	ΔB^1	ΔC^1	ΔD^1	ΔE^1	ΔF^1	ΔG^1	ΔH^1	Conditions
Type 0	0	0	e_{31}	0	0	0	e_{31}	0	$a^1_{31} = 0, b^1_{31} = 0, e^1_{31} = 1,$ $(a^{*1}_{31} = 0, b^{*1}_{31} = 0, e^{*1}_{31} = 1)$
Type 1	0	0	e_{31}	0	0	0	e_{31}	0	$a^1_{31} = 1, b^1_{31} = 1, e^1_{31} = 1$ $(a^{*1}_{31} = 1, b^{*1}_{31} = 1, e^{*1}_{31} = 1)$
Type 2	0	0	e_{31}	0	0	0	e_{31}	e_{31}	$a^1_{31} = 0, b^1_{31} = 0, e^1_{31} = 0$ $(a^{*1}_{31} = 0, b^{*1}_{31} = 0, e^{*1}_{31} = 0)$
Type 3	0	0	e_{31}	0	0	0	e_{31}	e_{31}	$a^1_{31} = 1, b^1_{31} = 1, e^1_{31} = 0$ $(a^{*1}_{31} = 1, b^{*1}_{31} = 1, e^{*1}_{31} = 0)$
Type 4	0	0	e_{31}	e_{31}	0	0	e_{31}	0	$a^1_{31} = 1, b^1_{31} = 0, e^1_{31} = 0$ $(a^{*1}_{31} = 1, b^{*1}_{31} = 0, e^{*1}_{31} = 0)$
Type 5	0	0	e_{31}	e_{31}	0	0	e_{31}	0	$a^1_{31} = 0, b^1_{31} = 1, e^1_{31} = 0$ $(a^{*1}_{31} = 0, b^{*1}_{31} = 1, e^{*1}_{31} = 0)$
Type 6	0	0	e_{31}	e_{31}	0	0	e_{31}	e_{31}	$a^1_{31} = 0, b^1_{31} = 1, e^1_{31} = 1$ $(a^{*1}_{31} = 0, b^{*1}_{31} = 1, e^{*1}_{31} = 1)$
Type 7	0	0	e_{31}	e_{31}	0	0	e_{31}	e_{31}	$a^1_{31} = 1, b^1_{31} = 0, e^1_{31} = 1$ $(a^{*1}_{31} = 1, b^{*1}_{31} = 0, e^{*1}_{31} = 1)$

Table 2. Eight types of input differences of round 1

Second, we describe how to construct the input difference of round 0, i.e., the difference of P and P^*. Since the values a^1_{31}, e^1_{31}, a^{*1}_{31} and e^{*1}_{31} are affected by round key W^0, and the values of them should be fixed as shown in Table 2, we can not construct the difference of P and P^* without considering W^0. But fortunately, by fixing many bits of P and P^*, we can control them to be affected by only the value w^0_{31}.

Assume that a pair (P, P^*) with the difference $(0, e_{31}, 0, 0, 0, e_{31}, 0, 0)$ satisfies the following conditions.

$$
\begin{aligned}
& A^0 = C^0 = 0, \ D^0 = 2^{31}, \ e^0_{31} = g^0_{31} = 0, \\
& H^0 = -(\Sigma_1(E^0) + Ch(E^0, F^0, G^0) + K^0) \\
& A^{*0} = C^{*0} = 0, \ D^{*0} = 2^{31}, \ e^{*0}_{31} = g^{*0}_{31} = 0, \\
& H^{*0} = -(\Sigma_1(E^{*0}) + Ch(E^{*0}, F^{*0}, G^{*0}) + K^{*0})
\end{aligned}
\tag{21}
$$

Then, in the case of $w^0_{31} = 0$, the difference of P^1 and P^{*1} is of Type 0 in Table 2. The reason is as follows. Clearly, we have $\Delta B^1 = \Delta D^1 = \Delta F^1 = \Delta H^1 = 0$ and $\Delta C^1 = \Delta G^1 = e^{31}$ by using the encryption process. From the above conditions, we can check $T^1_1 = T^{*1}_1 = W^0$, $T^1_2 = T^{*1}_2 = 0$ and $b^1_{31} = b^{*1}_{31} = 0$. By applying these values to Equations (6) and (10), we can also check E^1 and E^{*1} are equal to $W^0 + 2^{31}$, (i.e., $\Delta E^1 = 0$) and A^1 and A^{*1} are equal to W^0, (i.e., $\Delta A^1 = 0$). So, in the case of $w^0_{31} = 0$, we have $a^1_{31} = a^{*1}_{31} = 0$ and $e^1_{31} = e^{*1}_{31} = 1$. Therefore, the difference of P^1 and P^{*1} is of Type 0 in Table 2. We denote the difference of such a pair (P, P^*) by $\mathcal{D}_0$, i.e., $P \oplus P^* = \mathcal{D}_0$. See Table 3.

Assume that a pair (P, P^{**}) with the difference $(0, e_{31}, e_{31}, 0, 0, e_{31}, 0, 0)$ satisfies the following conditions.

$$A^0 = C^0 = 0, \ D^0 = 2^{31}, \ b^0_{31} = 1, e^0_{31} = g^0_{31} = 0,$$
$$H^0 = -(\Sigma_1(E^0) + Ch(E^0, F^0, G^0) + K^0) \tag{22}$$
$$A^{**0} = 0, \ C^{**0} = 2^{31}, \ D^{**0} = 2^{31}, \ b^{**0}_{31} = 0, e^{**0}_{31} = g^{**0}_{31} = 0,$$
$$H^{**0} = -(\Sigma_1(E^{**0}) + Ch(E^{**0}, F^{**0}, G^{**0}) + K^{**0})$$

Then, in the case of $w^0_{31} = 1$, the difference of P^1 and P^{**1} is of Type 4 in Table 2. This fact can be checked by the same method used in the previous paragraph. We denote the difference of such a pair (P, P^{**}) by $\mathcal{D}_1$, i.e., $P \oplus P^{**} = \mathcal{D}_1$. Note that the differences $\mathcal{D}_0$ and $\mathcal{D}_1$ include the conditions (21) and (22), respectively.

Difference	ΔA^0	ΔB^0	ΔC^0	ΔD^0	ΔE^0	ΔF^0	ΔG^0	ΔH^0	Conditions
$\mathcal{D}_0$	0	e_{31}	0	0	0	e_{31}	0	0	(21)
$\mathcal{D}_1$	0	e_{31}	e_{31}	0	0	e_{31}	0	0	(22)

Table 3. Two types of input differences of round 0

Hence, we have two types of 11-round impossible differential characteristics with respect to the value of w^0_{31}. These two types of 11-round impossible differential characteristics are described as follows.

Property 1. If $w^0_{31} = 0$, given any pair with difference $\mathcal{D}_0$ (i.e., $P \oplus P^* = \mathcal{D}_0$), the l.s.b. of difference in the eighth word after 11 rounds can not be of 1 (i.e., $h^{11}_0 \oplus h^{*11}_0 \neq 1$). If $w^0_{31} = 1$, given any pair with difference $\mathcal{D}_1$ (i.e., $P \oplus P^{**} = \mathcal{D}_1$), the l.s.b. of difference in the eighth word after 11 rounds can not be of 1 (i.e., $h^{11}_0 \oplus h^{**11}_0 \neq 1$).

3.2   14-Round Impossible Characteristics

In this subsection, we show how to combine a nonlinear equation of 3 rounds to the 11-round impossible differential characteristics. As stated earlier, we only concern the value of Δh^{11}_0 in Equation (20). Thus, we can combine a nonlinear equation of 3 rounds to them by representing h^{11}_0 as a nonlinear equation of some bits in $A^{14}, B^{14}, \cdots, H^{14}, K^{11}, K^{12}, K^{13}, W^{11}, W^{12}$ and W^{13}. For this work, we use the fact that the addition at the l.s.b's is the same as the exclusive or. By using Equation (12), we can represent h_0^{11} as a nonlinear equation of some bits in $A^{12}, B^{12}, \cdots, H^{12}, K^{11}$ and W^{11} as like Equation (23).

$$h_0^{11} = a_0^{12} \oplus (\neg(b_2^{12} \oplus b_{13}^{12} \oplus b_{22}^{12})) \oplus (\neg((b_0^{12}\&c_0^{12}) \oplus (b_0^{12}\&d_0^{12}) \oplus (c_0^{12}\&d_0^{12}))) \qquad (23)$$
$$\oplus (\neg(f_6^{12} \oplus f_{11}^{12} \oplus f_{25}^{12})) \oplus (\neg((f_0^{12}\&g_0^{12}) \oplus ((\neg f_0^{12})\&h_0^{12}))) \oplus (\neg k_0^{11}) \oplus (\neg w_0^{11})$$

Using the decryption process, we can also represent h_0^{11} as a nonlinear equation of some bits in $A^{13}, B^{13}, \cdots, H^{13}, K^{11}, K^{12}, W^{11}$ and W^{12} as like Equation (24).

$$h_0^{11} = b_0^{13} \oplus (\neg(c_2^{13} \oplus c_{13}^{13} \oplus c_{22}^{13})) \oplus (\neg((c_0^{13}\&d_0^{13}) \oplus (c_0^{13}\&(e_0^{13} \oplus t_{1,0}^{13})) \oplus (d_0^{13}\&(e_0^{13} \oplus t_{1,0}^{13}))))$$
$$\oplus (\ \neg(g_6^{13} \oplus g_{11}^{13} \oplus g_{25}^{13})) \oplus (\neg((g_0^{13}\&h_0^{13}) \oplus ((\neg g_0^{13})\&h_0^{12}))) \oplus (\neg k_0^{11}) \oplus (\neg w_0^{11}) \qquad (24)$$

The values $t_{1,0}^{13}$ and h_0^{12} in Equation (24) are represented as like Equations (25) and (26), which are induced by Equations (11) and (23), respectively.

$$t_{1,0}^{13} = a_0^{13} \oplus (\neg(b_2^{13} \oplus b_{13}^{13} \oplus b_{22}^{13})) \oplus (\neg((b_0^{13}\&c_0^{13}) \oplus (b_0^{13}\&d_0^{13}) \oplus (c_0^{13}\&d_0^{13}))) \qquad (25)$$

$$h_0^{12} = a_0^{13} \oplus (\neg(b_2^{13} \oplus b_{13}^{13} \oplus b_{22}^{13})) \oplus (\neg((b_0^{13}\&c_0^{13}) \oplus (b_0^{13}\&d_0^{13}) \oplus (c_0^{13}\&d_0^{13}))) \qquad (26)$$
$$\oplus (\neg(f_6^{13} \oplus f_{11}^{13} \oplus f_{25}^{13})) \oplus (\neg((f_0^{13}\&g_0^{13}) \oplus ((\neg f_0^{13})\&h_0^{13}))) \oplus (\neg k_0^{12}) \oplus (\neg w_0^{12})$$

Similarly, we can represent h_0^{11} as a nonlinear equation of some bits in $A^{14}, B^{14}, \cdots, H^{14}, K^{11}, K^{12}, K^{13}, W^{11}, W^{12}$ and W^{13} as like Equation (27).

$$h_0^{11} = c_0^{14} \oplus (\neg(d_2^{14} \oplus d_{13}^{14} \oplus d_{22}^{14})) \oplus (\neg((d_0^{14}\&(e_0^{14} \oplus t_{1,0}^{14})) \oplus (d_0^{14}\&(f_0^{14} \oplus t_{1,0}^{13}))$$
$$\oplus ((e_0^{14} \oplus t_{1,0}^{14})\&(f_0^{14} \oplus t_{1,0}^{13})))) \oplus (\neg(h_6^{14} \oplus h_{11}^{14} \oplus h_{25}^{14})) \qquad (27)$$
$$\oplus (\neg((h_0^{14}\&h_0^{13}) \oplus ((\neg h_0^{14})\&h_0^{12}))) \oplus (\neg k_0^{11}) \oplus (\neg w_0^{11})$$

The values $h_0^{12}, t_{1,0}^{13}, h_0^{13}$ and $t_{1,0}^{14}$ in Equation (27) are represented as like Equations (28), (29), (30) and (31), which are induced by Equations (24), (25), (26) and (25), respectively.

$$h_0^{12} = b_0^{14} \oplus (\neg(c_2^{14} \oplus c_{13}^{14} \oplus c_{22}^{14})) \oplus (\neg((c_0^{14}\&d_0^{14}) \oplus (c_0^{14}\&(e_0^{14} \oplus t_{1,0}^{14})) \oplus (d_0^{14}\&(e_0^{14} \oplus t_{1,0}^{14}))))$$
$$\oplus (\neg(g_6^{14} \oplus g_{11}^{14} \oplus g_{25}^{14})) \oplus (\neg((g_0^{14}\&h_0^{14}) \oplus (\neg g_0^{14}\&h_0^{13}))) \oplus (\neg k_0^{12}) \oplus (\neg w_0^{12}) \qquad (28)$$

$$t_{1,0}^{13} = b_0^{14} \oplus (\neg(c_2^{14} \oplus c_{13}^{14} \oplus c_{22}^{14})) \oplus (\neg((c_0^{14}\&d_0^{14}) \oplus (c_0^{14}\&(e_0^{14} \oplus t_{1,0}^{14})) \oplus (d_0^{14}\&(e_0^{14} \oplus t_{1,0}^{14}))))$$
$$\qquad (29)$$

$$h_0^{13} = a_0^{14} \oplus (\neg(b_2^{14} \oplus b_{13}^{14} \oplus b_{22}^{14})) \oplus (\neg((b_0^{14}\&c_0^{14}) \oplus (b_0^{14}\&d_0^{14}) \oplus (c_0^{14}\&d_0^{14})))$$
$$\oplus (\neg(f_6^{14} \oplus f_{11}^{14} \oplus f_{25}^{14})) \oplus (\neg((f_0^{14}\&g_0^{14}) \oplus (\neg f_0^{14}\&h_0^{14}))) \oplus (\neg k_0^{13}) \oplus (\neg w_0^{13}) \qquad (30)$$

$$t_{1,0}^{14} = a_0^{14} \oplus (\neg(b_2^{14} \oplus b_{13}^{14} \oplus b_{22}^{14})) \oplus (\neg((b_0^{14}\&c_0^{14}) \oplus (b_0^{14}\&d_0^{14}) \oplus (c_0^{14}\&d_0^{14}))) \qquad (31)$$

So, we can represent h_0^{11} as the output of nonlinear function $NF(A^{14}, B^{14}, \cdots,$ $H^{14}, K^{11}, K^{12}, K^{13}, W^{11}, W^{12}, W^{13})$, denoted NF^{14}. Thus we have the following property for 14-round SHACAL-2.

Property 2. Assume that a plaintext triples (P, P^*, P^{**}) satisfies the conditions $P \oplus P^* = \mathcal{D}_0$ and $P \oplus P^{**} = \mathcal{D}_1$.

1. If $w_{31}^0 = 0$, $NF^{14} \oplus NF^{*14} \neq 1$ with probability 1.
2. If $w_{31}^0 = 1$, $NF^{14} \oplus NF^{**14} \neq 1$ with probability 1.

Therefore, we have two types of 14-round impossible characteristics for rounds 0-13 with respect to the value w_{31}^0. (In the same method, we also have two types of 14-round impossible characteristics for rounds i-$(i + 13)$ with respect to the value w_{31}^i.)

4   Impossible Differential Attack on 30-Round SHACAL-2 with 512-Bit Key

In this section, we present a method to use two types of the 14-round impossible characteristics to find a master key of 30-round SHACAL-2.

Let (P_i, P_i^*, P_i^{**}) be a plaintext triple. Assume that 248 plaintext triples (P_i, P_i^*, P_i^{**}) satisfy the conditions $P_i \oplus P_i^* = \mathcal{D}_0$ and $P_i \oplus P_i^{**} = \mathcal{D}_1$, for all $i = 1, \cdots, 248$. Let (C_i, C_i^*, C_i^{**}) be the corresponding ciphertext triple of (P_i, P_i^*, P_i^{**}) and $(NF^{14} \oplus NF^{*14})_i$ (resp. $(NF^{14} \oplus NF^{**14})_i$) be the difference between the l.s.b. of 8th word of P_i^{11} and that of P_i^{*11} (resp. P_i^{**11}). Then, for each triple (P_i, P_i^*, P_i^{**}), the foregoing property 2 holds.

To know ΔNF^{14}, the key value w_0^{11} in Equation (27) is not required. Also to know not the value NF^{14} but the value ΔNF^{14} in Equation (27), we only need the key value $w_0^{14}, w_1^{14}, \cdots, w_{24}^{14}, w_0^{15}, w_1^{15}, \cdots, w_{25}^{15}, w_0^{16}, w_1^{16}, \cdots, w_{25}^{16}$ with assumption that the output of round 17 are known. Hence if we guess 495-bit key, $W^{29}, W^{28}, \cdots, W^{17}, w_0^{16}, w_1^{16}, \cdots, w_{25}^{16}, w_0^{15}, w_1^{15}, \cdots, w_{25}^{15}, w_0^{14}, w_1^{14}, \cdots, w_{24}^{14},$ w_0^{13}, and w_0^{12} with assumption that the output of round 30 are known, we can calculate $(NF^{14} \oplus NF_0^{*14})_i$ and $(NF^{14} \oplus NF^{**14})_i$ from the decryption process. However, by the key schedule we can get the value w_0^{13} by guessing W^{29}, W^{27}, W^{22}, w_3^{14}, w_7^{14}, w_{18}^{14}. Since there is no need to guess the value w_0^{13}, we will guess 494-bit key.

Let $k_1, k_2, \cdots, k_{2^{494}}$ be all possible 494-bit guessed key values. For each k_j, check whether $(NF^{14} \oplus NF^{*14})_i = 0$ or $(NF^{14} \oplus NF^{**14})_i = 0$ for all i. If k_j satisfies $(NF^{14} \oplus NF^{*14})_i = 0$ (resp. $(NF^{14} \oplus NF^{**14})_i = 0$) for all i, then keep k_j and $w_{31}^0 = 0$ (resp. $w_{31}^0 = 1$) and then for this key, do an exhaustive search for the 17 remaining key bits. Algorithm 1 describes our attack scenario.

The time complexity of Algorithm 1 is dominated by Step1. Because the suggested keys which are tested in Step 2 have a small portion of 2^{494} subkeys. Thus, the attack for 30-round SHACAL-2 requires $744(= 3 \cdot 248)$ chosen plaintexts and processing equivalent to about $\frac{16}{30} \cdot \sum_{i=0}^{247} \left(2^{494-2 \cdot i} \cdot 3\right) \simeq 2^{495.1}$ 30-round SHACAL-2 encryptions.

Input : 248 ciphertext triples (C_i, C_i^*, C_i^{**}) for $i = 1, \cdots, 248$

Output : $Masterkey$

---

1. For $j = 1, \cdots, 2^{494}$

   1.1 For $i = 1, \cdots, 248$

      1.1.1 Decrypt C_i and C_i^* using k_j and compute $a_i = (NF^{14} \oplus NF^{*14})_i$.

      1.1.2 If $(a_i = 1)$, goto 1.2

           else if $(a_i = 0$ and $i = 248)$, keep k_j and $w_{31}^0 = 0$

   1.2 For $i = 1, \cdots, 248$

      1.2.1 Decrypt C_i and C_i^{**} using k_j and compute $b_i = (NF^{14} \oplus NF^{**14})_i$.

      1.2.2 If $(b_i = 1)$, goto 1

           else if $(b_i = 0$ and $i = 248)$, keep k_j and $w_{31}^0 = 1$

2. For each suggested key, do an exhaustive search for the 17 remaining key bits using trial encryption.

Algorithm 1. Attacking 30-round SHACAL-2

Note: If given a wrong guess the value ΔNF^{14} is random, the expected value of 495-bit subkeys which pass Step 1 is $0.5(= 2^{495 - 2 \cdot 248})$. However, according to our simulation, the number of subkeys which pass Step 1 is more than this expectation. In case 495-bit subkeys are correctly guessed except for the 25-bit of W^{14}, about 52.6 subkeys passed Step 1, and in case 495-bit subkeys are correctly guessed except for the 26-bit of W^{15} (resp. W^{16}), about 9.3 subkeys (resp. 4.8 subkeys) passed Step 1. (Each number of passed subkeys is the average value over 10 tests). Following lemma illustrates the reason why the number of subkeys which pass Step 1 is more than the expectation.

Lemma 1. *Assume that $Z = X + Y$, $Z^* = X^* + Y^*$, $X \oplus X^* = e_j$, and $Y = Y^*$ where X, Y and X^*, Y^* be 32-bit words. Then it holds $Z \oplus Z^* = e_{j,j+1,\cdots,j+k-1}$ with probability $1/2^k$ $(j + k - 1 \leq 31)$.*

For example, let's consider the 495-bit subkey which is correctly guessed except for w_{12}^{14}. Then for this wrong subkey, we get $\Delta NF^{14} = 1$ with probability 2^{-13} from Lemma 1. However, given a subkey which is wrong guessed in some bits of $W^{17} \sim W^{29}$, the value ΔNF^{14} is random. Hence we can attack 30-round SHACAL-2 with time complexity of $2^{495.1}$ encryptions, because almost all the 495-bit wrong subkeys produce the random value ΔNF^{14}.

5   Conclusion

In this paper we showed how to use a nonlinear equation related to the least significant bit to extend impossible differential characteristics whose output differences are fixed in the position of the least significant bit. The method enabled us to extend 11-round impossible differential characteristics to 14-round distinguishers. Using the 14-round distinguishers, we attacked 30-round SHACAL-2 with data complexity of 744 chosen plaintexts and time complexity of $2^{495.1}$ encryptions.

References

1. E. Biham, A. Biryukov and A. Shamir, *Cryptanalysis of skipjack reduced to 31 rounds using impossible differentials*, Advances in Cryptology - EUROCRYPT'99, LNCS 1592, pp. 12–23, Springer-Verlag, 1999.
2. E. Biham, O. Dunkelman and N. Keller, *Rectangle Attacks on 49-Round SHACAL-1*, FSE 2003, Springer-Verlag, 2003.
3. E. Biham and A. Shamir, *Differential cryptanalysis of the full 16-round DES*, Advances in Cryptology - CRYPTO'92, LNCS 740, pp. 487–496, Springer-Verlag, 1992.
4. H. Handschuh and D. Naccache, *SHACAL : A Family of Block Ciphers*, Submission to the NESSIE project, 2002.
5. H. Handschuh, L. R. Knudsen and M. J. Robshaw, *Analysis of SHA-1 in Encryption Mode*, CT-RSA 2001, LNCS 2020, pp. 70–83, Springer-Verlag, 2001.
6. J.S. Kim, D.J. Moon, W.L. Lee, S.H. Hong, S.J. Lee and S.W. Jung, *Amplified Boomerang Attack against Reduced-Round SHACAL*, Advances in Cryptology, ASIACRYPT 2002, LNCS 2501, pp. 243–253, Springer-Verlag, 2002.
7. D.J. Moon, K.D. Hwang, W.I. Lee, S.J. Lee and J.I. Lim, *Impossible Differential Cryptanalysis of Reduced Round XTEA and TEA*, FSE 2002, LNCS 2365, pp.49–60, Springer-Verlag, 2002.
8. J. Nakahara Jr, *The Statistical Evaluation of the NESSIE Submission*, October 2001.
9. M.-J.O. Saarinen, *Cryptanalysis of Block Ciphers Based on SHA-1 and MD5*, advances in Cryptology, proc. of FSE 2003.
10. U.S. Department of Commerce.*FIPS 180-1*: Secure Hash Standard ,Federal Information Processing Standards Publication, N.I.S.T., April 1995.
11. U.S. Department of Commerce.*FIPS 180-2*: Secure Hash Standard ,Federal Information Processing Standards Publication, N.I.S.T., August 2002.

Construction of Perfect Nonlinear and Maximally Nonlinear Multi-output Boolean Functions Satisfying Higher Order Strict Avalanche Criteria
(Extended Abstract)

Kishan Chand Gupta and Palash Sarkar

Cryptology Research Group
Applied Statistics Unit
Indian Statistical Institute
203, B.T. Road
Kolkata 700108, India
{kishan_t,palash}@isical.ac.in

Abstract. We consider the problem of constructing perfect nonlinear multi-output Boolean functions satisfying higher order strict avalanche criteria (SAC). Our first construction is an infinite family of 2-ouput perfect nonlinear functions satisfying higher order SAC. This construction is achieved using the theory of bilinear forms and symplectic matrices. Next we build on a known connection between 1-factorization of a complete graph and SAC to construct more examples of 2 and 3-output perfect nonlinear functions. In certain cases, the constructed S-boxes have optimal trade-off between the following parameters: numbers of input and output variables, nonlinearity and order of SAC. In case the number of input variables is odd, we modify the construction for perfect nonlinear S-boxes to obtain a construction for maximally nonlinear S-boxes satisfying higher order SAC. Our constructions present the first examples of perfect nonlinear and maximally nonlinear multioutput S-boxes satisfying higher order SAC. Lastly, we present a simple method for improving the degree of the constructed functions with a small trade-off in nonlinearity and the SAC property. This yields functions which have possible applications in the design of block ciphers.
Keywords: S-box, SAC, bent function, bilinear form, symplectic matrix, nonlinearity, symmetric ciphers.

1 Introduction

A Boolean function is a map from $\{0,1\}^n$ to $\{0,1\}$ and by a multi-output Boolean function we mean a map from $\{0,1\}^n$ to $\{0,1\}^m$. Multi-output Boolean functions are usually called S-boxes and are used as basic primitives for designing symmetric ciphers. For example, the S-boxes used in DES have $n = 6$ and $m = 4$ and the S-box used in the design of AES has $n = m = 8$. We next describe some properties of S-boxes which have been studied previously.

T. Johansson and S. Maitra (Eds.): INDOCRYPT 2003, LNCS 2904, pp. 107–120, 2003.

Nonlinearity is one of the basic properties of an S-box. The nonlinearity of a Boolean function measures the distance of the function to the set of all affine functions. The nonlinearity of an S-box is a natural generalization of this notion. For even n, functions achieving the maximum possible nonlinearity are called *perfect nonlinear* S-boxes [10]. If $m = 1$, such functions are called *bent* functions [12]. For odd n and $m > 1$, functions achieving the maximum possible nonlinearity are called *maximally nonlinear* functions.

The concept of propagation characteristic was introduced in the cryptology literature in [11]. An S-box $f(x)$ is said to satisfy propagation characteristic of degree l and order k (PC(l) of order k) if the following holds: Let $g(y)$ be a function obtained from $f(x)$ by fixing at most k inputs to constant values and let α be a non zero vector of weight at most l. Then $g(y) \oplus g(y \oplus \alpha)$ is a balanced function.

If $k = 0$, then the function is simply said to satisfy PC(l). PC(l) of order k functions have been studied in [3,4] and constructions of Boolean functions and S-boxes satisfying PC(l) of order k are known [8,7,13]. S-boxes satisfying PC(1) of order k are said to satisfy strict avalanche criteria of order k (SAC(k)). If $k = 0$, then the S-box is said to satisfy SAC. The notion of SAC was introduced in [14]. It is known [9] that any bent function or any perfect nonlinear S-box satisfies PC(n). It is also possible to construct bent functions satisfying SAC($n - 2$). However, for $m > 1$, construction of perfect nonlinear S-boxes satisfying SAC(k) for $k > 0$ has been a open problem.

In this paper, we (partially) solve this problem by providing constructions of perfect nonlinear S-boxes with $m = 2, 3$ and satisfying SAC(k) for $k \geq 1$. Our contributions are the following.

- Construction of an infinite family of 2-output perfect nonlinear S-boxes satisfying higher order SAC. More precisely, for each even $n \geq 6$, we construct a 2-output perfect nonlinear S-box satisfying SAC($(n/2) - 2$).
- In an earlier paper [8], a 1-factorization of the complete graph on n-vertices was used to construct S-boxes satisfying higher order SAC. However, the S-boxes constructed in [8] did not satisfy perfect nonlinearity. We make a more detailed analysis of the connection between 1-factorization and higher order SAC to construct 2 and 3 output *perfect nonlinear* S-boxes satisfying higher order SAC.
- In certain cases, the functions that we construct achieve the best possible trade-off among the following parameters: number of input variables, number of output variables, nonlinearity and order of SAC. Hence for such functions, it is not possible to improve any one parameter without changing some other parameter.
- For small n, our constructions provide S-boxes which cannot be obtained from the currently known constructions [8,7,13]. Some examples of such functions are the following.
  - 8-input, 2-output perfect nonlinear S-box satisfying SAC(2).
  - 8-input, 3-output perfect nonlinear S-box satisfying SAC(1).
  - 10-input, 3-output perfect nonlinear S-box satisfying SAC(3).

The last example is also an example of an S-box achieving the best possible trade-off.

- Our constructions are based on bilinear forms and symplectic matrices used in the study of second order Reed-Muller code. We show that if n is odd, then the construction for $(n + 1)$ can be modified to obtain maximally nonlinear S-boxes satisfying higher order SAC.
- We provide a simple technique for improving the degree of an S-box with a small sacrifice in nonlinearity and the SAC property. This results in S-boxes which have possible applications in the design of symmetric ciphers

In this extended abstract we skip the proofs of some of the results. These are available in the full version [5].

2 Preliminaries

Let $F_2 = GF(2)$. We consider the domain of a Boolean function to be the vector space $(F_2^n, \oplus)$ over F_2, where $\oplus$ is used to denote the addition operator over both F_2 and the vector space F_2^n. The inner product of two vectors $u, v \in F_2^n$ will be denoted by $\langle u, v \rangle$. The weight of an n-bit vector u is the number of ones in u and will be denoted by $\mathsf{wt}(u)$. The (Hamming) distance between two vectors $x = (x_1, x_2, \cdots, x_n)$ and $y = (y_1, y_2, \cdots, y_n)$ is the number of places where they differ and is denoted by $d(x, y)$. The bitwise complement of a bit string x will be denoted by $\overline{x}$.

2.1 Boolean Functions

An n-variable Boolean function is a map $f : F_2^n \to F_2$. The weight of f, denoted by $\mathsf{wt}(f)$ is defined as $\mathsf{wt}(f) = |\{x : f(x) = 1\}|$. The function f is said to be balanced if $\mathsf{wt}(f) = 2^{n-1}$. The (Hamming) distance between two n-variable Boolean functions f and g is $d(f, g) = |\{x : f(x) \neq g(x)\}|$.

A parameter of fundamental importance in cryptography is the nonlinearity of a Boolean function. This quantity measures the distance of a Boolean function from the set of all affine functions. An n-variable affine function is of the form $l_{u,b}(x) = \langle u, x \rangle \oplus b$, where $u \in F_2^n$ and $b \in F_2$. Let A_n be the set of all n-variable affine functions. The nonlinearity $\mathsf{nl}(f)$ of an n-variable Boolean function is defined as $\mathsf{nl}(f) = \min_{l \in A_n} d(f, l)$. The maximum nonlinearity achievable by an n-variable Boolean function is $2^{n-1} - 2^{(n-2)/2}$. Functions achieving this value of nonlinearity are called bent and can exist only when n is even [12]. When n is odd, the maximum nonlinearity achievable by an n-variable Boolean function is not known. However, functions achieving a nonlinearity of $2^{n-1} - 2^{(n-1)/2}$ are easy to construct and are called almost optimally nonlinear [4].

An n-variable Boolean function f satisfies strict avalanche criteria (SAC) if $f(x) \oplus f(x \oplus \alpha)$ is balanced for any $\alpha \in F_2^n$ with $\mathsf{wt}(\alpha) = 1$ [14]. A function f satisfies SAC(k) if every subfunction obtained from $f(x_1, \cdots, x_n)$ by keeping at most k input bits constant satisfies SAC.

An n-variable Boolean function can be represented as a multivariate polynomial over F_2. The degree of this polynomial is called the degree of the function. Affine functions have degree one and functions of degree two are called quadratic.

2.2   S-boxes

An (n, m) S-box (or vectorial function) is a map $f : \{0,1\}^n \to \{0,1\}^m$. Let $f : \{0,1\}^n \to \{0,1\}^m$ be an S-box and $g : \{0,1\}^m \to \{0,1\}$ be an m-variable Boolean function. The composition of g and f, denoted by $g \circ f$ is an n-variable Boolean function defined by $(g \circ f)(x) = g(f(x))$.

Let f be an (n, m) S-box. The nonlinearity of f is defined to be $\mathsf{nl}(f) = \min\{\mathsf{nl}(l \circ f) : l$ is a non-constant m-variable linear function$\}$. The maximum achievable nonlinearity of an n-variable function is $2^{n-1} - 2^{(n-2)/2}$ and S-boxes achieving this value of nonlinearity are called perfect nonlinear S-boxes. Such S-boxes exist only if n is even and $m \leq (n/2)$ [10]. For odd n and $m = n$, the maximum possible nonlinearity achievable is $2^{n-1} - 2^{(n-1)/2}$ and S-boxes achieving this value of nonlinearity are called maximal nonlinear S-boxes. For odd n and $1 < m < n$, the maximum possible achievable nonlinearity is an open problem. However, for odd n, $1 < m < n$, and quadratic functions the maximum possible achievable nonlinearity is $2^{n-1} - 2^{(n-1)/2}$. We will also call such functions to be maximally nonlinear.

We define the degree of an (n, m) S-box f to be the minimum of the degrees of $l \circ f$, where l ranges over all non constant m-variable linear functions. This definition is more meaningful to cryptography than the definition where the degree of an S-box is taken to be the maximum of the degrees of all the component functions. The later definition has been used in [2].

An (n, m) S-box f is said to be SAC(k), if $l \circ f$ is SAC(k) for every nonconstant m-variable linear function l. By an (n, m, k) S-box we mean an (n, m) S-box which is SAC(k). We will be interested in (n, m, k) S-boxes with maximum possible nonlinearity. More specifically, we will be interested in (n, m, k) perfect nonlinear S-boxes if n is even and in (n, m, k) maximally nonlinear S-boxes if n is odd. Such S-boxes have important applications in the design of secure block ciphers.

2.3   Binary Quadratic Form

An n-variable Boolean function g of degree ≤ 2 can be written as (see [9, page 434]) $g(x) = xQx^T \oplus Lx^T \oplus b$ where $Q = (q_{ij})$ is an upper triangular $n \times n$ binary matrix, $L = (l_1, \cdots, l_n)$ is a binary vector and b is 0 or 1. The expression xQx^T is called a quadratic form and Lx^T is called a linear form. Let $B = Q \oplus Q^T$. Then B is a binary symmetric matrix with zero diagonal. Such a matrix is called a symplectic matrix (see [9, page 435]). Thus from a quadratic Boolean function we can define a symplectic matrix. *Conversely, given a symplectic matrix B we can construct a quadratic Boolean function by reversing the above steps. We denote this Boolean function by f_B.*

It is known that the rank of a symplectic matrix is always even [9, page 436]. The nonlinearity of the Boolean function g is related to the rank of B by the following result [9, page 441].

Proposition 1. *Let g be a quadratic n-variable Boolean function and B be its associated symplectic form. Then the nonlinearity of g is equal to $2^{n-1} - 2^{n-h-1}$, where the rank of B is $2h$.*

Consequently, a quadratic Boolean function is bent if and only if the associated symplectic matrix is of full rank.

3   Basic Results

We will be interested in nonlinear quadratic functions satisfying higher order SAC. From Proposition 1, a convenient way to study the nonlinearity of quadratic functions is through the rank of the associated symplectic matrix. We now develop the basic relationships between the nonlinearity and SAC property of a quadratic S-box and the symplectic matrices associated with the component functions.

Proposition 2. *Let f be a quadratic Boolean function and B its associated symplectic matrix. Then f satisfies $SAC(k)$ if and only if for all $1 \leq i \leq n$, we have $\mathsf{wt}(r^{(i)}) \geq k+1$, where $r^{(i)}$ is the ith row of B. (Since B is symmetric, a similar property holds for the columns of B.)*

Let $f = (f_1, \cdots, f_m)$ be an (n,m) quadratic S-box. Then each of the component functions f_i is an n-variable quadratic Boolean function. For $1 \leq i \leq m$, let B_i be the symplectic matrix associated with the component function f_i. Clearly, any linear combination of symplectic matrices is also a symplectic matrix. We have the following extension of Proposition 2.

Lemma 1. *Let f be an (n,m) S-box with quadratic component functions f_i and associated symplectic forms B_i for $1 \leq i \leq m$. Then f satisfies $SAC(k)$ if and only if the weight of every row in any non zero linear combination of the B_i's is at least $k+1$.*

A similar result for nonlinearity can be stated by extending Proposition 1.

Lemma 2. *Let f be an (n,m) S-box with quadratic component functions f_i and associated symplectic forms B_i for $1 \leq i \leq m$. The nonlinearity of f is $2^{n-1} - 2^{n-h-1}$, where $2h$ is the minimum of the ranks of any non zero linear combination of the B_i's. Consequently for even n, the S-box f is perfect nonlinear if and only if every non zero linear combination of the B_i's has full rank. Similarly, for odd n, the S-box f is maximally nonlinear if and only if every non zero linear combination of the B_i's has rank $(n-1)$.*

Lemmas 1 and 2 will be used in proving the correctness of our constructions in the next sections.

4   Construction of $(n, 2, \frac{n}{2} - 2)$ S-box

Our construction will be via symplectic matrices. Given any (n, r) quadratic S-box, it is clear from the above discussion that the symplectic matrices associated with the output component function defines the S-box. Thus to describe the construction, it is sufficient to define these symplectic matrices and use Lemmas 1 and 2 to prove the correctness of the construction.

In this section, we describe the construction of $(n, 2)$ S-boxes. Hence it is sufficient to define two symplectic matrices. We proceed to do this as follows. For each even $n \geq 6$, we define two sequences of $n \times n$ matrices and show that these matrices are the symplectic matrices required in the construction. For the rest of this paper, we will use the following notation.

- For each $n \geq 1$, define v_n to be a string of length n which is the alternating sequence of 0's and 1's starting with a 0. For example, $v_4 = 0101$ and $v_5 = 01010$. Define $w_n = 1\overline{v_{n-1}}$.
- For each even $n \geq 2$, define u_n as $u_n = \underbrace{1\ldots1}_{(n/2)}\underbrace{0\ldots0}_{(n/2)}$. For odd $n \geq 3$, define $x_n = 1\overline{u_{n-1}}$.

Define $M_4 = [0010, 0010, 1101, 0010]^T$ and $N_4 = [0101, 1011, 0101, 1110]^T$. Further, for even $n > 4$ define

$$
M_n = \begin{bmatrix} 0 & v_{n-2} & 0 \\ v_{n-2}^T & M_{n-2} & v_{n-2}^T \\ 0 & v_{n-2} & 0 \end{bmatrix}, F_n = \begin{bmatrix} 0 & v_{n-2} & 0 \\ v_{n-2}^T & M_{n-2} & u_{n-2}^T \\ 0 & u_{n-2} & 0 \end{bmatrix};
$$

$$
N_n = \begin{bmatrix} 0 & \overline{v_{n-2}} & 1 \\ v_{n-2}^T & N_{n-2} & v_{n-2}^T \\ 1 & \overline{v_{n-2}} & 0 \end{bmatrix}, G_n = \begin{bmatrix} 0 & \overline{v_{n-2}} & 1 \\ v_{n-2}^T & N_{n-2} & u_{n-2}^T \\ 1 & \overline{u_{n-2}} & 0 \end{bmatrix}. \tag{1}
$$

The following result is easy to prove by induction on even $n \geq 6$.

Lemma 3. F_n, G_n and $F_n \oplus G_n$ are symplectic matrices, where F_n and G_n are defined by equation 1.

The matrices F_n and G_n are our required symplectic matrices which define the two output component functions of the required $(n, 2)$ S-box. In particular, we have the following result.

Theorem 1. Let $n \geq 6$ be an even integer. The S-box $f : F_2^n \rightarrow F_2^2$ defined by $f(x) = (f_{F_n}(x), f_{G_n}(x))$ is a perfect nonlinear S-box satisfying $SAC(\frac{n}{2} - 2)$.

We now turn to the proof of correctness of Theorem 1. The proof is in two parts – in the first part we prove the statement about SAC and in the second part we prove the statement about nonlinearity.

Lemma 4. The S-box f defined in Theorem 1 satisfy $SAC(\frac{n}{2} - 2)$.

We next turn to the nonlinearity of the S-box defined in Theorem 1.

Lemma 5. *For even $n \geq 6$, the rank of F_n is n.*

Proof : First we prove that the rank of M_n is $n - 2$. It is easy to check that the rank of M_4 is 2. Assume that the rank of M_{n-2} is $n - 4$. It is clear that 1-st column and n-th column of M_n are identical. Likewise 1-st column and $n - 2$-th column of M_{n-2} are identical. Consider the matrix

$$M_n' = \begin{bmatrix} 0 & v_{n-2} \\ v_{n-2}^T & M_{n-2} \end{bmatrix}.$$

From the definition of v_n, we have that the first bit of v_{n-2} is 0 and $(n-2)$-th bit is 1. So v_{n-2} is linearly independent of rows of M_{n-2}. So rank of M_n' is at least $n - 4 + 1 = n - 3$. But M_n' is symplectic matrix and hence its rank must be even (see [9, page 436]). So the rank of M_n' (and hence M_n) is $n - 2$.

Now we turn to the rank of F_n. As M_n' has rank $n - 2$, the rank of F_n is at least $n - 2$. It is simple to verify by induction that $\frac{n}{2}$-th column and $(\frac{n}{2} + 2)$-th column of M_n are identical. From definition, the $\frac{n}{2}$-th bit of $0u_{n-2}0$ is 1 and the $(\frac{n}{2} + 2)$-th bit is 0. Hence the last row $0u_{n-2}0$ of F_n is linearly independent of the previous $(n - 1)$ rows. Thus the rank of F_n is at least $n - 2 + 1 = n - 1$. But F_n is a binary symplectic matrix and hence its rank must be even. Hence the rank of F_n is n. $\square$

Lemma 6. *For even $n \geq 6$, the rank of $F_n \oplus G_n$ is n.*

Proof : Note $M_4 \oplus N_4 = [0111, 1001, 1000, 1100]^T$ and hence the rank of $M_4 \oplus N_4$ is 4. Assume that the rank of $M_{n-2} \oplus N_{n-2}$ is $n - 2$. Note

$$F_n \oplus G_n = M_n \oplus N_n = \begin{bmatrix} 0 & J_{n-2} & 1 \\ J_{n-2}^T & M_{n-2} \oplus N_{n-2} & J_{n-2}^T \\ 1 & J_{n-2} & 0 \end{bmatrix},$$

where J_n is the all 1 vector. The row $1J_{n-2}0$ is linearly independent of rows of matrix $J_{n-2}^T(M_{n-2} \oplus N_{n-2})J_{n-2}^T$. So rank of

$$\begin{bmatrix} J_{n-2}^T & M_{n-2} \oplus N_{n-2} & J_{n-2}^T \\ 1 & J_{n-2} & 0 \end{bmatrix}$$

is at least $n - 2 + 1 = n - 1$ and hence the rank of $F_n \oplus G_n$ is at least $n - 1$. Again since $F_n \oplus G_n$ is a symplectic matrix its rank must be even. Hence its rank is n. $\square$

We define $T_5 = [01010, 10101, 01011, 10101, 01110]^T$,

$$T_n = \begin{bmatrix} 0 & \overline{v_{n-2}} & 0 \\ v_{n-2}^T & T_{n-2} & w_{n-2}^T \\ 0 & w_{n-2} & 0 \end{bmatrix} \text{ for odd } n > 5$$

and

$$H_n = \begin{bmatrix} T_{n-1} & x_{n-1}^T \\ x_{n-1} & 0 \end{bmatrix} \text{ for even } n \geq 6.$$

First we prove the following result.

Lemma 7. $G_n = H_n$ *for all even* $n \geq 6$.

Proof : We first prove the following statement by induction on n.

$$T_n = \begin{bmatrix} 0 & \overline{v}_{n-1} \\ v_{n-1}^T & N_{n-1} \end{bmatrix} \text{ for odd } n \geq 5 \text{ and } N_n = \begin{bmatrix} T_{n-1} & w_{n-1}^T \\ w_{n-1} & 0 \end{bmatrix} \text{ for even } n \geq 6. (2)$$

It is easy to verify that $T_5 = \begin{bmatrix} 0 & \overline{v}_4 \\ v_4^T & N_4 \end{bmatrix}$ and $N_6 = \begin{bmatrix} T_5 & w_5^T \\ w_5 & 0 \end{bmatrix}$. Assume that (2)

holds for $(n-1)$. By definition and using $\overline{v}_{n-1} = \overline{v}_{n-2}0$ we have that for

odd $n \geq 7$, $\begin{bmatrix} 0 & \overline{v}_{n-1} \\ v_{n-1}^T & N_{n-1} \end{bmatrix} = \begin{bmatrix} 0 & \overline{v}_{n-2} & 0 \\ v_{n-2}^T & T_{n-2} & w_{n-2}^T \\ 0 & w_{n-2} & 0 \end{bmatrix} = T_n$. Similarly, by defini-

tion and using $1\overline{v}_{n-2} = \overline{w}_{n-1}0$ we have that for even $n \geq 8$, $\begin{bmatrix} T_{n-1} & w_{n-1}^T \\ w_{n-1}^T & 0 \end{bmatrix} =$

$\begin{bmatrix} 0 & \overline{v}_{n-2} & 1 \\ v_{n-2}^T & N_{n-2} & v_{n-2}^T \\ 1 & \overline{v}_{n-2} & 0 \end{bmatrix} = N_n$. This completes the proof of (2). Now to prove $G_n =$

H_n it is sufficient to show $T_{n-1} = \begin{bmatrix} 0 & \overline{v}_{n-2} \\ v_{n-2}^T & N_{n-2} \end{bmatrix}$ and $x_{n-1}0 = 1\overline{u}_{n-2}0$. The first

statement follows from (2) and the second statement follows from the definition
of x_n.   □

Lemma 8. *For odd* $n \geq 5$, *the following statements hold for* T_n.
(1) The first column of T_{n-2} *is* v_{n-2}^T *and the second column is* v_{n-2}^T; *(2) The*
$\lfloor \frac{n}{2} \rfloor$*-th column and* $(\lfloor \frac{n}{2} \rfloor + 2)$*-th column of* T_n *are identical; (3) The rank of* T_n
is $(n-1)$.

Proof : All three statements are proved using induction on odd $n \geq 5$. We only
describe the proof for the third statement. For $n = 5$ it is easy to verify that
the rank of T_5 is 4. Assume that the rank of T_{n-2} is $n-3$. Consider the matrix

$A_n = \begin{bmatrix} v_{n-2}^T & T_{n-2} \\ 0 & w_{n-2} \end{bmatrix}$. By the first statement of the lemma, the first and third

columns of the matrix $[v_{n-2}^T \ T_{n-2}]$ are identical. At the same time the first and
third bits of the vector $0w_{n-2}$ are 0 and 1 respectively. So the last row of A_n is
linearly independent of other rows. Hence the rank of A_n is $n - 3 + 1 = n - 2$.
Consequently, T_n has rank at least $n-2$. Again since T_n is a symplectic matrix,
its rank must be even and hence must be $n-1$.   □

Now we are in a position to prove that G_n is of full rank.

Lemma 9. *The rank of* G_n *is* n.

Proof : Consider $G_n = H_n = \begin{bmatrix} T_{n-1} & x_{n-1}^T \\ x_{n-1} & 0 \end{bmatrix}$. Since the rank of T_{n-1} is $(n-2)$

the rank of H_n is at least $n-2$. Again from Lemma 8, the $\lfloor \frac{n-1}{2} \rfloor$-th column
and the $(\lfloor \frac{n-1}{2} \rfloor + 2)$-th column of T_{n-1} are identical. But the $\lfloor \frac{n-1}{2} \rfloor$-th and
the $(\lfloor \frac{n-1}{2} \rfloor + 2)$-th bits of x_{n-1} are 0 and 1 respectively. Hence x_{n-1} is linearly

independent of T_{n-1}. Thus the rank of G_n is at least $n - 2 + 1 = n - 1$. Again since G_n is a symplectic matrix its rank must be even and hence its rank is n. $\square$

Thus we have the following result which completes the proof of Theorem 1.

Lemma 10. *The S-box f defined in Theorem 1 is a perfect nonlinear S-box.*

Proof : Using Lemmas 5, 6 and 9, we know that F_n, G_n and $F_n \oplus G_n$ have full rank. Hence the Boolean functions f_{F_n}, f_{G_n} and $f_{F_n} \oplus f_{G_n} = f_{F_n \oplus G_n}$ are bent. Thus the function f defined in Theorem 1 is a perfect nonlinear function. $\square$

5   Relation with One Factorization of a Complete Graph

A one-factor of a graph G is a one-regular spanning subgraph of G. A one-factorization of G is a partition of the edges of G into one-factors.

Let K_n be the complete graph with n vertices. For even $n \geq 2$, it is well known that K_n can be decomposed into $(n - 1)$ edge disjoint, one-factors [1]. One such decomposition of K_n is described as follows. For even n and $1 \leq i \leq n - 1$, define

$$\mathcal{F}_i^n = \{(n, i)\} \cup \{((n - 2 - j + i) \bmod (n - 1) + 1, (i + j - 1) \bmod (n - 1) + 1)$$

$$: 1 \leq j \leq \frac{n}{2} - 1\}$$

The collection $\mathcal{T}_n = \{\mathcal{F}_1^n, \ldots, \mathcal{F}_{n-1}^n\}$ is a one factorization of K_n where the vertices are labeled by the integers $1, \ldots, n$. When n is clear from the context we will write $\mathcal{F}_i$ instead of $\mathcal{F}_i^n$.

In [8], one factorization of K_n was used as a tool for construction of S-boxes satisfying SAC. We point out the connection of the construction of Section 4 to the one factorization of K_n. This connection will be developed in later sections to obtain other constructions of perfect nonlinear S-boxes satisfying higher order SAC.

Suppose $\mathcal{S} \subseteq \mathcal{T}_n$. We use $\mathcal{S}$ to define a symplectic matrix $B_\mathcal{S}$ in the following manner: For $1 \leq k, l \leq n$, the entry $B_\mathcal{S}[k, l] = 1$ if and only if either (k, l) or (l, k) is in $\mathcal{F}_i^n$ for some $\mathcal{F}_i^n \in \mathcal{S}$.

Theorem 2. *Let $n \geq 4$ be an even integer, $\mathcal{S}_1 = \{\mathcal{F}_2, \ldots, \mathcal{F}_{\frac{n}{2}}\}$ and $\mathcal{S}_2 = \mathcal{T} \setminus \mathcal{S}_1$. Let $B_{\mathcal{S}_1}$ and $B_{\mathcal{S}_2}$ be the symplectic matrices associated with $\mathcal{S}_1$ and $\mathcal{S}_2$ respectively. Then*
1. F_n is obtained from $B_{\mathcal{S}_1}$ by changing the zeros in positions $(\frac{n}{2} + 1, \frac{n}{2})$ and $(\frac{n}{2}, \frac{n}{2} + 1)$ to ones.
2. G_n is obtained from $B_{\mathcal{S}_2}$ by changing the zeros in positions $(\frac{n}{2} + 1, \frac{n}{2} + 2)$ and $(\frac{n}{2} + 2, \frac{n}{2} + 1)$ to ones.

Theorem 2 shows the relationship between one factorization and two output S-boxes of Section 4. This can be generalized to more than two output S-boxes. In fact, the earlier work of [8] provides such a generalization. However, there is one major difficulty with the generalization. It becomes very difficult to ensure that the resulting S-box is a perfect nonlinear S-box. Thus while the generalization of [8] ensures the SAC property, it results in functions with quite weak nonlinearity. On the other hand, our motivation is to obtain *perfect nonlinear* S-boxes satisfying higher order SAC. The rest of the paper is devoted to identifying other perfect nonlinear S-boxes satisfying higher order SAC.

5.1 Improvements for Two Output S-boxes

We know from [8] that for an $(n, 2, k)$-SAC function, $k \leq \lfloor \frac{2(n-1)}{3} \rfloor - 1$. Thus the construction in Section 4 is suboptimal with respect to the SAC property. (However, it is optimal with respect to nonlinearity).

Here we provide some examples of two output S-boxes with higher order SAC. All these examples were obtained using experimental method. The constructions are based on the relationship between the symplectic matrices and one factorization described above. These examples are summarized in Table 1. The interpretation of the entries in Table 1 is as follows. Each row describes a construction for the particular value of n. The second column describes two subsets $\mathcal{S}_1$ and $\mathcal{S}_2$ of $\mathcal{T}_n$. Let $B_{\mathcal{S}_1}$ and $B_{\mathcal{S}_2}$ be the symplectic matrices associated with these two sets. We set $B_1 = B_{\mathcal{S}_1}$ and B_2 is $B_{\mathcal{S}_2}$ with the following modification: If (k, l) is in the third column, then $B_{\mathcal{S}_2}[k, l]$ and $B_{\mathcal{S}_2}[l, k]$ are changed from 0 to 1. The desired S-box $f : F_2^n \rightarrow F_2^2$ is given by $f(x) = (f_{B_1}(x), f_{B_2}(x))$. *Each of these S-boxes is a perfect nonlinear S-box.* The fourth column provides the order of SAC that is achieved by the corresponding S-box. The fifth column provides the maximum order of SAC that can be achieved by an $(n, 2)$ S-box. *In the situation where this maximum is equal to the achieved order of SAC, the construction provides optimal trade-off among the following parameters : nonlinearity, order of SAC, number of input variables, number of output variables. None of these parameters can be improved without changing some other parameter.*

Table 1. Improved and Optimal Constructions of Two Output S-boxes.

n	Description	Modification	k	max k
8	$\mathcal{S}_1 = \{\mathcal{F}_2, \mathcal{F}_3, \mathcal{F}_4, \mathcal{F}_5, \mathcal{F}_7\}$	–	3	3
	$\mathcal{S}_2 = \{\mathcal{F}_1, \mathcal{F}_4, \mathcal{F}_5, \mathcal{F}_6\}$	(5,6)		
10	$\mathcal{S}_1 = \{\mathcal{F}_1, \mathcal{F}_2, \mathcal{F}_4, \mathcal{F}_7, \mathcal{F}_8\}$	–	4	5
	$\mathcal{S}_2 = \{\mathcal{F}_3, \mathcal{F}_5, \mathcal{F}_6, \mathcal{F}_7, \mathcal{F}_8\}$	(6,9)		
12	$\mathcal{S}_1 = \{\mathcal{F}_1, \mathcal{F}_3, \mathcal{F}_5, \mathcal{F}_6, \mathcal{F}_7, \mathcal{F}_8, \mathcal{F}_{11}\}$	–	6	6
	$\mathcal{S}_2 = \{\mathcal{F}_1, \mathcal{F}_2, \mathcal{F}_4, \mathcal{F}_7, \mathcal{F}_8, \mathcal{F}_9, \mathcal{F}_{10}\}$	(2,7)		
14	$\mathcal{S}_1 = \{\mathcal{F}_1, \mathcal{F}_2, \mathcal{F}_3, \mathcal{F}_4, \mathcal{F}_9, \mathcal{F}_{10}, \mathcal{F}_{11}, \mathcal{F}_{12}, \mathcal{F}_{13}\}$	–	7	7
	$\mathcal{S}_2 = \{\mathcal{F}_5, \mathcal{F}_6, \mathcal{F}_7, \mathcal{F}_8, \mathcal{F}_9, \mathcal{F}_{10}, \mathcal{F}_{11}, \mathcal{F}_{12}\}$	(8,9)		
16	$\mathcal{S}_1 = \{\mathcal{F}_2, \mathcal{F}_3, \mathcal{F}_4, \mathcal{F}_5, \mathcal{F}_6, \mathcal{F}_7, \mathcal{F}_8, \mathcal{F}_9, \mathcal{F}_{15}\}$	–	8	9
	$\mathcal{S}_2 = \{\mathcal{F}_1, \mathcal{F}_7, \mathcal{F}_8, \mathcal{F}_9, \mathcal{F}_{10}, \mathcal{F}_{11}, \mathcal{F}_{12}, \mathcal{F}_{13}, \mathcal{F}_{14}\}$	(3,9)		

6   Construction of $(n, 3, k)$ S-boxes

We describe constructions of $(n, 3, k)$ perfect nonlinear S-boxes. These constructions were obtained by experimental trial and error methods. Some of the constructions seem to have a general pattern, though it has not been possible to prove a general result. There are several cases in the construction though the description of the constructions in all the cases is similar. We first identify three subsets S_1, S_2 and S_3 of $\mathcal{T}_n$. These three subsets define three symplectic matrices B_{S_1}, B_{S_2} and B_{S_3}. These matrices are then modified by changing a number of zeros to ones to obtain three other symplectic matrices B_1, B_2 and B_3. The positions where the changes are to be made are given by the third column. If (k, l) is in the third column, then $B_{S_j}[k, l]$ and $B_{S_j}[k, l]$ $(1 \leq j \leq 3)$ are changed from 0 to 1. The required $(n, 3)$ S-box $f : F_2^n \rightarrow F_2^3$ is obtained from these three matrices in the following manner: $f(x) = (f_{B_1}(x), f_{B_2}(x), f_{B_3}(x))$. There are three cases.

1. Table 2 describes several cases of constructions for $n \equiv 0 \bmod 8$. For $n > 8$, there is a general heuristic which provides the required construction. For $n = 8$, a special construction is required.

2. Table 3 describes constructions for $n \equiv 4 \bmod 8$. These constructions have a general pattern.

3. Table 4 describes several constructions for $n \equiv 2 \bmod 4$. There does not appear to be any general pattern for these constructions.

The constructions for $n = 10, 22$ provide optimal trade-off between the following parameters: numbers of input and output variables, nonlinearity and the order of SAC. Further, for $n = 12, 16, 20$ and 24 the achieved value of k is only one less than the upper bound on k.

Table 2. Constructions for $n \equiv 0 \bmod 8$.

n	Description	Modification	k	max k
8	$S_1 = \{\mathcal{F}_2, \mathcal{F}_3, \mathcal{F}_7\}$	$(4,5)$	1	2
	$S_2 = \{\mathcal{F}_3, \mathcal{F}_4, \mathcal{F}_5\}$	$-$		
	$S_3 = \{\mathcal{F}_1, \mathcal{F}_3, \mathcal{F}_6\}$	$(4,7)$		
16,	$S_1 = \{\mathcal{F}_2, \mathcal{F}_3, \ldots, \mathcal{F}_{\frac{n}{2}-1}, \mathcal{F}_{n-1}\}$	$(\frac{n}{2}, \frac{n}{2}+1)$	$\frac{n}{2}-2$	min(
24,	$S_2 = \{\mathcal{F}_{\frac{n}{4}+1}, \ldots, \mathcal{F}_{\frac{3n}{4}-1}\}$	$(\frac{n}{2}, \frac{n}{4}+1),$		$\lfloor \frac{4(n-1)}{7} \rfloor - 1,$
		$(\frac{n}{2}, \frac{3n}{4}+1)$		
32	$S_3 = \{\mathcal{F}_1, \mathcal{F}_{\frac{n}{4}+1}, \ldots, \mathcal{F}_{\frac{n}{2}-1}, \mathcal{F}_{\frac{3n}{4}}, \ldots, \mathcal{F}_{n-2}\}$	$(\frac{n}{2}, \frac{3n}{4})$		$2\lfloor \frac{2n}{7} \rfloor - 1)$

Table 3. Constructions for $n \equiv 4 \bmod 8$.

n	Description	Modification	k	max k
12,	$S_1 = \{\mathcal{F}_2, \mathcal{F}_3, \ldots, \mathcal{F}_{\frac{n}{2}-2}, \mathcal{F}_{n-1}\}$	$(\frac{n}{2}, \frac{n}{2}+1)$	$\frac{n}{2}-2$	min(
20,	$S_2 = \{\mathcal{F}_{\frac{n}{4}+1}, \ldots, \mathcal{F}_{\frac{3n}{4}-1}\}$	$(\frac{n}{2}, \frac{n}{4}+2)$		$\lfloor \frac{4(n-1)}{7} \rfloor - 1,$
28	$S_3 = \{\mathcal{F}_1, \mathcal{F}_{\frac{n}{4}+1}, \ldots, \mathcal{F}_{\frac{n}{2}-1}, \mathcal{F}_{\frac{3n}{4}}, \ldots, \mathcal{F}_{n-2}\}$	$(\frac{n}{2}, \frac{3n}{4}+1)$		$2\lfloor \frac{2n}{7} \rfloor - 1)$

Table 4. Constructions for $n \equiv 2 \bmod 4$.

n	Description	Modification	k	max k
10	$\mathcal{S}_1 = \{\mathcal{F}_3, \mathcal{F}_7, \mathcal{F}_8, \mathcal{F}_9\}$	(6,9)	3	3
	$\mathcal{S}_2 = \{\mathcal{F}_1, \mathcal{F}_2, \mathcal{F}_4, \mathcal{F}_7, \mathcal{F}_8\}$	–		
	$\mathcal{S}_3 = \{\mathcal{F}_5, \mathcal{F}_6, \mathcal{F}_7, \mathcal{F}_8\}$	(5,6)		
14	$\mathcal{S}_1 = \{\mathcal{F}_1, \mathcal{F}_3, \mathcal{F}_4, \mathcal{F}_{11}, \mathcal{F}_{12}, \mathcal{F}_{13}\}$	(1,6)	4	6
	$\mathcal{S}_2 = \{\mathcal{F}_1, \mathcal{F}_2, \mathcal{F}_6, \mathcal{F}_9, \mathcal{F}_{10}, \mathcal{F}_{11}, \mathcal{F}_{12}\}$	–		
	$\mathcal{S}_3 = \{\mathcal{F}_1, \mathcal{F}_5, \mathcal{F}_7, \mathcal{F}_8, \mathcal{F}_{11}, \mathcal{F}_{12}\}$	(1,9)		
18	$\mathcal{S}_1 = \{\mathcal{F}_3, \mathcal{F}_9, \mathcal{F}_{10}, \mathcal{F}_{11}, \mathcal{F}_{12}, \mathcal{F}_{13}, \mathcal{F}_{14}, \mathcal{F}_{17}\}$	(5,10)	7	8
	$\mathcal{S}_2 = \{\mathcal{F}_1, \mathcal{F}_2, \mathcal{F}_4, \mathcal{F}_{11}, \mathcal{F}_{12}, \mathcal{F}_{13}, \mathcal{F}_{14}, \mathcal{F}_{15}, \mathcal{F}_{16}\}$	–		
	$\mathcal{S}_3 = \{\mathcal{F}_5, \mathcal{F}_6, \mathcal{F}_7, \mathcal{F}_8, \mathcal{F}_{11}, \mathcal{F}_{12}, \mathcal{F}_{13}, \mathcal{F}_{14}\}$	(9,11)		
22	$\mathcal{S}_1 = \{\mathcal{F}_1, \mathcal{F}_3, \mathcal{F}_4, \mathcal{F}_5, \mathcal{F}_6, \mathcal{F}_{13}, \mathcal{F}_{14}, \mathcal{F}_{15}, \mathcal{F}_{16}, \mathcal{F}_{17}, \mathcal{F}_{21}\}$	(1,5)	9	9
	$\mathcal{S}_2 = \{\mathcal{F}_1, \mathcal{F}_2, \mathcal{F}_8, \mathcal{F}_{13}, \mathcal{F}_{14}, \mathcal{F}_{15}, \mathcal{F}_{16}, \mathcal{F}_{17}, \mathcal{F}_{18}, \mathcal{F}_{19}, \mathcal{F}_{20}\}$	–		
	$\mathcal{S}_3 = \{\mathcal{F}_7, \mathcal{F}_9, \mathcal{F}_{10}, \mathcal{F}_{11}, \mathcal{F}_{12}, \mathcal{F}_{13}, \mathcal{F}_{14}, \mathcal{F}_{15}, \mathcal{F}_{16}, \mathcal{F}_{17}\}$	(1,16)		

7   Maximally Nonlinear Functions

The constructions described so far hold when the number of input bits n is even. In case n is odd, there do not exist any perfect nonlinear S-boxes. The best nonlinearity achieved by an (n, m) quadratic S-box with $m > 1$ is $2^{n-1} - 2^{(n-1)/2}$ and S-boxes achieving this value of nonlinearity are called maximally nonlinear. In this section, we describe a simple modification of the previously described constructions which provide maximally nonlinear S-boxes.

Theorem 3. *Let f be a $(2r, m, k)$ perfect nonlinear quadratic S-box where the symplectic matrices associated with the component functions are $B_1, \ldots, B_m$. For $1 \leq i \leq m$, let B_i' be obtained from B_i by deleting the first row and column. Then the S-box $f' : F_2^{2r-1} \to F_2^m$ defined by $f'(x) = (f_{B_1'}(x), \ldots, f_{B_m'}(x))$ is a $(2r - 1, m, k - 1)$ maximally nonlinear quadratic S-box.*

8   Improving Algebraic Degree

The constructions described in the previous sections provide quadratic functions. In this section, we describe a method of improving the degree of the constructed functions with a small trade-off in the nonlinearity and the SAC property. We first need to relax the notion of SAC. (See [6] for the notion of almost PC(l) of order k functions.)

Definition 1. *An n-variable Boolean function f is said to be (ϵ, k)-SAC if the following property holds: Let g be an $(n - i)$-variable Boolean function obtained from f by fixing $i \leq k$ input variables to constants. Then $\left| \frac{wt(g(x) \oplus g(x \oplus \alpha))}{2^{n-i}} - \frac{1}{2} \right| \leq \epsilon$ for any α of weight 1. An (n, m) S-box is said to be (n, m, ϵ, k)-SAC if every nonzero linear combination of the component functions is an (ϵ, k)-SAC function.*

The next result shows how to convert an (n, m, k) S-box into an (n, m, ϵ, k) S-box for a small ϵ and with a small change in nonlinearity.

Theorem 4. *Let $f = (f_1, \ldots, f_m)$ be an (n, m, k) S-box where the degree of any f_i is less than $(n - 1)$. Then it is possible to construct an (n, m, ϵ, k) S-box g with algebraic degree $n - 1$, $\epsilon = \frac{m+1}{2^{n-k-1}}$ and $\mathsf{nl}(g) \geq \mathsf{nl}(f) - (m + 1)$ if m is odd; $\mathsf{nl}(g) \geq \mathsf{nl}(f) - m$ if m is even.*

Table 5 provides some examples to illustrate Theorem 4. The interpretation

Table 5. Values of k, ϵ and nonlinearity for 2 and 3 output S-boxes for different values of n (see Theorem 4).

n	degree	$m = 2$	$m = 3$
8	7	(3, 0.1875, 118)	(1, 0.0625, 116)
9	8	(3, 0.0938, 238)	(2, 0.0625, 236)
10	9	(4, 0.0938, 494)	(3, 0.0625, 492)
11	10	(5, 0.0938, 990)	(3, 0.0313, 988)
12	11	(6, 0.0938, 2014)	(4, 0.0313, 2012)

of Table 5 is as follows. Each entry is of the form (k, ϵ, x), where k is the order of SAC, ϵ is defined in Theorem 4 and x is the nonlinearity of the modified function. (When m is even, the value of nonlinearity is one more than the lower bound given in Theorem 4.) Note that in each case the algebraic degree is $n - 1$. The drop in nonlinearity is very small; for example for $n = 8$, the lower bound from Theorem 4 is 117 while the maximum possible nonlinearity is 120. Similarly, in each of the above cases, the value of ϵ is small. Hence the deviation from perfect nonlinearity and the (perfect) SAC property is small. On the other hand, the degree increases to the maximum possible. Thus such S-boxes are amply suited for use in the design of practical block cipher algorithms.

9 Conclusion

In this paper, we have considered the problem of constructing perfect nonlinear S-boxes satisfying higher order SAC. Previous work in this area [8] also provided constructions of S-boxes satisfying higher order SAC. However, the nonlinearity obtained was lower. To the best of our knowledge, we provide the first examples of S-boxes satisfying higher order SAC *and* perfect nonlinearity. Some of the constructed S-boxes also achieve optimal trade-off between the numbers of input and output variables, nonlinearity and the order of SAC. Our construction uses bilinear forms and symplectic matrices and yields quadratic functions. We show that the degree can be significantly improved by a small sacrifice in nonlinearity and the SAC property. This yields S-boxes which have possible applications in the design of block ciphers. Lastly, we would like to remark that more research is necessary to generalize our construction using symplectic matrices to more than 3 outputs and also to obtain direct constructions of higher degree S-boxes which satisfy higher order SAC and perfect nonlinearity.

Acknowledgements: We wish to thank the reviewers for detailed comments which helped to improve the quality of the paper.

References

1. J. A. Bondy and U. S. R. Murthy. *Graph Theory with Applications*. London, Macmillan Press, 1977.
2. A. Canteaut and M. Videau. Degree of composition of highly nonlinear functions and applications to higher order differential cryptanalysis. *Advances in Cryptology – Eurocrypt 2002*, LNCS 2332, pages 518-533.
3. C. Carlet. On cryptographic propagation criteria for Boolean functions. *Information and Computation*, 151:32–56, 1999.
4. A. Canteaut, C. Carlet, P. Charpin and C. Fontaine. Propagation Characteristics and Correlation-Immunity of Highly Nonlinear Boolean Functions. In *Advances in Cryptology – Eurocrypt 2000*, pages 507–522, Lecture Notes in Computer Science, Springer-Verlag, 2000.
5. K. C. Gupta and P. Sarkar. Construction of Perfect Nonlinear and Maximally Nonlinear Multi-Output Boolean Functions Satisfying Higher Order Strict Avalanche Criteria . Cryptology e-print archive, http://eprint.iacr.org/2003/198.
6. K. Kurosawa. Almost security of cryptographic Boolean functions. Cryptology e-print archive, http://eprint.iacr.org/2003/075.
7. K. Kurosawa and T. Satoh. Design of $SAC/PC(l)$ of order k Boolean functions and three other cryptographic criteria. In Advances in Cryptology - Eurocrypt'97, number 1233 in Lecture Notes in Computer Science Series, Springer-Verlag, 434-449, 1997.
8. K. Kurosawa and T. Satoh. Generalization of Higher Order SAC to Vector Output Boolean Functions. In *Advances in Cryptology - Asiacrypt 1996*, Lecture Notes in Computer Science, Springer-Verlag, 1996.
9. F. J. MacWillams and N. J. A. Sloane. *The Theory of Error Correcting Codes*. North Holland, 1977.
10. K. Nyberg. Perfect Nonlinear S-boxes. In *Advances in Cryptology - EUROCRYPT 1991*, pages 378–386, Lecture Notes in Computer Science, Springer-Verlag, 1991.
11. B. Preneel, W. Van Leekwijck, L. Van Linden, R. Govaerts and J. Vandewalle. Propagation Characteristics of Boolean Functions. In *Advances in Cryptology - EUROCRYPT 1990*, pages 161–173, Lecture Notes in Computer Science, Springer-Verlag, 1991.
12. O. S. Rothaus. On bent functions. *Journal of Combinatorial Theory, Series A*, 20:300–305, 1976.
13. P. Sarkar and S. Maitra. Construction of Nonlinear Boolean Functions with Important Cryptographic Properties. *In Advances in Cryptology - EUROCRYPT 2000*, pages 485–506, Lecture Notes in Computer Science, Springer-Verlag, 2000.
14. A. F. Webster and S. E. Tavares. On the Design of S-boxes. In *Advances in Cryptology - Crypto 1985*, pages 523–534, Lecture Notes in Computer Science, Springer-Verlag, 1986.

Improved Cost Function in the Design of Boolean Functions Satisfying Multiple Criteria

Selçuk Kavut and Melek D. Yücel

Department of Electrical and Electronics Engineering
Middle East Technical University-ODTÜ, 06531 Ankara, Türkiye
{kavut, melekdy}@metu.edu.tr

Abstract. We develop an improved cost function to be used in simulated annealing followed by hill-climbing to find Boolean functions satisfying multiple desirable criteria such as high nonlinearity, low autocorrelation, balancedness, and high algebraic degree. Using this cost function that does not necessitate experimental search for parameter tuning, the annealing-based algorithm reaches the desired function profiles more rapidly. Some Boolean functions of eight and nine variables have been found, which are unattained in the computer search based literature, in terms of joint optimization of nonlinearity and autocorrelation. Global characteristics of eight-variable Boolean functions generated by algebraic construction or computer search are compared with respect to the sum-of-squared-errors in their squared spectra, which is also proportional to the sum-of-squared-errors in their autocorrelation function, the term 'error' denoting the deviation from bent function characteristics. Preliminary results consisting of cryptographically strong Boolean functions of nine, ten and eleven variables obtained using a three-stage optimization technique are also presented.

Keywords: Simulated annealing, bent Boolean functions, nonlinearity, Walsh-Hadamard transforms, autocorrelation.

1 Introduction

In cryptographic applications, Boolean functions are required to satisfy various criteria, mainly high nonlinearity and low autocorrelation, to resist linear cryptanalysis and differential cryptanalysis particularly. Constructing Boolean functions with desirable cryptographic properties has received a lot of attention in the literature [1,5,8-10,18]. Some search techniques such as random search, hill-climbing, genetic algorithms, and hybrid approach have been investigated [11-13]; however, these techniques seem to be insufficient in designing *"near-the best"* Boolean functions. Recently, Clark et al have proposed [2-4] the use of simulated annealing, a heuristic optimization technique based on annealing process for metals, for finding Boolean functions with good cryptographic properties. The results obtained are promising; therefore, in this study, we investigate this optimization method in detail.

We develop an improved cost function, which does not necessitate experimental search for parameter tuning, to be used by the simulated annealing process. The optimization algorithm works fast, since Boolean functions with desired characteristics

T. Johansson and S. Maitra (Eds.): INDOCRYPT 2003, LNCS 2904, pp. 121-134, 2003.

are encountered more frequently and one does not have to tune parameters for each different number of variables as in [2-4]. Defining the '*profile*' as the (input length '*n*', nonlinearity '*nl*', autocorrelation coefficient with maximum magnitude '*ac*', degree '*d*'), we have encountered a balanced Boolean function profile (n, nl, ac, d) = (8, **114, 16**, 7), in which the bold entries are particularly significant. Such low autocorrelation and high nonlinearity have not appeared together for the same Boolean function [2-4, 11-13] in the related literature. For an input length of 9, we have obtained a balanced function with profile (9, 234, **32**, 8), where the maximum autocorrelation magnitude, 32, is equal to that of the construction in [9] and lowest possible value for *n*=9 according to some conjectured bounds for autocorrelation [19]. Table 1 given below compares Clark et al's best achieved results with ours. After the preliminaries given in the next section, we compare in Table 2, the results of computer search based approaches ([2-4] and ours) with some algebraic constructions [1, 10, 16, 18] of balanced Boolean functions, for *n*=8.

Table. 1 Comparison of the best achieved computer search results for (*nl, ac, d*)

Results	(*nl, ac, d*) for			
	n=8	*n*=9	*n*=10	*n*=11
Clark et.al.[2-4]	(**116**, 24, 7) (112, **16**, 5)	(**238**, 40, 8)	(**486**, 72, 9) (484, 56, 9)	(**984**, 96, 9) (982, **88**, 10)
Ours	(**116**, 24, 7) (114, **16**, 7)	(**238**, 40, 8) (234, **32**, 8) (236, **32**, 8)*	(**486**, **56**, 9)*	(**984**, **80**, 10)*

* These Boolean functions are obtained using the three-stage method described as a future work in Section 5.

2   Preliminaries

Let $f: F_2^n \rightarrow F_2$ be a Boolean function that maps each possible combination of *n*-bit variables to a single bit.

Balancedness. If the number of 0's are equal to the number of 1's in the truth table, then the Boolean function $f(\mathbf{x})$ is said to be balanced.

Affine and Linear Functions. A Boolean function $f(\mathbf{x})$ is called an affine function of $\mathbf{x} = (x_1, x_2, ..., x_n) \in F_2^n$, if it is in the form

$$f(\mathbf{x}) = a_1 \otimes x_1 \oplus a_2 \otimes x_2 \oplus ... \oplus a_n \otimes x_n \oplus c = \mathbf{w} \cdot \mathbf{x} \oplus c, \qquad (1)$$

where $a_1, ..., a_n, c \in F_2$, $\mathbf{w} = (a_1, ..., a_n) \in F_2^n$, and $\oplus, \otimes, \cdot$ respectively denote addition, multiplication and inner product operations in F_2. $f(\mathbf{x})$ is called linear if c=0.

Walsh-Hadamard Transform. For a function f, the Walsh-Hadamard transform (or spectrum) is defined as

$$F(\mathbf{w}) = \sum_{\mathbf{x} \in F_2^n} (-1)^{f(\mathbf{x})} (-1)^{\mathbf{w} \cdot \mathbf{x}}. \tag{2}$$

We denote the maximum absolute value by $WH_f = \max_{\mathbf{w} \in F_2^n} | F(\mathbf{w}) |$, which is closely related to the nonlinearity of $f(\mathbf{x})$.

Nonlinearity Measure. The nonlinearity of a Boolean function is defined as the minimum distance to the set of affine functions, and can be expressed as

$$nl_f = (2^n - WH_f) / 2. \tag{3}$$

Parseval's Theorem. It states that the sum of squared $F(\mathbf{w})$ values, over all $\mathbf{w} \in F_2^n$, is constant and equal to 2^{2n}, which has motivated the derivation of the cost functions in [2-4]:

$$\sum_{\mathbf{w} \in F_2^n} (F(\mathbf{w}))^2 = 2^{2n}. \tag{4}$$

For bent Boolean functions [15], the squared spectrum is flat, so $(F(\mathbf{w}))^2 = 2^n$ for all values of $\mathbf{w}$.

Autocorrelation Function. The autocorrelation function of a Boolean function is given by

$$r_f(\mathbf{d}) = \sum_{\mathbf{x} \in F_2^n} (-1)^{f(\mathbf{x})} (-1)^{f(\mathbf{x} \oplus \mathbf{d})}. \tag{5}$$

The maximum absolute value that we denote by $ac_f = \max_{\mathbf{d} \neq \mathbf{0} \in F_2^n} | r_f(\mathbf{d}) |$ is also known as the absolute indicator [18].

Another measure related to the autocorrelation function is commonly called the sum-of-squares indicator [18], given by the sum $\sum_{\mathbf{d} \in F_2^n} (r_f(\mathbf{d}))^2$. We prefer to use the sum-of-squared-errors (SSE_f), $\sum_{\mathbf{d} \neq \mathbf{0} \in F_2^n} (r_f(\mathbf{d}))^2$, instead of the sum-of-squares indicator, since SSE_f is proportional to the sum of squared spectrum deviations [17] from that of the bent functions, that is

$$\sum_{\mathbf{d} \neq \mathbf{0} \in F_2^n} (r_f(\mathbf{d}))^2 = 2^{-n} \sum_{\mathbf{w} \in F_2^n} \left[(F(\mathbf{w}))^2 - 2^n \right]^2. \tag{6}$$

If f is affine, the sum (6) of squared autocorrelation errors, i.e., the autocorrelation deviations from the autocorrelation of bent functions, is maximum and equal to $2^{3n} - 2^{2n}$. Hence, dividing (6) by $2^{3n} - 2^{2n}$, one obtains the useful measure of mean squared error (MSE_f), which takes rational values in the interval [0,1]. The mean squared error percentage $100MSE_f$ of the Boolean function f shows the percentage of total squared deviations of its autocorrelation function $r_f(\mathbf{d})$ and squared spectrum $F^2(\mathbf{w})$ respectively, from the autocorrelation and the squared spectrum of bent functions [17].

Algebraic Degree. The algebraic degree d or simply degree of f is defined as the degree of its algebraic normal form.

Table 2. Comparison for 8-variable functions obtained either by algebraic construction or computer search

Function f	nl_f	$\lvert r_f(\mathbf{d})\rvert_{\substack{\max \\ \mathbf{d}\neq 0}}$ (ac_f)	$\sum\limits_{all\ \mathbf{d}\neq 0} r_f^2(\mathbf{d})$ (SSE_f)	$\dfrac{\sum_{all\ \mathbf{d}\neq 0} r_f^2(\mathbf{d})}{167116.8}$ $(100MSE_f)$
Affine	0	256	16711680	100 %
Stanica, Sung [16]	112	256	196608	1.176471 %
Cauteaut et al [1], example 2, f_1	112	256	172032	1.029412 %
Cauteaut et al [1], example 2, f_2	112	256	196608	1.176471 %
Maitra [10]	116	128	55296	0.330882 %
Zhang, Zheng [18], Theorem 16	≥ 112	≤ 32	24576	0.147059 %
Ours	114	16	23424	0.140165 %
Ours and Clark's [2-4]	116	24	21120	0.126378 %
Bent	120	0	0	0 %

In Table 2, we compare the computer search based approaches in [2-4] and our results, with balanced Boolean function constructions [1, 10, 16, 18], for $n=8$. (We also include affine and bent –so unbalanced– functions, as reference.) The constructions in [1, 10, 16, 18] satisfy the SAC or can be made to satisfy the SAC by a change of basis; which is possible for most of our results as well. The functions are ranked in descending order of the sum-of-squared-errors and $100MSE_f$, shown in the last two columns of Table 2, respectively. Notice that the sum-of-squared-errors SSE_f, does not contain the extra and unnecessary constant $(r_f(\mathbf{0}))^2 = 2^{2n}$ (which equals 65536 for $n=8$), therefore the comparison of different functions can be done on a much fair basis.

In terms of the absolute indicator ac_f and sum-of-squared-errors SSE_f given by (6); although search based approaches seem to yield better results than theoretical constructions for $n=8$, yet it is not clear that they can be that much successful for higher values of n. For instance, for $n=9$, the construction in Theorem 17 of [18] yields (nl, ac) values equal to (240, 32), (which have not yet been encountered by any search based algorithm) and quite a small percentage of mean squared error ($100MSE_f = 0.195695\%$). Same (nl, ac) values of (240, 32) are also obtained by a different construction (Construction 0 and Theorem 9 of [9]). For odd values of $n \geq 15$, the construction in [10] is very promising and yields an extremely small ac value, equal to 0.635% of the maximum possible autocorrelation magnitude (2^n); in addition to a very small percentage of mean squared error, which is equal to $18.92/(2^n-1)\%$ (so, for $n=15$, $100MSE_f = 0.000577\%$).

3   Two-Stage Optimization

Clark et al employ [2-4] a two-stage optimization method, which basically consists of an annealing-based search followed by hill-climbing. As the first stage of the optimization, simulated annealing process [7] starts at some initial state ($S = S_0$) and initial temperature ($T = T_0$), and carries out a certain number of moves (**Moves in Inner Loop**: *MIL*) at each temperature (see Fig.1 taken from [3] adding the common stopping criterion). As the initial state, one selects a balanced function randomly, then disturbs the function just by complementing two randomly chosen bits to maintain the balancedness, and compares the cost for the disturbed function to that of the original.

```
T = T₀
IC = 0
MFC = 0
Generate f (x) randomly: S₀
while ( MFC<N & IC<M )
{
        nomove = 0
        for (i = 0; i < MIL; i++)
        {
                Generate g(x) randomly from the neighborhood of f (x): S
                Calculate the value of pre-defined cost function: δ
                if (δ<0)
                        f (x) = g(x): S₀ = S
                else
                {
                        Generate a value between (0,1) randomly: U
                        if ( U<exp(−δ /T ) )
                                f (x) = g(x): S₀ = S
                        else
                                nomove = nomove + 1

                }
        }
        T = T x α
        if (nomove = MIL)
                MFC = MFC + 1
        else
                MFC = 0
}
```

Fig. 1. Basic Simulated Annealing Process for Minimization Problems with Common Stopping Criteria

The algorithm always accepts cost-improving moves; however, it may also accept cost-worsening moves probabilistically, depending upon the temperature T of the search. The probability of accepting worsening moves is high at the beginning and the algorithm accepts any move virtually. The probability decreases gradually due to the *cooling parameter* α (generally ranging from 0.90 to 0.99), and eventually only improving moves are accepted.

We use the common stopping criteria; which stops the simulated annealing process when the number of **M**aximum **F**ailed **C**ycles (*MFC*) of consecutive temperature cycles reaches N, or upon completion of M **I**teration **C**ounts (*IC*). In our study, we use *MIL* = 400, M = 300, and N = 50, which are the same as in [2-4].

After the annealing-based search, hill-climbing with respect to nonlinearity or autocorrelation is used as the second stage of optimization.

4   Improved Cost Function

Since bent Boolean functions [15] are the best in terms of high nonlinearity and low autocorrelation, we prefer to use a cost function of the form

$$\text{Cost}_1 = \sum_{\mathbf{w}} \left| |F(\mathbf{w})| - 2^{n/2} \right|^3 + \sum_{\mathbf{d} \neq 0} |r_f(\mathbf{d})|^3 , \tag{7}$$

where the first term in the summation forces the Walsh-Hadamard transform, and the second term forces the autocorrelation of the searched function to those of the bent functions. Cost_1 is equal to zero for a bent function, and greater than zero for a balanced function; because, bent functions are not balanced.

The third power enhances the effect of any large deviation of a term from the optimum, more than that of a small deviation. Therefore, cubing gives higher priority and higher chance of being minimized, to the term which increases more rapidly than the others, whenever there is an unbalance among the terms of the summation.

The cost function in [2] minimizes the deviation of the absolute value of $F(\mathbf{w})$ from $(2^{n/2}+K)$, and it is defined as

$$\text{Cost}_2 = \sum_{\mathbf{w}} \left| |F(\mathbf{w})| - (2^{n/2} + K) \right|^3 . \tag{8}$$

Cost_2 is the same as the first term of Cost_1 given by Eq.(7), when K=0. Although Clark et al [2-4] have been motivated by Parseval's theorem, and started with the minimization of the deviation from $2^{n/2}$ of the absolute value of $F(\mathbf{w})$, which corresponds to Cost_2 for K=0, they have found better profile distributions for nonzero values of K. In [3, 4], a parameter X ranging from −16 to 30 is used instead of the term $(2^{n/2}+K)$ of Eq.(8).

The improved cost function Cost_1 does not contain any experimental parameter. To make fair comparisons between the two cost functions, Cost_2 is used with values of K as experimentally tuned by Clark et al (more specifically, K= −6 for *n*=8, K= −12 for *n*=9 and K= −16 for *n*=10).

5   Experimental Results

In this section, we detail the results of applying the technique with $Cost_1$ to the design of balanced functions. It is of particular interest to see, how two separate terms of $Cost_1$ proceed during the course of the simulated annealing algorithm; which is shown in Fig.2 for a typical run. A close view of Fig.2 reveals that the contribution of the first term to the overall cost is much less than that of the second term.

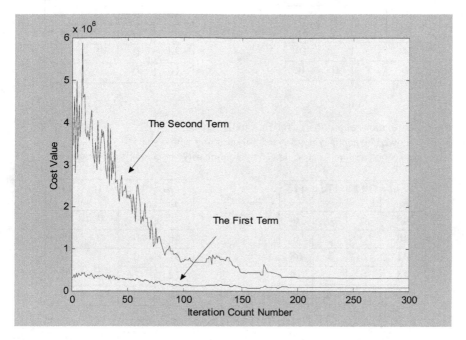

Fig.2. Change of the first term and second term of $Cost_1$, while minimizing by simulated annealing for $n=8$ and $\alpha=0.95$

To compare the final effects of these terms, profile distributions within 100 runs are found. Simulated annealing algorithm with a given cost function, followed by nonlinearity or autocorrelation targeted hill climbing is run 100 times to find a Boolean function for each run. Distribution of profiles in 100 runs are then tabulated. Table 3(b) shows the profile distributions obtained by choosing only the first term of $Cost_1$ as the cost function of the simulated annealing algorithm. Results of 100 independent runs using only the second term of $Cost_1$ as the cost function are tabulated in Table 4(b). In both cases, part (a) of the table is the profile distribution for the total cost, $Cost_1$, over a different set of 100 runs. In all tables, 'nl' stands for nl_f and 'ac' stands for ac_f.

Comparing 3(b) and 4(b) one can say that the second term of the cost function is more indispensable than the first one; because the distribution in Table 3(b) corresponding to the first term, is worse than that of 4(b). However; none of the distributions of part (b) is as good as the distributions in part (a), so the sum $Cost_1$ serves as a better cost function than any of the two terms in the sum.

Table 3. Distributions of profiles in 100 runs for $n=8$ using
(a) $Cost_1$ followed by nonlinearity targeted hill-climbing with $\alpha=0.95$
(b) Only the first term of $Cost_1$ followed by nonlinearity targeted hill-climbing with $\alpha=0.95$

nl / ac	112	114	116
56	0	0	0
48	0	0	0
40	0	2	6
32	0	3	68
24	0	1	19
16	1	0	0

(a)

nl / ac	112	114	116
56	2	0	0
48	1	11	0
40	9	35	3
32	14	17	8
24	0	0	0
16	0	0	0

(b)

Table 4. Distributions of profiles in 100 runs for $n=8$ using
(a) $Cost_1$ followed by nonlinearity targeted hill-climbing with $\alpha=0.95$
(b) Only the second term of $Cost_1$ followed by nonlinearity targeted hill-climbing with $\alpha=0.95$

nl / ac	112	114	116
48	0	0	0
40	0	2	6
32	0	3	68
24	0	1	19
16	1	0	0

(a)

nl / ac	112	114	116
48	0	0	0
40	0	0	2
32	0	5	46
24	2	2	16
16	27	0	0

(b)

Theoretically, the best achievable values for $n=8$ balanced functions are the nonlinearity of $nl=118$, and the absolute indicator of $ac=16$ [19]. Profiles of (118,16) have still not been encountered in our experiments and also in the related literature of algebraic construction and computer search. The values $(nl,ac)=(116,16)$ have also not appeared together, nowadays. On the other hand, the profiles (8, 114, **16**, 7) that appear in our experiments are new; and can be found with $Cost_1$ repeatably, in sufficiently large number of trials. Among the nine different (8, 114, **16**, 7) functions obtained, eight have a sum-of-squared-error, $SSE=23424$ and one has $SSE=23808$. The support of one (8, 114, **16**, 7) function having $SSE=23424$ is

f = 149016CDD1931F10860B4B8BECEF5557B8177A8565229B775E08F97B7692C32D.

For $n=9$, we have obtained balanced functions with profile (9, 234, **32**, 8); which is new for computer based search literature; however, algebraic constructions in [9, 18] yield better nonlinearity $nl=240$ and absolute indicator $ac=32$. To give some statistical insight, we summarize our experimental results. Table 5 shows distributions obtained after 100 runs using $Cost_1$, yielding our best achieved Boolean functions in terms of autocorrelation. We have also obtained a balanced function of profile (9, 236, **32**, 8) using the three-stage optimization method discussed in Section 5 (see Table 10(b)).

Table 5. Distributions of profiles in 100 runs, yielding our best achieved Boolean functions in terms of autocorrelation, using Cost$_1$
(a) followed by autocorrelation targeted hill-climbing with α=0.95, for n=8
(b) followed by autocorrelation targeted hill-climbing with α=0.95, for n=9

nl ac	110	112	114	116
24	3	79	7	2
16	0	8	1	0

(a)

nl ac	232	234	236
40	3	47	48
32	1	1	0

(b)

The following three tables are prepared to compare profile distributions of Cost$_1$, and Cost$_2$, both followed by nonlinearity targeted hill-climbing We compare our results (Cost$_1$), with Clark's results (Cost$_2$) reproduced from [2] in Table 6 for n=8, in Table 7 for n=9, and in Table 8 for n=10. Cost$_1$ generates better distributions than Cost$_2$, as can be seen by comparing parts (a) and (b) of each figure. All results taken from Clark et al, have been obtained by experimentally tuned values of the parameter (K) to attain the best profiles in terms of nonlinearity.

Table 6. Distributions of profiles in 100 runs for n=8 using
(a) Cost$_1$, followed by nonlinearity targeted hill-climbing with α=0.95
(b) Cost$_2$ with K=−6, followed by nonlinearity targeted hill-climbing with α=0.9
(taken from [2])

nl ac	112	114	116
48	0	0	0
40	0	2	6
32	0	3	68
24	0	1	19
16	1	0	0

(a)

nl ac	112	114	116
48	0	0	1
40	0	5	13
32	2	27	42
24	0	0	10
16	0	0	0

(b)

Table 7. Distributions of profiles in 400 runs for n=9 using
(a) Cost$_1$, followed by nonlinearity targeted hill-climbing with α=0.95
(b) Cost$_2$ with K=−12, followed by nonlinearity targeted hill-climbing with α=0.9
(taken from [2])

nl ac	236	238
72	0	0
64	2	1
56	28	12
48	168	77
40	91	21

(a)

nl ac	236	238
72	3	1
64	16	13
56	97	60
48	125	76
40	5	4

(b)

Table 8. Distributions of profiles in 100 runs for $n=10$ using
(a) $Cost_1$, followed by nonlinearity targeted hill-climbing with $\alpha=0.95$
(b) $Cost_2$ with K=−16, followed by nonlinearity targeted hill-climbing with $\alpha=0.9$
(taken from [2])

nl / ac	482	484
104	0	0
96	0	0
88	0	1
80	0	11
72	0	51
64	0	37

(a)

nl / ac	482	484
104	1	1
96	0	8
88	1	23
80	1	41
72	0	24
64	0	0

(b)

6   Conclusions and Future Work

We think that $Cost_1$ is a promising cost function for the annealing-based search of Boolean functions with desired properties. Since as the first step, we are not interested in the resiliency of the functions; there is no term in $Cost_1$, which forces spectral terms at small-weight frequencies to zero. We have obtained unattained Boolean functions with profiles (8, 114, 16, 7) and (9, 234, 32, 8). Our cost function gives rise to better distributions of the resulting Boolean functions, and it does not use any parameter which should be experimentally tuned. In fact, experimentation with this cost function demonstrates that there is a close relationship between the autocorrelation r_f (d) and Walsh-Hadamard transform $F(\mathbf{w})$; and one gets closer to the desirable Boolean function profiles when both of them are optimized.

We have also considered a three-stage optimization method using the below cost function $Cost_3$, which also confirms the close relationship between the autocorrelation r_f (d) and Walsh-Hadamard transform $F(\mathbf{w})$:

$$Cost_3 = \sum_{\mathbf{w}} |F(\mathbf{w})|^R + \sum_{\mathbf{d} \neq 0} |r_f(\mathbf{d})|^S . \tag{9}$$

Our three-stage optimization algorithm is as follows:

1. Apply simulated annealing setting R=S=3,
2. Use hill-climbing technique with respect to $Cost_3$ setting R=S=13,
3. Use hill-climbing technique with respect to nonlinearity or autocorrelation.

Although we have tested this heuristic optimization method only for a small number of runs, we have obtained some unattained Boolean functions like (9, **236**, **32**, 8), (10, **486**, **56**, 9), and (11, **984**, **80**, 10). Besides, we get better distributions in this case (see Table 9-12).

Table 9. Distributions of profiles in 100 runs for $n=8$ using
(a) $Cost_1$, followed by nonlinearity targeted hill-climbing with $\alpha=0.95$
(b) Three-stage optimization method with nonlinearity targeted hill-climbing as the third step with $\alpha=0.95$

nl / *ac*	112	114	116
40	0	2	6
32	0	3	68
24	0	1	19
16	1	0	0

nl / *ac*	112	114	116
40	0	0	0
32	0	0	41
24	0	0	59
16	0	0	0

(a) (b)

Table 10. Distributions of profiles in 100 runs for $n=9$ using
(a) $Cost_1$, followed by autocorrelation targeted hill-climbing with $\alpha=0.95$
(b) Three-stage optimization method with autocorrelation targeted hill-climbing as the third step with $\alpha=0.95$

nl / *ac*	232	234	236	238
40	3	47	48	0
32	1	1	0	0

nl / *ac*	232	234	236	238
40	0	1	90	4
32	0	2	3	0

(a) (b)

Table 11. Distributions of profiles in 100 runs for $n=10$ using
(a) $Cost_1$, followed by nonlinearity targeted hill-climbing with $\alpha=0.95$
(b) Three-stage optimization method with nonlinearity targeted hill-climbing as the third step with $\alpha=0.95$

nl / *ac*	482	484	486
88	0	1	0
80	0	11	0
72	0	51	0
64	0	37	0
56	0	0	0

nl / *ac*	482	484	486
88	0	0	0
80	0	0	0
72	0	0	0
64	0	38	3
56	0	57	2

(a) (b)

Table 12. Distributions of profiles in 10 runs for $n=11$ using
(a) Three-stage optimization method with autocorrelation targeted hill-climbing as the third step with $\alpha=0.95$
(b) Three-stage optimization method with nonlinearity targeted hill-climbing as the third step with $\alpha=0.95$

nl ac	980	982	984
96	0	0	0
88	0	0	0
80	1	8	1

(a)

nl ac	980	982	984
96	0	0	1
88	0	0	8
80	0	0	1

(b)

To examine the effect of hill-climbing with R=S=13 in the second stage of the three-stage optimization method, the cost function used in the first stage is recorded during the second stage (see Fig.3).

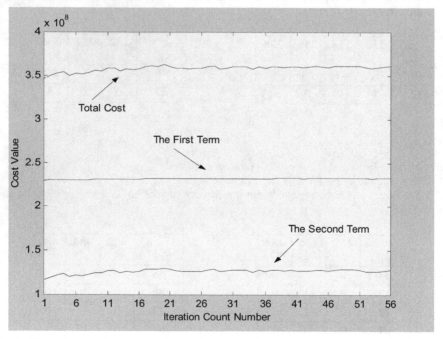

Fig. 3. Change of the first term and second term of $Cost_3$ with R=S=3 during the second stage of the three-stage optimization for $n=11$.

As one can see, the cost value with R=S=3 increases slightly, i.e., it doesn't have to be minimized as in the first stage, because the minimization is with respect to the cost function with R=S=13. A decrease in the value of a cost function doesn't necessarily mean an improvement in the nonlinearity or autocorrelation values, and the second stage allows some increments in the cost function of the first stage. On the other hand, the thirteenth power truly enhances the effects of deviations from optimum, and gives extremely higher priorities and higher chances of being minimized, to the terms which increase more rapidly than the others. To understand and control the true behaviour of this heuristic three-stage mechanism is a subject of future work.

References

[1]   Cauteaut, A., Carlet, C., Charpin, P., Fontaine, C., Propagation characteristics and correlation immunity of highly nonlinear Boolean functions. In Advances in Cryptology – EUROCRYPT'00, LNCS, Volume 1807, pages 507-522. Springer Berlin, 2000.

[2]   Clark, J.A. and Jacob, J.L., Two-Stage Optimisation in the Design of Boolean Functions. In 5th Australasian Conference on Information, Security and Privacy – ACISP 2000, LNCS, Volume 1841, pages 242-254. Springer Verlag, 2000.

[3]   Clark, J.A., Jacob, J.L., Stepney, S., Maitra, S., and Millan, W., Evolving Boolean Functions Satisfying Multiple Criteria. In 3rd International Conference on Cryptology in India – Indocrypt 2002, LNCS, Volume 2551, pages 246-259, Springer Verlag, December 2002.

[4]   Clark, J.A., Metaheuristic Search as a Cryptological Tool. D.Phil. Thesis. YCSY-2002-07. Department of Computer Science, University of York, York UK. December 2001. (http://www.cs.york.ac.uk/ftpdir/reports)

[5]   Dobbertin, H., Construction of bent functions and balanced functions with high nonlinearity. In Fast Software Encryption, 1994 Leuven Workshop, LNCS, Volume 1008, pages 61-74. Springer Verlag, Berlin, 1994.

[6]   Hou, X.D., On the Norm and Covering Radius of First-Order Reed-Muller Codes. IEEE Transactions on Information Theory, 43(3):1025-1027, May 1997.

[7]   Kirkpatrick, S., Jr. Gelatt, C. D., and Vecchi, M. P., Optimization by simulated annealing. Science, 220(4598):671-680, May 1983.

[8]   Maitra, S. and Pasalic, E., Further constructions of resilient Boolean functions with very high nonlinearity. IEEE Transactions on Information Theory, 48(7):1825-1834, July 2002.

[9]   Maitra, S., Highly nonlinear balanced Boolean functions with very good autocorrelation property. In Workshop on Coding and Cryptograhy – WCC 2001, Paris, January 8-12, 2001. Electronic Notes in Discrete Mathematics, Volume 6, Elsevier Science, 2001.

[10]   Maitra, S., Highly nonlinear balanced Boolean functions with good local and global avalanche characteristics. Information Processing Letters 83, pages 281-286, 2002.

[11]   Millan, W., Clark, A., and Dawson, E., An effective genetic algorithm for finding highly nonlinear Boolean functions. In First International Conference on Information and Communications Security, LNCS, Volume 1334, pages 149-158. Springer Verlag, 1997.

[12]   Millan, W., Clark, A., and Dawson, E., Heuristic Design of Cryptographically Strong Balanced Boolean Functions. In Advances in Cryptology EUROCRYPT'98, LNCS, Volume 1403, pages 489-499. Springer Verlag. 1998.

[13]   Millan, W., Clark, A., and Dawson, E., Boolean function design using hill climbing methods. In 4th Australasian Conference on Information, Security and Privacy, LNCS, Volume 1587, pages 1-11. Springer Verlag, April 1999.

[14]   Patterson, N.J. and Wiedemann, D.H., The covering radius of the $(2^{15}, 16)$ Reed-Muller code is at least 16276. IEEE Transactions on Information Theory, IT-29(3):354-356, 1983 (see correction IT-36(2):443, 1990).

[15]   Rothaus, O. S., On Bent Functions. Journal of Combinatorial Theory, pp.300-305, 1976.

[16] Stanica, P., Sung, S.H., Improving the nonlinearity of certain balanced Boolean functions with good local and global avalanche characteristics. Information Processing Letters 79 (4), pages 167-172, 2001.

[17] Yücel, M. D., Alternative Nonlinearity Criteria for Boolean Functions. Electrical and Electronics Engineering Departmental Memorandum, No.2001-1, Middle East Technical University (ODTÜ), Ankara, Türkiye, 20 pages, January 2001.

[18] Zhang, X. M. and Zheng, Y., GAC–the criterion for global avalanche characteristics of cryptographic functions, Journal for Universal Computer Science, 1(5), pages 316-333, 1995.

[19] Zheng, Y. and Zhang, X.M., Improved upper bound on the nonlinearity of high order correlation immune functions. In Selected Areas in Cryptography – SAC 2000, LNCS, Volume 2012, pages 264-274. Springer Verlag, 2000.

On Multiplicative Linear Secret Sharing Schemes

Ventzislav Nikov[1*], Svetla Nikova[2**], and Bart Preneel[2]

[1] Department of Mathematics and Computing Science,
Eindhoven University of Technology
P.O. Box 513, 5600 MB, Eindhoven, the Netherlands
v.nikov@tue.nl
[2] Department Electrical Engineering, ESAT/COSIC,
Katholieke Universiteit Leuven, Kasteelpark Arenberg 10,
B-3001 Heverlee-Leuven, Belgium
svetla.nikova,bart.preneel@esat.kuleuven.ac.be

Abstract. We consider both information-theoretic and cryptographic settings for Multi-Party Computation (MPC), based on the underlying linear secret sharing scheme. Our goal is to study the Monotone Span Program (MSP), that is the result of local multiplication of shares distributed by two given MSPs as well as the access structure that this *resulting* MSP computes. First, we expand the construction proposed by Cramer *et al.* for multiplying two different general access structures and we prove some properties of the resulting MSP. We prove that using two (different) MSPs to compute their resulting MSP is more efficient than building a multiplicative MSP. Next we define a (strongly) multiplicative resulting MSP and we prove that when one uses dual MSPs only all players together can compute the product. An analog of the algebraic simplification protocol of Gennaro *et al.* is presented. We show which conditions the resulting access structure should fulfill in order to achieve MPC secure against an adaptive, active adversary in the zero-error case in both the computational and the information-theoretic model.

1 Introduction

Background. The concept of *secret sharing* was introduced by Shamir [14] as a tool to protect a secret simultaneously from exposure and from being lost. It allows a so called *dealer* to share the secret among a set of entities, usually called *players*, in such a way that only certain specified subsets of the players are able to reconstruct the secret while smaller subsets have no information about it. Denote by P the set of participants in the scheme. The groups who are allowed to reconstruct the secret are called *qualified* (denoted by Γ), and

* The research has been supported by a Marie Curie Fellowship of the European Community Programme under contract number HPMT-CT-2000-00093.
** The author was partially supported by the IWT STWW project on Anonymity and Privacy in Electronic Services and the Concerted Research Action GOA-MEFISTO-666 of the Flemish Government; part of the work was done during the author's visit at the Ruhr University, Bochum.

T. Johansson and S. Maitra (Eds.): INDOCRYPT 2003, LNCS 2904, pp. 135–147, 2003.

the groups who should not be able to obtain any information about the secret are called *forbidden* (or curious) (denoted by Δ). Γ is *monotone increasing* and can be described by the set Γ^- consisting of its *minimal elements* (sets). Δ is *monotone decreasing* and similarly, the set Δ^+ consists of the *maximal elements* (sets) in Δ. The tuple (Γ, Δ) is called an *access structure* if $\Gamma \cap \Delta = \emptyset$. If $\Gamma = \Delta^c$ is the complement of Δ, then we say that (Γ, Δ) is *complete* and we denote it only by Γ. The *dual* $\Gamma^\perp$ of a monotone access structure Γ, defined on P, is the collection of sets $A \subseteq P$ such that $A^c \notin \Gamma$. An access structure Γ is *connected* if each player belongs to at least one minimal set.

It is common to model cheating by considering an *adversary* who may corrupt some subset of the players. One can distinguish between *passive* and *active* corruption, see Fehr *et al.* [6] for recent results. Passive corruption means that the adversary obtains the complete information held by the corrupt players, but the players execute the protocol correctly. Active corruption means that the adversary takes full control of the corrupt players. Thus in a so called *mixed adversary model* an adversary is characterized by a *privacy structure* Δ (the curious players) and an *adversary structure* $\Delta_A \subseteq \Delta$ (the corrupt players). Denote the complement $\Gamma_A = \Delta_A^c$ and call its dual access structure $\Gamma_A^\perp$ the *honest* (or *good*) players structure. Both passive and active adversaries may be *static*, meaning that the set of corrupt players is chosen once and for all before the protocol starts, or *adaptive* meaning that the adversary can at any time during the protocol choose to corrupt a new player based on all the information he has at the time, as long as the total set is in Δ_A.

A wide range of general approaches for designing Secret Sharing Schemes (SSS) is known, but most of these techniques result in *linear* SSS (LSSS). Since late 80's many efforts has been put into finding better presentations (algebraic, geometric, combinatorial) which allow to compute any monotone access structure. In this paper we will use an algebraic computational device introduced by Karchmer and Wigderson [10] called *Monotone Span Program*. It is well known that there is one-to-one correspondence between LSSS and MSPs and that MSPs can compute any complete monotone access structure.

Since an LSSS neither guarantees reconstructability when some shares are incorrect, nor verifiability of a shared value a stronger primitive called *verifiable secret sharing* (VSS) has been introduced in [1,5]. In a VSS a dealer distributes a secret value among the players, where the dealer and/or some of the players may be cheating. It is guaranteed that if the dealer is honest, then the cheaters obtain no information about the secret, and all honest players will later be able to reconstruct it, without the help of the dealer. Even if the dealer cheats, a unique value will be determined and is reconstructible without the help of the cheaters.

Secure *multi-party computation* (MPC) can be defined as follows: n players compute an agreed function of their inputs in a "secure" way, where "secure" means guaranteeing the correctness of the output as well as the privacy of the players' inputs, even when some players cheat. VSS is a key tool for secure MPC.

The Model. We will consider the standard *secure-channels model*, where the players are connected by bilateral, synchronous, reliable secure channels. We assume also the availability of a broadcast channel. By default, we consider *unconditional* security against an adaptive, active adversary (mixed adversary model) and error-free protocols.

Organization. In the first part of the next section we give some notations and linear algebra techniques, then we describe our results. In Section 3 we propose the main construction diamond $\diamond$ and investigate its properties. Then in Section 4 conditions for the existence of MPC based on LSSS, which are secure against adaptive, active adversaries are considered.

2 Preliminaries

Related Works. We briefly recall some definitions and observations. The following operation (called element-wise union) for monotone decreasing sets was introduced in [6,12].

Definition 1. *[6,12] We define the operation $\uplus$ for any monotone **decreasing** sets Δ_1, Δ_2 as follows: $\Delta_1 \uplus \Delta_2 = \{A = A_1 \cup A_2; A_1 \in \Delta_1, A_2 \in \Delta_2\}$ and the operation $\uplus$ for any monotone **increasing** sets Γ_1, Γ_2 as follows: $\Gamma_1 \uplus \Gamma_2 = \{A = A_1 \cup A_2; A_1 \notin \Gamma_1, A_2 \notin \Gamma_2\}^c$.*

For an arbitrary matrix M over a finite field $\mathbb{F}$, with m rows labelled by $1, \ldots, m$ let M_A denote the matrix obtained by keeping only those rows i with $i \in A$. Let M_A^T denote the transpose of M_A, and let $Im(M_A^T)$ denote the $\mathbb{F}$-linear span of the rows of M_A. We use $Ker(M_A)$ to denote the kernel of M_A, i.e., all linear combinations of the columns of M_A, leading to the 0. Let $\mathbf{v} = (v_1, \ldots, v_{t_1}) \in \mathbb{F}^{t_1}$ and $\mathbf{w} = (w_1, \ldots, w_{t_2}) \in \mathbb{F}^{t_2}$ be two vectors. By $\langle \mathbf{v}, \mathbf{w} \rangle$ we denote the standard inner product. The tensor vector product $\mathbf{v} \otimes \mathbf{w}$ is defined as a vector in $\mathbb{F}^{t_1 t_2}$ such that the j-coordinate in $\mathbf{v}$ (denoted by v_j) is replaced by $v_j \mathbf{w}$, i.e., $\mathbf{v} \otimes \mathbf{w} = (v_1 \mathbf{w}, \ldots, v_{t_1} \mathbf{w}) \in \mathbb{F}^{t_1 t_2}$. The tensor matrix product $\mathbf{v} \bar{\otimes} \mathbf{w}$ is defined as a $t_1 \times t_2$ matrix such that the j-column is equal to $v_j \mathbf{w}$.

Definition 2. *[3,10] A **Monotone Span Program** (MSP) $\mathcal{M}$ is a quadruple $(\mathbb{F}, M, \varepsilon, \psi)$, where $\mathbb{F}$ is a finite field, M is a matrix (with m rows and $d \leq m$ columns) over $\mathbb{F}$, $\psi : \{1, \ldots, m\} \to \{1, \ldots, n\}$ is a surjective function and ε is a fixed vector, called **target vector**, e.g., a column vector $(1, 0, ..., 0) \in \mathbb{F}^d$. The size of $\mathcal{M}$ is the number m of rows.*

As ψ labels each row with a number from $[1, \ldots, m]$ corresponding to a fixed player, we can think of each player as being the "owner" of one or more rows. For every player we consider a function φ which gives the set of rows owned by the player, i.e., φ is "inverse" of ψ. Note the difference between $M_{\varphi(G)}$ for $G \subseteq P$ and M_N for $N \subseteq \{1, \ldots, m\}$, but for the sake of simplicity we will write M_G instead of $M_{\varphi(G)}$.

An MSP is said to *compute* a (complete) access structure Γ when $\varepsilon \in Im(M_G^T)$ if and only if G is a member of Γ. Hence, the players can reconstruct the secret precisely if the rows they own contain in their linear span the target vector of $\mathcal{M}$, and otherwise they get no information about the secret, i.e., there exists a so called *recombination vector* $\mathbf{r}$ such that $M_G^T \mathbf{r} = \varepsilon$. Thus $\langle \mathbf{r}, M_G(s, \mathbf{c}) \rangle = \langle M_G^T \mathbf{r}, (s, \mathbf{c}) \rangle = \langle \varepsilon, (s, \mathbf{c}) \rangle = s$ for any secret s and any vector $\mathbf{c}$. It is well known that the vector $\varepsilon \notin Im(M_N^T)$ if and only if there exists a vector $\mathbf{k} \in \mathbb{F}^d$ such that $M_N \mathbf{k} = 0$ and $\mathbf{k}_1 = 1$.

Because of the linearity LSSSs provide it is easy to add secrets securely: it is sufficient for each player to add up the shares he holds. Therefore, to achieve general MPC, it suffices to implement multiplication of shared secrets. That is, we need a protocol where each player initially holds shared secrets s and s', and ends up holding a share of the product ss'. Several such protocols are known for the threshold case [1,2,7,8] and for general access structure [3].

We follow the approach proposed by Cramer *et al.* in [3] to build an MPC from any LSSS, provided that the LSSS is called *(strongly) multiplicative*. Loosely speaking, an LSSS is (strongly) multiplicative if each player i can, from his shares of secrets s and s', compute a value c_i, such that the product ss' can be obtained using all values (only values from honest players). Let Γ be an access structure, computed by the MSP $\mathcal{M} = (\mathbb{F}, M, \varepsilon, \psi)$. Given two m-vectors $\mathbf{x}$ and $\mathbf{y}$, Cramer *et al.* in [3] denote $\mathbf{x} \diamond \mathbf{y}$ to be the vector containing all the entries of the form $x_i y_j$, where $\psi(i) = \psi(j)$. Thus, if $m_i = |\varphi(i)|$ is the number of rows owned by a player i, then $\mathbf{x} \diamond \mathbf{y}$ has $\overline{m} = \sum_i m_i^2$ entries. So, if $\mathbf{x}$ and $\mathbf{y}$ contain shares resulting from sharing two secrets using $\mathcal{M}$, then the vector $\mathbf{x} \diamond \mathbf{y}$ can be computed using only local computations by the players, i.e., each component of the vector can be computed by one player. Denote by $\mathcal{M}_A$ the MSP obtained from $\mathcal{M}$ restricted to the set players A.

Definition 3. *[3] A* **multiplicative** *MSP is an MSP $\mathcal{M}$ for which there exists an $\overline{m}$-vector $\mathbf{r}$ called a* **recombination vector**, *such that for any two secrets s' and s'' and any random vectors $\mathbf{c}'$ and $\mathbf{c}''$, it holds that*

$$s's'' = \langle \mathbf{r}, M(s', \mathbf{c}') \diamond M(s'', \mathbf{c}'') \rangle .$$

It is said that $\mathcal{M}$ is **strongly multiplicative** *if for any subset A of honest players $\mathcal{M}_A$ is multiplicative.*

In the recent paper of Cramer *et al.* [4] this definition is rephrased.

Definition 4. *[4] The MSP $\mathcal{M}$ is called* **multiplicative** *if there exists a block-diagonal matrix $D \in \mathbb{F}^{m \times m}$ such that $M^T D M = \varepsilon \varepsilon^T$, where block-diagonal is to be understood as follows. Let the rows and columns of D be labelled by ψ, then the non-zero entries of D are collected in blocks $D_1, \ldots, D_n$ such that for every player $i \in P$ the rows and columns in D_i are labelled by i. $\mathcal{M}$ is called* **strongly multiplicative** *if, for any subset A of honest players $\mathcal{M}_A$ is multiplicative.*

Hirt and Maurer [9] call the adversary structure Q^2 (Q^3) if no two (three) sets in Δ_A cover the full player set P. Unconditional secure MPC for arbitrary Q^2 (in the passive case) and Q^3 (in the active case) access structures has been completely solved by Hirt and Maurer [9]. Efficient MPC (no error in the passive case and negligible error in the active case) from LSSS has been proposed by

Cramer *et al.* [3]. They have also proposed a LSSS with strong multiplication, but for this case both their solution and the solution of Hirt and Maurer are not efficient. Defining complexity measure for MPC is rather subtle. For that reason the complexity of the MSP is used, which is a measure of the complexity of its adversary structure.

Define $msp_\mathbb{F}(f)$ to be the size of the smallest MSP over $\mathbb{F}$ computing a monotone boolean function f. Next define $\mu_\mathbb{F}(f)$ to be the size of the smallest multiplicative MSP over $\mathbb{F}$ computing f. Similarly, define $\mu_\mathbb{F}^*(f)$ to be the size of the smallest strongly multiplicative MSP. In other words for a given adversary $\mathcal{A}$ with adversary structure Δ_A the requirement is for every set $B \in \Delta_A$ to have $B \notin \Gamma$, but $B^c \in \Gamma$. By definition, we have $msp_\mathbb{F}(f) \leq \mu_\mathbb{F}(f) \leq \mu_\mathbb{F}^*(f)$. In [3] Cramer *et al.* characterized the functions that (strongly) multiplicative MSPs can compute, and proved that the multiplication property for an MSP can be achieved without loss of efficiency. In particular, for the passive (multiplicative) case they proved that $\mu_\mathbb{F}(f) \leq 2\, msp_\mathbb{F}(f)$ provided that f is Q^2 function. Unfortunately there is no similar result for the strongly multiplicative case. Instead the authors in [3] proved that for an active adversary (strongly multiplicative) case $\mu_\mathbb{F}^*(f)$ is bounded by the so-called "formula complexity", provided that f is Q^3 function.

Recently Maurer [11] has proved that general unconditional information-theoretically MPC secure against a mixed (Δ_1, Δ_A)-adversary is possible if and only if $P \notin \Delta_1 \uplus \Delta_1 \uplus \Delta_A$ or equivalently if and only if $\Gamma_A^\perp \subseteq \Gamma_1 \uplus \Gamma_1$. Another important recent result, which gives necessary and sufficient conditions for the existence of an information-theoretically secure VSS, against a mixed (Δ_1, Δ_A)-adversary, has been proved by Fehr and Maurer in [6]: the robustness conditions for VSS are fulfilled if and only if $P \notin \Delta_1 \uplus \Delta_A \uplus \Delta_A$ or equivalently if and only if $(\Gamma_A \uplus \Gamma_A)^\perp \subseteq \Gamma_1$. We will refer to those two results as the MPC and VSS *conditions*.

Our Results. We will use the approach proposed by Cramer *et al.* in [3] for building General Secure Multi-Party Computation based on an underlying linear secret sharing scheme. First we expand the construction proposed by Cramer *et al.* in [3]. Let Γ_1 and Γ_2 be access structures, computed by MSPs $\mathcal{M}_1 = (\mathbb{F}, M1, \varepsilon1, \psi_1)$ and $\mathcal{M}_2 = (\mathbb{F}, M2, \varepsilon2, \psi_2)$. Let also $M1$ be an $m_1 \times d_1$ matrix, $M2$ be an $m_2 \times d_2$ matrix and φ_1, φ_2 are the "inverse" functions of ψ_1 and ψ_2. Given an m_1-vector $\mathbf{x}$ and an m_2-vector $\mathbf{y}$, we denote $\mathbf{x} \diamond \mathbf{y}$ to be the vector containing all entries of form $x_i y_j$, where $\psi_1(i) = \psi_2(j)$. Thus $\mathbf{x} \diamond \mathbf{y}$ has $m = \sum_i |\varphi_1(i)||\varphi_2(i)|$ entries (notice that $m < m_1 m_2$). So, if $\mathbf{x}$ and $\mathbf{y}$ contain shares resulting from sharing two secrets using $\mathcal{M}_1$ and $\mathcal{M}_2$, then the vector $\mathbf{x} \diamond \mathbf{y}$ can be computed using only local computation by the players, i.e., each component of the vector can be computed by one player. In other words we define the operation diamond $\diamond$ for vectors (and analogously for matrices) as concatenation of vectors (matrices), which are the tensor multiplication ($\otimes$) of the sub-vectors (sub-matrices) belonging to a fixed player. In order to better characterize the multiplicative property of an MSP we introduce a new notion *multiplicative resulting* MSP.

Definition 5. *Define MSP $\mathcal{M}$ to be $(\mathbb{F}, M = M1 \diamond M2, \varepsilon = \varepsilon 1 \diamond \varepsilon 2, \psi)$, where $\psi(i,j) = r$ if and only if $\psi_1(i) = \psi_2(j) = r$. Given two MSPs $\mathcal{M}_1$ and $\mathcal{M}_2$, the MSP $\mathcal{M}$ is called their* **multiplicative resulting MSP** *if there exists an m-vector $\mathbf{r}$ called a recombination vector, such that for any two secrets s' and s'' and any random vectors $\mathbf{c}'$ and $\mathbf{c}''$, it holds that*

$$s's'' = \langle \mathbf{r}, M1\,(s',\mathbf{c}') \diamond M2\,(s'',\mathbf{c}'') \rangle = \langle \mathbf{r}, M((s',\mathbf{c}') \otimes (s'',\mathbf{c}'')) \rangle.$$

An MSP $\mathcal{M}$ is called a **strongly multiplicative resulting MSP** *if for the access structure Γ computed by $\mathcal{M}$ we have $\{P\} \subset \Gamma$.*

This means that one can construct a multiplicative resulting MSP with which some subsets of players are able to compute the product of the secrets shared by MSPs $\mathcal{M}_1$ and $\mathcal{M}_2$; these subsets constitute a new access structure (called *resulting*) Γ. The difference between the multiplicative resulting MSP and the strongly multiplicative resulting MSP is that in the first one $\Gamma = \{P\}$.

Recall that in [3] the mixed adversary model is not considered, i.e. the authors consider access structures Γ_1 such that $\Delta_A = \Delta_1$ is $\mathcal{Q}^2$ $(\mathcal{Q}^3)$. The intuition behind this new definition is the following. In [3] two scenarios (ways) to build MPC are proposed:

1) For a given Q^2 (Q^3) access structure Γ_1 find (directly construct) a (strongly) multiplicative MSP computing Γ_1.
2) For an MSP $\mathcal{M}_1$ computing the Q^2 (Q^3) access structure Γ_1, construct a new (strongly) multiplicative MSP $\mathcal{M}_1'$ computing the same access structure.

It is shown in [3] that in the multiplicative case, for any MSP $\mathcal{M}_1$ one can efficiently construct multiplicative MSP $\mathcal{M}_1'$ computing the same access structure. Hence scenario 2) applies in that case. But for the strongly multiplicative case there is no efficient solution neither for scenario 1) nor for 2).

On the other hand we consider more grained mixed adversaries with Q^2, (Q^3) adversary structure. The adversary is called (Δ_1, Δ_A)-adversary if Δ_1 is its privacy structure and $\Delta_A \subseteq \Delta_1$ is its adversary structure. In our adversary model we have adversary with two privacy structures Δ_1, Δ_2 and with one adversary structure $\Delta_A \subseteq \Delta_1$, $\Delta_A \subseteq \Delta_2$, let us call it $(\Delta_1, \Delta_2, \Delta_A)$-adversary. In our MPC model there are also two scenarios.

A) Find conditions for the MSPs $\mathcal{M}_1$ and $\mathcal{M}_2$, computing Γ_1 and Γ_2 respectively, such that the access structure Γ (Γ is the (strongly) multiplicative resulting access structure) fulfills certain conditions.
B) For a MSP $\mathcal{M}_1$ computing Γ_1, find second MSP $\mathcal{M}_2$, computing Γ_2, such that the resulting access structure Γ fulfills certain conditions.

We will discuss later which conditions Γ should fulfil, in order to obtain secure MPC. Note that our main goal is to investigate the properties of the access structure Γ and the MSP $\mathcal{M}$ and how these properties depend on the initial MSPs, while the approaches in [3] are focused on the constructions. A partial answer for scenario A) is given in Proposition 1, stating that for the resulting access structure Γ we have $\Gamma \subseteq \Gamma_1 \uplus \Gamma_2$. Unfortunately, we still do not know when the equality holds. But solving this problem we will yield an efficient solution for the strongly multiplicative case. Our second main result Theorem 1 shows that the access structure Γ computed by the resulting MSP $\mathcal{M}$ of MSPs $\mathcal{M}_1$ and $\mathcal{M}_1^{\perp}$ is in fact the whole set of players P. Theorem 1 implies that only all players together can compute the product of the secrets, hence $\mathcal{M}$ is the multiplicative resulting MSP, but not the strongly multiplicative resulting MSP. Hence for the multiplicative case scenario B) holds, for any $\mathcal{M}_1$ with its dual $\mathcal{M}_1^{\perp} = \mathcal{M}_2$. Unfortunately this result also means that the construction proposed by Cramer *et al.* in [3] is not applicable in the strongly multiplicative case, i.e. even if we apply it for the Q^3 access structure. Let us define $\nu_{\mathbb{F}}(f)$ to be the size of the smallest multiplicative resulting MSP over $\mathbb{F}$ computing f and respectively $\nu_{\mathbb{F}}^*(f)$ to be the size of the smallest strongly multiplicative resulting MSP. In fact by the definition of the operation $\diamond$ (see Definition 5) this size depends on the sizes of the two initial MSPs, thus it is more accurate to denote it by $\nu_{\mathbb{F}}(f_1, f_2)$ $(\nu_{\mathbb{F}}^*(f_1, f_2))$. Denote by f^* the function which is the dual of f. The third main result, Theorem 2, shows that $msp_{\mathbb{F}}(f) = \nu_{\mathbb{F}}(f, f^*) \leq \nu_{\mathbb{F}}(f, f) = \mu_{\mathbb{F}}(f)$ and $\nu_{\mathbb{F}}^*(f, \bar{f}) \leq \nu_{\mathbb{F}}^*(f, f) = \mu_{\mathbb{F}}^*(f)$. The relations mean that when we use a (strongly) multiplicative MSP to compute the multiplicative resulting MSP the efficiency is the same. However, if we have an MSP without the (strongly) multiplicative property the usage of specific pair of MSPs (e.g. the given one and its dual in the multiplicative case) we gain better efficiency. The knowledge of the access structure Γ allows us to find which recombination vector corresponds to each qualified group. In the adversary model we consider, for a given adversary $\mathcal{A}$ with adversary structure Δ_A the requirement is for every set $B \in \Delta_A$ to have $B \notin \Gamma_1$, $B \notin \Gamma_2$ but $B^c \in \Gamma$. Recently Maurer [11] gave necessary and sufficient conditions for the existence of secure MPC in the mixed adversary model. Since Maurer considers general SSS, it was not clear whether using only LSSS these conditions still hold. In our model these conditions correspond to the conditions Γ should fulfill. And as we prove in Theorems 3 and 4 in both settings (unconditional information-theoretic and computational) for secure general MPC we have similar to those of Maurer conditions.

3   Enhanced Construction

3.1   The Diamond $\diamond$ Construction and Its Properties

A natural construction for the resulting MSP is the well known Kronecker product (construction $\otimes$) of matrices. The problem with this construction is that we do not know whom each new row belongs to, since we multiply a row owned by one player to a row owned by another player, hence local computation is not applicable. To avoid the inherent problem of the construction $\otimes$, we study the *diamond* $\diamond$ construction.

Some useful properties of the matrices $M = M1 \otimes M2$ and $M = M1 \diamond M2$ are given in an earlier version of this paper [13].

Consider the vector $\mathbf{x}$. Let us collect the coordinates in $\mathbf{x}$, which belong to the player t in a sub-vector $\bar{x}_t$ or $\mathbf{x} = (\bar{x}_1, \ldots, \bar{x}_n)$. Hence $\bar{x}_t \in \mathbb{F}^{|\varphi(t)|}$. The operation diamond $\diamond$ for vectors could be defined as: $\mathbf{x} \diamond \mathbf{y} = (\bar{x}_1 \otimes \bar{y}_1, \ldots, \bar{x}_n \otimes \bar{y}_n)$. We define an operation diamond for matrices and we denote the new matrix by $M = M1 \diamond M2$. We construct the new matrix M as follows. Denote by $M1_t$ the matrix formed by rows of $M1$ owned by player t and correspondingly by $M2_t$ the matrix formed by rows of $M2$ owned by player t. Thus the construction diamond $\diamond$ for $M = M1 \diamond M2$ is the concatenation of matrices $M1_t \otimes M2_t$ for $t = 1, \ldots, n$. First we show that the construction is symmetric regarding to the MSPs $\mathcal{M}_1$ and $\mathcal{M}_2$.

Lemma 1. *The MSPs $\mathcal{M} = \mathcal{M}_1 \diamond \mathcal{M}_2$ and $\widetilde{\mathcal{M}} = \mathcal{M}_2 \diamond \mathcal{M}_1$ compute the same access structure Γ.*

Lemma 2. *Let $M1$ be an $m_1 \times d_1$ matrix, and $M2$ be an $m_2 \times d_2$ matrix. Construct the matrix M following the construction $\diamond$ (i.e., $M = M1 \diamond M2$ is $m \times d_1 d_2$ matrix), then for arbitrary column vectors $\lambda1 \in \mathbb{F}^{d_1}$, $\lambda2 \in \mathbb{F}^{d_2}$ the following equality holds: $(M1 \diamond M2)(\lambda1 \otimes \lambda2) = (M1\ \lambda1) \diamond (M2\ \lambda2)$.*

Note that the construction diamond $\diamond$ confirms our intuitive expectations that the players could locally compute their new shares, as shown in the following lemma.

Lemma 3. *Let us denote by $\mathbf{s1} = M1\ (s_1, \mathbf{a})$ and $\mathbf{s2} = M2\ (s_2, \mathbf{b})$ the shares distributed by MSPs $\mathcal{M}_1$ and $\mathcal{M}_2$, for the secrets s_1 and s_2 resp. Then MSP $\mathcal{M}$ actually distributes shares $\mathbf{s} = \mathbf{s1} \diamond \mathbf{s2}$ for the secret $s_1 s_2$.*

Note that we have $\mathbf{s} = (M1 \diamond M2)((s_1, \mathbf{a}) \otimes (s_2, \mathbf{b}))$ and that the vector $(s_1, \mathbf{a}) \otimes (s_2, \mathbf{b})$ is no longer random. Now we are in position to prove our first main result using the operation $\uplus$ and the construction $\diamond$ we have introduced.

Proposition 1. *Let Γ_1 and Γ_2 be the access structures computed by the MSPs $\mathcal{M}_1$ and $\mathcal{M}_2$. Let the MSP $\mathcal{M}$ be the strongly multiplicative result of MSPs $\mathcal{M}_1$ and $\mathcal{M}_2$, and let the access structure Γ be computed by the MSP $\mathcal{M}$. Then $\Gamma \subseteq \Gamma_1 \uplus \Gamma_2$. (Notice that Γ may be trivial, e.g. $\emptyset$.)*

Proof: Let $A1 \notin \Gamma_1$. Hence there exists a vector $\mathbf{k} \in Ker(M1_{A1})$ such that $\mathbf{k}_1 = 1$. Analogously, let $A2 \notin \Gamma_2$. Hence there exists a vector $\mathbf{r} \in Ker(M2_{A2})$ such that $\mathbf{r}_1 = 1$. Notice that $\mathbf{k} \in \mathbb{F}^{d_1}$ and $\mathbf{r} \in \mathbb{F}^{d_2}$. Let $A = A1 \cup A2$, so we have $A \notin \Gamma_1 \uplus \Gamma_2$. Form a new vector $\mathbf{k} \otimes \mathbf{r} \in \mathbb{F}^{d_1 d_2}$. It is easy to check that the vector $\mathbf{k} \otimes \mathbf{r} \in Ker(M_A)$ and $(\mathbf{k} \otimes \mathbf{r})_1 = 1$. Hence $A \notin \Gamma$, thus $\Gamma \subseteq \Gamma_1 \uplus \Gamma_2$. $\square$

3.2   Properties of the Resulting MSP

An interesting open question is when the "equality" holds? One can see from the examples given in [13] that "equality" does not always hold. Consider for example the threshold case. Denote by $T_{s,n}$ the s-out-of-n threshold access structure, then it is easy to verify that $T_{l,n} \uplus T_{s,n} = T_{l+s-1,n}$. On the other hand each player t holds vectors $\mathbf{w} = (1, \alpha_t, \ldots, \alpha_t^{s-1})$ and $\mathbf{v} = (1, \alpha_t, \ldots, \alpha_t^{l-1})$ from MSPs computing $T_{s,n}$ and $T_{l,n}$ correspondingly. Thus the construction proposed above gives $\mathbf{v} \otimes \mathbf{w} = (1, \alpha_t, \ldots, \alpha_t^{s-1}, \alpha_t, \alpha_t^2, \ldots, \alpha_t^s, \ldots \ldots \ldots, \alpha_t^{l-1}, \ldots, \alpha_t^{s+l-2})$.

Using the fact that without changing the access structure we can always replace the 2nd up to the last column of M by any set of vectors that generates the same space we obtain that $\mathbf{v} \otimes \mathbf{w}$ is equivalent to $(1, \alpha_t, \ldots, \alpha_t^{s+l-2})$. But this is exactly the row owned by the player t in MSP computing $T_{l+s-1,n}$. This means that in the threshold case we have equality in Proposition 1. That is why we believe that for an MSP $\mathcal{M}_1$ there should exist another MSP $\mathcal{M}_2$ such that for their strongly multiplicative resulting MSP $\mathcal{M}$, computing the access structure Γ, we have $\Gamma = \Gamma_1 \uplus \Gamma_2$. The first step in this direction is [3, Theorem 7], where $\mathcal{M}_1$ and $\mathcal{M}_2$ are dual, i.e., $\Gamma_2^{\perp} = \Gamma_1$. Cramer *et al.* have proved in [3, Theorem 7] that $\varepsilon = \varepsilon 1 \diamond \varepsilon 1$ belongs to the linear span of the rows of $M = M1 \diamond M1^{\perp}$, when the matrices $M1$ and $M1^{\perp}$ satisfy the condition $M1^T M1^{\perp} = \overline{E}$. Here $\overline{E} = \varepsilon 1 \, \varepsilon 1^T$ is the matrix with zeros everywhere, except in its upper-left corner where the entry is 1. It is known [3] how to derive the matrix $M1^{\perp}$ from $M1$ such that they satisfy $M1^T M1^{\perp} = \overline{E}$.

One of the key results in [3] is a method to construct, from any MSP $\mathcal{M}_1$ with Q^2 access structure Γ_1, a multiplicative MSP $\mathcal{M}'$ with the same access structure and with twice bigger size (hence with twice bigger complexity). Unfortunately no similar result is known for the strongly multiplicative case. It is natural to ask what happens if $\mathcal{M}_1$ computes the Q^3 access structure Γ_1 instead. We are ready to prove our second main result, which gives an answer to this question.

Theorem 1. *Let Γ_1 and $\Gamma_1^{\perp}$ be the connected access structures computed by the MSPs $\mathcal{M}_1$ and $\mathcal{M}_1^{\perp}$ and $M^T M\perp = \overline{E}$ holds. Let the MSP $\mathcal{M}$ be the strongly multiplicative result of MSPs $\mathcal{M}_1$ and $\mathcal{M}_1^{\perp}$, and let the access structure Γ be computed by the MSP $\mathcal{M}$. Then $\Gamma = \Gamma_1 \uplus \Gamma_1^{\perp} = \{P\}$.*

Proof: It is known that $\{P\} \in \Gamma$. On the other hand from Proposition 1 we have $\Gamma \subseteq \Gamma_1 \uplus \Gamma_1^{\perp}$, thus it is sufficient to prove that there is no other sets in $\Gamma_1 \uplus \Gamma_1^{\perp}$ except $\{P\}$.

For any set $A \in \Delta_1^+$ and any player $i \in P$, $i \notin A$ we have $(A \cup i) \in \Gamma_1$. Set $B^c = A \cup i$ and hence $B = P \setminus B^c \in \Delta_1^{\perp}$. Therefore $A \cup B = (P \setminus i) \in (\Delta_1 \uplus \Delta_1^{\perp})$. Let us assume that there exists a player j such that $(P \setminus j) \notin (\Delta_1 \uplus \Delta_1^{\perp})$. So, $j \in A$ for every set $A \in \Delta_1^+$, because otherwise using the construction given above we arrive at a contradiction. Hence the access structure Γ_1 has the star topology for the forbidden sets, i.e., there exists a player j such that for any set $A \in \Delta^+$, $j \in A$. Hence Γ_1 is not connected – contradiction and we are done. $\square$

As an example let us consider again the threshold case. Taking into account that $(T_{l,n})^{\perp} = T_{n-l+1,n}$, we have $T_{l,n} \uplus (T_{l,n})^{\perp} = T_{n,n} = \{P\}$, which is in accordance with Theorem 1.

3.3   Relations with Multiplicative MSPs

Lemma 4.   *Let $\mathcal{M}$ be a multiplicative MSP computing Γ and satisfying $M^T D M = \overline{E}$ for some block-diagonal matrix D (Definition 4). Define MSP $\overline{\mathcal{M}}$ computing $\overline{\Gamma}$ by $\overline{M} = DM$. Then $\Gamma^{\perp} \subseteq \overline{\Gamma} \subseteq \Gamma$ holds.*

Note that as a consequence we obtain that $M^T D M = \overline{E}$ imply $\Gamma^{\perp} \subseteq \Gamma$, i.e. the Q^2 property. Let $\mathcal{M}$ be multiplicative MSP and D be a block-diagonal matrix satisfying the condition from Definition 4. Then for any invertible block-diagonal matrix $\widetilde{D}$ the matrices $\overline{M} = \widetilde{D}M$ and $\overline{D} = (\widetilde{D}^{-1})^T D \widetilde{D}^{-1}$ satisfy also the condition from Definition 4.

Corollary 1. *For any self-dual access structure Γ there exist MSPs $\mathcal{M}$ and $\mathcal{M}^\perp$ and block-diagonal matrix D such that the following relations hold $M^T M^\perp = \overline{E}$ and $DM = M^\perp$.*

Theorem 2. *For any (strongly) multiplicative MSP $\mathcal{M}$ computing Γ (f) and its dual MSP $\mathcal{M}^\perp$ computing $\Gamma^\perp$ (f*) we have*

$$msp_{\mathbb{F}}(f) = \nu_{\mathbb{F}}(f, f^*) \leq \nu_{\mathbb{F}}(f, f) = \mu_{\mathbb{F}}(f) \quad and \quad \nu_{\mathbb{F}}^*(f, \bar{f}) \leq \nu_{\mathbb{F}}^*(f, f) = \mu_{\mathbb{F}}^*(f).$$

Proof: For the sake of simplicity we will prove only the multiplicative case, since the strongly multiplicative case is a straightforward consequence. Let M be an $m \times d$ matrix, thus D is an $m \times m$ matrix. Let's compute $M^T DM$ denoting by $d_{i,j}$ the element in i-th row and j-th column in D. Thus $M^T DM = \sum_{i,j=1}^m d_{i,j} M_i \overline{\otimes} M_j$, where $\overline{\otimes}$ is the tensor matrix product. But $\sum_{i,j=1}^m d_{i,j} M_i \overline{\otimes} M_j = \overline{E}$ is equivalent to $\sum_{i,j=1}^m d_{i,j} M_i \otimes M_j = \varepsilon$, since the only difference is the way the tensor product is presented in matrix or in vector form. Thus the condition $M^T DM = \overline{E}$ for some block-diagonal matrix D is equivalent to the condition of the existence of a recombination vector $\mathbf{r}$ for the resulting matrix $M \diamond M$. In fact the block-diagonal matrix D is the recombination vector $\mathbf{r}$ written in matrix block form. Thus we prove the right equality, namely $\nu_{\mathbb{F}}(f, f) = \mu_{\mathbb{F}}(f)$.

Revisiting the construction for multiplicative MSP given in [3], we notice that the matrix $\widetilde{M}$ from the multiplicative MSP consists of two separate parts (matrices) M and $M^\perp$. Thus sharing a secret s by $\widetilde{M}$ with random vector of the form $(\mathbf{a}, \mathbf{b})$, where $\mathbf{a}, \mathbf{b} \in \mathbb{F}^{d-1}$ we have $\widetilde{s} = \widetilde{M}(s, \mathbf{a}, \mathbf{b})$. Define $\mathbf{s} = M(s, \mathbf{a})$ and $\mathbf{s}^\perp = M^\perp(s, \mathbf{b})$. Notice that $\widetilde{s} = (\mathbf{s}, \mathbf{s}^\perp)$. Therefore using the construction of Cramer *et al.* for multiplicative MSP we have two shares of the secret s: one corresponding to M and one corresponding to its dual $M^\perp$. Now considering a multiplication gate we have as input two secrets s_1 and s_2 sharing them with $\widetilde{M}$ gives us shares for s_i ($i = 1, 2$) for both M and $M^\perp$. On the other hand using the resulting MSP of M and $M^\perp$ we need only shares of s_1 shared by M and s_2 shared by $M^\perp$. Thus we need twice less shares to be distributed. Therefore we have $msp_{\mathbb{F}}(f) = msp_{\mathbb{F}}(f^*) = \nu_{\mathbb{F}}(f, f^*)$ and that is the best possible, since we always need to share the two inputs to a given multiplication gate.   □

The fact that we use two different MSP to share the inputs in every multiplication gate make the computation of a given arithmetic circuit more complicated compared to the case when all inputs are shared just by one MSP. Let consider some examples:

- If the function we want to compute is $s_1 s_2$, then as we proved we need to share s_1 by M and s_2 by $M^\perp$ and we are twice more efficient, note that this is the best possible improvement.

- On the other hand if the function we want to compute is s^2, then in fact sharing s by M and $M^\perp$ gives us the same as sharing it by $\widetilde{M}$ thus the efficiency here is the same.
- Another indicative example is when the function we want to compute is $s_1 s_2 + s_2 s_3 + s_3 s_1$, then we share s_1 by M and s_2 by $M^\perp$, but then we are forced to share s_3 by both M and $M^\perp$, (i.e. by $\widetilde{M}$). Thus we are $\frac{3}{2}$ times more efficient.

Since the function we want to compute is public it is required in our model to figure out in advance for each multiplication gate which MSP we will use (M, $M^\perp$ or $\widetilde{M}$). We can compare this to coloring coloring a graph with two colors (for M and $M^\perp$), but some nodes could be colored by both colors. Thus the following question arises: classify the functions by the criterion whether the inputs and all nodes could be "colored" only by the two colors, i.e. there are no nodes colored by both colors.

4   Adaptive, Active Adversary: The Zero-Error Case

Recall that in our adversary model we have adversary with two privacy structures Δ_1, Δ_2 and with one adversary structure $\Delta_A \subseteq \Delta_1$, $\Delta_A \subseteq \Delta_2$. To build secure MPC protocol we employ the error-free commitment protocols [3], provided that the MSP we have is strongly multiplicative.

The use of strongly multiplicative LSSS allows to compute the product of two secrets without interaction between the players. Unfortunately in the general case the picture coincides with the threshold case. As Ben-Or et $al.$ note in their seminal paper [1] the new shares computed after local multiplication correspond to a higher (double) degree polynomial which is not random. To overcome this problem they introduced a degree reduction and randomization protocols. Later Gennaro et $al.$ [7] achieve both tasks in a single step, which they call an algebraic simplification for the multiplication protocol. As we noticed in the case of general access structures we have the same problem as described by Ben-Or et $al.$ The new shares computed after local multiplication correspond to a much "smaller" access structure Γ and the shares are computed using a non-random vector. On the other hand the knowledge of the access structure Γ allows us to build an analog of the algebraic simplification protocol of Gennaro et $al.$ [7], which we will describe in the next subsection.

4.1   Algebraic Simplification for the Multiplication Protocol on a General Access Structure

Let the two secrets s_1 and s_2 are shared using the MSPs $\mathcal{M}_1$ and $\mathcal{M}_2$ (computing Γ_1 and Γ_2 respectively). Denote their resulting MSP as usual by $\mathcal{M}$ with access structure Γ. Let us choose another MSP $\mathcal{M}_3$ computing Γ_3 to which we want to reduce Γ. Then the simplified multiplication protocol is as follows:

1. Each player i multiplies locally his shares (for simplicity let they own one share from each of the access structures and denote them by) $\mathbf{s1}_i$ and $\mathbf{s2}_i$.
2. Then the player i chooses a random vector $\mathbf{h(i)}$ such that its first coordinate is the product, (i.e., $\mathbf{s1}_i\,\mathbf{s2}_i = \mathbf{s}_i$.)
3. With the MSP $\mathcal{M}_3$ and applying the VSS protocol the i-th player re-shares its product $\mathbf{s}_i$, i.e. using vector $\mathbf{h(i)}$.
4. In this way every player k receives from player i a temporary share, denoted by $\mathbf{ts(i)}_k$.
5. For some set of "good" players $A \in \Gamma$ with recombination vector λ, each player k calculates its new-share $\mathbf{ns}_k = \sum_{i \in A} \mathbf{ts(i)}_k\,\lambda_i$.
6. Finally the new-shares have the property that any set of "good" players $B \in \Gamma_3$ could restore the secret $s_1 s_2$.

For a proof that this protocol is correct and secure we refer to [13].

4.2   Information-Theoretic Settings

In order to build an MPC protocol secure against active, adaptive adversary in the non-computational model it is sufficient for the MSPs $\mathcal{M}_1$, $\mathcal{M}_2$ and $\mathcal{M}_3$ to satisfy the VSS conditions from [6] and Γ to be the strongly multiplicative result of MSPs computing Γ_1 and Γ_2. Using the algebraic simplification protocol, and the homomorphic commitments (information-theoretic secure VSS) [3] we could "reduce" the access structure Γ to any access structure Γ_3, which we call "reduced", provided Γ_3 satisfies the VSS conditions. Hence combining Proposition 1 and the VSS conditions of Fehr and Maurer our fourth main result follows.

Theorem 3. *Let Γ be the access structure computed by the strongly multiplicative resulting MSP $\mathcal{M} = \mathcal{M}_1 \diamond \mathcal{M}_2$ and Γ_3 be the "reduced" access structure. Then the sufficient conditions for the existence of general unconditional information-theoretically secure MPC, secure against $(\Delta_1, \Delta_2, \Delta_A)$-adversary are:*

$$\Gamma_A^{\perp} \subseteq \Gamma \subseteq \Gamma_1 \uplus \Gamma_2, \quad (\Gamma_A \uplus \Gamma_A)^{\perp} \subseteq \Gamma_i, \quad \text{for } i = 1,2,3.$$

Note that from Theorem 3 it follows that we have $P \notin \Delta_1 \uplus \Delta_2 \uplus \Delta_A$, which corresponds to the condition of Maurer [11].

4.3   Computational Settings

In order to build an MPC protocol secure against an active adversary in the computational model it is sufficient for the MSPs $\mathcal{M}_1$, $\mathcal{M}_2$, $\mathcal{M}_3$ to satisfy the VSS conditions and for Γ to be the strongly multiplicative result of MSPs computing Γ_1 and Γ_2. Again using the algebraic simplification protocol, and the homomorphic commitments (computational secure VSS plus one-way trapdoor permutations) [3,7] we could "reduce" the access structure Γ to any access structure Γ_3, provided Γ_3 satisfy VSS conditions. Hence we obtain our next result.

Theorem 4. *Let Γ be the access structure computed by the strongly multiplicative resulting MSP $\mathcal{M} = \mathcal{M}_1 \diamond \mathcal{M}_2$ and Γ_3 be the "reduced" access structure. If a trapdoor one-way permutation exists, then the sufficient conditions for the existence of general unconditional secure MPC in the cryptographic scenario, secure against $(\Delta_1, \Delta_2, \Delta_A)$-adversary are:*

$$\Gamma_A^\perp \subseteq \Gamma \subseteq \Gamma_1 \uplus \Gamma_2, \quad \Gamma_A^\perp \subseteq \Gamma_i, \quad \text{for} \quad i = 1, 2, 3.$$

Note again the similarity of the conditions for existence of MPC.

Acknowledgements. The authors would like to thank Ronald Cramer and Ivan Damgård for the helpful discussions and comments.

References

1. M. Ben-Or, S. Goldwasser, A. Wigderson, Completeness theorems for Non- Cryptographic Fault-Tolerant Distributed Computation, *STOC 1988*, 1988, pp. 1-10.
2. D. Chaum, C. Crepeau, I. Damgård, Multi-Party Unconditionally Secure Protocols, *STOC 1988*, 1988, pp. 11-19.
3. R. Cramer, I. Damgård, U. Maurer, General Secure Multi-Party Computation from any linear secret sharing scheme, *EUROCRYPT 2000*, LNCS 1807, pp. 316-334.
4. R. Cramer, S. Fehr, Y. Ishai, E. Kushilevitz, Efficient Multi-Party Computation over Rings, *EUROCRYPT 2003*, LNCS 2656, pp. 596-613.
5. B. Chor, S. Goldwasser, S. Micali, B. Awerbuch, Verifiable secret sharing and achieving simultaneity in the presence of faults, *FOCS 1985*, pp. 383-395.
6. S. Fehr, U. Maurer, Linear VSS and Distributed Commitments Based on Secret Sharing and Pairwise Checks, *CRYPTO 2002*, LNCS 2442, pp. 565-580.
7. R. Gennaro, M. Rabin, T. Rabin, Simplified VSS and Fast-Track Multi-party Computations with Applications to Threshold Cryptography, *PODC'98*, pp. 101-111.
8. O. Goldreich, S. Micali, A. Wigderson, How to Play Any Mental Game or a Completeness Theorem for Protocols with Honest Majority, *STOC'87*, pp. 218-229.
9. M. Hirt, U. Maurer, Complete characterization of Adversaries Tolerable in General Multiparty Computations, *PODC'97*, pp. 25-34.
10. M. Karchmer, A. Wigderson. On Span Programs, *Proc. of 8-th Annual Structure in Complexity Theory Conference*, 1993, pp. 102-111.
11. U. Maurer, Secure Multi-Party Computation Made Simple, *3rd Conference on Security in Communication Networks* 2002, LNCS 2576, pp. 14-28, 2003.
12. V. Nikov, S. Nikova, B. Preneel, J. Vandewalle, Applying General Access Structure to Proactive Secret Sharing Schemes, *Proc. of the 23rd Symposium on Information Theory in the Benelux*, May 29-31, 2002, Universite Catolique de Lovain (UCL), Lovain-la-Neuve, Belgium, pp. 197-206, *Cryptology ePrint Archive*: Report 2002/141.
13. V. Nikov, S. Nikova, B. Preneel. Multi-Party Computation from any Linear Secret Sharing Scheme Secure against Adaptive Adversary: The Zero-Error Case, *Cryptology ePrint Archive:* Report 2003/006.
14. A. Shamir. How to share a secret, *Commun. ACM* 22, 1979, pp. 612-613.

A New $(2, n)$-Visual Threshold Scheme for Color Images

Avishek Adhikari[1] and Somnath Sikdar[2]

[1] Applied Statistics Unit, Indian Statistical Institute,
203, B T Road, Calcutta 700 035, INDIA
avishek_r@isical.ac.in
[2] The Institute of Mathematical Sciences,
C.I.T Campus, Tharamani, Chennai 600 113, INDIA
somnath@imsc.ernet.in

Abstract. In this paper we propose a new scheme for a $(2, n)$-visual threshold scheme (VTS) for color images. Our scheme achieves a better color ratio than the schemes proposed by Koga *et al* in 1998 and 2001. The pixel expansion of our scheme is reasonably good and we also give a lower bound on the color ratio of our scheme.

Keywords: secret sharing scheme, visual secret sharing scheme, visual cryptography.

1 Introduction

Visual cryptography was introduced by Naor and Shamir [9]. It is a new cryptographic paradigm that enables a secret image to be split into n shares, each share being printed on a transparency. The shares are distributed among n participants of whom only some are qualified to recover the original image. The secret image is reconstructed by stacking a certain number k ($2 \leq k \leq n$) of these transparencies from the set of qualified participants. If fewer than k transparencies are superimposed, then it is impossible to decode the original image. The resulting cryptographic scheme is called a $(k, n)-$visual threshold scheme (VTS). Since the reconstruction is done by the human visual system, no computations are involved during decoding unlike traditional cryptographic schemes where a fair amount of computation is needed to reconstruct the plain text.

The schemes proposed by Naor and Shamir [9] involved black and white images. Further research [4], [6], [7], [8] extended the idea to gray-scale and color images. There are several efficient ways to encrypt black and white images but gray-scale and color images present problems. Color images are specially difficult to encrypt because of two reasons. Firstly, there seems to be no way of encrypting a color image using a reasonably small pixel expansion (a scheme has pixel expansion m if each pixel of the original image is encoded as m subpixels on each transparency) .

T. Johansson and S. Maitra (Eds.): INDOCRYPT 2003, LNCS 2904, pp. 148–161, 2003.

The second problem with encrypting color images relates to the brightness of colors in the reconstructed image. In all the schemes proposed on color visual cryptography so far, [6], [8], the brightness of the reconstructed image is poor. This is because, on stacking the requisite number of transparencies, the number of subpixels that have the color of the original encoded pixel is very small compared to the total number of subpixels. Since, the ratio

$$\frac{\text{number of subpixels that possess the true color}}{\text{the total number of subpixels}} \quad (1)$$

influences the brightness of the reconstructed image, it deserves a separate definition.

Definition 1. *Let the (i, j)th pixel of the secret image have color c and suppose that we are working in a $(k, n)-VTS$ for color images. The* **color ratio** *of the (i, j)th pixel in the reconstructed image is the ratio (1) above, when k transparencies from qualified participants are stacked together.*

It is conceivable that the encryption strategy is such that the color ratios of the different pixels in the reconstructed image are different. In this case,we could define the color ratio of a scheme to be the minimum value of the ratio defined in (1), the minimum being taken over all possible different colored pixels. On the other hand, the encryption strategy could be so regular that each pixel, irrespective of its color, has the same color ratio. In this case, we need not define the color ratio separately for each pixel of the reconstructed image. In such a case, if the color ratio of each pixel is R, we will say that the encoding scheme attains a color ratio R.

Koga *et al* ([8]) have proposed a construction of an $(n, n)-VTS$ with colors $\{c_1, c_2, \ldots, c_k\}$. Their construction is defined over a bounded upper semilattice. From the basis matrices of a $(t, t)-VTS$ for color images, they have constructed basis matrices of a $(t, n)-VTS$ for color images. The disadvantage of their scheme is that when the number of shares (and also the number of colors) increases, the pixel expansion shoots up; the color ratio decreases correspondingly. Since their scheme has no positive lower bound on the color ratio, the reconstructed image becomes progressively darker as the number of shares increases.

In this paper, we propose a new scheme for a $(2, n)-VTS$ for color images. Our scheme has the following advantages:

1. Our scheme provides a lower bound on the color ratio. For instance, if the secret image is composed of the colors cyan, yellow and green, then the color ratio of our scheme is lower bounded by $1/4$. This lower bound depends only on the number of colors in the secret image and is *independent* of the number of shares. Other schemes dealing with color images can provide no such lower bound ([6], [8]).

2. The color ratio of our scheme is very high compared to what other schemes provide. The result is that our images are brighter. If the secret image has the colors $\mathcal{C} = \{c_1, c_2, \ldots, c_k\}$ such that no two colors c_i and $c_j \in \mathcal{C}$ can be combined to produce a third color $c_l \in \mathcal{C}$, then the color ratio of our scheme

is lower bounded by $1/2k$. Compare this with the color ratio $2/k \cdot n(n-1)$, which is what the scheme proposed by Koga ([6]) achieves. The scheme proposed by Koga in 2001, ([8]), achieves a color ratio of $2/kn!$

3. The pixel expansion of our scheme is reasonably good. If the secret image consists of colors cyan, yellow and green, then the pixel expansion is $4m$ where $m = \binom{n}{\lfloor n/2 \rfloor}$. In case, the secret image is composed of the colors red, green, blue, cyan, magenta and green, the pixel expansion is $6m$. Compare this with the pixel expansion given in [8], which for these same two color sets is $2n!$ and $3n!$ respectively. It can be easily seen that: $2n! > 4m$, and $3n! > 6m$, for $n > 3$

This paper is organized as follows. Section 2 deals with some of the definitions and results on visual cryptography that we have made use of in this paper. In section 3, we construct a $(2, n)$−color VTS with three base colors. Finally, in section 4 we show how to construct a $(2, n)$−color VTS with six base colors. We also show how our scheme can be extended to an arbitrary number of colors.

2   Preliminaries

2.1   The Model

The model that we describe here is taken nearly verbatim from Blundo, De Santis, and Stinson [3]. Let $\mathcal{P} = \{1, \ldots, n\}$ be a set of elements called *participants*, and let $2^{\mathcal{P}}$ denote the set of all subsets of $\mathcal{P}$. Let Γ_{Qual} and Γ_{Forb} be subsets of $2^{\mathcal{P}}$, where $\Gamma_{Qual} \cap \Gamma_{Forb} = \emptyset$. We will refer to members of Γ_{Qual} as *qualified sets* and the members of Γ_{Forb} as *forbidden sets*. The pair $(\Gamma_{Qual}, \Gamma_{Forb})$ is called the *access structure* of the scheme.

We assume that the secret image consists of a collection of black and white pixels, each pixel being encrypted separately. To understand the encryption process consider the case where the secret image consists of just a single black or white pixel. On encryption, this pixel appears in the n shares distributed to the participants. However, in each share the pixel is subdivided into m *subpixels* (m is the pixel expansion), each of which is either black or white. It is important to note that the shares are printed on transparencies, and that a "white" subpixel is actually an area where nothing is printed, and therefore left transparent. We assume that the subpixels are sufficiently small and close enough so that the eye averages them to some shade of grey. We can represent this with an $n \times m$ boolean matrix $S[i, j]$, where $S[i, j] = 1$ if and only if the jth subpixel in the ith share is black. When the shares are stacked together, the perceived grey level is proportional to the number of 1's in the boolean OR of the m−vectors representing the shares of each participant. When the secret image consists of more than one pixel, we encrypt each pixel separately.

We give the following definition of a visual cryptography scheme for a general access structure. The phrasing is taken directly from Atienese, Blundo, De Santis, and Stinson [1]. We use $OR\ V$ to denote the boolean operation OR of a set of vectors with result V. The *Hamming weight* $w(V)$ is the number of 1's in the boolean vector V.

Definition 2. *Let* $(\Gamma_{Qual}, \Gamma_{Forb})$ *be an access structure on a set of* n *partici-pants. Two collections (multisets) of* $n \times m$ *Boolean matrices* C_0 *and* C_1 *consti-tute a visual cryptography scheme* $(\Gamma_{Qual}, \Gamma_{Forb}, m) - VCS$ *if there exists values* $\alpha(m)$ *and* $\{t_X\}_{X \in \Gamma_{Qual}}$ *satisfying:*

1. *Any qualified set* $X = \{i_1, i_2, \dots, i_p\} \in \Gamma_{Qual}$ *can recover the shared image by stacking their transparencies.*
   Formally, for any $M \in C_0$, *the "or"* V *of rows* $i_1, \dots, i_p$ *satisfies* $w(V) \le t_X - \alpha(m) \cdot m$; *whereas, for any* $M \in C_1$ *it results that* $w(V) \ge t_X$.
2. *Any forbidden set* $X = \{i_1, i_2, \dots, i_p\} \in \Gamma_{Forb}$ *has no information on the shared image.*
   Formaly, the two collections of $p \times m$ *matrices* D_t, *with* $t \in \{0, 1\}$, *obtained by restricting each* $n \times m$ *matrix in* C_t *to rows* $i_1, \dots, i_p$ *are indistinguishable in the sense that they contain the same matrices with the same frequencies.*

In order that the recovered image is clearly discernible, it is important that the grey level of a black pixel be darker than that of a white pixel. Informally, the difference in the grey levels of the two pixel types is called *contrast*. We want the contrast to be as large as possible. Three variables control the perception of black and white regions in the recovered image: a threshold value (t), a relative difference (α), and the pixelexpansion (m). The *threshold value* is a numeric value that represents a grey level that is perceived by the human eye as the color black. The value $\alpha \cdot m$ is the contrast, which we want to be as large as possible. We require that $\alpha \cdot m \ge 1$ to ensure that black and white areas will be distinguishable.

Each pixel of the original image will be encrypted into n pixels, each of which consist of m subpixels. To share a white (resp. black) pixel, the dealer randomly chooses one of the matrices in C_0 (resp. C_1), and distributes row i to participant i. Thus the chosen matrix defines the m subpixels in each of the n transparencies. Note that in the definition above we allow a matrix to appear more than once in C_0 (C_1). Finally, note that the size of the collections C_0 and C_1 need not be the same.

2.2 Basis Matrices

Instead of working with the collections C_0 and C_1, it is convenient (in terms of memory requirements) to consider only two $n \times m$ boolean matrices, S^0 and S^1 called *basis matrices* which satisfy the following definition.

Definition 3. *Let* $(\Gamma_{Qual}, \Gamma_{Forb})$ *be an access structure on a set* $\mathcal{P}$ *of* n *par-ticipants. A* $(\Gamma_{Qual}, \Gamma_{Forb}, m)$-*VCS with relative difference* $\alpha(m)$ *and a set of thresholds* $\{t_X\}_{X \in \Gamma_{Qual}}$ *is realized using the* $n \times m$ *basis matrices* S^0 *and* S^1 *if the following two conditions hold:*

1. If $X = \{i_1, i_2, \ldots, i_p\} \in \Gamma_{Qual}$, then the OR V of the rows $i_1, i_2, \ldots, i_p$ of S^0 satisfies $w(V) \leq t_X - \alpha(m) \cdot m$; whereas, for S^1 it results that $w(V) \geq t_X$.
2. If $X = \{i_1, i_2, \ldots, i_p\} \in \Gamma_{Forb}$, the two $p \times m$ matrices obtained by restricting S^0 and S^1 to rows $i_1, i_2, \ldots, i_p$ are equal up to a column permutation.

The collections $\mathcal{C}_0$ and $\mathcal{C}_1$ are obtained by permuting the columns of the corresponding basis matrix (S^0 for $\mathcal{C}_0$ and S^1 for $\mathcal{C}_1$) in all possible ways. Note that, in this case, the sizes of the collections $\mathcal{C}_0$ and $\mathcal{C}_1$ are the same.

2.3   Share Distribution Algorithm

Now that we will be working with the basis matrices S^0 and S^1, we need to modify the encryption process slightly as described below.

For each pixel P, do the following:

1. Generate a random permutation π of the set $\{1, 2, \ldots, m\}$.
2. If P is a black pixel, then apply π to the columns of S^0; else apply π to the columns of S^1. Call the resulting matrix T.
3. For $1 \leq i \leq n$, row i of T comprises the m subpixels of P in the ith share.

2.4   Threshold Schemes

A (k, n)−threshold structure is any access structure $(\Gamma_{Qual}, \Gamma_{Forb})$ in which $\Gamma_0 = \{B \subseteq \mathcal{P} : |B| = k\}$ and $\Gamma_{Forb} = \{B \subseteq \mathcal{P} : |B| \leq k - 1\}$. In any (k, n)−threshold VCS, the image is visible if any k of the n participants stack their transparencies, but totally invisible if fewer than k transparencies are stacked together or analyzed by any other method. In a strong (k, n)−threshold VCS, the image remains visible if more than k participants stack their transparencies.

2.5   A $(2, n)$−Threshold VCS with Optimal Contrast

We now present a $(2, n)$−threshold VCS due to Blundo, De Santis, and Stinson [3] in which the relative difference is optimal.

The $n \times m$ basis matrices S^0 and S^1 are constructed as follows: The columns of S^1 consist of all binary n−vectors of weight $\lfloor n/2 \rfloor$. Hence, $m = \binom{n}{\lfloor n/2 \rfloor}$ and any row in S^1 has weight equal to $\binom{n-1}{\lfloor n/2 \rfloor - 1}$. S^0 is constructed from n identical row vectors of length m, and of weight $\binom{n-1}{\lfloor n/2 \rfloor - 1}$.

Theorem 1. *Let $n \geq 2$. In any $(2, n)$−threshold visual cryptography scheme with pixel expansion $m = \binom{n}{\lfloor n/2 \rfloor}$, $\alpha(m) \leq \alpha^*(n) = \lfloor n/2 \rfloor \lceil n/2 \rceil / n(n-1)$.*

Theorem 2. *For any $n \geq 2$, there exists a strong $(2, n)$−threshold visual cryptography scheme with pixel expansion $m = \binom{n}{\lfloor n/2 \rfloor}$ and $\alpha(m) = \alpha^*(n)$.*

2.6   Lattice-Based Color VTS

In this section we present constructions for lattice based visual cryptography schemes for color images due to Koga *et al* [6,7,8].We begin with a few definitions.

A poset $(L, \leq)$ is called a *lattice* if for all $a, b \in L$, the set $\{a, b\}$ has both a l.u.b and a g.l.b. A poset $(L, \leq)$ is called an *upper semi-lattice* if for all $a, b \in L$ the set $\{a, b\}$ has a l.u.b. If $(L, \leq)$ is an upper semi-lattice then the *join* (denoted by $\sqcup$) of L is a binary operation on L defined as follows: For $x, y \in L$, $x \sqcup y = $ least upper bound of $\{x, y\}$. An upper semi-lattice is said to be *bounded* if it contains the least element 0 and the greatest element 1 such that $x \sqcup 0 = x$ and $x \sqcup 1 = 1$ for all $x \in L$.

Let L be a bounded upper semi-lattice of colors and let $m \geq 2$ be an integer. Define $\mathcal{C} = \{c_1, \dots, c_K\} \subseteq L$ to be the set of colors that the secret image contains. This subset need not be a sublattice of L. Let $\mathcal{P}$ be a set of n participants. We can view an element of $(L^m)^n$ as an $n \times m$ matrix S whose entries are elements of L. For $1 \leq p \leq n$ and $A = \{i_1, \dots, i_p\} \subseteq \mathcal{P}$ define $S[A]$ to be the $p \times m$ matrix obtained by restricting S to rows $i_1, \dots, i_p$. For such a p and A define the mapping $h : (L^m)^n \rightarrow L^m$ as $h(S[A]) = \bar{s}_{i_1} \sqcup_m \bar{s}_{i_2} \sqcup_m \dots \sqcup_m \bar{s}_{i_p}$ where $\bar{s}_{i_j}$ denotes the i_j-th row of matrix S and $\sqcup_m$ denotes the join of L^m defined previously. Note that h describes the physical operation of stacking the shares of the i_1th, $\dots, i_p$th participants.

A lattice-based visual cryptography scheme for an access structure Γ is defined as follows:

Definition 4. *Let $\Gamma = (\Gamma_{Qual}, \Gamma_{Forb})$ be an access structure on a set $\mathcal{P}$ of participants. Let L be a bounded upper semi-lattice of colors and $\mathcal{C} = \{c_1, \dots, c_K\}$ a subset of L. Let $\mathcal{X}_{c_i}$ $(1 \leq i \leq K)$ denote a collection of $n \times m$ matrices with elements from L. The set $\{\mathcal{X}_{c_i}\}_{i=1}^K$ is called a lattice-based visual cryptography scheme for the access structure Γ with colors $\mathcal{C}$ and pixel expansion m if the following two properties hold:*

1. *For each $1 \leq i \leq K$, if $A \in \Gamma_{Qual}^*$ and $S \in \mathcal{X}_{c_i}$, then $h(S[A]) \in L^m$ contains only 1s and at least one c_i. Γ_{Qual}^* is a minimum qualified set of Γ_{Qual}.*
2. *If $A \in \Gamma_{Forb}^*$, then the sets $\mathcal{X}_{c_i}[A]$ $(1 \leq i \leq K)$ are indistinguishable in the sense that they contain the same elements with the same frequencies. $\mathcal{X}_{c_i}[A]$ is defined as $\mathcal{X}_{c_i}[A] = \{S[A] : S \in \mathcal{X}_{c_i}\}$. Γ_{Forb}^* refers to the maximal forbidden sets of Γ_{Forb}.*

2.7   (2,n)-Lattice-Based VTS

Let us first present a few notations. For $j = 1, 2, \dots, n$, define $M_{n,n-j}(x)$ as the matrix obtained by permuting a column containing one x, $n - j$ 0's, and $j - 1$ 1's in all possible ways.

If M is a matrix and $\alpha \geq 1$ an integer, $M^{[\alpha]}$ denotes the matrix obtained by concatenating M with itself α times i.e $M^{[\alpha]} = \underbrace{M \circ M \circ \dots \circ M}_{\alpha \text{ times}}$.

In the paper [6] if the color set is $\mathcal{C} = \{c_1, \ldots, c_K\}$, the basis matrices S_{c_i} ($c_i \in \{c_1, \ldots, c_K\}$) are defined as follows: $S_{c_i} = A(c_i) \circ D(c_1) \circ \ldots \circ D(c_{i-1}) \circ D(c_{i+1}) \circ \ldots \circ D(c_K)$ where $A(x) = M_{n,1}(x)$ and $D(x) = M_{n,0}^{[n-1]}(x)$. Here the pixel expansion is $kn(n-1)$ and the color ratio is $2/kn(n-1)$.

In the paper [8] ,they first construct the basis matrices for $(2,2)$-color VTS. If the color set is $\mathcal{C} = \{c_1, \ldots, c_K\}$,then the basis matrices X_{c_i} ($c_i \in \{c_1, \ldots, c_K\}$) of $(2,2)$-color VTS are defined as follows: $X_{c_i} = A(c_i) \circ D(c_1) \circ \ldots \circ D(c_{i-1}) \circ D(c_{i+1}) \circ \ldots \circ D(c_K)$ where $A(X) = M_2(X, 0)$ and $D(X) = M_2(X, 1)$ and $M_n(a_1, \ldots, a_n)$ is an $n \times n!$ matrix consisting of all permutations of the vector $(a_1, \ldots, a_n)^t$. To construct the basis matrices S_{c_i} of a $(2,n)$-color VTS with a set of colors $\mathcal{C} = \{c_1, \ldots, c_K\}$,they just replace $M_2(X,Y)$ of X_{c_i} by $M_n(X, Y, 1, \ldots, 1)$. So they have pixel expansion $kn!$ and color ratio $2/kn!$.

3   A New Construction of a $(2,n)$−Color VTS with Three Base Colors

Before we describe our scheme, we define what we mean by a color visual threshold scheme.

Definition 5. *Let us suppose we can generate all the colors in the secret image using the color set $\mathcal{C} = \{c_1, c_2, \ldots, c_J\}$. A set of J $n \times m$ matrices G_i with entries from the set $\{0, 1, c_1, c_2, \ldots, c_J\}$ form a (k,n) visual threshold scheme if the following properties are satisfied:*

1. *For any i, ($1 \leq i \leq J$), the m−vector obtained by "superimposing" any k rows of G_i has at least L_i entries which are c_i; each of the remaining colors c_j appear at most U_i^j times in this m−vector.*
2. *For any subset $\{i_1, i_2, \ldots, i_j\} \subset \{1, \ldots, n\}$, the submatrices G_i' obtained by restricting each G_i to the rows $i_1, \ldots, i_j$ are identical up to a column permutation.*

Here the L_i's signify lower bounds and the U_i^j's signify upper bounds. The true color c_i must appear at least L_i times when two shares for a pixel of color c_i are combined; other colors c_j may appear on superimposing two arbitrary shares for c_i, but they do so at most U_i^j times. The second property is related to security and it ensures that no set of $k - 1$ participants or fewer can decipher the secret image.

suppose that the original image is composed of only three colors, namely Cyan (C), Yellow (Y) and Green (G). We will first construct a $(2,2)$−VTS with the color set $\mathcal{C} = \{C, Y, G\}$. To do this, it is sufficient to construct the basis matrices S_C, S_Y, S_G corresponding to the colors C, Y and G respectively. These basis matrices are given below:

$$S_C = \begin{bmatrix} C & 0 & Y & 1 \\ 0 & C & 1 & Y \end{bmatrix}, \; S_Y = \begin{bmatrix} Y & 0 & C & 1 \\ 0 & Y & 1 & C \end{bmatrix} \; and \; S_G = \begin{bmatrix} C & Y & 1 & 0 \\ Y & C & 0 & 1 \end{bmatrix}$$

To construct a $(2,n)$−color VTS ($n \geq 3$) with base colors C, Y, and G, we first consider the basis matrix S^1 of the $(2,n)$−VTS for black and white images as described in [3]. S^1 is an $n \times m$ Boolean matrix which is realized by considering all binary $n-$vectors of weight $\lfloor n/2 \rfloor$, where $m = \begin{pmatrix} n \\ \lfloor n/2 \rfloor \end{pmatrix}$.

To construct our $(2, n)-$ color VTS $(n \geq 3)$, we will use the basis matrices of the $(2,2)$ scheme as templates in the construction of the basis matrices of the $(2, n)$ scheme. From here on, we will denote the basis matrix S_{col} of the $(2,2)$ scheme by X_{col}. Therefore, S_C, S_Y, S_G defined above will be denoted as X_C, X_Y, and X_G respectively. S_C, S_Y and S_G, the basis matrices for the $(2, n)$ scheme, are defined below.

S_C is an $n \times 4m$ matrix defined in terms of S^1 as follows:
Replace a 0 in S^1 by the first row of X_C and a 1 in S^1 by the second row of X_C.

S_Y and S_G are constructed in a similar fashion. One can verify that the ith rows $(1 \leq i \leq n)$ of the matrices S_C, S_Y and S_G thus obtained are identical up to a column permutation.

3.1 An Example of a $(2, 4)-$Color VTS with Base Colors C, Y, and G

To construct a $(2, 4)-$color VTS with base colors C, Y, and G we first look at the basis matrix S^1 of a $(2, 4)-$VTS for black and white images. The matrix S^1 is shown below.

$$\begin{bmatrix} 1 & 0 & 0 & 1 & 0 & 1 \\ 1 & 1 & 0 & 0 & 1 & 0 \\ 0 & 1 & 1 & 1 & 0 & 0 \\ 0 & 0 & 1 & 0 & 1 & 1 \end{bmatrix}.$$

Since $X_C = \begin{bmatrix} C & 0 & Y & 1 \\ 0 & C & 1 & Y \end{bmatrix}$, $X_Y = \begin{bmatrix} Y & 0 & C & 1 \\ 0 & Y & 1 & C \end{bmatrix}$ and $X_G = \begin{bmatrix} C & Y & 1 & 0 \\ Y & C & 0 & 1 \end{bmatrix}$ we have

S_C:

$$\begin{bmatrix} 0 & C & 1 & Y & C & 0 & Y & 1 & C & 0 & Y & 1 & 0 & C & 1 & Y & C & 0 & Y & 1 & 0 & C & 1 & Y \\ 0 & C & 1 & Y & 0 & C & 1 & Y & C & 0 & Y & 1 & C & 0 & Y & 1 & 0 & C & 1 & Y & C & 0 & Y & 1 \\ C & 0 & Y & 1 & 0 & C & 1 & Y & 0 & C & 1 & Y & C & 0 & Y & 1 & C & 0 & Y & 1 & C & 0 & Y & 1 \\ C & 0 & Y & 1 & C & 0 & Y & 1 & 0 & C & 1 & Y & C & 0 & Y & 1 & 0 & C & 1 & Y & 0 & C & 1 & Y \end{bmatrix}$$

S_Y:

$$\begin{bmatrix} 0 & Y & 1 & C & Y & 0 & C & 1 & Y & 0 & C & 1 & 0 & Y & 1 & C & Y & 0 & C & 1 & 0 & Y & 1 & C \\ 0 & Y & 1 & C & 0 & Y & 1 & C & Y & 0 & C & 1 & Y & 0 & C & 1 & 0 & Y & 1 & C & Y & 0 & C & 1 \\ Y & 0 & C & 1 & 0 & Y & 1 & C & 0 & Y & 1 & C & Y & 0 & C & 1 & Y & 0 & C & 1 & Y & 0 & C & 1 \\ Y & 0 & C & 1 & Y & 0 & C & 1 & 0 & Y & 1 & C & Y & 0 & C & 1 & 0 & Y & 1 & C & 0 & Y & 1 & C \end{bmatrix}$$

S_G:

$$\begin{bmatrix} Y & C & 0 & 1 & C & Y & 1 & 0 & C & Y & 1 & 0 & Y & C & 0 & 1 & C & Y & 1 & 0 & Y & C & 0 & 1 \\ Y & C & 0 & 1 & Y & C & 0 & 1 & C & Y & 1 & 0 & C & Y & 1 & 0 & Y & C & 0 & 1 & C & Y & 1 & 0 \\ C & Y & 1 & 0 & Y & C & 0 & 1 & Y & C & 0 & 1 & Y & C & 0 & 1 & C & Y & 1 & 0 & C & Y & 1 & 0 \\ C & Y & 1 & 0 & C & Y & 1 & 0 & Y & C & 0 & 1 & C & Y & 1 & 0 & Y & C & 0 & 1 & Y & C & 0 & 1 \end{bmatrix}$$

Here $R_C = 10/24$, $R_Y = 10/24$ and $R_G = 8/24$. Note that the color ratio of the above scheme is $1/3$.

3.2 Share Distribution Algorithm

We use the following algorithm to encode the secret image.

For each pixel P in the secret image do the following:

1. Generate a random permutation π of the set $\{1, 2, \ldots, 4m\}$.
2. If P is a cyan pixel, apply π to the columns of S_C. Call the resulting matrix T_C. If P is Y or G, apply π to the columns of the matrices S_Y or S_G respectively to form T_Y or T_G.
3. For $1 \leq i \leq n$, row i of T_X, where $X \in \{C, Y, G\}$, describes the color distribution among the $4m$ subpixels of the ith share.

If $n = 2$, we do the same thing as above by taking $m = 1$.

We next show that the color ratio attained by this encryption scheme is bounded below by $1/4$. To prove this result we need the following lemmas:

Lemma 1. *For any distinct $i, j = 1, 2, \ldots, n$ let $S^1\{i, j\}$ denote the $2 \times m$ matrix obtained by restricting S^1 to rows i and j. Then the submatrix $S^1\{i, j\}$ has equal number of patterns of the forms:* $\begin{bmatrix} 0 \\ 1 \end{bmatrix}$ *and* $\begin{bmatrix} 1 \\ 0 \end{bmatrix}$ *and this number is equal to $m \cdot \alpha(m)$. Also the total number of patterns of the forms $\begin{bmatrix} 0 \\ 0 \end{bmatrix}$ and $\begin{bmatrix} 1 \\ 1 \end{bmatrix}$ is given by $m - 2 \cdot m \cdot \alpha(m)$. Here $\alpha(m)$ is given by* $\alpha(m) = \begin{cases} 1/4 + 1/4(n-1), & \text{if } n \text{ is even} \\ 1/4 + 1/4n, & \text{if } n \text{ is odd} \end{cases}$

Proof. Choose two distinct rows i and j and suppose that $i < j$. The number of $\begin{bmatrix} 1 \\ 0 \end{bmatrix}$ patterns in $S^1\{i, j\}$ is equal to the number of ways of distributing the remaining $\lfloor n/2 \rfloor - 1$ 1's among the remaining $n - 2$ rows. This can be done in $\binom{n-2}{\lfloor n/2 \rfloor - 1}$ ways. It so happens that the contrast $\alpha(m) \cdot m$ of the $(2, n)$ scheme of which S^1 is a basis matrix is equal to the above number. By a similar argument, there is an equal number of $\begin{bmatrix} 0 \\ 1 \end{bmatrix}$ patterns in $S^1\{i, j\}$. Since S^1 has m columns, the total number of patterns of the forms $\begin{bmatrix} 0 \\ 0 \end{bmatrix}$ and $\begin{bmatrix} 1 \\ 1 \end{bmatrix}$ must be $m - 2 \cdot m \cdot \alpha(m)$. $\qquad\square$

Note. From here on whenever we use the term $\alpha(m)$, we will use it as defined above in Lemma 1. To see how $\alpha(m)$ is obtained, consult reference [3].

Lemma 2. *For each cyan (or yellow) pixel in the original image, the total number of cyan (or yellow) colored subpixels in the image formed by superimposing two arbitrary shares is given by*

$$m[1 + 2 \cdot \alpha(m)].$$

For each green pixel in the original image, the total number of green colored subpixels in the superimposed image is given by $4m \cdot \alpha(m)$.

Proof. Recall that we defined X_C as follows: $X_C = \begin{bmatrix} C & 0 & Y & 1 \\ 0 & C & 1 & Y \end{bmatrix}$. First note that each one of the patterns $\begin{bmatrix} 0 \\ 1 \end{bmatrix}, \begin{bmatrix} 1 \\ 0 \end{bmatrix}$ in S^1 contributes two C's in S_C. Each one of the patterns $\begin{bmatrix} 0 \\ 0 \end{bmatrix}, \begin{bmatrix} 1 \\ 1 \end{bmatrix}$ in S^1 contributes one C in S_C. So for each cyan colored pixel in the original image, the total number of cyan colored subpixels in the image obtained by superimposing two arbitrary shares, is given by (by using lemma 1) $2m \cdot \alpha(m) + 2m \cdot \alpha(m) + m - 2m \cdot \alpha(m)$ which simplifies to $m[1 + 2 \cdot \alpha(m)]$ One can verify that this result holds for a yellow pixel also.

Next we will work this out for a green pixel. Recall that we defined X_G as follows $X_G = \begin{bmatrix} C & Y & 1 & 0 \\ Y & C & 0 & 1 \end{bmatrix}$ Note that each one of the patterns $\begin{bmatrix} 0 \\ 1 \end{bmatrix}, \begin{bmatrix} 1 \\ 0 \end{bmatrix}$ in S^1 contributes two G's in S_G. The patterns $\begin{bmatrix} 0 \\ 0 \end{bmatrix}$ and $\begin{bmatrix} 1 \\ 1 \end{bmatrix}$ contribute no G's in S_G. Thus for each green pixel in the original image, the total number of green colored subpixels in the superimposed image is given by (by using lemma 1)
$$2m \cdot \alpha(m) + 2m \cdot \alpha(m) + 0 + 0 = 4m \cdot \alpha(m).$$ □

Now we come to the main theorem.

Theorem 3. *For any $n \geq 3$, there exists a $(2, n)-$color VTS with base colors Cyan (C), Yellow(Y), and Green (G) with pixel expansion $4m$ and color ratios*

$$R_C = \begin{cases} 3/8 + 1/8(n-1), & \text{if } n \text{ is even} \\ 3/8 + 1/8n, & \text{if } n \text{ is odd} \end{cases}$$

$$R_Y = \begin{cases} 3/8 + 1/8(n-1), & \text{if } n \text{ is even} \\ 3/8 + 1/8n, & \text{if } n \text{ is odd} \end{cases}$$

$$R_G = \begin{cases} 1/4 + 1/4(n-1), & \text{if } n \text{ is even} \\ 1/4 + 1/4n, & \text{if } n \text{ is odd} \end{cases}$$

For $n = 2$, the color ratio of the $(2, n)-$color VTS is $1/2$ and the pixel expansion is 4.

Proof. That for $n = 2$, the color ratio is $1/2$ and the pixel expansion is 4 can be seen by examining the basis matrices of the (2,2) scheme.

Therefore, consider the case when $n \geq 3$. The number of columns in each S_X, $X \in \{C, Y, G\}$ is $4m$. To find out the color ratios R_C, R_Y and R_G we need to examine the construction of the basis matrix S^1. The color ratio depends on the patterns $\begin{bmatrix} 0 \\ 1 \end{bmatrix}, \begin{bmatrix} 1 \\ 0 \end{bmatrix}, \begin{bmatrix} 0 \\ 0 \end{bmatrix}, \begin{bmatrix} 1 \\ 1 \end{bmatrix}$ that appear in the matrix $S^1\{i, j\}$ for any $i, j = 1, 2, \ldots, n$. Now by using lemma 2, the color ratio R_C is given by

$$R_C = [m \cdot (2 \cdot \alpha(m) + 1)]/4m = 1/4 + \alpha(m)/2$$

From [3] we know that $\alpha(m)$ is given by,

$$\alpha(m) = \begin{cases} 1/4 + 1/4(n-1), & \text{if } n \text{ is even} \\ 1/4 + 1/4n, & \text{if } n \text{ is odd} \end{cases}$$

Therefore, R_C is given by

$$R_C = \begin{cases} 3/8 + 1/8(n-1), & \text{if } n \text{ is even} \\ 3/8 + 1/8n, & \text{if } n \text{ is odd} \end{cases}$$

R_Y can be derived in a similar fashion. One can verify that R_Y is given by:

$$R_Y = \begin{cases} 3/8 + 1/8(n-1), & \text{if } n \text{ is even} \\ 3/8 + 1/8n, & \text{if } n \text{ is odd} \end{cases}$$

The color ratio R_G is given by

$$R_G = 4m \cdot \alpha(m)/4m = \alpha(m) = \begin{cases} 1/4 + 1/4(n-1), & \text{if } n \text{ is even} \\ 1/4 + 1/4n, & \text{if } n \text{ is odd} \end{cases}$$

$\square$

Observe that each share contains the original base colors C, Y, and G along with white (0) and black (1). If the encoded pixel is colored cyan then on superimposing any two shares, the resulting reconstructed pixel will have the colors cyan, yellow, black, and white. Note that the color yellow is unwanted here, because too many yellow subpixels may fool the visual system into thinking this as a yellow pixel. We will call such unwanted colors *nuisance colors*.

Suppose that the encoded pixel has color c_i, and on superimposing two arbitrary shares, we find some subpixels with color c_j ($c_j \neq$ black, white). Then we define c_j to be a nuisance color for c_i and we denote the *number* of such subpixels in a reconstructed pixel for c_i by $N_{c_i}^{c_j}$. In this case, yellow is a nuisance color for cyan and we denote the number of yellow subpixels in a cyan pixel by N_C^Y. Similarly for a yellow pixel, the nuisance color is cyan and we denote the number of cyan subpixels by N_Y^C. Finally, a green pixel has as nuisance colors both cyan and yellow and we denote their numbers by N_G^C and N_G^Y respectively.

Note that we have not defined white or black as nuisance colors. This is because the presence of white subpixels serve to make the image lighter, but they do not hinder the visual system from discerning the true color of a pixel. Black subpixels darken the image, but again, they do not hinder the visual system from discerning the color of a pixel.

We will next determine the numbers N_C^Y, N_Y^C, N_G^Y, and N_G^C. Note that in the (2,2) color VTS with color set $\{C, Y, G\}$ there are no nuisance colors. Let the encoded pixel be a cyan colored one. When the ith and jth ($1 \leq i < j \leq n$) shares are superimposed, the nuisance color yellow appears due to the patterns $\begin{bmatrix} 0 \\ 0 \end{bmatrix}$ and $\begin{bmatrix} 1 \\ 1 \end{bmatrix}$ in $S^1\{i, j\}$. Each of the above patterns contributes one yellow subpixel in the superimposed image. Recall that the total number of such patterns is given by $m - 2 \cdot m \cdot \alpha(m)$. Hence the number of yellow subpixels N_C^Y for each cyan pixel in the original image is given by

$$N_C^Y = m - 2 \cdot m \cdot \alpha(m) = \begin{cases} m[\frac{1}{2} - \frac{1}{2(n-1)}], & \text{if } n \text{ is even} \\ m[\frac{1}{2} - \frac{1}{2n}], & \text{if } n \text{ is odd} \end{cases}$$

Note that $N_C^Y < m/2$. The pixel expansion, in this case, is $4m$. Out of these $4m$ subpixels, less than $m/2$ subpixels are yellow. That is, less than $1/8$th of the subpixels will be nuisance colored. One can show that $N_Y^C = N_C^Y$. Therefore, these observations hold for cyan colored pixels also. For each green pixel in the secret image, the number of cyan and yellow colored subpixels are each less than $m/2$. Hence, $N_G^C + N_G^Y < m$. That is, for each green pixel in the reconstructed image the total number of nuisance colored subpixels $< m$ and the fraction of such subpixels $< m/4m = 1/4$.

Although, the reconstructed image has nuisance colors, they will not affect the quality of the reconstructed image significantly. Nuisance colors do not produce any diminution of brightness, as the overwhelming majority of black subpixels do in [6] and [8].

4   Construction of a $(2, n)$−Color VTS with Six Base Colors

Suppose that the original image is made up of exactly six colors namely Red (R), Green (G), Blue (B), Cyan (C), Yellow (Y) and Magenta (M). To construct a $(2, n)$−color VTS with base colors R, G, B, Y, C, and M we first consider the basis matrix S^1 of the $(2, n)$− VTS for black and white images that was described earlier. Recall that S^1 is an $n \times m$ Boolean matrix which is realized by considering all binary n-vectors of weight $\lfloor n/2 \rfloor$, where $m = \binom{n}{\lfloor n/2 \rfloor}$. Now for each of the six colors we construct a 2×6 matrix as shown below.

$$X_Y = \begin{bmatrix} Y & 0 & C & M & 1 & 1 \\ 0 & Y & 1 & 1 & C & M \end{bmatrix}, X_C = \begin{bmatrix} C & 0 & M & Y & 1 & 1 \\ 0 & C & 1 & 1 & M & Y \end{bmatrix}, X_M = \begin{bmatrix} M & 0 & Y & C & 1 & 1 \\ 0 & M & 1 & 1 & Y & C \end{bmatrix}$$

$$X_R = \begin{bmatrix} Y & M & C & 1 & 1 & 0 \\ M & Y & 1 & C & 0 & 1 \end{bmatrix}, X_G = \begin{bmatrix} Y & C & M & 1 & 1 & 0 \\ C & Y & 1 & M & 0 & 1 \end{bmatrix}, X_B = \begin{bmatrix} M & C & Y & 1 & 1 & 0 \\ C & M & 1 & Y & 0 & 1 \end{bmatrix}$$

Note that C + Y = G, C + M = B and Y + M = R. To construct a $(2, n)$−color VTS it is sufficient to construct the basis matrices S_R, S_G, S_B, S_C, S_Y, and S_M for the colors R, G, B, C, Y, and M respectively. For $n = 2$, we have $S_P = X_P$ where P = C, Y, M, R, G, B. Now we consider our $(2, n)$−color VTS with $n \geq 3$.

We define S_R as an $n \times 6m$ matrix that is constructed as follows:

Replace each occurrence of a 0 in S^1 by the first row of X_R and that of a 1 by the second row of X_R.

S_G, S_B, S_C, S_Y and S_M are constructed in a similar fashion. Note that the ith rows of the matrices S_R, S_G, S_B, S_C, S_Y, and S_M thus obtained are identical up to a column permutation.

4.1   Share Distribution Algorithm

The share distribution algorithm is exactly the same as the one given for three colors. The proof is similar to the proof of Theorem 3.

Theorem 4. *For any $n \geq 3$, there exists a $(2, n)-color$ VTS with base colors Red (R), Green (G), Blue (B), Cyan (C), Yellow (Y), and Magenta (M) with pixel expansion 6m and color ratios*

$$R_Y = \begin{cases} 1/4 + 1/12(n-1), & \text{if } n \text{ is even} \\ 1/4 + 1/12n, & \text{if } n \text{ is odd} \end{cases}$$

$$R_C = \begin{cases} 1/4 + 1/12(n-1), & \text{if } n \text{ is even} \\ 1/4 + 1/12n, & \text{if } n \text{ is odd} \end{cases}$$

$$R_M = \begin{cases} 1/4 + 1/12(n-1), & \text{if } n \text{ is even} \\ 1/4 + 1/12n, & \text{if } n \text{ is odd} \end{cases}$$

$$R_R = \begin{cases} 1/6 + 1/6(n-1), & \text{if } n \text{ is even} \\ 1/6 + 1/6n, & \text{if } n \text{ is odd} \end{cases}$$

$$R_G = \begin{cases} 1/6 + 1/6(n-1), & \text{if } n \text{ is even} \\ 1/6 + 1/6n, & \text{if } n \text{ is odd} \end{cases}$$

$$R_B = \begin{cases} 1/6 + 1/6(n-1), & \text{if } n \text{ is even} \\ 1/6 + 1/6n, & \text{if } n \text{ is odd} \end{cases}$$

If $n = 2$, then the pixel expansion is 6 and the color ratio of the scheme is 1/3.

5   Generalization to an Arbitrary Number of Colors

Our scheme can be generalized to an arbitrary number of colors. If the secret image has the colors $\mathcal{C} = \{c_1, c_2, \ldots, c_k\}$ such that no two colors c_i and $c_j \in \mathcal{C}$ can be combined to produce a third color $c_l \in \mathcal{C}$, then for each color $c_i \in \mathcal{C}$ we define matrix X_{c_i} as follows :

$$X_{c_i} = \begin{bmatrix} c_i & 0 & c_1 & \ldots & c_{i-1} & c_{i+1} & \ldots & c_k & 1 & \ldots & 1 & 1 & \ldots & 1 \\ 0 & c_i & 1 & \ldots & 1 & 1 & \ldots & 1 & c_1 & \ldots & c_{i-1} & c_{i+1} & \ldots & c_k \end{bmatrix}$$

Note that X_{c_i} has $2k$ columns. We define S_{c_i} in terms of X_{c_i} and S^1 as follows:
S_{c_i} is an $n \times 2 \cdot m \cdot k$ matrix obtained by replacing the 0's of S^1 by the first row of X_{c_i} and the the 1's of S^1 by the second row of X_{c_i}.
On combining any two rows of S_{c_i}, the number of columns with color c_i is $m[1 + 2 \cdot \alpha(m)]$. The proof of this is similar to the proof of Lemma 2. The color ratio X_{c_i} is therefore:

$$X_{c_i} = \frac{m[1 + 2 \cdot \alpha(m)]}{2mk}$$

$$= \frac{1}{2k} + \frac{\alpha(m)}{k}$$

This lower bound is independent of the number of shares and depends only on the number of colors in the secret image. In case colors can be combined, we will get a better lower bound for the color ratio.

6   Conclusion

In this paper we have proposed a new construction for a $(2, n)$−VTS with the color sets $\mathcal{C}_1 = \{C, Y, G\}$ and $\mathcal{C}_2 = \{C, Y, M, R, G, B\}$. For the first color set $\mathcal{C}_1$, our scheme achieves a color ratio that is lower bounded by $1/4$; for the second color set $\mathcal{C}_2$, the lower bound on the color ratio is $1/6$. The color ratio that we obtain is also much higher than those obtained from the schemes in [6] and [8]. The pixel expansion is quite good, and is certainly better than the one in [8]. We have also shown that our scheme can be extended to an arbitrary number of colors and that the pixel expansion of our scheme does not depend on the number of shares but only on the number of base colors. Extending this scheme to a (k, n)−VTS remains an open problem.

References

1. G. Ateniese, C. Blundo, A. De Santis, and D.R. Stinson, *Visual Cryptography for General Access Structures* Information and Computation, vol.129, pp.86-106, 1996.
2. G. Ateniese, C. Blundo, A. De Santis, and D.R. Stinson, *Constructions and Bounds for Visual Cryptography*, in "23rd International Colloquim on Automata, Languages and Programming" (ICALP '96), F.M. auf der Heide and B. Monien Eds., Vol. 1099 of "Lecture Notes in Computer Science", Springer-Verlag, Berlin, pp. 416-428,1996.
3. C. Blundo, A.De Santis, and D.R. Stinson, *On the Contrast in Visual Cryptography Schemes*, J. Cryptology, vol.12, no.4, pp.261-289, 1999.
4. C. Blundo, A.De Santis and M. Naor, *Visual Cryptography for gray Level Images*, Inf. Process. Lett., Vol. 75, Issue. 6, pp.255-259, 2001.
5. S. Droste, *New Results on Visual Cryptography*, Advance in Cryptography-CRYPT'96, Lecture Notes in Computer Science, 1109, pp. 401-415, Springer-Verlag, 1996.
6. H. Koga and H. Yamamoto, *Proposal of a Lattice-Based Visual Secret Sharing Sceme for Color and Gray-Scale Images*, IEICE Trans. Fundamentals, Vol. E81-A, No. 6 June 1998.
7. H. Koga and T. Ishihara, *New Constructions of the Lattice-Based Visual Secret Sharing Scheme Using Mixture of Colors*, IEICE Trans. Fundamentals, Vol. E85-A, No. 1 January 2002.
8. H. Koga, M. Iwamoto and H. Yamamoto, *An Analytic Construction of the Visual Secret Scheme for Color Images*, IEICE Trans. Fundamentals, Vol. E84-A, No.1 January 2001.
9. M. Naor and A. Shamir, *Visual Cryptography*, Advance in Cryptography, Eurocrypt'94, Lecture Notes in Computer Science 950, pp. 1-12, Springer-Verlag, 1994.
10. V. Rijmen and B. Preneel, *Efficient Colour Visual Encryption or "Shared Colors of Benetton"*, presented at EUROCRYPT '96 Rump Session. Available at http://www.iacr.org/conferences/ec96/rump/preneel.ps.

On the Power of Computational Secret Sharing

V. Vinod[1][*], Arvind Narayanan[2], K. Srinathan[2][**], C. Pandu Rangan[2][***], and Kwangjo Kim[3]

[1] Laboratory for Computer Science, Massachusetts Institute of Technology, Cambridge, MA 02139, USA.
vinodv@mit.edu
[2] Department of Computer Science and Engineering, Indian Institute of Technology, Madras, Chennai - 600036, India.
arvindn@meenakshi.iitm.ernet.in, ksrinath@cs.iitm.ernet.in, rangan@iitm.ernet.in
[3] IRIS (International Center for Information Security), ICU (Information and Communications University) 58-4, Hwaam-Dong, Yusong-gu, Taejon 305-732, South Korea.
kkj@icu.ac.kr

Abstract. Secret sharing is a very important primitive in cryptography and distributed computing. In this work, we consider computational secret sharing (CSS) which provably allows a smaller share size (and hence greater efficiency) than its information-theoretic counterparts. Extant CSS schemes result in succinct share-size and are in a few cases, like threshold access structures, optimal. However, in general, they are not *efficient* (share-size not polynomial in the number of players n), since they either assume efficient perfect schemes for the given access structure (as in [10]) or make use of exponential (in n) amount of public information (like in [5]). In this paper, our goal is to explore other classes of access structures that admit of efficient CSS, without making any other assumptions. We construct efficient CSS schemes for every access structure in monotone P. As of now, most of the efficient information-theoretic schemes known are for access structures in algebraic NC^2. Monotone P and algebraic NC^2 are not comparable in the sense one does not include other. Thus our work leads to secret sharing schemes for a new class of access structures. In the second part of the paper, we introduce the notion of secret sharing with a semi-trusted third party, and prove that in this relaxed model efficient CSS schemes exist for a wider class of access structures, namely monotone NP.

Keywords: Secret sharing, computationally bounded players, access structures, monotone P.

---

[*] Work done when the author was at the Indian Institute of Technology, Madras.
[**] Financial support from Infosys Technologies Limited, India is acknowledged.
[***] Financial support from the Ministry of Information Technology, Government of Korea is acknowledged.

T. Johansson and S. Maitra (Eds.): INDOCRYPT 2003, LNCS 2904, pp. 162–176, 2003.

1   Introduction

Secret sharing schemes protect the secrecy and integrity of information by distributing the information over different locations (not necessarily geographical). This forces the *adversary* to attack multiple locations in order to learn or destroy the information. In a secret sharing protocol, the *dealer* shares his secret among n players. In the so called *threshold* model, the sharing is done so that subsets of $t + 1$ (or more) players can later correctly reconstruct the secret, while subsets of t (or fewer) players cannot reconstruct it. This notion can be generalized by specifying a family of authorized subsets of the n players, called *access structures*. The dealer shares the secret in such a way that only authorized subsets of players can reconstruct the secret while the players in non-authorized subsets cannot.

1.1   Why Computational Secret Sharing ?

CSS schemes allow us to achieve a better *information rate* than is possible with information theoretic schemes. The information rate of a secret sharing scheme is the maximum length of a (player's) share per unit size of the secret. This measure is of interest when the size of the message to be shared is large. For instance, it is possible ([10]) to t-share a secret, in the threshold model, with the share of each player being only $\frac{1}{t}$ as long as the secret (which is clearly optimal). Indeed, Beguin and Cresti [1] have shown that it is possible to achieve the optimal information rate for *any* access structure provided that a fixed length secret can be shared on the same access structure. The above schemes build on the idea of *information dispersal algorithms* [13]. CSS schemes which make use of a *bulletin board* on which an arbitrary amount of information can be published have also been proposed ([5]).

Clearly, a secret sharing scheme can be called *efficient* only if the information rate is polynomially bounded as a function of the number of players n. Furthermore, we posit that the amount of information published on the bulletin board be polynomial in n. Note that the CSS of [1] is not applicable when there is no known efficient secret sharing scheme for the corresponding access structure. Similarly, the CSS scheme of [5] becomes inefficient when the size of the access structure is not polynomial in n. Our interest in studying CSS is due to the fact that there may exist access structures that have efficient CSS schemes but do not permit any other (information theoretic) efficient secret sharing scheme.

Most of the results in extant literature on secret sharing schemes deal with information-theoretic secret sharing. We know that efficient information-theoretic perfect linear secret sharing schemes exist only if the access structure is in algebraic $NC^2 \cap mono$, the class of monotone languages which can be computed by algebraic circuits of logarithmic size and log-squared depth. Non-linear schemes appear to be more powerful [2]. However, there remains a large gap between the known upper bounds and lower bounds in the case of information theoretic secret sharing.

1.2   Boolean Formulas and Boolean Circuits

The contribution of the first part of this paper is the construction of efficient CSS schemes for all access structures which can be computed by monotone Boolean circuits of polynomial size. This complexity class is known as mP, for monotone P. Monotone Boolean circuits differ from monotone Boolean formulas in that in the former, the output of a node can serve as the input of more than one node. We say that gates in a monotone Boolean circuit can have a *fanout* of more than one. Another way of understanding the difference is that a monotone Boolean circuit is a *directed acyclic graph* where each node represents an AND gate or an OR gate whereas a monotone Boolean formula is a *tree* with the same property.

This makes monotone Boolean circuits much more powerful than formulas. In fact, it was believed that monotone Boolean circuits could simulate any (deterministic) Turing machine accepting a monotone set (thus making mP equivalent to $P \cap mono$, the class of monotone languages in P), but Razborov ([14]) proved that this is not the case.[4]

Benaloh and Leichter [3], in their landmark result on the existence of linear perfect secret sharing schemes for any monotone access structure (as first defined in [8]), showed how to combine secret sharing schemes across AND and OR gates (in other words, how to realize secret sharing schemes for the union and intersection of two access structures) and recursively applied this result to the Boolean formula computing the access structure. Our construction is similar in spirit. We represent the Boolean circuit computing the access structure as a graph whose nodes are AND, OR, or $FANOUT$ gates. (The $FANOUT$ gate takes a single input and produces multiple copies of the input. We do this in order that the $FANOUT$ is the only gate with more than one output.) With each edge of this graph, we associate a (virtual) secret, which we call the share of that edge. The shares of the input wires of the circuit form the shares of the players. The sharing scheme has the property that a subset S of players can compute the share of some edge E if the wire corresponding to E evaluates to 1 when the circuit is given (the encoding of) S as input. We show how to associate shares with edges in such a way that the above property is carried across AND, OR and $FANOUT$ gates.

Our techniques are similar in spirit to Yao's landmark garbled circuit construction ([15]), but very different in application since in the case of secret sharing, non-interactivity is essential. Thus our result does not follow from Yao's since secret sharing protocols cannot be expressed as special case(s) of interactive secure function evaluation protocols.

1.3   Semi-trusted Third Party

Our second result deals with secret sharing using a semi-trusted third party. The use of this construct is to introduce a limited amount of interactivity into

---

[4]   Razborov showed superpolynomial lower bounds for the monotone circuit complexity of the *matching* function.

the protocol, and thus increase its power. Just as the relaxation of the security requirement from information theoretic to computational security allows us to give protocols for a broader class of access structures, the relaxation of the non-interactivity requirement results in a further broadening. We prove that, using a semi-trusted third party, efficient CSS schemes exist for any class of access structures in monotone NP (denoted mNP). This is the class of languages accepted by monotone non-deterministic Turing machines in polynomial time, which also turns out to be the class of monotone languages in NP. Clearly, this includes all access structures that could possibly be interesting in practice.

The notion of a semi-trusted third party has been made use of in protocols for *fair exchange* ([6]). It allows two parties to exchange a secret in such a way that neither party can gain an unfair advantage by aborting the protocol at any point. To the best of our knowledge, however, secret sharing with a semi-trusted third party has not been considered.

A semi-trusted third party may try to deviate from the protocol, but it cannot collude with any of the players. It cannot be trusted with any of the private information of the other players. Therefore, in the case of secret sharing, we have the restriction that the semi-trusted third party should neither gain knowledge of the secret nor be able to identify the access set of players that tries to determine the secret. We make these notions formal in the next section.

Semi-trusted third parties are worthy of study because, unlike trusted third parties, they are readily realizable in practice. Indeed, [6] have suggested that in networks such as the internet, a *random* player can be chosen as a semi-trusted third party. In such a scenario, the third party is both geographically and logically separated from the players, and thus the possibility of both the third party and some of the other players coming under the control of a common adversary is remote. Another practical possibility is for a bank to play the role of a semi-trusted third party.

2   Preliminaries and Definitions

To begin with, we formally define the concept of Computational Secret Sharing for general access structures. A Secret Sharing scheme is a protocol between the set of players $\mathcal{P} = \{P_1, P_2, \ldots, P_n\}$ and a dealer D, where we assume $D \notin \mathcal{P}$. An access structure $\mathcal{A} \subseteq 2^{\mathcal{P}}$ consists of sets of players qualified to recover the secret. It is natural to consider only monotone access structures $\mathcal{A}$, that is, if $A \in \mathcal{A}$ and $A \subseteq A' \subseteq \mathcal{P}$, then $A' \in \mathcal{A}$. The set $\bar{\mathcal{A}} = 2^{\mathcal{P}} - \mathcal{A}$ is called the *adversary structure*. A set of players $A \in \mathcal{A}$ is called an *access set* or a *qualified subset*. A set of players $A \notin \mathcal{A}$ is called an *adversary set* or a *non-qualified subset*. We associate a class of access structures $\{\mathcal{A}_n\}$ with a language

$$L_{\mathcal{A}} = \{x = x_1 x_2 \ldots x_n : x_i \in \{0,1\}, \{P_i | x_i = 1\} \in \mathcal{A}_n\}$$

We make statements such as "the access structure $\mathcal{A}$ is in monotone P", when we actually mean to say that $L_{\mathcal{A}} \in$ monotone P.

The set of all possible secrets is called the *secret domain* (denoted by $\mathbb{S}$) and the set of all possible shares is called the *share domain* (denoted by $\mathbb{S}'$). Now, we formally define a computational secret sharing scheme.

Definition 1 (Computational Secret Sharing). *A computational secret sharing scheme is a protocol π between D and $\mathcal{P}$ to share a secret $S \in \mathbb{S}$, respective to an access structure $\mathcal{A}$ such that*

- *The dealer D transmits a share $S_i \in \mathbb{S}'$ to the player P_i, for $i = 1, 2, \ldots, n$. D retires from the protocol immediately afterwards.*
- *There is a polynomial-time algorithm π_{REC} such that $\pi_{REC}(S_{i_1}, S_{i_2}, \ldots, S_{i_m})$ $= S$ with probability 1 if $\{P_{i_1}, P_{i_2}, \ldots, P_{i_m}\} \in \mathcal{A}$.*
- *For any set of players $\{P_{i_1}, P_{i_2}, \ldots, P_{i_m}\} \notin \mathcal{A}$ and any (possibly randomized) polynomial-time algorithm π_{ADV}, $Prob[\pi_{ADV}(S_{i_1}, S_{i_2}, \ldots, S_{i_m}) = S] \leq \frac{1}{|\mathbb{S}|^c}$ for all constants c and suitably chosen $|\mathbb{S}|$.*

In our secret-sharing schemes, the domains of the secret and of the shares are the same, and this common domain is a finite field, which we denote by $\mathcal{F}$.

Definition 2 (Secret sharing using a semi-trusted third party). *A computational secret sharing scheme using a semi-trusted third party is a pair of protocols σ and ρ between a dealer D, the set of players $\mathcal{P}$ and a third party T, to share a secret $S \in \mathbb{S}$, respective to an access structure $\mathcal{A}$ such that*

- *In the sharing protocol σ, the dealer D transmits a share $S_i \in \mathbb{S}'$ to the player P_i, for $i = 1, 2, \ldots, n$, and a share S_0 to T. D retires from the protocol immediately afterwards.*
- *The reconstruction protocol ρ is an interactive protocol between T and some subset $A \subset \mathcal{P}$, represented by the virtual player P, at the end of which:*
  - *T should not obtain any information about S or about A.*
  - *If $A \in \mathcal{A}$ then P should be able to compute the secret S with certainty in polynomial time.*
  - *If $A \notin \mathcal{A}$, then the probability that P can compute S in polynomial time should be negligible.*

Monotone Boolean circuit. A monotone Boolean circuit is a Boolean circuit consisting of *AND*, *OR* and *FANOUT gates* connected by *wires*. Both *AND* and *OR* gates have two inputs and one output; *FANOUT* gates have one input and two outputs. *AND* and *OR* gates perform Boolean multiplication and addition respectively on their inputs, while the *FANOUT* gates propagate their input to both outputs. The circuit has n input wires (which are not the output of any gate) and one output wire (which is not the input of any gate).

There are two values associated with each wire W of the circuit - the Boolean value of W obtained by the evaluation of the circuit on some input assignment and the share value associated with W during the sharing and reconstruction process. The Boolean value of W corresponding to an input assignment $x_1, x_2, \ldots, x_n$ is denoted by $Eval(W, A)$ where $(x_1, x_2, \ldots x_n)$ is the encoding of A. We abuse notation and denote $Eval(W, A)$ by $Eval(W)$ when it is clear which set of players A we are referring to. Given a wire W in the circuit, we denote by $V(W)$ the share-value associated with W.

Definition 3 (Nondeterministic Boolean circuit). *A nondeterministic circuit for a Boolean function* $f(x_1, x_2, \ldots x_n)$ *is a circuit* C *with standard inputs* $x_1, x_2, \ldots x_n$ *and auxiliary inputs* $y_1, y_2, \ldots y_m$*, where* $m = poly(n)$*, such that if* $f(x_1, x_2, \ldots x_n) = 1$*, then there is a assignment for the inputs* y *such that* $C(x_1, x_2, \ldots x_n, y_1, y_2, \ldots y_m) = 1$ *and if* $f(x_1, x_2, \ldots x_n) = 0$*, then there is no such assignment.*

We model such a circuit as a directed acyclic graph whose nodes are AND, OR, NOT, or $FANOUT$ gates.

Definition 4 (Monotone nondeterministic Boolean circuit). *A monotone nondeterministic Boolean circuit is a nondeterministic Boolean circuit in which a gate that (transitively) depends on a standard input[5] cannot be a* NOT *gate.*

Given an access structure $\mathcal{A}$ in mNP, we associate with it the monotone nondeterministic Boolean circuit for the characteristic function of the language $\mathcal{L}_{\mathcal{A}}$.

Lemma 1 ([7]). *A language* $\mathcal{L}$ *is in* mNP *if and only if the monotone nondeterministic Boolean circuit computing* $\mathcal{L}$ *has polynomial size.*

Definition 5 (Oblivious transfer). *An oblivious transfer (or OT) protocol is a protocol between a sender* S *and a receiver* R *in which*

- *S's input is* (s_1, s_2) *which are elements of the secret domain* $\mathbb{S}$.
- *R's input is an index* $\alpha \in \{0, 1\}$.
- *At the end of the protocol* R *must obtain* s_α *but should not get any information about* $s_{1-\alpha}$.
- *S should not obtain any information about* α.

The definition above refers to "1-out-of-2 OT", or OT_1^2. There are more general notions of oblivious transfer, but we will not require them.

3   Our Computational Secret Sharing Scheme

We assume that the players are provided a monotone Boolean circuit C that accepts the access structure $\mathcal{A}$ (the circuit can be individually given to all the players or published on a bulletin board). We consider the circuit as composed of AND, OR and $FANOUT$ gates. Each wire of the circuit will be associated with a value during the sharing and reconstruction phases. In the sharing phase, the problem is to compute the values corresponding to the input wires (the shares of the corresponding players) from the value of the output wire (the secret). We perform the reverse of this during reconstruction. Our computational secret sharing scheme is as follows.

---

[5] In other words, there is a directed path from a standard input to that gate.

Let $ENC_K : \mathcal{F} \to \mathcal{F}$ be a family of trapdoor one-way functions[6] on $\mathcal{F}$ with the index K varying over $\mathcal{F}$, and $DEC_K : \mathcal{F} \to \mathcal{F}$ the corresponding inverses.

Algorithm Share

1. Let W be the output wire. Assign $V(W) = s$, where s is the secret.
2. Choose a gate G whose output wire has been assigned a value.[7]
   - G is an AND gate : Let W be the output wire of G and W_1 and W_2 be the input wires. Pick a random x in $\mathcal{F}$. Assign: $V(W_1) = x \oplus V(W)$ and $V(W_2) = x$.
   - G is an OR gate : Let W be the output wire of G and W_1 and W_2 be the input wires. Assign: $V(W_1) = V(W)$ and $V(W_2) = V(W)$.
   - G is a $FANOUT$ gate : Let W_1, W_2 be the outputs of G and W be the input. Pick a random key K from $\mathcal{F}$. Publish: $ENC_K(V(W_i))$ for $i = 1, 2$. Assign: $V(W) = K$.
3. Repeat step 2 until all gates are considered. The values at each input wire form the shares of the corresponding players.

Algorithm Reconstruct

1. Let W_P be the input wire corresponding to player P. Assign $V(W_P)$ to the share of player P. Choose a gate G whose input wires have been assigned values.
   - G is an AND gate : Let W be the output wire of G and W_1 and W_2 be the input wires. Assign $V(W) = V(W_1) \oplus V(W_2)$.
   - G is an OR gate : Let W be the output wire of G and W_1 and W_2 be the input wires. Choose the input wire $W_i (i = 1, 2)$ such that $Eval(W_i) = 1$. Assign $V(W) = V(W_i)$.
   - G is a $FANOUT$ gate : Let W be the input wire of G and W_1, W_2 be the outputs. By applying DEC, compute $V(W_i)$ from $V(W)$ and $ENC_{V(W)}(V(W_i))$.
2. Repeat step 1 until $V(W_O)$ for the output wire W_O is constructed. $V(W_O)$ is the secret.

3.1   Correctness and Security

Correctness. It is easy to show that any access set A can recover the secret with probability 1. We prove the stronger statement that A can recover $V(W)$ for any wire W with $Eval(W, A) = 1$. We show this by induction on the depth of the W: it is clearly true for the input wires. If W is any other wire, it must be the output of an AND, OR or $FANOUT$ gate G. In each of these cases algorithm reconstruct shows how to obtain $V(W)$ from the V-values at those input wires of G for which $Eval$ is 1 (the V-values at the input wires, which have a smaller depth than W, are assumed to be known by the induction hypothesis).   □

---

[6] We have used trapdoor functions for clarity of presentation, but the protocol will work with minor modifications even when ENC is a one-way function.

[7] If G is a $FANOUT$ gate, both its outputs should have been assigned a value.

Theorem 1. *The above computational secret sharing scheme is secure, i.e, for any $A \notin \mathcal{A}$, A cannot recover the secret.*

Proof. We have shown that the access set A can recover $V(W)$ for any wire W with $Eval(W, A) = 1$. The converse of this assertion is not true: A may be obtain $V(W)$ even when $Eval(W, A) = 0$. To see this consider an OR gate whose inputs W_1 and W_2 evaluate to 1 and 0 respectively. Then A can obtain $V(W_2) = V(W_1)$. To get around this difficulty, we introduce the concept of a *fanout-free region*. It will turn out that the converse statement indeed holds on those wires that connect fanout-free regions.

A subcircuit C' of a circuit C is a connected subgraph of C induced by some of the nodes (gates). A fanout free region (FFR) of C is a maximal subcircuit of C having no $FANOUT$ gates. Note that C can be considered to be a *directed acyclic graph* of fanout free regions connected by $FANOUT$ gates. We call this the *fanout graph* of C. Also note that any FFR is a tree of AND and OR gates.

With every FFR F, we associate a virtual adversary A_F. A_F is given as input the circuit F, the share values of those input wires W of F for which $Eval(W) = 1$, and nothing else. Our goal is to prove that the players' view of the FFR F is indistinguishable from the view of A_F on F.

Let $\pi(W)$ be the property that if $Eval(W)$ is 0 then $V(W)$ is computationally indistinguishable from the uniform distribution on $\mathcal{F}$.

We begin with a restatement of Benaloh and Leichter's result [3]:

Lemma 2. *For the adversary A_F, π holds on the output wire of F.*

The next lemma states that if the input share value of a $FANOUT$ gate is not known, then the public value associated with that gate gives no new information about any of its output share values. In particular this means that the property π is carried across a $FANOUT$ gate.

Lemma 3. *Let W be the input wire and W' an output wire of a $FANOUT$ gate. If $V(W)$ is computationally indistinguishable from the uniform distribution on $\mathcal{F}$, then so is $V(W')$.*

Proof: Since $DEC_K(.)$ is a family of pseudo-random permutations (from $\mathcal{F}$ to $\mathcal{F}$, indexed by the key K), uniform distribution on the key space, given the ciphertext, implies that the distribution on the plaintext space is computationally indistinguishable from uniform distribution. If the distribution of $V(W')$ is computationally distinguishable from the uniform distribution, it means that $V(W)$ is computationally distinguishable from uniform distribution, which is by assumption, false. $\square$

The next lemma formalizes the notion that a $FANOUT$ gate does not "leak any information" in the reverse direction.

Lemma 4. *Let W be the input wire and W' be an output wire of a $FANOUT$ gate. Then the distributions $V(W)$ and $V(W)|V(W')$ are computationally indistinguishable.*

We observe that the sharing algorithm fixes $V(W)$ randomly and independently of $V(W')$. Therefore, if the lemma is false it would mean that the knowledge of an arbitrary $(plaintext, ciphertext)$ pair gives information about the key, which contradicts the assumption that ENC is secure. $\square$

A simple generalization of the above lemma to any pair of wires with the property that any path connecting them must pass through a $FANOUT$ gate is:

Lemma 5. *In the fanout graph of C, let F be an FFR of depth d and F' an FFR of depth d', $d' > d$. Let W and W' be wires in F and F' respectively. Then $V(W)$ and $V(W)|V(W')$ are computationally indistinguishable.* $\square$

The above lemma allows us to apply induction on the depth on the fanout graph, at every step ignoring all FFRs at a greater depth than the current FFR.

Lemma 6. *Let F be an FFR at depth d. Assume that π holds on the outputs of all FFRs of depth $< d$. Then π holds on the output of F.*

Proof. By lemma 5, we can ignore the effect of all FFRs of depth $> d$. By assumption, π holds on all wires that feed any of the inputs of F. Applying lemma 3 to each $FANOUT$ gate feeding F, we find that the players' view of the inputs of F is identical to that of A_F. Therefore by lemma 2, π holds on the output of F. $\square$

The rest of the proof is straightforward. By applying induction on the depth d of the FFRs, we find that π holds on the output of every FFR. In particular, π holds on the output wire of the circuit. $\square$

3.2   Efficiency

Theorem 2. *The above scheme is efficient for all access structures $\mathcal{A} \in mP$.*

Proof. The total number of shares given to the players is $O(n)$ since each player gets exactly one share, corresponding to one of the input wires in the circuit. The number of published share values is twice the number of $FANOUT$ gates, which is polynomial in n when the circuit is poly-size. Therefore, for all access structures $\mathcal{A}$ having a polynomial-size circuit (i.e, $\mathcal{A} \in mP$), this scheme is efficient. $\square$

The scheme is also computationally efficient for all $\mathcal{A} \in mP$ since the computational effort required by D is equivalent to that of evaluating the circuit on some input assignment and performing a polynomial number of encryptions. Reconstruction of the secret can be naturally parallelized and the parallel time complexity of reconstructing the secret by an access set $A \in \mathcal{A}$ is proportional to the *depth* of the circuit.

4   Secret Sharing with Semi-trusted Third Party

Our goal is to explore the limits on the access structures for which we can give secret sharing schemes by relaxing the requirements. Thus, even though secret sharing as such is a non-interactive protocol, we wish to make it more powerful by allowing a limited amount of interaction. We do this by introducing a third party T who is allowed to interact with the players. However, at the end of the protocol T should be no wiser about the inputs of the dealer and the players than before the beginning of the protocol.

The algorithms for sharing and reconstruction are similar to the first protocol. The main difference is that the circuit consists of NOT gates also. Therefore, we need to associate *two* share values with each wire: one corresponding to the evaluation of the wire being 1 and the other corresponding to the evaluation of the wire being 0. (We denote these by $V(W, 1)$ and $V(W, 0)$ respectively, and call them the *1-share* and the *0-share* respectively of W.) Propagating the values across AND, OR and $FANOUT$ gates is done as in the previous protocol. In the case of NOT gates, the share value of the input wire with evaluation 0 is related to the share value of the output wire with evaluation 1, and vice versa.

Role of the third party. In monotone circuits, $Eval(W, A) \geq Eval(W, B)$ whenever $A \supset B$. Therefore, the dealer need not worry about a set of players obtaining some share values by evaluating the circuit (i.e, invoking the reconstruction algorithm) with some input x_i set to 0 even though is possible to evaluate the circuit with $x_i = 1$. In the case of a general circuit (which includes NOT gates), however, this is not true, and therefore it is possible that the players might obtain both the 0-share and the 1-share of some wire. The role of the third party is to ensure that this cannot happen. It is enough to ensure that the players cannot get both the 0-share and the 1-share of any auxiliary input wire of the circuit. This is done by executing an *Oblivious Transfer* protocol [12] for each auxiliary input wire of the circuit.

Let $\mathcal{A}$ be an access structure in mNP. By lemma 1, there exists a monotone nondeterministic Boolean circuit C of polynomial size that computes $\mathcal{A}$. Using this circuit we will construct a CSS scheme for $\mathcal{A}$.

4.1   Protocol

Sharing The sharing algorithm is essentially the sharing algorithm of the previous section invoked twice, once for the 0-shares and once for the 1-shares.

1. Let W be the output wire of C. Assign $V(W, 1) = s$, where s is the secret.
2. Choose a gate G whose output wire has been assigned a 1-share.
   - G is an AND gate : Let W be the output wire of G and W_1 and W_2 be the input wires. Pick a random x in $\mathcal{F}$. Assign: $V(W_1, 1) = x \oplus V(W, 1)$ and $V(W_2, 1) = x$.
   - G is an OR gate : Let W be the output wire of G and W_1 and W_2 be the input wires. Assign: $V(W_1, 1) = V(W, 1)$ and $V(W_2, 1) = V(W, 1)$.
   - G is a NOT gate: Assign $V(W', 0) = V(W, 1)$ where W is the output wire and W' is the input wire of G.
   - G is a $FANOUT$ gate : Choose a random key K from $\mathcal{F}$. Let W_1, W_2 be the outputs of G and W be the input. Publish: $ENC_K(V(W_i, 1))$ for $i = 1, 2$. Assign: $V(W, 1) = K$.

Choose a gate G whose output wire has been assigned a 0-share.
- G is an AND gate : Let W be the output wire of G and W_1 and W_2 be the input wires. Assign: $V(W_1, 0) = V(W, 0)$ and $V(W_2, 0) = V(W, 0)$.
- G is an OR gate : Let W be the output wire of G and W_1 and W_2 be the input wires. Pick a random x in $\mathcal{F}$. Assign: $V(W_0, 0) = x \oplus V(W, 0)$ and $V(W_2, 0) = x$.
- G is a NOT gate: Assign $V(W', 1) = V(W, 0)$ where W is the output wire and W' is the input wire of G.
- G is a $FANOUT$ gate : Choose a random key K from $\mathcal{F}$. Let W_1, W_2 be the outputs of G and W be the input. Publish: $ENC_K(V(W_i, 0))$ for $i = 1, 2$. Assign: $V(W, 0) = K$.
3. Repeat steps 2 and 3 until all gates are considered.
4. For every (W, b) that has not been assigned a share, assign a random value to $V(W, b)$.
5. The 1-shares of the input wires form the shares of the corresponding players. The 0-shares of the input wires are published.
6. For every auxiliary input wire W, $\{(V(W, 0), V(W, 1)\}$ is sent to the third party.

Reconstruction The reconstruction consists of two stages: in the first stage the players interact with the third party; in the second the players perform some local computations to recover the secret (if they form an access set).

Stage 1. The players cannot interact individually with the third party through separate channels, because of the requirement that the third party should not gain any information about the set of players involved in the reconstruction procedure. Therefore we consider all the players as constituting one virtual player P.

Let A be the access set of players participating in the reconstruction algorithm. For each auxiliary input wire W:

- T and P execute a OT_1^2 protocol with $V(W, 0)$ and $V(W, 1)$ as T's secrets. Since A is a qualified subset, there exists an assignment of values to the auxiliary inputs y such that the circuit evaluates to 1. For each wire W, P chooses the corresponding value of y as its index.

Stage 2. The stage 2 is similar to the reconstruction phase of the first protocol. The goal is to compute $V(W, Eval(W, A))$ for each wire W. For the input wires these values are already known from the share values and the public information.

1. Choose a gate G whose input wires have been assigned shares.
   - G is an AND gate with output 1: Let W be the output wire of G and W_1 and W_2 be the input wires. Assign $V(W, 1) = V(W_1, 1) \oplus V(W_2, 1)$.
   - G is an OR gate with output 1: Let W be the output wire of G and W_1 and W_2 be the input wires. Choose the input wire $W_i (i = 1, 2)$ such that $Eval(W_i) = 1$. Assign $V(W, 1) = V(W_i, 1)$.

- G is an AND gate with output 0: Let W be the output wire of G and W_1 and W_2 be the input wires. Choose the input wire $W_i (i = 1, 2)$ such that $Eval(W_i) = 0$. Assign $V(W, 0) = V(W_i, 0)$.
- G is an OR gate with output 0: Let W be the output wire of G and W_1 and W_2 be the input wires. Assign $V(W, 0) = V(W_1, 0) \oplus V(W_2, 0)$.
- G is a NOT gate: Let Let W be the output wire of G and W' the input wire. Assign $V(W, b) = V(W', 1 - b)$ where $b = Eval(W)$.
- G is a $FANOUT$ gate : Let W be the input wire of G and W_1, W_2 be the outputs. Apply DEC to compute $V(W_i, b)$ from $V(W, b)$ and $ENC_{V(W,b)}(V(W_i, b))$, where $b = Eval(W)$.

2. Repeat step 1 until $V(W_O, 1)$ for the output wire W_O is constructed. $V(W_O, 1)$ is the secret.

4.2   Correctness and Security

To prove the correctness we first note if A is a qualified subset then the circuit evaluates to 1 on the input $(x_1, x_2, \ldots x_n, y_1, y_2, \ldots y_n)$ as chosen in the stage 1 of the reconstruction protocol. Next, we prove by induction that if the wire W evaluates to b then the players can compute $V(W, b)$. Clearly this is true of the input wires. If W is any other wire, it must be the output of an AND, OR, NOT, or $FANOUT$ gate G. In each of these cases, stage 2 of the reconstruction algorithm shows how the players can obtain the $V(W, b)$ from the relevant V-values of the input wires of G.

Security. Suppose $A \notin \mathcal{A}$. Then for every $P_i \notin A$, P has no way of knowing the 1-share of P_i's input wire. Therefore when P evaluates the circuit in stage 2, the input x_i must be 0. Hence from the definition of C there is no assignment $(y_1, y_2, \ldots y_n)$ which will make C evaluate to 1. Further, since C computes a monotone function, C will evaluate to zero even if some of the inputs x_i with $P_i \in A$ are set to zero.

It remains to prove the correctness of stage 2, i.e, that P cannot find the secret if C evaluates to 0. The proof of this is very similar to the proof of security of the reconstruction algorithm of the first protocol, and is hence omitted.

Security against the third party. In the sharing protocol, the T gets no information about the 1-shares of the input wires. Further, from the definition of OT, T gets no new information in stage 2 of the reconstruction protocol. Therefore, T can only evaluate the circuit with all inputs 0, which means that T cannot get the 1-share of the output wire.

Again, since T learns nothing at all during interaction with the virtual player, T cannot identify the access set A.

4.3   Efficiency

As with the previous protocol, this one is also efficient when the circuit is of polynomial size (i.e, $\mathcal{A} \in mNP$) since the total size of the shares is linear in n and the amount of published information is proportional to the number of $FANOUT$ gates. The question of computational complexity is somewhat tricky.

Strictly speaking, the protocol is computationally efficient since the sharing and reconstruction algorithms involve only a constant amount of computation for each gate of the circuit. However, the players need to determine if the set A is a qualified subset before they can start the reconstruction algorithm. This computation is, by definition, a general problem in mNP. The implications of this are discussed in the next section.

The *round complexity* of the interactive protocol between T and P is the same as the round complexity of the OT protocol used, since all the $m + n$ OTs can be invoked in parallel. If we use a scheme like the ones in [11], this complexity is 2. It might appear at first glance that if we use non-interactive OT schemes like [4], then the need for a third party would disappear. However, non-interactive OT schemes are not applicable in this context since the receiver needs to choose the index after the start of the protocol. Using non-interactive OT would require the access set of players to be known beforehand.

5   Discussion and Future Work

Theorem 3 (A simple upper bound). *Efficient CSS schemes cannot exist for an access structure $\mathcal{A}$ not in co-RP.*

Proof. To show this, we construct a deterministic algorithm to solve the problem "Does $A \in \mathcal{A}$" using the algorithms *share* and *reconstruct* as oracles. The algorithm chooses a random secret, shares it and uses *reconstruct* with the shares corresponding to A as inputs to see if it gets back the secret which it chose. If it is the same it decides that $A \in \mathcal{A}$. Else, it decides that $A \notin \mathcal{A}$. We note that if indeed $A \in \mathcal{A}$, it decides correctly with probability 1, while if $A \notin \mathcal{A}$, there is a small probability of error. Since we decide L_A with a deterministic algorithm, if *share* and *reconstruct* are poly-time then $\mathcal{A}$ must be in co-RP.   □

Implications of Our Results

- It was not known whether it is possible to construct efficient secret sharing schemes for access structures outside (algebraic $NC^2 \cap$ mono), though [2] provided evidence that it is possible. Our result shows that computational secret sharing is possible over the entire class mP which contains access structures not in algebraic NC^2.
- Combining our result with that of [1], it is possible to achieve the optimal information rate (for large secret length) for every access structure in mP.
- As we have remarked earlier, to carry out the reconstruction algorithm the players need to determine if they form an access set, and this computation could lie outside P in the third party case. This does not mean, however, that the third party result is purely of theoretical significance, for two reasons: when the players are probabilistic algorithms, the class of access structures that can be decided in poly-time is co-$RP \cap$ mono, as shown above, and this class is bigger than mP. Further, even for access structures admitting of deterministic poly-size circuits, it might be more efficient to use a randomized algorithm, in which case the protocol using nondeterministic Boolean circuits must be used.

Our result must be understood more as an existence result for efficient protocols, rather than as a method to construct such protocols. For instance, it is likely that using threshold gates as building blocks in addition to AND and OR would give more efficient protocols. One direction for future work in this area is to construct particularly efficient CSS schemes for interesting special cases of access structures in monotone P.

The most important open question is to determine the exact power of efficient CSS schemes; in particular, do there exist efficient CSS schemes for access structures outside monotone P.

Another direction for future work is to investigate models for CSS which do not have the co-RP upper bound. One way of doing this would be to allow the dealer to be computationally unbounded. This would have the effect of the language L_A being decidable in poly-time by a Turing machine with advice strings. However it is unreasonable to assume the dealer alone to be computationally unbounded, since secret sharing is usually a part of a larger protocol.

Another approach has been explored by Cachin [5], who uses a bulletin board on which an arbitrary amount of information may be published. Our construction using a semi-trusted third party also, as we have remarked, sidesteps the problem. We propose yet another approach. Since the trouble with access structures outside co-RP is the infeasibility of determining whether or not a given set is in the access structure, we provide the players with an oracle which can answer precisely that question: the players can make queries to the oracle with a subset of $\mathcal{P}$ as input, and the oracle decides whether it is a qualified subset or not. We note that this is a weaker assumption than that of [5]. It would be interesting to see if it is possible to construct efficient CSS schemes for the class mNP under this model.

References

1. P. Beguin and A. Cresti. General short computational secret sharing schemes. In *Advances in Cryptology - EUROCRYPT '95*, volume 921 of *LNCS*, pp. 194–208, 1995.
2. A. Beimel and Y. Ishai. On the Power of Non-Linear Secret Sharing. newblock *In proceedings of 16th IEEE Structure in Complexity Theory* , 2001.
3. J. Benaloh and J. Leichter. Generalized secret sharing and monotone functions. *In proceedings of CRYPTO 88*, volume 403 of *LNCS*, pp. 27–36. Springer-Verlag, 1988.
4. M. Bellare and S. Micali. Non-interactive oblivous transfer and applications. *Advances in Cryptology - CRYPTO'89*, pp. 547–557, 1989.
5. Cachin. On-line secret sharing. In *IMA: IMA Conference on Cryptography and Coding, LNCS lately (earlier: Cryptography and Coding II, Clarendon Press, 1992)*, 1995.
6. M. K. Franklin and M. K. Reiter. Fair exchange with a semi-trusted third party. In *ACM Conference on Computer and Communications Security*, pp. 1–5, 1997.
7. Grigni and Sipser. Monotone Complexity. In *PATBOOL: Boolean Function Complexity*, London Mathematical Society Lecture Note Series 169, Cambridge University Press, 1992.

8. M. Ito, A. Saito, and T. Nishizeki. Secret sharing scheme realizing general access structure. In *Proceedings of IEEE Globecom 87*, pp. 99–102. IEEE, 1987.
9. M. Karchmer and A. Wigderson. On span programs. In *Proceedings of the 8th Annual IEEE Structure in Complexity Theory*, pp. 102–111, 1993.
10. H. Krawczyk. Secret sharing made short. In *CRYPTO'93*, LNCS, pp. 136–146, 1993.
11. M. Naor and B. Pinkas. Efficient Oblivious Transfer Protocols. In *SODA 01*, pp. 448–457, 2001.
12. M. Rabin. How to exchange secrets by oblivious transfer. *Technical Report TR-81, Aiken Computation Laboratory, Harvard University*, 1981.
13. M. Rabin. Efficient dispersal of information for security, load-balancing and fault-tolerance. *JACM*, volume 36, pp. 335–348, 1989.
14. A. Razborov. A lower bound on the monotone network complexity of the logical permanent, (in russian). In *Mat. Zametki*, volume 37(6), pp. 887 – 900, 1985.
15. A. C. Yao. How to generate and exchange secrets. *In Proc. of STOC*, pp. 162–167, 1986.

Identity-Based Broadcasting

Yi Mu[1], Willy Susilo[1], and Yan-Xia Lin[2]

[1] School of IT and Computer Science
University of Wollongong, Wollongong, NSW 2522, Australia
{ymu, wsusilo}@uow.edu.au
[2] School of Mathematics and Applied Statistics
University of Wollongong, Wollongong, NSW 2522, Australia
yanxia@uow.edu.au

Abstract. In this paper, we introduce a new concept called "Identity-Based Broadcasting Encryption" (IBBE), which can be applied to dynamic key management in secure broadcasting. Based on this new concept, in the proposed system a broadcaster can *dynamically add* or *remove* a user to or from the receiver group without any involvement of users. We classify our systems into three different scenarios and give three provably secure and elegant constructions of IBBE system based on the pairing. Our system naturally suits *multi-group broadcasting*, where a message can be selectively broadcasted to certain groups of users.

Keywords: Broadcasting, Encryption, Pairing.

1 Introduction

Recently, a number of closely related models and constructions with the aim of securing electronic distribution of digital content have been proposed. An example of such services that rely on this kind of distribution is a pay TV system where a broadcaster needs assurance that only paid customers will receive the service and can only become viable if security of the distribution can be guaranteed.

The protection of a pay TV program is normally based on a symmetric-key cryptographic algorithm. That is, a broadcaster and all its users in a group share a secret key that is used by the broadcaster to encrypt a TV signal and is then used by users to decrypt the signal. The major disadvantage of such a scheme is that it is difficult for the broadcaster to stop an illegal user who has a forged secret key to receive pay TV programs. Changing a secret shared key requires updating all decoding boxes of users. This is infeasible and costly.

The key management problem above can be solved with a hybrid model. That is, the session key distribution relies on an efficient public-key algorithm that allows multiple decryption keys and the actual message is encrypted using the session key. With a dynamic key management, a broadcaster can *dynamically add* or *remove* a user to or from a receiver group without involvement of

T. Johansson and S. Maitra (Eds.): INDOCRYPT 2003, LNCS 2904, pp. 177–190, 2003.

users. Conceptually, in such systems, an encryption key maps to multiple decryption keys (forming a set $\mathbb{K}$). A new decryption key can arbitrarily be added to $\mathbb{K}$ and an existing decryption key can arbitrarily be removed from $\mathbb{K}$ without involvement of users.

The concept of *secure broadcasting* was introduced by Fiat-Naor[5] for solving the problem of multi-message encryption, which is known as the *broadcast encryption*. Conceptually, a broadcast encryption is based on a single symmetric cipher equipped with a number of affine substitution boxes, where n messages can be converted into n ciphertexts that are broadcasted to the other end of a communication channel. The ciphertexts are then decrypted with the same key(s). We need to highlight that the broadcast encryption is completely different from our broadcasting concept that will be studied in this paper.

Another important related area is the work on *secure multicasting* [12]. In this concept, the multicast group must share a common key to enable the multicast communication. This problem is also known as the *re-keying* problem, which requires an algorithm to securely and efficiently update the group key whenever needed. Several constructions have been proposed (e.g. [12,1,10]) that consider the group's dynamic. However, we must point out that this system is not suitable for our purpose. In this system, each user needs to update his/her secret key whenever there is a dynamic in the system, for example due to the addition of a new user or removal of a user in the system. This solution is also not practical, since the user needs to update his/her secret key which might not be doable in some scenario (for example, consider a black box that is used to receive a pay TV broadcast channel).

In this paper, we introduce a new concept called "Identity-Based Broadcasting Encryption" (IBBE). Our schemes have the following distinct properties. (1) Users can be *dynamically* divided into groups with no involvement of users. (2) User groups can be dynamically *updated* by the broadcaster without any involvement of users. (3) It is *identity-based* so that the broadcaster can easily broadcast a message or messages to a group in term of the ID of the group. Furthermore, a group in our system can be divided into subgroups where each subgroup has a unique ID. The obvious application of sub-grouping is that a pay TV series is sold to a group while each subgroup could view a different program.

The primary reason to use the pairing in our system is that it allows most of computations to be done in elliptic curves and presents a promise for identity based encryption. The pairing such as the Weil pairing suggests that two points in an elliptic curve can be mapped to a point in a finite field. The Weil pairing was originally considered to be a bad thing, since it can be used for attacking elliptic curves[8]. Recently, it has been showed that the Weil pairing can be used to construct a protocol for three party one round Diffie-Hellman key exchange[7]. Boneh-Franklin have recently proposed a concrete identity based encryption protocol[3] and a short signature scheme based on the Weil pairing[4]. There have been a number of publications in the applications of the pairing. For example, Verheul has found the Weil pairing is useful for credential pseudonymous certificate systems[11]; Gentry and Silverberg introduced the concept of hierarchical ID-based cryptography using the Weil pairing in [6]; and Zhang and Kim proposed an ID-based signatures from pairing in [13].

We find that it is possible to use the Weil pairing to construct a mapping such that a public key can map into multiple private keys. These private keys can be dynamically split into multiple groups; each with a unique identity. Our schemes are instances of the elliptic curve discrete logarithm (ECDL) problems which are believed to be intractable.

It should be pointed out that a secure broadcasting scheme has been proposed to handle the dynamic user update issue [9]. In that system, a broadcaster can add/remove a user to/from any group dynamically. However, it is not identity-based and does not allow a group to have a subgroup in a broadcasting scenario. Moreover, the underlying assumption in that scheme is based on the intractability of the discrete logarithm problem and the overhead of the initial encryption key computation is proportional to the maximum number of users (although the computational overhead of encryption/decryption/update required is very low). Compared with [9], our new scheme shows better computational efficiency in the construction of initial encryption keys. This is due to the computation that is performed to "future" users that might join the system later on in [9]. In our schemes, the encryption key can be computed dynamically when a new user joins the system.

The rest of this paper is organized as follows. Section 2 gives the basic definitions and models of our systems. Section 3 provides some preliminaries that will be used in construction of our protocols. Section 4 describes the first IBBE scheme, where each group has a unique system setting and a group can be dynamically divided into subgroups. Section 5 presents the second scheme, where one set of cryptographic keys is required to broadcast a message to multiple groups, where each group has a group ID. Section 6 is devoted to a new protocol that is secure against exhaustive search attacks by insiders. Section 7 provides complete security proofs for all three protocols. Section 8 concludes the paper.

2   Definitions and Models

In this section, we give the definitions and models of our systems.

Definition 1. *A designated group has an* ID *that is an arbitrary string* $\{0,1\}^*$.

Definition 2. *IBBE is a system that consists of a broadcaster* $\mathcal{T}$, *a set of m users* $\mathcal{U} = \{U_1, U_2, \cdots, U_m\}$, *and a set of $\hat{k}$ user groups* $\{\mathbb{U}^{\mathsf{ID}_1}, \mathbb{U}^{\mathsf{ID}_2}, \cdots, \mathbb{U}^{\mathsf{ID}_{\hat{k}}}\}$. *Each group* $\mathbb{U}^{\mathsf{ID}_i}$ *can contain $\tilde{k}$ subgroups* $\{\mathbb{U}_1^{\mathsf{ID}_i}, \mathbb{U}_2^{\mathsf{ID}_i}, \cdots, \mathbb{U}_{\tilde{k}}^{\mathsf{ID}_i}\}$ *or no subgroups. Each group* $\mathbb{U}^{\mathsf{ID}_i}$ *contains several users* $\subseteq \mathcal{U}$ *where each user U_j may belong to several group.*

The organization of groups is illustrated in Figure 1.

$\mathcal{T}$ has a private encryption key. Each user has a private decryption key. The management of keys varies in terms of group structures.

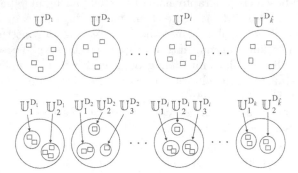

Fig. 1. The first row shows a set of user groups, where a user is represented by a square. Each group has a unique ID. In the second row, the user groups are divided into subgroups, where all subgroups have their parent's ID.

Definition 3. *Given a unique encryption key $\mathcal{E}$ (for $\mathcal{T}$) and users' decryption keys $\mathcal{D}_i$, $\forall i \in \{1, 2, \cdots, m\}$, the IBBE is referred to as the following scenarios:*

(1) *A group without subgroups. There exists a mapping Ψ_1 between a unique encryption key $\mathcal{E}_1$ and a group $\mathbb{U}^{\mathsf{ID}_i}$, $i \in \{1, ..., \hat{k}\}$. There exists a mapping Ψ_2 between the encryption key $\mathcal{E}_1$ and decryption keys $\mathcal{D}_i$, $i = 1, ..., m$, of users in $\mathbb{U}^{\mathsf{ID}_i}$, $i \in \{1, ..., \hat{k}\}$.*

(2) *A group with subgroups. There exists a mapping φ_1 between a unique encryption key $\mathcal{E}_2$ and a subgroup $\mathbb{U}_j^{\mathsf{ID}_i} \subset \mathbb{U}^{\mathsf{ID}_i}$, $j \in \{1, ..., \tilde{k}\}$. There exists a mapping φ_2 between the encryption key $\mathcal{E}_2$ and decryption keys $\mathcal{D}_i$, $i = 1, ..., k'$, of users in $\mathbb{U}_j^{\mathsf{ID}_i}$, where k' denotes the number of users in the subgroup.*

(3) *$\hat{k}$ groups without subgroups. There exists a mapping ω_1 between a unique encryption key $\mathcal{E}_3$ and groups $\mathbb{U}^{\mathsf{ID}_i}, \forall i \in \eta$, where η is a subset of $\{1, ..., \hat{k}\}$. There exists a mapping ω_2 between the encryption key $\mathcal{E}_3$ and decryption keys $\mathcal{D}_i$ of users in $\mathbb{U}^{\mathsf{ID}_i}, \forall i \in \eta$.*

In terms of the three scenarios, broadcasting can be classified as:

Scenario 1: A message encrypted with $\mathcal{E}_1$ can be decrypted by users in a single group $\mathbb{U}^{\mathsf{ID}_i}$, where ID_i is the ID of the corresponding group.

Scenario 2: A message encrypted with $\mathcal{E}_2$ can be decrypted by user in a subgroup $\mathbb{U}_j^{\mathsf{ID}_i}$ for the corresponding i and j.

Scenario 3: A message encrypted with $\mathcal{E}_3$ can be decrypted by users in groups $\mathbb{U}^{\mathsf{ID}_i}, \forall i \in \eta$.

Definition 4. *An IBBE is specified by five algorithms:* Setup, KeyGen, Encrypt, Decrypt, Update.

Setup : A randomized algorithm that takes as input a security parameter $\ell \in \mathbb{Z}$ and outputs system parameters (params1, params2). That is,

$$(\text{params1}, \text{params2}) \leftarrow \text{Setup}(\ell).$$

params1 is known only to the broadcaster $\mathcal{T}$. params2 is public.

KeyGen : A randomized algorithm that takes as input $(\ell, \text{params1}, \text{params2})$, and outputs an encryption key tuple $\mathcal{E}$ and decryption keys $\mathcal{D}_i$ $(i = 1, ..., m)$. That is, $(\mathcal{E}, \{\mathcal{D}_1, \mathcal{D}_2, \cdots, \mathcal{D}_m\}) \leftarrow \text{KeyGen}(\ell, \text{params1}, \text{params2})$.

In Scenario 1, $\mathcal{E} = \mathcal{E}_1$, which is assigned to a user group $\mathbb{U}^{\text{ID}_i}$.

$$\Psi_1 : \text{ID}_i \mapsto \mathcal{E}_1, \quad \Psi_2 : \{\mathcal{D}_1, \cdots, \mathcal{D}_m\} \mapsto \mathcal{E}_1.$$

In Scenario 2, $\mathcal{E} = \{\mathcal{E}_{2,1}, \cdots, \mathcal{E}_{2,\tilde{k}}\}$, which are assigned to subgroups $\mathbb{U}_j^{\text{ID}_i} \subset \mathbb{U}^{\text{ID}_i}$, $j = 1, \cdots, \tilde{k}$.

$$\varphi_1 : \text{ID}_i \mapsto \mathcal{E}, \quad \varphi_2 : \{\mathcal{D}_1, \cdots, \mathcal{D}_{k'}\} \mapsto \mathcal{E}_{2,j}, j \in \{1, \cdots, \tilde{k}\}.$$

In Scenario 3, $\mathcal{E} = \mathcal{E}_3$, which is assigned to groups $\mathbb{U}^{\text{ID}_i}, i \in \eta$.

$$\omega_1 : \text{ID}_i \mapsto \mathcal{E}_3, \forall i \in \eta, \quad \omega_2 : \mathcal{D}_i \mapsto \mathcal{E}_3, \forall i \in \eta.$$

Encrypt : An algorithm that uses params1, params2, an encryption key $\mathcal{E}$, an ID, and a message M as its inputs and outputs a ciphertext tuple, c.

$$c \leftarrow \text{Encrypt}(\text{params1}, \text{params2}, \mathcal{E}, \text{ID}, M).$$

For clarity, in Scenario 1, the Encrypt algorithm is defined as

$$c \leftarrow \text{Encrypt}(\text{params1}, \text{params2}, \mathcal{E}_1, \text{ID}_i, M),$$

in Scenario 2, it is defined as

$$c \leftarrow \text{Encrypt}(\text{params1}, \text{params2}, \mathcal{E}_{2,j}, \text{ID}_i, M), \quad \forall j \in \{1, \cdots, \tilde{k}\}$$

and in Scenario 3, it is defined as

$$c \leftarrow \text{Encrypt}(\text{params1}, \text{params2}, \mathcal{E}_3, \text{ID}_{\forall i \in \eta}, M).$$

Decrypt : An algorithm that takes as input: params2, one of decryption keys $\mathcal{D}_i$, $i = 1, 2, \cdots, m$, and a ciphertext c, and outputs the corresponding plaintext M, if $\mathcal{D}_i$ is valid. It outputs $\perp$ otherwise.

$$\text{Decrypt}(\text{params2}, \mathcal{D}_i, c) = \begin{cases} M & \text{if } \mathcal{D}_i \text{ is valid} \\ \perp & \text{otherwise} \end{cases}$$

Update : An algorithm that takes as input the encryption key tuple, and outputs a new encryption key tuple. That is,

$$\hat{\mathcal{E}} \leftarrow \text{Update}(\mathcal{E})$$

The basic setup of our schemes are based on the pairing. The basic system parameters for our systems are described as follows. Let E denote an elliptic curve over a field K with characteristic > 0, and $E[n]$ be its group of n-torsion points.

Definition 5. *Let $n \in \mathbb{Z}_{\geq 2}$ denote an integer, coprime to the characteristic of K with characteristic > 0. The Weil pairing is a mapping*

$$\hat{e} : E[n] \times E[n] \to \mu_n$$

where μ_n is the group of nth roots of unity in $\bar{K}$.

Under the definition of the Weil pairing, if $\hat{e}(P, Q)$ is not the unit in μ_n, then $\hat{e}(aP, bQ) = \hat{e}(P, Q)^{ab}$ for $P, Q \in E[n]$ and all $a, b \in \mathbb{Z}$. Please refer to Page 43 of [2] for details of the Weil pairing.

Group $E[n]$ is a cyclic additive group, now denoted $\mathbb{G}_1$, which maps to a cyclic multiplicative group $\mathbb{G}_2$ by the Weil pairing. If n is prime, then both G_1 and G_2 have a prime order. From Definition 5, n is not necessary to be prime. Thus, the order of $\mathbb{G}_1$ and $\mathbb{G}_2$ is not necessarily prime. In this paper, we consider the case where the order q of $\mathbb{G}_1$ and $\mathbb{G}_2$ is a product of some primes. q is kept secret by the broadcaster, since users do not need it in decryption. For simplicity, we will omit modulus n in the presentation.

For convenience of the presentation, hereafter we will denote the pairing $\hat{e}$ by $\langle ., . \rangle$.

3   Preliminaries

Before describing our schemes, in this section we give some basic results to support the validity and practicability of our schemes.

Definition 6. *An integer u_i is called an Identity Element associated with u'_i and q, defined by $I(u'_i, q)$, if the following property is held: $u_i u'_i = u'_i \bmod q$.*

Lemma 1. *Assume $q = p_1^{k_1} p_2^{k_2} \cdots p_T^{k_T}$, where p_i are primes and k_i are integers ($p_i \neq p_{i'}$ for $i \neq i'$). Set $u'_i \leftarrow p_i^{k_i}$, $i = 1, ..., t$ and $t < T$. Set $u_i \leftarrow \prod_{j \neq i} p_j^{k_j} + 1$. Then, u_i is an $I(u'_i, q)$. Also, there exists no u'_{j_0} for $j_0 \neq i$, such that u_i is an $I(u'_{j_0}, q)$.*

Proof. Proving that u_i is an $I(u'_i, q)$ is equivalent to proving $(u_i - 1)u'_i = kq$ for certain k. By noting that $(u_i - 1)u'_i = (\prod_{i \neq j} p_j^{k_j})p_i^{k_i} = q$, it is obvious that $u_i u'_i = u'_i \bmod q$ is held.

We prove the second statement by contradiction. If there is u'_{j_0} such that $u_i u'_{j_0} = u'_{j_0} \bmod q$, $i \neq j_0$, then there exists an integer $k_0 \neq 0$ such that $(u_i - 1)u'_{j_0} = k_0 q$. This implies $\prod_{j \neq i} p_j^{k_j} p_{j_0}^{k_{j_0}} = k_0 \prod_{j=1}^{T} p_j^{k_j}$. That is, $p_{j_0}^{k_{j_0}} = k_0 p_i^{k_i}$. It is contradictory to which p_{j_0} is prime, as $p_{j_0} \neq p_i$ and $k_0 \neq 0$. $\square$

Remark 1: Given u'_i as above, there may exist more than one $I(u'_i, q)$. For example, $\prod_{j \neq i} p_j^{k_j} p_i + 1$ is also an $I(u'_i, q)$. However, using u_i defined in the lemma above, we ensure that u_i is unique to $I(u'_i, q)$.

Definition 7. *The doublet (u_i, u'_i) given in Lemma 1 is defined as a "qualified pair."*

Definition 8. *Let $v \in Z_q$ be prime and $\gcd(v, q) = 1$. An integer $v_i \in Z_q$ is called the image of u_i associated with v if $vv_i = u_i \bmod q$.*

Lemma 2. *The mapping from u_i to its image v_i is unique.*

Proof. Since $\gcd(q, v) = 1$, v has an inverse, v^{-1}. v_i can then be computed from

$$v_i = v^{-1} u_i \bmod q, \qquad i = 1, 2, \cdots, m.$$

Thus,

$$vv_i \bmod q = (vv^{-1}u_i) \bmod q = u_i.$$

It is trivial that $v_i \neq v_j$ if $u_i \neq u_j$. $\qquad\qquad\qquad\qquad\qquad\qquad\qquad\square$

4   Identity Based Broadcasting

This scheme fits into Scenario 1 given in Section 2. We assume that users are assigned to multiple groups. Each group $\mathbb{U}^{\mathsf{ID}_i}$, $i \in \{1, ..., \hat{k}\}$, has a unique identity (ID). In terms of security, we require the broadcaster $\mathcal{T}$ to have a separate ID-based encryption key for each group. Each user in a group is assigned a unique decryption key. An encrypted message broadcasted to a targeted group can be decrypted by any of users in the group. Group members can be dynamically added to or removed from a group by $\mathcal{T}$ without involvement of any existing users. We here use the words "encryption key" to replace "public key", because the encryption key (tuple) is only known to $\mathcal{T}$.

4.1   The Protocol

Setup : $\mathcal{T}$ inputs $\ell \in \mathbb{Z}$ as a security parameter to generate private params1 $\leftarrow (P \in \mathbb{G}_1, q)$ and public params2$\leftarrow (\mathbb{G}_1, \mathbb{G}_2, \hat{e})$ as output. He then constructs two strong hash functions, $H_1 : \{0, 1\}^l \to \mathbb{G}_1$, $H_2 : \mathbb{G}_2 \to \{0, 1\}^l$. H_1 is only known to $\mathcal{T}$ and H_2 is publicly available.

KeyGen : $\mathcal{T}$ inputs $(\ell, \mathsf{params1}, \mathsf{params2})$ to KeyGen and obtains

- $u \leftarrow \prod_{i=1}^{m} u'_i \bmod q$, where $u_i u'_i = u'_i \bmod q$. Namely, (u_i, u'_i) is a qualified pair defined earlier.
- a prime v such that $\gcd(v, q) = 1$,
- $\{v_i\}$ as images of $\{u_i\}$ such that $vv_i \bmod q = u_i$, and

- a set of keys described as follows:
  - Set $d_i \leftarrow (xu_i + 1)v_i \bmod q$, where x, the master key, is a prime selected from $\mathbb{Z}_q$.
  - Set $E_1 \leftarrow uP$ and $E_2 \leftarrow uvP$.
  - Extract the group ID from the group identifier $\mathsf{ID} \in \{0,1\}^l$: $Q_{\mathsf{ID}} \leftarrow H_1(\mathsf{ID})$. The decryption key for a user is $D_i \leftarrow d_i Q_{\mathsf{ID}}$.

The outputs are the encryption key triplet (E_1, E_2, x) and decryption keys D_i, $i = 1, \cdots, m$. We note that the only information for U_i will be his decryption key D_i, and the public modulus n along with $\hat{e}, \mathbb{G}_1, \mathbb{G}_2$. All other data including P, Q_{ID} are known to $\mathcal{T}$ only.

Encrypt : $\mathcal{T}$ inputs a message $M \in \{0,1\}^l$ and the encryption key tuple (E_1, E_2, x), selects a random $r \in \mathbb{Z}_q$ and then computes $R \leftarrow rE_2$, and $b \leftarrow \langle E_1, (x + 1)Q_{\mathsf{ID}} \rangle$. The ciphertext is obtained from a bitwise XOR operation $c \leftarrow M \oplus H_2(b^r)$. The output from Encrypt is the tuple: (c, R), which is then broadcasted to users in the group with the group ID: ID_i. Note that M could be a session key and the real message is encrypted with this session key.

Decrypt : $U_i \in \mathbb{U}^{\mathsf{ID}_i}$ inputs (c, R) and his decryption key D_i and computes

$$\langle R, D_i \rangle = \langle rvuP, (xu_i + 1)v_i Q_{\mathsf{ID}} \rangle$$
$$= \langle ruP, (xu_i + 1)u_i Q_{\mathsf{ID}} \rangle$$
$$= \langle rE_1, (x + 1)Q_{\mathsf{ID}} \rangle$$
$$= b^r.$$

Upon obtaining this value, he can decrypt the message $M \leftarrow c \oplus H_2(b^r)$.

Remark 2. The broadcaster $\mathcal{T}$ can arbitrarily construct user groups at a runtime so that each group can receive a single pay TV program they entitle to watch (Scenario 2). It is actually trivial to achieve it in our protocol. What $\mathcal{T}$ has to do is to construct a new u that is the product of u_j for $j \in \{selected\ users\}$. That is, $u \leftarrow \prod_j u'_j \bmod q$.

Lemma 3. *The collusion of t users in the system, $t \le m$, cannot produce a valid decryption key.*

Proof. We are interested to see what can be gained by a collusion of t users in the system. It is clear to see that t users cannot gain a new decryption key. Without loss of generality, we assume that there are two malicious users who hold two legitimate decryption keys D_A and D_B respectively. By addition, they can obtain $D' \leftarrow D_A + D_B$, hoping that D' is a new decryption key. However, a "decryption" with D' will produce $(b^r)^2$. Since q is unknown to users, it is infeasible for users to compute the inverse of 2 in order to remove 2. It is easy to check other cases; therefore, we here omit the details.   □

The complete security proof of this scheme is given in Section 7.

4.2   Dynamic Update

The idea is to allow $\mathcal{T}$ to add a new user U_z to or remove an existing user $U_{z'}$ from the system without current users' involvement. Formally, we allow

$$\hat{\mathbb{U}}^{\mathsf{ID}_i} \leftarrow \mathbb{U}^{\mathsf{ID}_i} \cup \{U_z\}$$

and

$$\hat{\mathbb{U}}^{\mathsf{ID}_i} \leftarrow \mathbb{U}^{\mathsf{ID}_i} \setminus \{U_{z'}\}$$

without any involvement of $\forall U_j \in \mathbb{U}^{\mathsf{ID}_i}$.

When a user $U_{z'} \in \mathbb{U}^{\mathsf{ID}_i}$ is to be removed from the current system, $\mathcal{T}$ simply updates the encryption key by recomputing his encryption key as $\hat{E}_1 \leftarrow u_{z'}^{-1} E_1$, $\hat{E}_2 \leftarrow u_{z'}^{-1} E_2$ ($\hat{E}_1, \hat{E}_2$ are now the new encryption key tuple).

Adding a new user $U_z \notin \mathbb{U}^{\mathsf{ID}_i}$ into a group can be done with a similar fashion: $\hat{E}_1 \leftarrow u_z E_1$, $\hat{E}_2 \leftarrow u_z E_2$. The update scheme mentioned in this section is also applicable to the next two schemes that will be discussed in the next sections.

5   Broadcast to Multiple Groups

In the preceding scheme, a message can be broadcasted to multiple groups, but it requires a separate encryption key for each group and a message has to be encrypted several times. In this section, we will describe a new approach (Scenario 3) that has the following important features.

- A message can be broadcasted to multiple groups without multiple encryptions.
- $\mathcal{T}$ can use a specified ID for each group in the broadcast message.
- $\mathcal{T}$ can still use the original encryption key defined in the previous section.
- The protocol naturally has chosen-ciphtertext security.

We assume that the total number of users is m. They are divided into $\hat{k}$ groups, namely $\mathbb{U}^{\mathsf{ID}_1}, \mathbb{U}^{\mathsf{ID}_2}, \cdots, \mathbb{U}^{\mathsf{ID}_k}$. As in the preceding protocol, the secret decryption key for a user $U_i \in \mathbb{U}^{\mathsf{ID}_j}$ is denoted by $D_i^{\mathsf{ID}_j}$.

The encryption scheme is similar to that in the preceding protocol. The only required change is to construct the encryption key with respect to all group IDs. For clarity, for each group in $\{\mathbb{U}^{\mathsf{ID}_1}, \mathbb{U}^{\mathsf{ID}_2}, \cdots, \mathbb{U}^{\mathsf{ID}_k}\}$, we rewrite b (as defined in the preceding scheme) as $b_1, b_2, \cdots b_{\hat{k}}$, respectively. They are still computed from the the formula: $b_X \leftarrow \langle E_1, (x+1)Q_X \rangle$ for $\mathbb{U}^X$.

Without loss of generality, suppose that a pay TV program is intended to broadcast to groups $\mathbb{U}^{\mathsf{ID}_1}$ and $\mathbb{U}^{\mathsf{ID}_2}$, then $X \in \{\mathsf{ID}_1, \mathsf{ID}_2\}$. $\mathcal{T}$ needs to compute a new parameter as part of key distribution for each group as follows. Select $b'_{\mathsf{ID}_1}, b'_{\mathsf{ID}_2}$ such that they satisfy $b_X b'_X = \hat{b}_X$ for $\mathbb{U}^X$.

Encrypt : $\mathcal{T}$ carries out the following procedures:

- inputs a message $M \in \{0,1\}^l$, the encryption key E_1, E_2, x, and the session key $\hat{b}_X$ for $X \in \{\mathsf{ID}_1, \mathsf{ID}_2\}$,

- selects three additional cryptographic hash functions: $H_3 : \{0,1\}^l \times \{0,1\}^l \to \mathbb{Z}_q$, $H_4 : G_2 \to G_2$, and $H_5 : G_2 \to \{0,1\}^l$, where H_4 is publicly available and H_3, H_5 are known to $\mathcal{T}$ only.
- sets $r \leftarrow H_3(\sigma, M)$, where σ is selected at random,
- computes $R \leftarrow rE_2$, and $b_X \leftarrow \langle E_1, (x+1)Q_X \rangle$, where $X \in \{\mathsf{ID}_1, \mathsf{ID}_2\}$,
- computes $c_X \leftarrow b'^r_X H_4(b^r_X)$, and
- computes the ciphertext from a bitwise XOR operation $c \leftarrow M \oplus H_2(\hat{b}^r_X)$. Output: (c_X, c, R), which is then broadcasted.

Decrypt : A user $U_i \in \mathbb{U}^{\mathsf{ID}_1}$ or $\mathbb{U}^{\mathsf{ID}_2}$ inputs $(c, c_X, R, D_i^{(X)})$ for $X \in \{\mathsf{ID}_1, \mathsf{ID}_2\}$ to the decryption algorithm and computes

$$\langle R, D_i^{(X)} \rangle = \langle rvuP, (xu_i + 1)v_i Q_X \rangle$$
$$= \langle ruP, (x+1)Q_X \rangle$$
$$= b^r_X,$$

and obtains $c_X (H_4(b^r_X))^{-1} = b'^r_X$ and $b'^r_X b^r_X = \hat{b}^r_X$. Revealing this value, he can obtain the message $M \leftarrow c \oplus H_2(\hat{b}^r)$.

This protocol is chosen-ciphertext secure, since we have used the technique due to Boneh-Franklin [3]. The security proof is omitted. The reader is referred to [3] for details. The security against the chosen plaintext attacks is given in Section 7.

6   A Protocol against Exhaustive Search

We now describe an IBBE that is secure against exhaustive search by a legitimate user who wishes to find another decryption key pair, which is different from his own. This scheme is suitable for both cases described in Sections 4 and 5. We take the first one in the following description.

KeyGen: The basic setup is the same as that of the first protocol. $\mathcal{T}$ needs to reconstruct decryption key pairs. For a user $U_i \in \mathbb{U}^{\mathsf{ID}}$, a private key pair is constructed as $D_i \leftarrow x(u_i + y_i)v_i Q_{\mathsf{ID}}$, $d_i \leftarrow (1 + y_i)^{-1} \bmod q$, where $x \in \mathbb{Z}_q$ is the master key and y_i is an appropriate integer in $\mathbb{Z}_q$ such that $(1 + y_i)$ has an inverse in $\mathbb{Z}_q^*$. y_i, u_i, v_i are unique to U_i and x is unique to the system.

Encrypt: To encrypt a message $M \in \{0,1\}^l$, $\mathcal{T}$ chooses a number $r \in \mathbb{Z}_q$ and sets $R \leftarrow rE_2$. The ciphertext c is constructed from $M \oplus H_2(b^r)$, where $b = \langle E_1, xQ_{\mathsf{ID}} \rangle$.

Decrypt : $U_i \in \mathbb{U}^{\mathsf{ID}}$ computes

$$\langle R, D_i \rangle = \langle ruvP, x(u_i + y_i)v_i Q_{\mathsf{ID}} \rangle = \langle uP, xQ_{\mathsf{ID}} \rangle^{r(1+y_i)} \equiv \hat{b}_i,$$

$$\hat{b}_i^{d_i} = b^r.$$

Hence, he can reveal M by computing $M \leftarrow c \oplus H_2(b^r)$.

Lemma 4. *It is infeasible for $U_i \in \mathbb{U}^{\mathsf{ID}}$ to find a decryption key tuple that is different from his own decryption key, in terms of a ciphertext and his own decryption key.*

Proof. We prove it by contradiction. Assume that U_i is able to create a different decryption key tuple. Assume U_i has found a tuple of decryption key, $D \in \mathbb{G}_1$ and $d \in Z_q$, where D, d are independent of U_i's personal decryption key (D_i, d_i). Since $c = M \oplus H_2(b^r)$ and R are public, D_i and d_i must satisfy the following equations: $\langle R, D_i \rangle = \hat{b}_i$ and $\hat{b}_i^{d_i} \bmod q = b^r$, where $\hat{b}_i$ and b^r are known since U_i can use his own key pair to obtain the information of $\hat{b}_i$ and b^r.

Observing $\langle uP, xQ_{\mathsf{ID}} \rangle^{r(1+y_i)} = \hat{b}_i$, in order to obtain

$$\hat{b}_i^d \bmod q = \langle uP, xQ_{\mathsf{ID}} \rangle^{r(1+y_i)d} = b^r = \langle uP, xQ_{\mathsf{ID}} \rangle^r$$

d must satisfy $(1 + y_i)d = 1 \bmod q$. This suggests that $d = (1 + y_i)^{-1} \bmod q$ and U_i's key tuple is found. We obtain the contradiction. $\qquad\square$

Lemma 5. *The collusion of t users in the system, $t \leq m$, cannot produce a valid decryption key that can be given to a malicious user.*

Proof. Similar to the proof of Lemma 3.

7   Security Consideration

Consider an IBBE with encryption algorithm $\mathcal{S}$ wrt $M \oplus H_2(b^r)$, where M is a true message. We denote a ciphertext by c_S. We will omit the subscript S when it is clear from the context. Define a system $(K_{pub}, K_{pri}, \mathcal{L}, H_2)$, where K_{pub} contains all public information; K_{pri} consists of all private keys (encryption and decryption keys) and private hash functions; $\mathcal{L}$ is an operator mapping from (K_{pub}, K_{pri}) into $\mathbb{G}_2$; H_2 is a random oracle from $\mathbb{G}_2$ to $\{0, 1\}^l$.

Definition 9. *If the output $O \leftarrow \mathcal{L}(K_{pub}, K_{pri})$ is independent of K_{pri} and, for any c_S, $c_S \oplus H_2(O)$ outputs the true message M, we will call the system wrt $(K_{pub}, K_{pri}, \mathcal{L}, H_2)$ a **Valid** IBBE system associated to $\mathcal{S}$ through the decryption procedure $c_S \oplus H_2(O)$.*

(1) Consider the first scheme in Section 4. Let K_{pub} be a set consisting of $ruvP$ and $ruy'P$; K_{pri} consist of D and D' and operation $\mathcal{L}$ be defined as

$$\mathcal{L}(K_{pub}, K_{pri}) = \langle ruvP, D \rangle.$$

Then $(K_{pub}, K_{pri}, \mathcal{L}, H_2)$ forms a **Valid** system, because for any valid private key D, we have $\mathcal{L}(K_{pub}, K_{pri}) = b^r$ which is independent of any valid private key D and, for any encrypted message C given by the scheme, we have $c \oplus H_2(b^r) = M$, which is the true message.

(2) Consider the second scheme in Section 5. Let K_{pub} be a set consisting of (R, H_4, c_X); K_{pri} be $D^{(X)}$; and operation $\mathcal{L}$ be defined as

$$\mathcal{L}(K_{pub}, K_{pri}) = c_X(H_4(\langle R, D^{(X)} \rangle))^{-1} \langle R, D^{(X)} \rangle$$

Then $(K_{pub}, K_{pri}, \mathcal{L}, H_2)$ forms a Valid system, because for any valid private key $D^{(X)}$, $\mathcal{L}(K_{pub}, K_{pri}) = b^r$ which is independent of any valid private key $D^{(X)}$ and, for any ciphertext C of the scheme, we have $C \oplus H_2(b^r) = M$, which is the true message.

(3) Consider the third scheme in Section 6. Let K_{pub} be a set of (R, H_2, c); K_{pri} be a set of D and d, and $\mathcal{L}$ be defined as:

$$\mathcal{L}(K_{pub}, K_{pri}) = \langle R, D \rangle^d,$$

then, $(K_{pub}, K_{pri}, \mathcal{L}, H_2)$ forms a Valid system, because for any valid private key pair (D, d), $\mathcal{L}(K_{pub}, K_{pri}) = b^r$, and for any ciphertext c of the scheme, $c \oplus H_2(b^r) = M$ gives the true message.

Lemma 6. *Let $\mathcal{S}$ be an encryption algorithm wrt $M \oplus H_2(b^r)$ that maps a ciphertext string into $\{0,1\}^l$. Assume that $(K_{pub}, K_{pri}, \mathcal{L}, H_2)$ is a Valid IBBE system associated with $\mathcal{S}$. If there is an adversary $\mathcal{A}$ with advantage ε_{N_h} against $\mathcal{S}$ after making a total of $N_h > 0$ queries to H_2, then there is an algorithm $\mathcal{B}$ with advantage at least $\frac{2^l \varepsilon_{N_h} - 1}{2^l - 1}$ for identifying a valid decryption key with the running time is $\mathcal{O}(time\ (\mathcal{A}))$.*

Proof. For a given Valid IBBE system $(K_{pub}, K_{pri}, \mathcal{L}, H_2)$ with the encryption algorithm $\mathcal{S}$, we first define the algorithm $\mathcal{B}$ and, then, prove that the advantage of $\mathcal{B}$ taking into account the advantage of $\mathcal{A}$.

The input to $\mathcal{B}$ is K_{pub}. $\mathcal{B}$ picks a random string $\tilde{c}$ from $\{0,1\}^l$ and assumes that $\tilde{c}$ is an encrypted message. That is, there is a true message M such that $\tilde{c} = M \oplus H_2(\mathcal{L}(K_{pub}, \tilde{K}_{pri}))$. Let $\mathbb{K}_{pri}$ be a set contains all potential $\tilde{K}_{pri}$ for the underlying IBBE system. Obviously, the size of this set is very large and the likelihood that randomly picking up an element from $\mathbb{K}_{pri}$ that is a valid K_{pri} for the system is negligible. Otherwise, the following study is meaningless.

Challenge: $\mathcal{B}$ randomly chooses an element from $\mathbb{K}_{pri}$ to form a $\tilde{K}_{pri}$, and then, sends K_{pub}, $\tilde{K}_{pri}$, and $\tilde{c}$ to $\mathcal{A}$. $\mathcal{B}$ wants to utilize $\mathcal{A}$'s knowledge to make a decision if this $\tilde{K}_{pri}$ can be accepted as a valid key. For convenience, we call $\tilde{K}_{pri}$ as the candidate of a valid decryption key.

H_2-**queries:** $\mathcal{B}$ will independently repeat the above challenge N_h times and obtain a list of candidate keys, say $\{\tilde{K}_{pri,i}\}$. At the same time, $\mathcal{A}$ independently quires H_2 for N_h times based on $\mathcal{B}$'s requirement and observes the outputs

$$\tilde{c} \oplus H_2(\mathcal{L}(K_{pub}, \tilde{K}_{pri,i})) = \hat{M}_i, \qquad i = 1, 2, \cdots, N_h.$$

$\mathcal{B}$ will establish a list with all these outputs. The list is denoted by H_{list} having elements $\{(\tilde{K}_{pri,i}, \hat{M}_i)\}$.

Guess: After the N_h queries, $\mathcal{A}$ makes a guess on the true message M, say $\hat{M}$. If $\hat{M}$ coincides with some $\hat{M}_i$, say a $\tilde{M}_{i_0}$, in the list H_{list}, $\mathcal{B}$ will consider $(\tilde{K}_{pri,i_0})$ as a valid key; if $\hat{M}$ does not appear in H_{list}, $\mathcal{B}$ then randomly picks a key from $\mathbb{K}_{pri}$ and assigns it as K_{pri}.

We now show that, if $\mathcal{B}$ uses the above procedure to guess a valid K_{pri}, the probability of obtaining a really valid K_{pri} is at least $\frac{2^l \epsilon_{N_h}-1}{2^l-1}$. For convenience, we also denote by H_{list} the event that at least one valid $\tilde{K}_{pri}$ appears in the list H_{list}. Denote $\mathcal{B}_M$ the event that, after N_h quires, a valid $\tilde{K}_{pri}$ is identified through the above procedure. Then, we have

$$P(\mathcal{B}_M) = P(H_{list})P(\mathcal{B}_M|H_{list}) + (1 - P(H_{list}))P(\mathcal{B}_M|H_{list}^c)$$

$$\geq P(H_{list})P(\mathcal{B}_M|H_{list}) = P(H_{list}),$$

where H_{list}^c denotes the complement of H_{list}.

Since after N_h quires, the probability of $\mathcal{A}$ obtaining the true message is at least ϵ_{N_h},

$$P(\hat{M} = M) > \epsilon_{N_h}.$$

Thus

$$\varepsilon_{N_h} < P(\hat{M} = M) = P(H_{list})P(\hat{M} = M|H_{list}) + (1 - P(H_{list}))P(\hat{M} = M|H_{list}^c)$$

$$= P(H_{list}) + \frac{1}{2^l}(1 - P(H_{list})),$$

and

$$P(H_{list}) > \frac{2^l \epsilon_{N_h} - 1}{2^l - 1}.$$

This gives $P(\mathcal{B}_M) > \frac{2^l \epsilon_{N_h}-1}{2^l-1}$. □

Corollary 1 *If there is an adversary $\mathcal{A}$ with advantage ϵ_{N_h} against $\mathcal{S}$ after making a total of $N_h > 0$ queries to H_2, then for Scheme 1, there is an algorithm $\mathcal{B}$ such that finding valid pair keys D and D' with advantage at least $\frac{2^l \epsilon_{N_h}-1}{2^l-1}$ and the running time is $\mathcal{O}(time\ (\mathcal{A}))$.*

Corollary 2 *If there is an adversary $\mathcal{A}$ with advantage ϵ_{N_h} against $\mathcal{S}$ after making a total of $N_h > 0$ queries to H_2, then for Scheme 2, there is an algorithm $\mathcal{B}$ such that finding valid pair keys $D^{(X)}$ and $D'^{(X)}$ with advantage at least $\frac{2^l \epsilon_{N_h}-1}{2^l-1}$ and a running time is $\mathcal{O}(time\ (\mathcal{A}))$.*

Corollary 3 *If there is an adversary $\mathcal{A}$ with advantage ϵ_{N_h} against $\mathcal{S}$ after making a total of $N_h > 0$ queries to H_2, then for Scheme 3, there is an algorithm $\mathcal{B}$ such that finding valid pair keys D and d with advantage at least $\frac{2^l \epsilon_{N_h}-1}{2^l-1}$ and the running time is $\mathcal{O}(time\ (\mathcal{A}))$.*

8   Conclusion

We formalized the model of IBBE scheme. We proposed three identity based broadcast schemes that meet the requirement of IBBE. In these systems, a user group can be dynamically updated by the broadcaster without any involvement of any other users. The algorithm for updating the group is simple and efficient, since the broadcaster is not required to recompute the entire encryption key. These schemes are proven to be secure; especially the third protocol that is secure against exhaustive research attacks. We provided a complete security proof for our schemes.

References

1. J. Anzai, N. Matsuzaki, and T. Matsumoto. A Quick Group Key Distribution Scheme with "Entity Revocation". *Advances in Cryptology - Asiacrypt '99, Lecture Notes in Computer Science 1716*, pages 333 – 347, 1999.
2. I. Blake, G. Seroussi, and N. Smart. *Elliptic Curves in Cryptography.* Cambridge Unversity Press, 2001.
3. D. Boneh and M. Franklin. Identity-based encryption from the weil pairing. In *Advances in Cryptology, Crypto 2001,* Lecture Notes in Computer Science 2139, pages 213–229. Springer Verlag, 2001.
4. D. Boneh, B. Lynn, and H. Shacham. Short signatures from the weil pairing. In *Advances in Cryptology, Asiacrypt 2001, Lecture Notes in Computer Science 2248,* pages 514–532. Springer Verlag, 2001.
5. A. Fiat and M. Naor. Broadcast encryption. In *Advances in Cryptology, Crypto '93,* Lecture Notes in Computer Science 773, pages 480–491. Springer Verlag, 1994.
6. C. Gentry and A. Silverberg. Hierarchical ID-based Cryptography. *Advances in Cryptology, Asiacrypt 2002, Lecture Notes in Computer Science 2501,* pages 548 – 566, 2002.
7. A. Joux. A One Round Protocol for Tripartite Diffie-Hellman. In W. Bosma, editor, *ANTS-IV, Lecture Notes in Computer Science,* pages 385–394. Springer Verlag, 2000.
8. A. Menezes, T. Okamoto, and S. Vanstone. Reducing elliptic curve logarithms to logarithms in a finite field. *IEEE Transaction on Information Theory,* 39:1639–1646, 1993.
9. Y. Mu and V. Varadharajan. Robust and secure broadcasting. In *Indocrypt 2001, Lecture Notes in Computer Science,* pages 223 – 231. Springer, 2001. (the revised version: www.uow.edu.au/~ymu).
10. M. Naor and B. Pinkas. Efficient Trace and Revoke Schemes. *Financial Cryptography 2000, Lecture Notes in Computer Science 1962,* pages 1– 20, 2001.
11. E. R. Verheul. Self-blindable credential certificates from the weil pairing. In *Advances in Cryptology–Asiacrypt 2001,* Lecture Notes in Computer Science 2248, pages 533–551. Springer Verlag, 2001.
12. D. M. Wallner, E. J. Harder, and R. C. Agee. Key management for multicast: Issues and architectures. Internet Draft (draft-wallner-key-arch-01.txt), ftp:// ftp.ietf.org/ internet-drafts/ draft-wallner-key-arch-01.txt.
13. F. Zhang and K. Kim. ID-based Blind Signature and Ring Signature from Pairings. *Advances in Cryptology, Asiacrypt 2002, Lecture Notes in Computer Science 2501,* pages 33– 547, 2002.

Efficient Verifiably Encrypted Signature and Partially Blind Signature from Bilinear Pairings

Fangguo Zhang, Reihaneh Safavi-Naini, and Willy Susilo

School of Information Technology and Computer Science
University of Wollongong, NSW 2522 Australia
{fangguo, rei, wsusilo}@uow.edu.au

Abstract. Verifiably encrypted signatures are used when Alice wants to sign a message for Bob but does not want Bob to possess her signature on the message until a later date. Such signatures are used in optimistic contact signing to provide fair exchange. Partially blind signature schemes are an extension of blind signature schemes that allows a signer to sign a partially blinded message that include pre-agreed information such as expiry date or collateral conditions in unblinded form. These signatures are used in applications such as electronic cash (e-cash) where the signer requires part of the message to be of certain form. In this paper, we propose a new verifiably encrypted signature scheme and a partially blind signature scheme, both based on bilinear pairings. We analyze security and efficiency of these schemes and show that they are more efficient than the previous schemes of their kind.

Key words: Verifiably encrypted signature, partially blind signature, Bilinear pairings.

1 Introduction

When Alice wants to sign a message for Bob but does not want Bob to possess her signature on the message immediately. Alice can achieve this by encrypting her signature using the public key of a trusted third party (adjudicator), and sending the result to Bob along with a proof that she has given him a valid encryption of her signature. Bob can verify that Alice has signed the message but cannot deduce any information about her signature. At a later stag, Bob can either obtain the signature from Alice or resort to the adjudicator who can reveal Alice's signature. There are many applications of such verifiably encrypted signature scheme, such as online contract signing [3,4]. Boneh *et al.* [10] gave a verifiably encrypted signature scheme as an application of their aggregate signature. Their scheme is based on a short signature due to Boneh, Lynn, and Shacham (BLS) [11] constructed from bilinear pairings.

Blind signatures were first introduced by Chaum [13] and play a central role in cryptographic protocols such as e-cash or e-voting that require user anonymity. However, when we use blind signature to design e-cash schemes, there are two

T. Johansson and S. Maitra (Eds.): INDOCRYPT 2003, LNCS 2904, pp. 191–204, 2003.

obvious shortcomings: (1) To prevent a customer from double-spending his e-cash, the bank has to keep a database which stores all spent e-cash to check whether a specified e-cash has been spent or not by searching this database. This operation is referred to as the freshness checking (or the double-spending checking) of e-cash. Certainly, the database kept by the bank may grow unlimitedly. (2) The bank cannot inscribe the value on the blindly issued e-cash. To believe the face value of e-cash, there are two conventional solutions: First, the bank uses different public keys for different coin values. In this case, the shops and customers must always carry a list of those public keys in their electronic wallet, which is typically a smart card whose memory is very limited. Second solution, the bank can use the cut-and-choose algorithm [13] in the withdraw phase. But this is very inefficient.

Partially blind signatures were introduced by Abe and Fujisaki [1] to allow the signer to explicitly include some agreed information in the blind signature. Using partially blind signatures in e-cash system, we can prevent the bank's database from growing unlimitedly. Because the bank assures that each e-cash issued by it contains the information it desires, such as the date information. By embedding an expiration date into each e-cash issued by the bank, all expired e-cash recorded in the bank's database can be removed. At the same time, each e-cash can be embedded the face value, the bank can know the value on the blindly issued e-cash. A number of partially blind signature schemes using different assumptions have been proposed. Abe and Fujisaki's scheme is based on RSA [1]. Abe and Okamoto's scheme is based on discrete logarithm problem [2] and Fan and Lei's scheme is based on quadratic residues problem [14].

In this paper, we propose a new verifiably encrypted signature scheme and a partially blind signature scheme, both based on bilinear pairings. We analyze security of these schemes and show that they are more efficient than previous schemes.

The rest of the paper is organized as follows. In the next section we give a brief introduction to bilinear pairings and describe two signature schemes from bilinear pairings. Section 3 gives the definition and security properties of verifiably encrypted signature schemes and partially blind signature schemes. In Section 4 we describe our proposed verifiably encrypted signature schemes and in Section 5 analye its security. Sections 6 and 7 give our proposed partially blind signature scheme and its analysis, respectively. Section 8 concludes the paper.

2   Preliminaries

In recent years, bilinear pairings have been used to construct numerous new cryptographic primitives [9,11,12,16,17,19,21,22,23,24,25]. We recall the basic concept and properties of bilinear pairings.

Let $\mathbb{G}_1$ be a cyclic additive group generated by P, whose order is a prime q, and $\mathbb{G}_2$ be a cyclic multiplicative group with the same order q. Let $e : \mathbb{G}_1 \times \mathbb{G}_1 \to \mathbb{G}_2$ be a bilinear pairing with the following properties:

1. **Bilinearity:** $e(aP, bQ) = e(P, Q)^{ab}$ for all $P, Q \in \mathbb{G}_1, a, b \in Z_q$
2. **Non-degeneracy:** There exists $P, Q \in \mathbb{G}_1$ such that $e(P, Q) \neq 1$, in other words, the map does not send all pairs in $\mathbb{G}_1 \times \mathbb{G}_1$ to the identity in $\mathbb{G}_2$;
3. **Computability:** There is an efficient algorithm to compute $e(P, Q)$ for all $P, Q \in \mathbb{G}_1$.

Three well-known problems in groups that is commonly used in Cryptography are, Discrete Logarithm Problem (DLP), Decision Diffie-Hellman Problem (DDHP)and Computational Diffie-Hellman Problem (CDHP). For the sake of brevity we do not state the problems here and refer the reader to [8,18]. Two variations of CDHP are:

- **Inverse Computational Diffie-Hellman Problem (Inv-CDHP):** For $a \in Z_q^*$, given P, aP, to compute $a^{-1}P$..
- **Square Computational Diffie-Hellman Problem (Squ-CDHP):** For $a \in Z_q^*$, given P, aP, to compute $a^2 P$.

Generalizing these two problems, we can obtain the following problems:

Definition 1 (k-wCDHP (k-weak Computational Diffie-Hellman Problem)[19]). *Given $k + 1$ values $< P, yP, y^2 P, \ldots, y^k P >$, to compute $\frac{1}{y} P$.*

Definition 2 (k+1 Exponent Problem [26]). *Given $k + 1$ values $< P, yP, y^2 P, \ldots, y^k P >$, to compute $y^{k+1} P$.*

The following theorem due to [26], gives the relationship between the two problems:

Theorem 1. *k-wCDHP and k+1EP are polynomial time equivalent.*

Assumptions: We assume that DLP, CDHP, Inv-CDHP, Squ-CDHP and k+1 Exponent Problem are hard, which mean there are no polynomial time algorithm to solve them with non-negligible probability.

When the DDHP is easy but the CDHP is hard on the group G, we call G a *Gap Diffie-Hellman (GDH) group*. From bilinear pairing, we can obtain the GDH group. Such groups can be found on supersingular elliptic curves or hyperelliptic curves over finite field, and the bilinear parings can be derived from the Weil or Tate pairing. More details can be found in [9,12,16].

Schemes in this paper can work on any GDH group. Throughout this paper, we define the system parameters in all schemes are as follows: Let P be a generator of $\mathbb{G}_1$ with order q, the bilinear pairing is given by $e : \mathbb{G}_1 \times \mathbb{G}_1 \to \mathbb{G}_2$. These system parameter can be obtained using a **GDH Parameter Generator** $\mathcal{IG}$ [9]. Define two cryptographic hash function $H : \{0, 1\}^* \to \{0, 1\}^\lambda$, in general, $|q| \geq \lambda \geq 160$, and $H_0 : \{0, 1\}^* \to \mathbb{G}_1^*$. Denote $params = \{\mathbb{G}_1, \mathbb{G}_2, e, q, \lambda, P, H, H_0\}$.

2.1   The Basic Signature Scheme

A signature scheme is described by the following four algorithms : a parameter generation algorithm Generate, a key generation algorithm KeyGenparam, a signature generation algorithm Sign and a signature verification algorithm Ver.

We recall a basic signature scheme from bilinear pairings proposed in [26].

1. Generate. Generate the system parameters: $params$.
2. KeyGenparam. Pick random $x \in_R Z_q^*$, and compute $P_{pub} = xP$. The public key is P_{pub}. The secret key is x.
3. Sign. Given a secret key x, and a message m. Compute $S = \frac{1}{H(m)+x}P$.
4. Ver. Given a public key P_{pub}, a message m, and a signature S, verify if $e(H(m)P + P_{pub}, S) = e(P, P)$.

This signature scheme was proposed at [26], it can be regarded as being derived from Sakai-Kasahara's new ID-based encryption scheme with pairing [22]. In [26], the authors proved that this signature scheme was secure against existential forgery on adaptive chosen-message attacks (in the random oracle model) assuming the "$k + 1$ Exponent Problem" is hard in $\mathbb{G}_1$.

2.2   The Blind GDH Signature

We introduce a blind GDH signature scheme as follows:

- Generate. Generate the system parameters: $params$.
- KeyGenparam. Pick random $x \in_R Z_q^*$, and compute $P_{pub} = xP$. The public key is P_{pub}. The secret key is x.
- Blind signature issuing. The user wants a message $m \in \{0,1\}^*$ to be signed.
  - (Blinding) The user randomly chooses a number $r \in_R Z_q^*$, computes $M' = r \cdot H_0(m)$, and sends M' to the signer.
  - (Signing) The signer sends back σ', where $\sigma' = x \cdot M'$.
  - (Unblinding) The user then computes the signature $\sigma = r^{-1} \cdot \sigma'$ and outputs (m, σ).
- Ver. Given a public key P_{pub}, a message m, and a signature σ, verify if $e(P_{pub}, H_0(m)) = e(P, \sigma)$ holds.

This blind signature scheme can be regarded as the blind version of BLS signature scheme [11], that was firstly mentioned in [24]. In [7], Boldyreva gave a security proof of this blind signature scheme, they showed that this blind signature scheme was secure against one-more forgery under the "Chosen-target CDH" [7] assumption.

3   Definitions

In this section, we introduce the definitions and security properties of verifiably encrypted signature and partially blind signature.

Definition 3 (Verifiably Encrypted Signature [10]). *A verifiably encrypted signature scheme consists of three entities: signer, verifier and adjudicator. There are seven algorithms. Three,* **KeyGen**, **Sign**, *and* **Verify**, *are analogous to those in ordinary signature schemes. The others,* **AdjKeyGen**, **VESigCreate**, **VESigVerify**, *and* **Adjudicate**, *provide the verifiably encrypted signature capability.*

- **KeyGen, Sign, Verify:** *These are key generation, signing and verification of the signer, they are same as in standard signature schemes.*
- **AdjKeyGen:** *This is generating a public-private key pair* (APK, ASK) *for the adjudicator.*
- **VESigCreate:** *Given a secret key* SK, *a message* m, *and an adjudicator public key* APK, *compute a verifiably encrypted signature* ν *on* m.
- **VESigVerify:** *Given a public key* PK, *a message* m, *an adjudicator public key* APK, *and a verifiably encrypted signature* ν, *verify that* ν *is a valid verifiably encrypted signature on* m *under key* PK.
- **Adjudicate:** *Given an adjudicator keypair* (APK, ASK), *a certified public key* PK, *and a verifiably encrypted signature* σ *on some message* m, *extract and output* ν, *an ordinary signature on* m *under* PK.

Besides the ordinary notions of signature security in the signature component, we require three security properties of verifiably encrypted signatures:

Validity: This requires that **VESigVerify**(m,**VESigCreate**(m)) and **Verify**(m, **Adjudicate**(**VESigCreate**(m))) hold for all m and for all properly-generated keypairs and adjudicator keypairs.

Unforgeability: This requires that it be difficult to forge a valid verifiably encrypted signature.

Opacity: This requires that it be difficult, given a verifiably encrypted signature, to extract an ordinary signature on the same message.

Partially blind signatures were introduced by Abe and Fujisaki [1]. In [2], Abe and Okamoto presented a formal definition of partially blind signature schemes. The following definition is based on Abe-Okamoto's definition.

Definition 4 (Partially Blind Signature). *A Partially blind signature scheme consists of three participants: signer, user and verifier. There are three algorithms:* **Key Generation** *algorithm,* **Partially blind signature issuing** *algorithm and* **Verification** *algorithm.*

- **Key Generation** *is a probabilistic polynomial-time algorithm that takes security parameter* k *and outputs a public and secret key pair* (pk, sk).
- **Partially blind signature issuing** *is a interactive protocol between the signer and the user. The public input of the user contains* pk *and the public information* info. *The public input of the signer contains the public information* info. *The private input tape of the the signer contains* sk, *and that for the user contains message* m. *When they stop,the public output of the user contains either* completed *or* not completed, *the private output of the user contains either "fail" or* $(info, m, \sigma)$.

– **Verification** *is a (probabilistic) polynomial-time algorithm that takes (pk, info, m, σ) and outputs either accept or reject.*

Security of a partially blind signature scheme is in terms of three requirements: *completeness, partial blindness* and *non-forgeability*. Partial blindness must satisfy the following two properties: (1). The signer assures that an issued signature contains the information that it desires, and none can remove the embedded information from the signature. (2). For the same embedded information, the signer cannot link a signature to the instance of the signing protocol that produces the corresponding blind signature. The most powerful attack on a blind signature is *one-more signature forgery* introduced by Pointcheval and Stern in [20]. A partially blind signature scheme is called unforgeable against one-more forgery under chosen message attack, that means for each info, for some integer l, there is no probabilistic polynomial-time adversary $\mathcal{A}$ that can compute, after l interactions with the signer, $l+1$ signatures with non-negligible probability. A partially blind signature scheme is called secure if it satisfies these requirements.

4   A New Verifiably Encrypted Signature Scheme

At Eurocrypt 2003, Boneh, Gentry, Lynn and Shacham [10] proposed a verifiably encrypted signature scheme as an application of their aggregate signatures. Their scheme is based on a short signature due to Boneh, Lynn, and Shacham (BLS) [11] constructed from bilinear pairings. To get the verifiably encrypted signature, they used the ElGamal encryption algorithm. In this section, we propose a new verifiably encrypted signature scheme from bilinear pairings. This new scheme does not require the ElGamal encryption.

The new verifiably encrypted signature scheme uses the basic signature scheme in 2.1 and works as follows.

Key Generation. KeyGen and **AdjKeyGen** are the same as the key generation algorithm in the basic signature scheme, i.e., the system parameters are $\{\mathbb{G}_1, \mathbb{G}_2, e, q, \lambda, P, H\}$, the signer and adjudicator have the public-secret key pair (P_{pub}, x) and (P_{pubAd}, x_a), respectively.

Signing, Verification. *Sign* and *Verify* are the same as in the basic signature scheme, i.e., for a message m, the signature is $\sigma = \frac{1}{H(m)+x}P$, the verification is $e(H(m)P + P_{pub},\ \sigma) = e(P,\ P)$.

VESig Creation. Given a secret key $x \in Z_p$ a message m, and an adjudicator's public key P_{pubAd}, compute $\nu = \frac{1}{H(m)+x}P_{pubAd}$. The verifiably encrypted signature for message m is ν.

VESig Verification. Given a public key P_{pub}, a message m, an adjudicator's public key P_{pubAd}, and a verifiably encrypted signature ν, accept ν if and only if the following equation holds:

$$e(H(m)P + P_{pub}, \ \nu) = e(P, \ P_{pubAd}).$$

Adjudication. Given an adjudicator's public key P_{pubAd} and corresponding private key $x_a \in Z_q$, a certified public key P_{pub}, and a verifiably encrypted signature ν on some message m, ensure that the verifiably encrypted signature is valid; then output $\sigma = x_a^{-1}\nu$.

5   Analysis of the Verifiably Encrypted Signature Scheme

5.1   Security

We show that the proposed signature scheme satisfies properties of a verifiably encrypted signature scheme.

Validity. ν is the verifiably encrypted signature for message m. Since we have

$$e(H(m)P + P_{pub}, \ \nu) = e((H(m) + x)P, \ \frac{1}{H(m) + x}P_{pubAd}) = e(P, \ P_{pubAd}),$$

this means **VESigVerify**$(m,$**VESigCreate**$(m))$ holds, and

$$e(H(m)P + P_{pub}, \ x_a^{-1}\nu) = e((H(m) + x)P, \ x_a^{-1} \cdot \frac{1}{H(m) + x} \cdot x_aP) = e(P, P)$$

this means **Verify**$(m,$ **Adjudicate**$(**VESigCreate**(m))$, so *Validity* holds.

Unforgeability: For the unforgeability, we have the following claim:

Claim 1. If the basic signature scheme is secure against existential forgery, then the new verifiably encrypted signature scheme is secure against existential forgery.

To prove this claim, we show that if the new verifiably encrypted signature scheme is forgeable against existential forgery, then the basic signature scheme is forgebale too. That is if there is a probabilistic polynomial time forger algorithm $\mathcal{F}$ with a non-negligible probability ϵ under an adaptive chosen message attack for the verifiably encrypted signature scheme, then using $\mathcal{F}$, we can construct a new probabilistic polynomial time forger algorithm $\mathcal{F}'$ such that $\mathcal{F}'$ can forge a signature of the basic signature scheme with non-negligible probability. Because the basic signature scheme is secure against existential forgery using adaptive chosen-message attacks (in the random oracle model) and assuming the $k + 1$ Exponent Problem [26] is hard in $\mathbb{G}_1$, then the new verifiably encrypted signature scheme is unforgeable.

We adopt the security model of Boneh et al. [10]. We assume that the basic signature scheme is given as in 2.1 : $\{\mathbb{G}_1, \mathbb{G}_2, e, q, \lambda, P, H\}$, and the public-secret key pair of the signer is (P_{pub}, x). The forger algorithm $\mathcal{F}'$ sets up a verifiably encrypted signature scheme $\mathcal{V}$ based on the basic signature scheme: $\mathcal{F}'$ generates a key, $(x_0, P_0) \leftarrow$ **KeyGen**, which serves as the adjudicator's key. Suppose a probabilistic polynomial time forger algorithm $\mathcal{F}$ for the verifiably encrypted signature scheme $\mathcal{V}$ is given. Now, $\mathcal{F}'$ runs $\mathcal{F}$ on $\mathcal{V}$ and if $\mathcal{F}$ generates a forged verifiably encrypted signature ν' for a message m', then $\mathcal{F}'$ produces a forged signature σ' of the basic signature scheme for this message m', where $\sigma' = x_0^{-1}\nu'$.

Opacity: For the opacity, we have the following claim:

Claim 2. If the basic signature scheme is secure against existential forgery and the DLP is hard, then the new verifiably encrypted signature scheme is secure against extraction.

Suppose given a verifiably encrypted signature ν for a message m, an adversary $\mathcal{A}$ wants to compute the signature σ of the signer on the message m. $\mathcal{A}$ either forge a signature of the signer for message m (under the signer's public key P_{pub}), directly, or extract a signature σ' from ν, such that $e(H(m)P + P_{pub}, \sigma') = e(P, P)$.

Assuming that the basic signature scheme is secure against existential forgery, then it is impossible to forge a signature of the signer for message m. We show that extracting a signature σ' from ν such that $e(H(m)P + P_{pub}, \sigma') = e(P, P)$, is equivalent to solving DLP. Since ν satisfies

$$e(H(m)P + P_{pub},\ \nu) = e(P,\ P_{pubAd}) = e(P, P)^{x_a},$$

and due to the *bilinearity* property of the pairing, we have

$$e(H(m)P + P_{pub},\ x_a^{-1}\nu) = e(P,\ P).$$

Due to the $non - degeneracy$ of bilinear pairing, we have $\sigma' = x_a^{-1}\nu$. So, to get σ', the adversary $\mathcal{A}$ should know x_a which is the discrete logarithm of P_{pubAd} in base P.

5.2   Efficiency

We compare our verifiably encrypted signature scheme with Boneh et al.'s scheme [10] from the view point of computation overhead. We denote Pa the pairing operation, Pm the point scalar multiplication on $\mathbb{G}_1$, Ad the point addition on $\mathbb{G}_1$, Mu the multiplication on $\mathbb{G}_2$, Inv the inversion in Z_q and MTP the Map-ToPoint hash operation in BLS scheme [11]. We summarize the result in Table 1(we ignore the general hash operation).

Schemes	VESig Creation	VESig Verification	Adjudication
Proposed	$1Inv + 1Pm$	$2(or\ 1)Pa + 1Pm + 1Ad$	$1Inv + 1Pm$
Boneh et al.'s	$1MTP + 3Pm + 1Ad$	$1MTP + 3Pa + 1Mu$	$1Pm + 1Ad$

Table 1. Comparison of our scheme and the Boneh et al.'s scheme

We note that the computation of the pairing is the most time-consuming. Although there have been many papers discussing the complexity of pairings and how to speed up the pairing computation [5,6,15], the computation of the pairing still remains time-consuming. In our scheme, we can precompute $e(P, P_{pubAd})$ and publish it as part of the adjudicator's public keys, therefore, there is only one pairing operation in **VESig verification**, but there are three pairing operations in Boneh et al.'s scheme. On the other hand, our scheme does not require special hash functions but a general cryptographic hash functions such as SHA-1 or MD5. In Boneh et al.'s scheme, there is a special hash operation: MapToPoint, there is at least one quadratic or cubic equation over finite field need to be solved. Hence, our verifiably encrypted signature scheme is much more efficient than Boneh et al.'s scheme. Finally, our signature is shorter than Boneh et al.'s signature and consists of one element of $\mathbb{G}_1$ in our scheme, but two elements in Boneh et al.'s scheme.

6 A New Partially Blind Signature Scheme

We propose a new partially blind signature scheme from bilinear pairings. The proposed scheme can be regarded as the combination of the basic signature scheme in 2.1 and the blind GDH signature in 2.2.

The system parameters are: $params = \{\mathbb{G}_1, \mathbb{G}_2, e, q, \lambda, P, H, H_0\}$.

[**Key generation:**]
The signer picks random $x \in_R Z_q^*$, and computes $P_{pub} = xP$. The public key is P_{pub}. The secret key is x.

[**Partially blind signature issuing protocol:**]
Suppose that m be the message to be signed and c be the public information. The protocol is shown in Fig. 1.

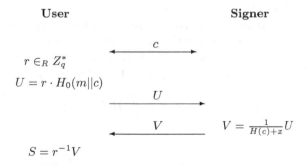

Fig. 1. The partially blind signature issuing protocol

- (Generation of the public information) The user and signer generate the public information c together.
- (Blinding) The user randomly chooses a number $r \in_R Z_q^*$, computes $U = r \cdot H_0(m\|c)$, and sends U to the signer.
- (Signing) The signer sends back V, where $V = (H(c) + x)^{-1}U$.
- (Unblinding) The user computes $S = r^{-1}V$.

Then (S, m, c) is the partially blind signature of the message m and public information c.

[**Verification:**]
A verifier can accept this partially blind signature if and only if

$$e(H(c)P + P_{pub}, \ S) = e(P, \ H_0(m||c)).$$

7   Analysis of the New Partially Blind Signature Scheme

7.1   Completeness

The completeness can be justified by the following equations:

$$
\begin{aligned}
& e(H(c)P + P_{pub}, \ S) \\
&= e((H(c) + x)P, \ r^{-1}V) \\
&= e((H(c) + x)P, \ r^{-1} \cdot (H(c) + x)^{-1}U) \\
&= e((H(c) + x)P, \ r^{-1} \cdot (H(c) + x)^{-1} \cdot r \cdot H_0(m||c)) \\
&= e(P, \ H_0(m||c))
\end{aligned}
$$

7.2   Partial Blindness

In the Blinding phase, r is chosen randomly from $\mathbb{Z}_q^*$ and so $U = r \cdot H_0(m||c)$ is a random element of the group $\mathbb{G}_1$. The signer receives this random information and the public information which he already knows and so no information about the message will be leaked.

The signer is assured that a signature issued by him contains the public information that he has agreed on and this information cannot be removed from the signature. This is true because if a malicious user could generate c' and replace c from the signer's signature (S, m, c) to obtain a signature with c'. Then we have

$$e(H(c')P + P_{pub}, \ S) = e(P, \ H_0(m||c')),$$

that means

$$e((H(c') - H(c))P, \ S) = e(P, \ H_0(m||c') - H_0(m||c)),$$

so, c and c' should satisfy $(H(c') - H(c))S = H_0(m||c') - H_0(m||c)$. This is unlikely, because H, H_0 are cryptographic hash functions.

Finally due to the randomness introduced during the Blinding phase and the fact that the public information is independent of the message, even if the same embedded information be used for two messages, the signer cannot link a signature to the corresponding instance of signature issuing protocol.

Hence the partially blind signature scheme satisfies the partial blindness property.

7.3 Unforgeability

To show that the proposed partially blind signature scheme is unforgeable, we first transform it to a fully blind signature scheme and then prove that the fully blind signature scheme is unforgeable.

For any public information c, the signer sets up the system parameters and public key as: $params = \{\mathbb{G}_1, \mathbb{G}_2, e, q, \lambda, P, H, H_0\}$ and Q. Here $Q = H(c)P + xP = sP$, $x \in_R Z_q^*$. The secret key is $s^{-1} = (H(c) + x)^{-1}$. Let m be the message to be signed. The blind signature issuing protocol of this fully blind signature scheme is shown as follows:

- (Blinding) The user randomly chooses a number $r \in_R Z_q^*$, computes $U = r \cdot H_0(m)$, and sends U to the signer.
- (Signing) The signer sends back V, where $V = s^{-1}U = (H(c) + x)^{-1}U$.
- (Unblinding) The user computes $S = r^{-1}V$.

The verification this blind signature is

$$e(Q, \ S) = e(P, \ H_0(m)).$$

We call above fully blind signature scheme FuBS. FuBS is derived from the proposed partially blind signature scheme. It is easy to see that if a massage-signature pair (m, c, S) can be forged for the proposed partially blind signature scheme, then a blind signature on the message $m' = m\|c$ for the corresponding FuBS can be forged. So, we have the following lemma.

Lemma 1. *If FuBS is secure against one-more forgery under chosen message attack. Then the proposed partially blind signature scheme is secure against one-more forgery under chosen message attack.*

Next, we show that FuBS is secure against one-more forgery under chosen message attack. It is easy to see that FuBS is very similar to the blind GDH signature in section 2.2. We will use similar technique in [7], where the author defined "Chosen target CDH" assumption and proved that their blind signature scheme is secure assuming the hardness of the chosen-target CDH problem. First, we give a variations of chosen-target CDH problem, named "Chosen target Inverse CDH" problem. We propose the problem and assumption as follows:

Definition 5. *Let $\mathbb{G}_1$ be GDH group of prime order q and P is a generator of $\mathbb{G}_1$. Let s be a random element of $\mathbb{Z}_q^*$ and $Q = sP$. Let $H_0 : \{0,1\}^* \to \mathbb{G}_1$ be a cryptographic hash function. The adversary $\mathcal{A}$ is given input (q, P, Q, H) and has access to the target oracle $T_{\mathbb{G}_1}$ that returns a random point U_i in $\mathbb{G}_1$ and the helper oracle $\texttt{Inv-cdh-}s(\cdot)$ (compute $s^{-1} \cdot (\cdot)$). Let q_T and q_H be the number of queries $\mathcal{A}$ made to the target oracle and the helper oracle respectively. The advantage of the adversary attacking the chosen-target inverse CDH problem $\text{Adv}_{\mathbb{G}_1}^{ct-icdh}(\mathcal{A})$ is defined as the probability of $\mathcal{A}$ to output a set of l pairs $((V_1, j_1), (V_2, j_2), \dots, (V_l, j_l))$, for all $i = 1, 2, \dots, l \; \exists \; j_i = 1, 2, \dots, q_T$ such that $V_i = s^{-1}U_{j_i}$ where all V_i are distinct and $q_H < q_T$.*
The chosen-target inverse CDH assumption states that there is no polynomial-time adversary $\mathcal{A}$ with non-negligible $\text{Adv}_{\mathbb{G}_1}^{ct-icdh}(\mathcal{A})$.

The following theorem shows that FuBS is secure assuming the chosen-target inverse CDH problem is hard.

Theorem 2. *If the chosen-target inverse CDH assumption is true in the group* $\mathbb{G}_1$ *then FuBS is secure against one-more forgery under chosen message attack.*

Proof. (sketch). If there is a probabilistic polynomial time one-more forger algorithm $\mathcal{F}$ with a non-negligible probability ϵ for FuBS under an chosen message attack, then using $\mathcal{F}$, we can construct an algorithm $\mathcal{A}$ such that $\mathcal{A}$ can solve the chosen-target inverse CDH problem with a non-negligible probability.

Suppose that a probabilistic polynomial time forger algorithm $\mathcal{F}$ is given. Suppose that $\mathcal{A}$ is given a challenge as in Definition 5. Now $\mathcal{F}$ has access to a blind signing oracle $s(\cdot)$ and the random hash oracle $H_0(\cdot)$. First, $\mathcal{A}$ provides $(\mathbb{G}_1, \mathbb{G}_2, e, q, P, H_0, Q)$ to $\mathcal{F}$ and $\mathcal{A}$ has to simulate the random hash oracle and the blind signing oracle for $\mathcal{F}$.

Each time $\mathcal{F}$ makes a new hash oracle query which differs from previous one, $\mathcal{A}$ will forward to its target oracle and returns the reply to $\mathcal{F}$. $\mathcal{A}$ stores the pair query-reply in the list of those pairs. If $\mathcal{F}$ makes a query to blind signing oracle, $\mathcal{A}$ will forward to its helper oracle Inv-cdh-$s(\cdot)$ and returns the answer to $\mathcal{F}$.

Eventually $\mathcal{F}$ halts and outputs a list of message-signature pairs $((m_1, S_1),$ $(m_2, S_2), \ldots, (m_l, S_l))$. $\mathcal{A}$ can find m_i in the list stored hash oracle query-reply for $i = 1, 2, \ldots, l$. Let j_i be the index of the found pair, then $\mathcal{A}$ can output its list as $((S_1, j_1), (S_2, j_2), \ldots, (S_l, j_l))$. Then this list is a solution to the problem in Definition 5.   □

From Lemma 1 and Theorem 2, we have the following theorem.

Theorem 3. *The proposed partially blind signature scheme is unforgeable under the chosen-target inverse CDH assumption in the group* $\mathbb{G}_1$.

7.4   Advantages

Comparing with previous partially blind signature schemes, such as [1], [2], [14], etc, the new partially blind signature scheme has a number of advantages:

A1. **Short signature.** In the proposed partially blind signature scheme, the signature only consists of an element in $\mathbb{G}_1$. In practice, the size of the element in $\mathbb{G}_1$ (elliptic curve group or hyperelliptic curve Jacobians) can be reduced by a factor of 2 with compression techniques. So, like BLS short signature scheme [11], our signature scheme can provide the short signature, the signature length is half the size of a DSA signature for a similar level of security. Short signatures are needed in low-bandwidth communication environments. An important application of partially blind signature is in e-cash system. E-coins are stored in users' electronic wallets which are typically implemented in smart cards with limited memory. The short length of the proposed signature makes the system much more practical.

A2. **Efficient**. The scheme can be implemented using elliptic curve cryptosystem, and is very efficient from the view point of the user and the bank. In the partially blind signature issuing protocol, the user only needs to perform *1MTP*, *2Pm* and *1Inv*, the bank only needs to perform *1Inv* and *1Pm*. In the verification, two pairing operations are needed (As we noted in 5.2, the computation of the pairing is the most time-consuming). In e-cash systems the verification will be done by the shop that can be assumed to have more computation power.

A3. **Batch verify** (For the same public information c). The efficiency of the system is of paramount importance when the number of verifications is considerably large (*e.g.*, when a bank issues a large number of electronic coins and the customer wishes to verify the correctness of the coins). The proposed partially blind signature scheme is very efficient when we consider the batch verification for the same public information c. Assuming that $S_1, S_2, \cdots, S_n$ are partially blind signatures on messages $m_1, m_2, \cdots, m_n$ with the same public information c. The batch verification is then to test if the following equation holds:

$$e(H(c)P + P_{pub}, \sum S_i) = e(P, \sum H_0(m_i\|c)).$$

8 Conclusion

Verifiably encrypted signature and partially blind signature are very important and useful cryptographic primitives. We proposed a new verifiably encrypted signature scheme and a partially blind signature scheme, both based on bilinear pairings. We analyzed the security and efficiency of them and showed that they are efficient than the previous schemes.

References

1. M. Abe and E. Fujisaki, *How to date blind signatures*, Advances in Cryptology - Asiacrypt 1996. LNCS 1163, pp. 244-251, Springer-Verlag, 2002.
2. M. Abe and T. Okamoto, *Provably secure partially blind signatures*, Advances in Cryptology - CRYPTO 2000. LNCS 1880, pp. 271-286, Springer-Verlag, 2000.
3. N. Asokan, V. Shoup and M. Waidner, *Optimistic fair exchange of digital signatures*. IEEE J. Selected Areas in Comm., 18(4):593-610, April 2000.
4. F. Bao, R. Deng and W. Mao. *Efficient and practical fair exchange protocols with offline TTP*. In Proceedings of IEEE Symposium on Security and Privacy, pp. 77-85, 1998.
5. P.S.L.M. Barreto, H.Y. Kim, B.Lynn, and M.Scott, *Efficient algorithms for pairing-based cryptosystems*, Advances in Cryptology-Crypto 2002, LNCS 2442, pp.354-368, Springer-Verlag, 2002.
6. P.S.L.M. Barreto, B.Lynn, and M.Scott, *On the selection of pairing-friendly groups*, to appear in the Workshop on Selected Areas in Cryptography (SAC) 2003.
7. A. Boldyreva, *Efficient threshold signature, multisignature and blind signature schemes based on the Gap-Diffie-Hellman -group signature scheme*, Public Key Cryptography - PKC 2003, LNCS 2139, pp.31-46, Springer-Verlag, 2003.

8. D. Boneh, *The decision Diffie-Hellman problem*, Proceedings of the Third Algorithmic Number Theory Symposium, LNCS 1423, pp. 48-63, Springer-Verlag, 1998.

9. D. Boneh and M. Franklin, *Identity-based encryption from the Weil pairing*, Advances in Cryptology-Crypto 2001, LNCS 2139, pp.213-229, Springer-Verlag, 2001.

10. D. Boneh, C. Gentry, B. Lynn and H. Shacham, *Aggregate and verifiably encrypted signatures from bilinear maps*, Eurocrypt 2003, LNCS 2656, pp.272-293, Springer-Verlag, 2003.

11. D. Boneh, B. Lynn, and H. Shacham, *Short signatures from the Weil pairing*, In C. Boyd, editor, Advances in Cryptology-Asiacrypt 2001, LNCS 2248, pp.514-532, Springer-Verlag, 2001.

12. J.C. Cha and J.H. Cheon, *An identity-based signature from gap Diffie-Hellman groups*, Public Key Cryptography - PKC 2003, LNCS 2139, pp.18-30, Springer-Verlag, 2003.

13. D. Chaum, *Blind signatures for untraceable payments*, Advances in Cryptology-Crypto 82, Plenum, NY, pp.199-203, 1983.

14. C.I. Fan and C.L. Lei, *Low-computation partially blind signatures for electronic cash*, IEICE Transactions on Fundamentals ofElectronics, Communications and Computer Sciences E81- A(5) (1998) 818-824.

15. S. D. Galbraith, K. Harrison, and D. Soldera, *Implementing the Tate pairing*, ANTS 2002, LNCS 2369, pp.324-337, Springer-Verlag, 2002.

16. F. Hess, *Efficient identity based signature schemes based on pairings*, SAC 2002, LNCS 2595, pp.310-324, Springer-Verlag, 2002.

17. A. Joux, *A one round protocol for tripartite Diffie-Hellman*, ANTS IV, LNCS 1838, pp.385-394, Springer-Verlag, 2000.

18. U. Maurer, *Towards the equivalence of breaking the Diffie-Hellman protocol and computing discrete logarithms*, Advances in Cryptology-Crypto 94, LNCS 839, pp.271-281, Springer-Verlag, 1994.

19. S. Mitsunari, R. Sakai and M. Kasahara, *A new traitor tracing*, IEICE Trans. Vol. E85-A, No.2, pp.481-484, 2002.

20. D. Pointcheval and J. Stern, *Security arguments for digital signatures and blind signatures*, Journal of Cryptology, Vol.13, No.3, pp.361-396, 2000.

21. R. Sakai, K. Ohgishi and M. Kasahara, *Cryptosystems based on pairing*, SCIS 2000-C20, Jan. 2000. Okinawa, Japan.

22. R. Sakai and M. Kasahara, *Cryptosystems based on pairing over elliptic curve*, SCIS 2003, 8C-1, Jan. 2003. Japan. This paper is available at Cryptology ePrint Archive, http://eprint.iacr.org/2003/054/.

23. N.P. Smart, *An identity based authenticated key agreement protocol based on the Weil pairing*, Electron. Lett., Vol.38, No.13, pp.630-632, 2002.

24. E. Verheul, *Self-blindable credential certificates from the Weil pairing*, Advances in Cryptology – Asiacrypt'2001, LNCS 2248, pp. 533–551, Springer-Verlag, 2001.

25. F. Zhang and K. Kim, *ID-based blind signature and ring signature from pairings*, Proc. of Asiacrpt2002, LNCS 2501, pp. 533-547, Springer-Verlag, 2002.

26. F. Zhang, R. Safavi-Naini and W. Susilo, *An efficient signature scheme from bilinear pairings and its applications*, Manuscript, 2003.

Extending Joux's Protocol to Multi Party Key Agreement
(Extended Abstract)

Rana Barua, Ratna Dutta, and Palash Sarkar

Cryptology Research Group
Stat-Math and Applied Statistics Unit
203, B.T. Road, Kolkata
India 700108
{rana,ratna_r,palash}@isical.ac.in

Abstract. We present a secure unauthenticated as well as an authenticated multi party key agreement protocol. The unauthenticated version of our protocol uses ternary trees and is based on bilinear maps and Joux's three party protocol. The number of rounds, computation/ communication complexity of our protocol compares favourably with previously known protocols. The authenticated version of our protocol also uses ternary trees and is based on public IDs and Key Generation Centres. The authenticated version of our protocol is more efficient than all previously known authenticated key agreement protocols.

Keywords : group key agreement, authenticated key agreement, pairing based cryptography, ID based cryptography.

1 Introduction

Key agreement is one of the fundamental cryptographic primitives. This is required in situations where two or more parties want to communicate securely among themselves. The situation where three or more parties share a secret key is often called *conference keying*. In this situation, the parties can securely send and receive messages from each other. An adversary not having access to the secret key will not be able to decrypt the message.

Key agreement protocols fall naturally into two classes – authenticated and unauthenticated. The first two party key agreement protocol was introduced by Diffie-Hellman in their seminal paper [6]. This is an unauthenticated protocol in the sense that an adversary who has control over the channel can use the man-in-the-middle attack to agree upon two separate keys with the two users without the users being aware of this. This situation is usually tackled by adding some form of authentication mechanism to the protocol.

In this paper, we present a secure multi-party key agreement protocol. The protocol can be used for both authenticated and unauthenticated key agreement. For n parties, $n > 2$, the number of rounds required for our protocol is $\lceil \log_3 n \rceil$ for both authenticated and unauthenticated key agreement. For unauthenticated

T. Johansson and S. Maitra (Eds.): INDOCRYPT 2003, LNCS 2904, pp. 205–217, 2003.

key agreement, the total number of messages exchanged and the total number of scalar multiplications (over a suitable elliptic curve) is less than $\frac{5}{2}(n-1)$. Additionally, $n\lceil\log_3 n\rceil$ pairings have to be computed. For authenticated version of our protocol, the combine message size is two times more than the unauthenticated version, the number of scalar multiplications is at most $9(n-1)$ and pairings is around five times more than the unauthenticated version.

Our protocol is based on bilinear maps. In one of the breakthroughs in key agreement protocols, Joux [7] proposed a three party, single round key agreement protocol. The basic Joux protocol is at the heart of our (unauthenticated) protocol.

Our multi party protocol is essentially a combination of Joux's tripartite Diffie-Hellman protocol and tree based group key agreement using ternary tree structure. Even though the idea of combining Joux's protocol and tree based group key agreement is a natural extension of Joux's original protocol, it is nontrivial to obtain a security proof for the protocol against passive adversary. One of our major contribution is to provide such a proof using techniques from [3], [9], [12]. In fact, any secure two or three party protocol can be used with the ternary tree structure to obtain a secure multi party protocol. Our security analysis against passive adversaries is also a kind of reduction. We argue that if the underlying two and three party protocols are secure, then our multi party protocol is also secure.

The remainder of the paper is organized as follows. Section 2 briefly explains the cryptographic bilinear map and the basic requirements of our protocol. Section 3 describes the protocol. The security analysis is provided in Section 4. Section 5 discusses the efficiency. Section 6 compares with other key agreement protocols. Finally Section 7 concludes the paper. Due to lack of space we omit some of the discussion and proofs. We refer the reader to [2] for the full version of the paper.

2   Preliminaries

2.1   Cryptographic Bilinear Maps

Let G_1, G_2 be two groups of the same prime order q. We view G_1 as an additive group and G_2 as a multiplicative group. Let P be an arbitrary generator of G_1. Assume that discrete logarithm problem (DLP) is hard in both G_1 and G_2. A mapping $e : G_1^2 \to G_2$ satisfying the following properties is called a bilinear map from a cryptographic point of view :

Bilinearity : $e(aP, bQ) = e(P, Q)^{ab}$ for all $P, Q \in G_1$ and $a, b \in Z_q^*$.

Non-degeneracy : If P is a generator of G_1, then $e(P, P)$ is a generator of G_2. In other words, $e(P, P) \neq 1$.

Computable : There exists an efficient algorithm to compute $e(P, Q)$ for all $P, Q \in G_1$.

Modified Weil Pairing [4] and Tate Pairing [1] are the examples of cryptographic bilinear maps.

2.2   Diffie Hellman Assumption

In this subsection we specify the version of the Diffie-Hellman problem which we will require. Consider $\langle G_1, G_2, e \rangle$ where G_1, G_2 are two cyclic subgroups of a large prime order q and $e : G_1^2 \rightarrow G_2$ is a cryptographic bilinear map. We take G_1 as an additive group and G_2 as a multiplicative group. (*By* $a \in_R Z_q^*$, *we mean a is randomly chosen from* Z_q^*.)

Decisional Hash Bilinear Diffie-Hellman (DHBDH) problem in $\langle G_1, G_2, e \rangle$:

Instance : (P, aP, bP, cP, r) for some $a, b, c, r \in Z_q^*$ and a one way hash function $H : G_2 \rightarrow Z_q^*$.

Solution : Output *yes* if $r = H(e(P, P)^{abc})$ mod q and output *no* otherwise.
The advantage of any probabilistic, polynomial time, 0/1-valued algorithm $\mathcal{A}$ in solving DHBDH problem in $\langle G_1, G_2, e \rangle$ is defined to be :
$Adv_{\mathcal{A}}^{DHBDH} = |Prob[\mathcal{A}(P, aP, bP, cP, r) = 1 : a, b, c, r \in_R Z_q^*]$
$-Prob[\mathcal{A}(P, aP, bP, cP, H(e(P, P)^{abc})) = 1 : a, b, c \in_R Z_q^*]|$
DHBDH assumption : There exists no polynomial time algorithm which can solve the DHBDH problem with non-negligible probability of success. In otherwords, for every probabilistic, polynomial time, 0/1-valued algorithm $\mathcal{A}$, $Adv_{\mathcal{A}}^{DHBDH} < \frac{1}{m^l}$ for every fixed $l > 0$ and sufficiently large m.

2.3   Protocol Requirements

Consider the n users who wish to agree upon a conference key to be the set $\{1, 2, \ldots, n\}$. Let $s_1, s_2, \ldots, s_n \in Z_q^*$, be their respective private keys. Let U be a subset of $\{1, 2, \ldots, n\}$ consisting of consecutive integers. We call U a user set. Let $Rep(U)$ stand for the representative of the set U. To be specific, let $Rep(U) = \min(U)$. We use the notation $A[1, \ldots, n]$ for an array of n elements $A_1, \ldots, A_n$ and write $A[i]$ and A_i interchangeably.
We take G_1 to be a cyclic subgroup of an elliptic curve group of some large prime order q and the bilinear map $e : G_1^2 \rightarrow G_2$ to be either a modified Weil pairing or a Tate pairing [1]. Let P be an arbitrary generator of G_1. Choose a hash function $H : G_2 \rightarrow Z_q^*$. The system parameters for the unauthenticated protocol are $params = \langle G_1, G_2, e, q, P, H \rangle$.

For ID-based authenticated key agreement, we will additionally require the followings:

1. Hash functions $\widehat{H} : G_1 \rightarrow Z_q^*$, $H_1 : \{0, 1\}^* \rightarrow G_1$.
2. A key generation centre (KGC) which chooses a random $s \in Z_q^*$ and sets $P_{pub} = sP$. It publishes P_{pub} as a system parameter and keeps s as secret which it treates as the master key. Each user i has an identity $ID_i \in \{0, 1\}^*$ and long term public key $Q_i = H_1(ID_i)$. User i sends Q_i to KGC and KGC sends back the long term private key $S_i = sQ_i$ to user i.
3. The keys $s_1, s_2, \ldots, s_n$ are short term private keys.

The system parameters for the authenticated protocol are
$params = \langle G_1, G_2, e, q, P, P_{pub}, H, \widehat{H}, H_1 \rangle$.

3   Protocol

In this section, we present a three-group and a two-group Diffie-Hellman key agreement protocol CombineThree and CombineTwo respectively together with an n-party recursive algorithm KeyAgreement which makes use of CombineThree and CombineTwo. *The boxed portions are executed for the authenticated version.*

The three-group Diffie-Hellman key agreement for unauthenticated as well as ID-based authenticated versions are jointly given by the subroutine CombineThree as described below. In this subroutine, when cardinality of each of these three groups is one, then the ID-based authenticated version is simply the three-party protocol proposed by Zhang, Liu, Kim in [13] while the unauthenticated version is the Joux [7] three-party key agreement protocol using bilinear map.

procedure CombineThree($U[1, 2, 3], s[1, 2, 3]$)
$i = 1$ to 3 **do**
 $Rep(U_i)$ computes $P_i = s_i P$
 $\boxed{\text{and } T_{Rep(U_i)} = \widehat{H}(P_i)S_{Rep(U_i)} + s_i P_i;}$
 Let $\{j, k\} = \{1, 2, 3\}\backslash\{i\}$;
 $Rep(U_i)$ sends P_i, $\boxed{T_{Rep(U_i)}}$ to all members of both U_j, U_k;
end do

$i = 1$ to 3 **do**
 Let $\{j, k\} = \{1, 2, 3\}\backslash\{i\}$;
each member of U_i
$\boxed{\begin{array}{l}\text{verifies}: e(T_{Rep(U_j)} + T_{Rep(U_k)}, P) = e(\widehat{H}(P_j)Q_{Rep(U_j)} + \widehat{H}(P_k)Q_{Rep(U_k)}, P_{pub}) \\ e(P_j, P_j)e(P_k, P_k) \text{ and}\end{array}}$
computes $H(e(P_j, P_k)^{s_i})$;
end do

end CombineThree

This subroutine does a key agreement among three user sets U_1, U_2, U_3 with s_1, s_2, s_3 respectively as their private keys (short term for ID-based) with common key $H(e(P, P)^{s_1 s_2 s_3})$.

Similarly, the two-group Diffie-Hellman key agreement for unauthenticated as well as ID-based authenticated versions are jointly given by the subroutine CombineTwo as described below. This subroutine reduces to a two-party Diffie-Hellman key agreement protocol when each of the two groups has cardinality one.

procedure CombineTwo($U[1, 2], s[1, 2]$)
$i = 1$ to 2 **do**
 $Rep(U_i)$ computes $P_i = s_i P$
 $\boxed{\text{and } T_{Rep(U_i)} = \widehat{H}(P_i)S_{Rep(U_i)} + s_i P_i;}$
end do

$Rep(U_1)$ generates $\overline{s} \in_R Z_q^*$ and sends $\overline{s}P$

and $\overline{T}_{Rep(U_1)} = \widehat{H}(\overline{s}P)S_{Rep(U_1)} + \overline{s}^2 P$ to the rest of the users;

each member of U_1, U_2 except $Rep(U_1)$ verifies :
$e(\overline{T}_{Rep(U_1)}, P) = e(\widehat{H}(\overline{s}P)Q_{Rep(U_1)}, P_{pub})e(\overline{s}P, \overline{s}P)$;

$Rep(U_1)$ sends P_1, $\boxed{T_{Rep(U_1)}}$ to all members of U_2;

$Rep(U_2)$ sends P_2, $\boxed{T_{Rep(U_2)}}$ to all members of U_1;

each member of U_1

verifies : $e(T_{Rep(U_2)}, P) = e(\widehat{H}(P_2)Q_{Rep(U_2)}, P_{pub})e(P_2, P_2)$ and
computes $H(e(P_2, \overline{s}P)^{s_1})$;

each member of U_2

verifies : $e(T_{Rep(U_1)}, P) = e(\widehat{H}(P_1)Q_{Rep(U_1)}, P_{pub})e(P_1, P_1)$ and
computes $H(e(P_1, \overline{s}P)^{s_2})$;

end CombineTwo

This subroutine does a key agreement among two user sets U_1, U_2 with s_1, s_2 respectively as their (short term) private keys with common key $H(e(P, P)^{s_1 s_2 \overline{s}})$ where $\overline{s}$ is generated randomly by the representative of the user set U_1. Thus, this subroutine is essentially the Joux's protocol invoked for two user sets.

Next we describe the tree structure KeyAgreement as a top down recursive procedure which uses the above two subroutines CombineTwo and CombineThree.

procedure KeyAgreement$(m, U[i+1, \ldots, i+m])$
    if $(m = 1)$ **then**
        $KEY = s[i+1]$;
    end if
    if $(m = 2)$ **then**
        call CombineTwo$(U[i+1, i+2], s[i+1, i+2])$;
        Let KEY be the agreed key between user sets U_{i+1}, U_{i+2};
    end if
    $n_0 = 0; n_1 = \lfloor \frac{m}{3} \rfloor; n_3 = \lceil \frac{m}{3} \rceil; n_2 = m - n_1 - n_3$;
    $j = 1$ to 3 **do**
        call KeyAgreement$(n_j, U[i + n_{j-1} + 1, \ldots, i + n_{j-1} + n_j])$;
        $\widehat{U}_j = U[i + n_{j-1} + 1, \ldots, i + n_{j-1} + n_j]$; $\widehat{s}_j = KEY$; $n_j = n_{j-1} + n_j$;
    end do;
    call CombineThree$(\widehat{U}[1, 2, 3], \widehat{s}[1, 2, 3])$;
    Let KEY be the agreed key among user sets $\widehat{U}_1, \widehat{U}_2, \widehat{U}_3$;
end KeyAgreement

The start of the recursive protocol KeyAgreement is made by the following two statements:

1. $U_j = j$ for $1 \le j \le n$;
2. **call** KeyAgreement$(n, U[1, \ldots, n])$;

The algorithm is recursive and goes through several levels starting with level 0. In each level, there are sets of users who have (or agree upon) a common secret key. In level 0, each user is in a set by himself/ herself and the (short term) secret key of the singleton set is the secret key of the concerned user. For n users, let the levels be numbered $0, \ldots, R(n)$. In level i, let the number of user sets be n_i. Thus $n_0 = n$, and $n_k = 1$, where $k = R(n)$. We identify the rounds of the algorithm as follows. There are $R(n)$ rounds of computation. The ith round of computation takes the state of the algorithm from level $i - 1$ to level i for $i = 1, \ldots, R(n)$. We introduce some notations for convenience of analysing the algorithm.

$-U_j^{(i)}$: the j-th user set at level $i, 0 \le i \le R(n), 1 \le j \le n_i$,

$-s_j^{(i)}$: common (short term) secret key agreed upon by users in the user set $U_j^{(i)}$,

$-P_j^{(i)}$: i-th level j-th public key, i.e. $P_j^{(i)} = s_j^{(i)} P$.

Let $p = \lfloor \frac{n}{3} \rfloor$ and $r = n \bmod 3$. We partition the set of users $U_1^{(k)} = \{1, \ldots, n\}$ into three user sets $U_1^{(k-1)}, U_2^{(k-1)}, U_3^{(k-1)}$ with cardinality p, p, p respectively if $r = 0$, with cardinality $p, p, p + 1$ respectively if $r = 1$ or with cardinality $p, p+1, p+1$ respectively if $r = 2$. We use this top down recursive procedure for each user set $U_j^{(i)}$ to split it into three user sets for $1 < i < k, 1 \le j \le n_i$ and for $i = 1$, each such user set is partioned into either one user set, or two user sets or three user sets depending on n. Note that with this tree structure, an user set with two users appears only at level 1 and so CombineTwo is never invoked for round ≥ 2. For $n = 10$, the working of the algorithm (unauthenticated version) is shown in figure 1.

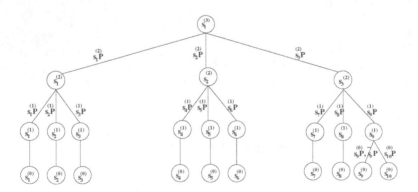

Fig. 1. Key agreement among $n = 10$ parties, user 9 generates $\bar{s}_9$ randomly and sends $\bar{s}_9 P$ to user 10 at the first level.

Lemma 31 *The final agreed key KEY among n users in the subroutine* KeyAgreement *is*

$$KEY = s_1^{(k)} = H(e(P,P)^{s_1^{(k-1)} s_2^{(k-1)} s_3^{(k-1)}})$$

where $k = R(n)$, $n > 2$.

Lemma 32 *Each member of $U_j^{(i)}$ can compute $s_j^{(i)}$ for $1 \leq j \leq n_i$, $i \leq k$ where $k = R(n)$. Consequently, all users are able to compute the common key $KEY = s_1^{(k)}$.*

4 Security Analysis

4.1 Security Against Passive Adversary

A secure key agreement protocol should withstand both passive and active attacks. In this subsection we define the **Decisional ternary tree group key agreement (DTGKA) problem** for our unauthenticated protocol and show that this problem is hard for the passive adversary by reducing the hardness of this problem to the hardness of DHBDH problem following the techniques used in [3], [9], [12]. The authentication is introduced in our unauthenticated protocol using a special signature scheme to get security against an active adversary.

Given a ternary tree T of height at most k with n leaf nodes ($n > 2$) and $X = (s_1, s_2, \ldots, s_n)$ for $s_i \in Z_q^*$, the public and secret values are collectively defined as follows :
$-v(k, X, T) := \{P_j^{(i)}$ where j and i are defined according to $T\}$
$-K(k, X, T) := \Psi = H(e(P,P)^{s_1^{(k-1)} s_2^{(k-1)} s_3^{(k-1)}})$

$v(k, X, T)$ is exactly the view of the passive adversary in the ternary tree T where final key is $K(k, X, T)$. We call the key $K(k, X, T)$ a DHBDH key. Our goal is to show that this DHBDH key generated by the unauthenticated protocol can not be distinguished by a polynomial time algorithm from a random number if all the transmitted values during a protocol run are known.

Suppose $\mathcal{T}_k$ is the set of all ternary trees of height k having structure of KeyAgreement. A tree T of height k is chosen randomly from $\mathcal{T}_k$ and let $X \in_R (Z_q^*)^n$ ($n \leq 3^k$) be the labels of (short term private keys associated with) the leaf nodes of T. Then k is the number of rounds with n users. Let us define two random variables $A_k, \widehat{A}_k$ as follows :
$-A_k := (v(k, X, T), y), y \in_R Z_q^*$
$-\widehat{A}_k := (v(k, X, T), K(k, X, T))$

Let $\mathcal{S}_k = \{(T, X) : T \in_R \mathcal{T}_k$ and $X \in_R (Z_q^*)^n$, where n is the number of leaf level nodes in $T\}$. Let B_T be the number of edges in the ternary tree T. For $(T, X) \in$

$\mathcal{S}_k$, define $\Gamma(T, X)$ to be the ordered tuple of all public information along the arcs of T. Clearly, $\Gamma(T, X) \subseteq (G_1)^{B_T}$. Then the random variable A_k takes values from the sample space $(G_1)^{B_T} \times Z_q^*$ according to the uniform probability distribution and $\widehat{A}_k$ takes values from the sample space $\Gamma(T, X) \times Z_q^* \subseteq (G_1)^{B_T} \times Z_q^*$ with the uniform probability distribution.

Definition 41 *Consider* $\langle G_1, G_2, e \rangle$. *Let* $n > 2$ *be a positive integer,* $X = (s_1, s_2, \ldots, s_n)$ *for* $s_i \in Z_q^*$ *and* T *be a ternary tree of height* k *with* n *leaf nodes labeled by* X, *and* $A_k, \widehat{A}_k$ *are defined as above. A* **Decisional ternary tree group key agreement (DTGKA)** *algorithm* $\mathcal{F}$ *for* $\langle G_1, G_2, e \rangle$ *is a probabilistic polynomial time algorithm that outputs either* 0 *or* 1, *satisfying, for some fixed* $l > 0$ *and sufficiently large* m *:*

$$|Prob[\mathcal{F}(A_k) = 1] - Prob[\mathcal{F}(\widehat{A}_k) = 1]| > \frac{1}{m^l}.$$

We call $\mathcal{F}$ *a polynomial time distinguisher that distinguishes* A_k *and* $\widehat{A}_k$.

The **DTGKA problem** with height k is to find a polynomial time distinguisher $\mathcal{F}$ for A_k and $\widehat{A}_k$ defined above.

Theorem 42 *If the DHBDH problem in* $\langle G_1, G_2, e \rangle$ *is hard, then* A_k *and* $\widehat{A}_k$ *are polynomially indistinguishable.*

Proof : First let us provide a plan of the proof. The proof is by induction on k.

Base Step : $k = 1$: Distinguishing A_1 and $\widehat{A}_1$ implies "solving" DHBDH problem in $\langle G_1, G_2, e \rangle$. Thus it is not possible to distinguish A_1 and $\widehat{A}_1$ assuming DHBDH problem is hard in $\langle G_1, G_2, e \rangle$.

Induction Hypothesis : Assume that for some $k \geq 2$, it is not possible to distinguish A_{k-1} and $\widehat{A}_{k-1}$.

Induction Step : We show that the ability to distinguish A_k and $\widehat{A}_k$ implies either (a) "solving" DHBDH problem in $\langle G_1, G_2, e \rangle$ or (b) ability to distinguish A_{k-1} and $\widehat{A}_{k-1}$. Since (a) is given to be hard and (b) is hard by induction hypothesis, it follows that it is not possible to distinguish between A_k and $\widehat{A}_k$.

Now we turn to a proof of the induction step. Let $T \in_R \mathcal{T}_k$ be a ternary tree of height k with n leaf nodes. Let $X \in_R (Z_q^*)^n$ be the labels of the leaf nodes of T. Let T_1, T_2, T_3 respectively be the left, middle and right subtree of height at most $k - 1$ of the tree T. This implies that at least one of these subtrees has height exactly $k-1$. If $X_1 = (s_1, \ldots, s_l), X_2 = (s_{l+1}, \ldots, s_m)$ and $X_3 = (s_{m+1}, \ldots, s_n)$, where s_1 through s_l are associated with T_1, s_{l+1} through s_m with T_2 and s_{m+1} through s_n with T_3, then A_k and $\widehat{A}_k$ can be rewritten as :
$A_k := (v(k, X, T), y)$ for random $y \in Z_q^*$
$= (v(k - 1, X_1, T_1), v(k - 1, X_2, T_2), v(k - 1, X_3, T_3), P_1^{(k-1)}, P_2^{(k-1)}, P_3^{(k-1)}, y)$
$= (v(k - 1, X_1, T_1), v(k - 1, X_2, T_2), v(k - 1, X_3, T_3), s_1^{(k-1)}P, s_2^{(k-1)}P, s_3^{(k-1)}P, y)$

$$\widehat{A}_k := (v(k, X, T), K(k, X, T))$$
$$= (v(k-1, X_1, T_1), v(k-1, X_2, T_2), v(k-1, X_3, T_3), P_1^{(k-1)}, P_2^{(k-1)}, P_3^{(k-1)}, \Psi)$$
$$= (v(k-1, X_1, T_1), v(k-1, X_2, T_2), v(k-1, X_3, T_3), s_1^{(k-1)}P, s_2^{(k-1)}P, s_3^{(k-1)}P, \Psi)$$

Let us consider the following random variables:
$$A_k := (v(k-1, X_1, T_1), v(k-1, X_2, T_2), v(k-1, X_3, T_3), P_1^{(k-1)}, P_2^{(k-1)}, P_3^{(k-1)}, y)$$
$$B_k := (v(k-1, X_1, T_1), v(k-1, X_2, T_2), v(k-1, X_3, T_3), r_1P, P_2^{(k-1)}, P_3^{(k-1)}, y)$$
$$C_k := (v(k-1, X_1, T_1), v(k-1, X_2, T_2), v(k-1, X_3, T_3), r_1P, r_2P, P_3^{(k-1)}, y)$$
$$D_k := (v(k-1, X_1, T_1), v(k-1, X_2, T_2), v(k-1, X_3, T_3), r_1P, r_2P, r_3P, y)$$
$$\widehat{D}_k := (v(k-1, X_1, T_1), v(k-1, X_2, T_2), v(k-1, X_3, T_3), r_1P, r_2P, r_3P, K_1)$$
$$\widehat{C}_k := (v(k-1, X_1, T_1), v(k-1, X_2, T_2), v(k-1, X_3, T_3), r_1P, r_2P, P_3^{(k-1)}, K_2)$$
$$\widehat{B}_k := (v(k-1, X_1, T_1), v(k-1, X_2, T_2), v(k-1, X_3, T_3), r_1P, P_2^{(k-1)}, P_3^{(k-1)}, K_3)$$
$$\widehat{A}_k := (v(k-1, X_1, T_1), v(k-1, X_2, T_2), v(k-1, X_3, T_3), P_1^{(k-1)}, P_2^{(k-1)}, P_3^{(k-1)}, \Psi)$$
where $r_1, r_2, r_3 \in_R Z_q^*$ and $K_1 = H(e(P, P)^{r_1 r_2 r_3})$, $K_2 = H(e(P, P)^{r_1 r_2 s_3^{(k-1)}})$ and $K_3 = H(e(P, P)^{r_1 s_2^{(k-1)} s_3^{(k-1)}})$.

Claim : If $A_k, \widehat{A}_k$ are distinguishable in polynomial time, then at least one of the followings can be distinguished : $(A_k, B_k), (B_k, C_k), (C_k, D_k), (D_k, \widehat{D}_k), (\widehat{D}_k, \widehat{C}_k), (\widehat{C}_k, \widehat{B}_k)$ or $(\widehat{B}_k, \widehat{A}_k)$.
Proof of the Claim : Let $a_1 = Prob[\mathcal{F}(A_k) = 1]$, $a_2 = Prob[\mathcal{F}(B_k) = 1]$, $a_3 = Prob[\mathcal{F}(C_k) = 1]$, $a_4 = Prob[\mathcal{F}(D_k) = 1]$, $a_5 = Prob[\mathcal{F}(\widehat{D}_k) = 1]$, $a_6 = Prob[\mathcal{F}(\widehat{C}_k) = 1]$, $a_7 = Prob[\mathcal{F}(\widehat{B}_k) = 1]$, $a_8 = Prob[\mathcal{F}(\widehat{A}_k) = 1]$. Since A_k and $\widehat{A}_k$ are distinguishable in polynomial time, $|a_1 - a_8| > \frac{1}{m^l}$ for sufficiently large m and for a fixed $l > 0$. Now we will show that at least one of the followings must hold : $|a_i - a_{i+1}| > \frac{1}{m^{l+1}}$ for $i = 1, \ldots, 7$. If not, let $|a_i - a_{i+1}| \leq \frac{1}{m^{l+1}}$ for all $i = 1, \ldots, 7$.
Then $|a_1 - a_8| \leq |a_1 - a_2| + \cdots + |a_7 - a_8| \leq \frac{7}{m^{l+1}} \leq \frac{1}{m^l}$, a contradiction, if $m \geq 7$.

We shall show that the ability to distinguish any one of $(A_k, B_k), (B_k, C_k), (C_k, D_k), (\widehat{D}_k, \widehat{C}_k), (\widehat{C}_k, \widehat{B}_k)$ or $(\widehat{B}_k, \widehat{A}_k)$ reduces to solving DTGKA problem with height $k-1$ and ability of distinguishing $(D_k, \widehat{D}_k)$ reduces to solving the DHBDH problem in $\langle G_1, G_2, e \rangle$. The proof of the two cases : (A_k, B_k) and $(D_k, \widehat{D}_k)$ are discussed here in details. A proof similar to the case (A_k, B_k) follows for others.

Case : Distinguish (A_k, B_k) : Suppose $\mathcal{F}_{AB_k}$ is a polynomial time distinguisher that can distinguish A_k and B_k in polynomial time. We will show that $\mathcal{F}_{AB_k}$ can be used to solve DTGKA problem with height $k-1$. We construct a polynomial time distinguisher $\mathcal{F}_{A\widehat{A}_{k-1}}$ that distinguishes A_{k-1} and $\widehat{A}_{k-1}$ in polynomial time as follows :
Let $V_{k-1}^* = (v(k-1, X^*, T^*), r^*)$ where T^* is a ternary tree of height $k-1$ with $|X^*| = n$ having structure of KeyAgreement. The distinguisher $\mathcal{F}_{A\widehat{A}_{k-1}}$ first constructs two ternary trees $\widetilde{T}$ and $\overline{T}$ with leaf level secret key distribution $\widetilde{X}$ and

$\overline{X}$ respectively in the following manner : if $n = 3^k$, then take $|\widetilde{X}| = |\overline{X}| = n$, else take either $|\widetilde{X}| = |\overline{X}| = n$ or $|\widetilde{X}| = n, |\overline{X}| = n + 1$ or $|\widetilde{X}| = |\overline{X}| = n + 1$. Then $\mathcal{F}_{A\widehat{A}_{k-1}}$ constructs a tree of height k with $|X^*| + |\widetilde{X}| + |\overline{X}|$ users and $T^*, \widetilde{T}, \overline{T}$ as the left, middle and right subtree respectively. The resulting tree is clearly a random member of $\mathcal{T}_k$. Next $\mathcal{F}_{A\widehat{A}_{k-1}}$ sets :

$V_k^* = (v(k - 1, X^*, T^*), v(k - 1, \widetilde{X}, \widetilde{T}), v(k - 1, \overline{X}, \overline{T}), r^*P, \widetilde{K}P, \overline{K}P, y)$, where $\widetilde{K} = K(k - 1, \widetilde{X}, \widetilde{T}), \overline{K} = K(k - 1, \overline{X}, \overline{T}), y \in_R Z_q^*$ and runs $\mathcal{F}_{AB_k}$ on input V_k^*. Now $Prob[\mathcal{F}_{AB_k}(A_k = V_k^*) = 1] = Prob[\mathcal{F}_{A\widehat{A}_{k-1}}(\widehat{A}_{k-1} = V_{k-1}^*) = 1]$ and $Prob[\mathcal{F}_{AB_k}(B_k = V_k^*) = 1] = Prob[\mathcal{F}_{A\widehat{A}_{k-1}}(A_{k-1} = V_{k-1}^*) = 1]$.

Consequently, $|Prob[\mathcal{F}_{A\widehat{A}_{k-1}}(\widehat{A}_{k-1} = V_{k-1}^*) = 1] - Prob[\mathcal{F}_{A\widehat{A}_{k-1}}(A_{k-1} = V_{k-1}^*) = 1]| = |Prob[\mathcal{F}_{AB_k}(A_k = V_k^*) = 1] - Prob[\mathcal{F}_{AB_k}(B_k = V_k^*) = 1]|$. Hence if $\mathcal{F}_{AB_k}$ can distinguish between A_k and B_k, then $\mathcal{F}_{A\widehat{A}_{k-1}}$ can distinguish between A_{k-1} and $\widehat{A}_{k-1}$.

Case : Distinguish $(D_k, \widehat{D}_k)$: Suppose $\mathcal{F}_{D\widehat{D}_k}$ is a polynomial time distinguisher that can distinguish D_k and $\widehat{D}_k$ in polynomial time. We shall show that $\mathcal{F}_{D\widehat{D}_k}$ can be used to construct a polynomial time algorithm $\mathcal{A}$ that solves the DHBDH problem in $\langle G_1, G_2, e \rangle$. Note that r_1P and r_2P are independent variables from $v(k - 1, X_1, T_1)$ and $v(k - 1, X_2, T_2)$.

Given $V_{k-1}^* = (P, r^*P, \widetilde{r}P, \overline{r}P, r))$, the algorithm $\mathcal{A}$ has to decide whether $r = H(e(P, P)^{r^*\widetilde{r}\overline{r}}) \bmod q$ ($\mathcal{A}$ outputs yes in this case) or r is random ($\mathcal{A}$ outputs no in this case) where $r^*, \widetilde{r}, \overline{r} \in_R Z_q^*$. For this, $\mathcal{A}$ first generates a tree of height k having structure of KeyAgreement with three subtrees T^*, $\widetilde{T}$ and $\overline{T}$. The leaf level secret key distribution of these subtrees are X^*, $\widetilde{X}$ and $\overline{X}$ respectively. Then $\mathcal{A}$ sets :

$V_k^* = (v(k - 1, X^*, T^*), v(k - 1, \widetilde{X}, \widetilde{T}), v(k - 1, \overline{X}, \overline{T}), r^*P, \widetilde{r}P, \overline{r}P, r))$ and runs $\mathcal{F}_{D\widehat{D}_k}$ on input V_k^*. Now $Prob[\mathcal{A}$ outputs no on input $V_{k-1}^*] = Prob[\mathcal{F}_{D\widehat{D}_k}(D_k = V_k^*) = 1]$

and $Prob[\mathcal{A}$ outputs yes on input $V_{k-1}^*] = Prob[\mathcal{F}_{D\widehat{D}_k}(\widehat{D}_k = V_k^*) = 1]$.

Consequently, $|Prob[\mathcal{A}$ outputs no on input $V_{k-1}^*] - Prob[\mathcal{A}$ outputs yes on input $V_{k-1}^*]| = |Prob[\mathcal{F}_{D\widehat{D}_k}(D_k = V_k^*) = 1] - Prob[\mathcal{F}_{D\widehat{D}_k}(\widehat{D}_k = V_k^*) = 1]|$. Hence if $\mathcal{F}_{D\widehat{D}_k}$ can distinguish between D_k and $\widehat{D}_k$, then $\mathcal{A}$ solves DHBDH problem in $\langle G_1, G_2, e \rangle$.   □

4.2   Security Against Active Adversary

This is discussed in the full version.

5   Efficiency

This involves the communication and computation efficiency. In each round, a user may have to transmit publicly an element of G_1 to some or all the other users. Also it has to perform some operations like scalar multiplications, pairing computations. The number of rounds, total group elements sent, total messages exchanged provides the communication overhead in the protocol whereas total pairing computation, total scalar multiplications used incurs the computation costs.

First consider the unauthenticated version. Let $R(n)$ denote the total number of rounds, $P(n)$ the total pairings computed, $B(n)$ the combined message size and $E(n)$ the total number of scalar multiplications. If an user sends publicly an element of G_1 to some or all remaining users, then this counts one to $B(n)$. Thus $B(n)$ is the total number of such group elements. Proofs of the following results will be provided in the full version of the paper.

Lemma 51 *For $n > 2$, the following recursions and bounds hold for $R(n), B(n)$, $E(n)$ and $P(n)$:*

1. $R(3n) = 1 + R(n); R(3n+1) = 1 + R(n+1); R(3n+2) = 1 + R(n+1);$ with initial conditions $R(1) = 0, R(2) = 1$. Consequently, $R(n) = \lceil \log_3 n \rceil$ for all n.

2. $B(3n) = 3 + 3B(n); B(3n+1) = 3 + 2B(n) + B(n+1); B(3n+2) = 3 + B(n) + 2B(n+1);$ with initial conditions $B(1) = 0, B(2) = 3$. Consequently, $B(n) < \frac{5}{2}(n-1)$ for $n > 2$.

3. $E(3n) = 3 + 3E(n); E(3n+1) = 3 + 2E(n) + E(n+1); E(3n+2) = 3 + E(n) + 2E(n+1);$ with initial conditions $E(1) = 0, E(2) = 3$. Consequently, $E(n) < \frac{5}{2}(n-1)$ for $n > 2$.

4. $P(3n) = 3n + 3P(n); P(3n+1) = 3n + 1 + 2P(n) + P(n+1); P(3n+2) = 3n + 2 + P(n) + 2P(n+1);$ with initial conditions $P(1) = 0, P(2) = 2$. Consequently, $P(n) \leq n \lceil \log_3 n \rceil$.

For authenticated version, let $R_a(n), B_a(n), E_a(n)$ and $P_a(n)$ be the corresponding terms of the unauthenticated version.

Lemma 52 *For $n > 2$, the following relations hold for $R_a(n), B_a(n), E_a(n)$ and $P_a(n)$:*

1. $R_a(n) = R(n)$. Consequently, $R_a(n) \leq n \lceil \log_3 n \rceil$.

2. $B_a(n) = 2B(n)$. Consequently, $B_a(n) < 5(n-1)$ for $n > 2$.

3. $E_a(3n) = 9 + E_a(n); E_a(3n+1) = 9 + 2E_a(n) + E_a(n+1); E_a(3n+2) = 9 + E_a(n) + 2E_a(n+1);$ with $E_a(1) = 0, E_a(2) = 12$. Consequently, $E_a(n) \leq 9(n-1)$ for $n > 2$.

4. $P_a(3n) = 5(3n) + 3P_a(n)$ for $n > 2; P_a(3n+1) = 5(3n+1) + 2P_a(n) + P_a(n+1); P_a(3n+2) = 5(3n+2) + P_a(n) + 2P_a(n+1);$ with $P_a(1) = 0, P_a(2) = 11$. Also $P_a(n) \leq 5P(n) + 3$. Consequently, $P_a(n) \leq 5n \lceil \log_3 n \rceil + 3$.

6   Comparison

In this section, we compare our protocol to some of the previously known protocols. Burmester and Desmedt present in [5] an efficient multi-party protocol that can be executed only in two rounds. A class of generic n-party Diffie-Hellman protocols ($n > 2$) is defined in [12]. The entire protocol class is shown to be secure against passive adversaries based on the intractability of the DDH problem. One group key distribution protocols introduced in [12] is GDH-3. A tree based Diffie-Hellman group key agreement protocol TGDH has been proposed by Kim, Perrig and Tsudik in [9] which is shown to be secure against passive adversaries. We note that the security assumptions behind the various protocols are different and hence in a strict sense an efficiency comparison might not be meaningful. However, we believe that the discussion presented below does provide some idea of the relative efficiency of the various protocols. Table 1 compares the unauthenticated version of our protocol with these protocols. (Inv stands for total number of modular inversions and Mul stands for Total number of multiplications used).

	$R(n)$	$B(n)$	$E(n)$	$P(n)$	Inv
BD [5]	2	$2n$	$n(n+1)$	–	n
GDH-3 [12]	$n+1$	$3(n-1)$	$5n-6$	–	–
TGDH [9]	$\lceil \log_2 n \rceil$	$n\lceil \log_2 n \rceil$	$n\lceil \log_2 n \rceil$	–	–
Our Protocol	$\lceil \log_3 n \rceil$	$< \frac{5}{2}(n-1)$	$< \frac{5}{2}(n-1)$	$\leq n\lceil \log_3 n \rceil$	–

Table 1: Protocol Comparison (unauthenticated versions).

Points to note for unauthenticated protocols :

1. The underlying group of GDH-3 and BD protocol is a multiplicative subgroup of Z_p^* of order q where p and q both are prime.
2. The communication complexity is measured by $R(n)$ and $B(n)$ and our protocol achieves the minimum for both among all known protocols, except BD protocol. The computation complexity of our protocol consists of two parts – exponentiation and pairing. The number of exponentiation is less than all other protocols, but additionally $n\lceil \log_3 n \rceil$ pairings are required. Assuming that each pairing computation is approximately equal to three exponentiations [1] the TGDH and GDH-3 algorithms are more efficient than ours. This is based on the current state of the art in the algorithm for computing pairings. Any improvement in pairing computation algorithms will improve the efficiency of our protocol with respect to both TGDH and GDH-3.
3. Moreover, all the above protocols give DH-key except BD protocol.

The comparison of our authenticated protocol with the existing authenticated protocols are discussed in the full version [2] of the paper.

7   Conclusion

We have described an unauthenticated as well as a ID-based authenticated multiparty key agreement protocol using pairing. In fact, our protocol can use any secure two and three party protocol and provides all the desirable security attributes possessed by both of them. A regorous proof of security against passive adversaries is provided for our unauthenticated protocol. As for security proof of our authenticated protocol, we refer the reader to the full version of our paper [2]. The computation and communication complexity of our protocol compares favourably with other known protocols.

Note : After completion of this work, we came to know of the result of Katz and Yung [8] which modifies the BD protocol to provide a provably secure constant round authenticated protocol.

References

1. P. S. L. M. Barreto, H. Y. Kim and M. Scott. *Efficient algorithms for pairing-based cryptosystems.* Advances in Cryptology - Crypto '2002, LNCS 2442, Springer-Verlag (2002), pp. 354-368.
2. R. Barua, R. Dutta and P. Sarkar. *Extending Joux's Protocol to Multi Party Key Agreement,* Report 2003/062, http://eprint.iacr.org, 2003.
3. K. Becker and U. Wille. *Communication Complexity of Group Key Distribution.* ACMCCS '98.
4. D. Boneh and M. Franklin. *Identity-Based Encryption from the Weil Pairing.* In Advances in Cryptology - CRYPTO '01, LNCS 2139, pages 213-229, Springer-Verlag, 2001.
5. M. Burmester and Y. Desmedt. *A Secure and Efficient Conference Key Distribution System.* In A. De Santis, editor, Advances in Cryptology EUROCRYPT '94, Workshop on the theory and Application of Cryptographic Techniques, LNCS 950, pages 275-286, Springer-Verlag, 1995.
6. W. Diffie and M. Hellman. *New Directions In Cryptography.* IEEE Transactions on Information Theory, IT-22(6) : 644-654, November 1976.
7. A. Joux. *A One Round Protocol for Tripartite Diffie-Hellman.* ANTS IV, LNCS 1838, pp. 385-394, Springer-Verlag, 2000.
8. J. Katz and M. Yung. *Scalable Protocols for Authenticated Group Key Exchange.* In Advances in Cryptology - CRYPTO 2003.
9. Y. Kim, A. Perrig, and G. Tsudik. *Tree based Group Key Agreement,* Report 2002/009, http://eprint.iacr.org, 2002.
10. D. Nalla and K. C. Reddy. *Identity Based Authenticated Group Key Agreement Protocol.* In Proceedings of INDOCRYPT-2002, LNCS , Springer-Verlag, 2002.
11. A. Shamir. *Identity-based Cryptosystems and Signature Schemes.* In Advances in Cryptology - CRYPTO '84, LNCS 196, pages 47-53, Springer-Verlag, 1984.
12. M. Steiner, G. Tsudik, M. Waidner. *Diffie-Hellman Key Distribution Extended to Group Communication,* ACM Conference on Computation and Communication Security, 1996.
13. F. Zhang, S. Liu and K. Kim. *ID-based One Round Authenticated Tripartite Key Agreement Protocol with Pairings.* Avalable at http://eprint.iacr.org, 2002.

Public Key Cryptosystems Based on Free Partially Commutative Monoids and Groups

P.J. Abisha, D.G. Thomas, and K.G. Subramanian

Department of Mathematics
Madras Christian College
Tambaram, Chennai - 600 059, India
dgthomasmcc@yahoo.com

Abstract. A public key cryptosystem based on free partially commutative monoids is constructed. The encryption of a message to create the cryptotext uses a Thue system which is formed from the free partially commutative monoid with the help of a trapdoor morphism. The decidability of the word problem for free partially commutative monoids can be used for decryption. Finding the trapdoor morphism of this system is shown to be NP-hard. But, a zero - knowledge protocol to convince a verifier that there is such a trapdoor morphism is provided. A related but different public key cryptosystem based on free partially commutative groups is also proposed.

Keywords. Public Key Cryptosystem, Finitely Presented Groups, Free Partially Commutative Monoids, Thue Systems, Word Problem, Zero Knowledge Protocol.

1 Introduction

In recent years, due to the need for security and secrecy in storage and transmission of sensitive data, there has been a lot of study in cryptography. After Diffie and Hellman introduced the concept of Public Key Cryptosystem (PKC), an increasing number of PKCs have been introduced. Salomaa [9] introduced the first PKC based on formal language theory using an ingenious technique for the construction of such systems. Subsequently, many PKCs based on formal language theory and semi-groups have been constructed [11,12,13,14]. For a compact survey of the language theoretic aspects of cryptology, we refer to [7].

Free partially commutative monoids have been studied in the literature. In computer science, they have served as an abstract model to describe concurrency [4]. A free partially commutative monoid generated by an alphabet Σ with respect to a concurrency relation θ is the quotient of Σ^* by the congruence relation $\equiv_\theta$. In fact for $(a, b) \in \theta$, a and b can be commuted which means an occurrence of $ab(respy.ba)$ in a word u can be replaced by $ba(respy.ab)$, yielding v, so that $u \equiv_\theta v$. The word problem for a free partially commutative monoid is the following : given $u, v \in \Sigma^*$, is $u \equiv_\theta v$? This problem is decidable in linear time [1].

T. Johansson and S. Maitra (Eds.): INDOCRYPT 2003, LNCS 2904, pp. 218–227, 2003.
© Springer-Verlag Berlin Heidelberg 2003

In this paper we present a PKC where the decryption is done using the decidability of the word problem of a free partially commutative monoid. The encryption is done using a Thue system which is obtained by applying a trapdoor morphism. In fact a Thue system T on an alphabet Σ is a finite subset of $\Sigma^* \times \Sigma^*$. The word problem is in general undecidable for Thue systems [2].

We prove that finding the trapdoor morphism from the encryption key is NP-hard. The proof technique relies on that of Kari [5]. We give a zero-knowledge protocol [10] to convince a verifier that there is a morphism which maps the known Thue system into a free partially commutative monoid. Here the morphism itself is not revealed by the protocol.

We present another related but different PKC which uses a free partially commutative group [15] for decryption and a finitely presented group for encryption. The advantage of finitely presented group G over that of finitely presented monoid (Thue system) is the existence of inverse for each letter. This helps to simplify the word problem in the following manner. For given two words $x, y \in \Sigma^*$, checking whether x is equal to y in G can be reduced to checking whether xy^{-1} is equal to λ (empty word) in G. The PKC presented uses this property to simplify the decryption process.

2   A PKC Based on Free Partially Commutative Monoids

We first give the relevant definitions [1,2].

Definition 1. *Let Σ be an alphabet and $\theta = \{(a,b)/ \text{ for some } a, b \in \Sigma\}$ be a concurrency relation on Σ. It means that a and b can be commuted. In other words an occurrence of ab can be replaced by ba and ba by ab. If a word $u \in \Sigma^*$ is obtained from a word $v \in \Sigma^*$ by such a sequence of replacements then we say that u and v are equivalent with respect to the relation θ, and denote it by $u \equiv v(mod\theta)$ or $u \equiv_\theta v$. Here $\equiv_\theta$ is a congruence relation on Σ^*. The free partially commutative monoid generated by Σ with respect to the relation θ is the quotient of Σ^* by $\equiv_\theta$.*

The word problem for the free partially commutative monoid is the following: given $u, v \in \Sigma^*$ is $u \equiv_\theta v$? This problem is decidable in linear time [1]. In fact, if $B \subseteq \Sigma^*$, we can define a projection mapping $\pi_B : \Sigma^* \longrightarrow B^*$ by $\pi_B(a) = a$ if $a \in B$ and $\pi_B(a) = \lambda$ otherwise. This π_B satisfies the following result [3] :
For any $u, v \in \Sigma^*, u \equiv_\theta v$ iff

1. $\pi_{\{a\}}(u) = \pi_{\{a\}}(v), \quad \forall a \in \Sigma$
2. $\pi_{\{a,b\}}(u) = \pi_{\{a,b\}}(v), \forall (a,b) \notin \theta$

Thus, the projection mapping can be used for checking whether $u \equiv_\theta v$ or not.

A Thue system T on Σ is a finite subset of $\Sigma^* \times \Sigma^*$. Each member of T is called a rule. The Thue congruence $\overset{*}{\longleftrightarrow}_T$ generated by T is the reflexive transitive closure of the symmetric relation $\longleftrightarrow_T$ defined as follows : for any

u, v such that $(u, v) \in T$ or $(v, u) \in T$ and any $x, y \in \Sigma^*$, $xuy \longleftrightarrow_T xvy$. Two strings $w, z \in \Sigma^*$ are congruent with respect to T iff $w \overset{*}{\longleftrightarrow}_T z$. The word problem for the Thue system on Σ is as follows: given any two words x and y is Σ^*, is $x \overset{*}{\longleftrightarrow}_T y$? The word problem is in general undecidable for Thue systems [2]. For example, G.C.Tzeitin (1958) has shown that the Thue system

$$T = \{(d_1 d_3, d_3 d_1), (d_1 d_4, d_4 d_1), (d_2 d_3, d_3 d_2), (d_2 d_4, d_4 d_2), (d_5 d_3 d_1, d_3 d_5),$$
$$(d_5 d_4 d_2, d_4 d_5), (d_3 d_3 d_1, d_3 d_3 d_1 d_5)\}$$

on $\Sigma = \{d_1, d_2, d_3, d_4, d_5\}$ has an undecidable word problem [6]. Similarly we have other examples of Thue systems with undecidable word problem in the literature.

We now construct a PKC based on free partially commutative monoid as follows:

Let Σ be an alphabet with a concurrency relation θ. Two words $x_0, x_1 \in \Sigma^*$ are chosen such that $x_0 \not\equiv_\theta x_1$. Another alphabet Δ whose cardinality is sufficiently greater than that of Σ is considered. Let g be a morphism from Δ to $\Sigma \cup \{\lambda\}$ where λ is the empty word. Define a Thue system T over Δ with rules (u, v) such that $(g(u), g(v))$ is either (x, x) where $x \in \Sigma^*$ or one of (ab, ba) or (ba, ab) where $(a, b) \in \theta$. Select two words y_0 and y_1 over Δ such that $g(y_0) \equiv_\theta x_0$ and $g(y_1) \equiv_\theta x_1$. (T, y_0, y_1) is the encryption key, g is the trapdoor morphism and (θ, x_0, x_1) is the decryption key.

Encryption of the bit 0 is done by considering the word y_0 and getting any word z which is congruent to y_0 with respect to the Thue system T. Decryption of the cryptotext z is done by first getting $g(z)$ and checking whether $g(z) \equiv_\theta x_0$ or not. If so, the plaintext is 0, otherwise it is 1. Computing $g(z)$ takes time linear in size of the cryptotext z and checking whether $g(z) \equiv_\theta x_0$ or not takes time linear in size of x_0.

But deciding whether z is congruent to y_0 with respect to the Thue system is the word problem for Thue systems which is undecidable.

We illustrate the technique with a simple example.

$$\Sigma \quad = \quad \{a, b, c\}, \theta = \{(a, b), (a, c)\}$$
$$\Delta \quad = \quad \{d_1, d_2, d_3, d_4, d_5, d_6, d_7, d_8, d_9\}$$
$$g \quad : \quad \Delta \to \Sigma \cup \{\lambda\} \text{ is given by}$$
$$g(d_1) \quad = \quad g(d_2) = g(d_3) = g(d_4) = g(d_5) = g(d_9) = \lambda$$
$$g(d_6) \quad = \quad a, g(d_7) = b, g(d_8) = c$$
$$x_0 \quad = \quad abbcc, x_1 = aabcc. \text{ Clearly } x_0 \not\equiv_\theta x_1$$
$$T \quad = \quad \{(d_1 d_3, d_3 d_1), (d_1 d_4, d_4 d_1), (d_2 d_3, d_3 d_2), (d_2 d_4, d_4 d_2), (d_5 d_3 d_1, d_3 d_5),$$
$$(d_5 d_4 d_2, d_4 d_5), (d_3 d_3 d_1, d_3 d_3 d_1 d_5), (d_6 d_7, d_7 d_6), (d_6 d_8, d_8 d_6),$$
$$(d_7 d_9, d_9 d_7), (d_8 d_9, d_9 d_8)\}$$
$$y_0 \quad = \quad d_1 d_2 d_2 d_4 d_3 d_3 d_1 d_6 d_7 d_5 d_7 d_9 d_1 d_2 d_8 d_3 d_4 d_8 d_3 d_9$$
$$y_1 \quad = \quad d_1 d_2 d_2 d_4 d_6 d_3 d_4 d_6 d_7 d_5 d_1 d_5 d_8 d_4 d_3 d_8 d_9$$

To encrypt 0,

$$y_0 \quad = \quad d_1 d_2 d_2 d_4 d_3 d_3 d_1 d_6 d_7 d_5 d_7 d_9 d_1 d_2 d_8 d_3 d_4 d_8 d_3 d_9$$
$$\longleftrightarrow \quad d_1 d_2 d_2 d_4 d_3 d_3 d_1 d_5 d_6 d_7 d_5 d_7 d_9 d_1 d_2 d_8 d_3 d_4 d_8 d_3 d_9$$
$$\longleftrightarrow \quad d_1 d_2 d_2 d_4 d_3 d_3 d_1 d_5 d_6 d_7 d_5 d_9 d_7 d_1 d_2 d_8 d_3 d_4 d_8 d_3 d_9$$
$$\longleftrightarrow \quad d_1 d_2 d_2 d_4 d_3 d_3 d_1 d_5 d_7 d_6 d_5 d_9 d_7 d_1 d_2 d_8 d_3 d_4 d_8 d_3 d_9 = z$$

To decrypt, $g(z) = babcc \equiv_\theta abbcc = x_0$.

Hence, the plain text bit is 0.

It can be noted that the word problem for T is undecidable as $T = R \cup S$ where R is the Thue system given by Tzeitin and S is a set of partially commuting rules over a concurrency alphabet. It is possible to add a finite number of rules (all need not be commuting rules) in T over Δ, thus enlarging T with the word problem for the new T still undecidable. It helps to increase some complication in analysing T even though T is made public.

We next prove that the problem of finding the trapdoor morphism g from the encryption key is NP-hard.

However, breaking the system is not equivalent to finding the trapdoor morphism as it is just one cryptanalytic attack on the cryptosystem.

Theorem 1. *Given a Thue system T on an alphabet Δ and two words y_O and y_1, the problem of constructing a nontrivial interpretation morphism $g : \Delta^* \longrightarrow \Sigma^*$ so that g maps T into a free partially commutative monoid on Σ with a concurrency relation θ is NP - hard. Here g maps the words y_0 and y_1 respectively to the words x_0 and x_1 with $x_0 \not\equiv_\theta x_1$ satisfying the following conditions:*

1. *for each letter $d \in \Delta, g(d)$ is either a letter in Σ or the empty word λ.*
2. *there exists a letter d in Δ such that $g(d)$ is a letter in Σ.*
3. *for every two words u and v in Δ^* with $u \xleftrightarrow{*}_T v, g(u) \equiv_\theta g(v)$.*
4. *$g(y_0) = x_0, g(y_1) = x_1$.*

Proof. Given a formula in conjunctive normal form, the problem of constructing a truth value assignment such that exactly one literal is true in every clause is NP-hard when we are given that such an assignment exists.

Let $X = \{x_1, x_2, ..., x_n\}$ be a set of variables and $L = C_1 \wedge C_2 \wedge ... \wedge C_m$ be a formula of propositional calculus where each clause C_i is a disjunction of some variables in X. Assume that there exists a truth value assignment $\alpha : X \rightarrow \{t, f\}$ that yields exactly one true variable in each clause C_i. We construct a Thue system T over Δ such that each interpretation morphism on T directly produces a solution to the satisfiability problem of L and the existence of the solution of α guarantees that there is a morphism g that maps T into a free partially commutative monoid over Σ with a concurrency relation θ satisfying the conditions of the above described PKC.

With each clause $C_j = x_{i_1} \vee x_{i_2} \vee ... \vee x_{i_p}, x_{i_j} \in X (1 \le j \le p)$, we associate a word $w(C_j)$ in X^* given by $w(C_j) = x_{i_1} x_{i_2} ... x_{i_p}$.

The Thue system T is now defined as follows:

The Thue system T has the alphabet $\Delta = X \cup P_X \cup Q_X$ where $P_X = \{p_x | x \in X\}$ and $Q_X = \{q_x | x \in X\}$.

The rules of the Thue system can now be defined :

$$w(C_i)p_x \longleftrightarrow p_x w(C_i)$$
$$w(C_i)q_x \longleftrightarrow q_x w(C_i), i = 1, 2, ..., m, x \in X$$
$$\text{If } C_1 = x_{i_1} \vee x_{i_2} \vee ... \vee x_{i_k} \text{ then}$$
$$y_0 = w(C_1)p_{x_{i_1}}p_{x_{i_1}}p_{x_{i_2}}p_{x_{i_2}}...p_{x_{i_k}}p_{x_{i_k}}$$
$$y_1 = w(C_1)w(C_1)p_{x_{i_1}}p_{x_{i_2}}...p_{x_{i_k}}$$

We prove that there exists an interpretation morphism g mapping the Thue system T into a free partially commutative monoid on Σ with a concurrency relation θ:

$$\Sigma = \{a, b, c\}, \quad \theta = \{(a, b), (a, c)\}$$

The morphism g is defined as follows:

For $x \in X$, if $\alpha(x) = t$, then $g(x) = a, g(p_x) = b, g(q_x) = c$. Otherwise, $g(x) = g(p_x) = g(q_x) = \lambda$.

Because in every clause C_i of L, there is exactly one true variable under the truth value assignment α, we have $g(w(C_i)) = a$.

$g(y_0) = abb; \quad g(y_1) = aab$

Clearly $abb \neq_\theta aab$ and hence $x_0 \neq_\theta x_1$.

Hence g is an interpretation morphism.

Now we show that every interpretation morphism of T yields a solution to the satisfiability problem for L.

Let Σ be any alphabet and θ be a concurrency relation on Σ. Let $g : \Delta^* \longrightarrow \Sigma^*$ be a morphism satisfying the requirements of the theorem. If $g(d) = \lambda$ for all $d \in \Delta$, then the condition(2) is violated. Therefore $g(d) \neq \lambda$ for some letter d in Δ. Again $g(y_0) \neq \lambda$. Hence if $g(w(C_1)) = a$, then $g(w(C_i)) = a$ for all i.

If we define the truth value assignment β on X by $\beta(x) = t$ iff $g(x) = a$, then β makes exactly one variable true in each clause C_i. So β constitutes a solution of the satisfiability problem.   □

3   Zero - Knowledge Protocol

We now give a zero-knowledge protocol [10, Chapter 6] to convince a Verifier that there is a morphism which maps the known Thue system into a free partially commutative monoid. Here the morphism itself is not revealed.

Assume that P, the Prover knows some information and there is an effective procedure for checking the validity of the information. P wants to convince V, the Verifier that he has the information. But P does not want to reveal anything of the information to V. A protocol for this set up is called zero-knowledge if and only if the following are true:

1. The Prover probably cannot cheat the Verifier. If the prover does not know a proof of the theorem, his chances of convincing the Verifier that he knows a proof are negligible.

2. The Verifier cannot cheat the Prover. She gets not a slightest hint of the proof, apart from the fact that the Prover knows the proof. In particular, the Verifier cannot prove the theorem to anyone else without proving it herself from the scratch.
3. Verifier learns nothing from the Prover that she could not learn by herself without the Prover.

In the above PKC based on free partially commutative monoids, there is a trapdoor morphism g which maps a Thue system T into a concurrency alphabet. We assume that P knows g. The Verifier V wants to verify that P really has such a $'g'$. But P does not want to reveal any bit of information about the morphism g. This can be achieved by the following protocol.

Step 1 : P constructs six different types of boxes

1. Letter Boxes : If T is a Thue system over Δ and the cardinality of $\Delta = n$, then P constructs $2n$ letter boxes B_i and each letter of Δ is locked in two boxes.
2. Dummy Boxes : P constructs $2n$ dummy boxes D_i containing 0 and 1 and there is an i where the letter in B_i has the correct value in D_i. The correct value is 1, if the letter is a dummy, 0 if it is not.
3. Equality Boxes : P constructs $4n^2$ equality boxes E_{ij} such that each equality box contains 1 if the letters in the boxes B_i and B_j are equal or equal according to g. Otherwise it contains 0.
4. Left and Right Boxes : P constructs $2m$ boxes l_i and r_i where m is the number of rules in T. Each box $l_i(r_i)$ contains one left (right) side of the rules.
5. Rule Boxes : P constructs m^2 rule boxes R_{ij}. Each rule box contains 1 if there is a rule between the contents of l_i and r_j. Otherwise, it contains 0.
6. Commute Boxes: P constructs $4n^2$ commute boxes C_{ij}. Each commute box contains 0 if B_i contains a letter x and B_j contains a letter y and $g(xy) \neq g(yx)$. Otherwise it contains 1.

Step 2 : V flips a coin.
Step 3 :

1. If head, P opens all letter boxes, left and right boxes and rule boxes.
2. If tail, P opens all the other boxes.

Step 4 :

1. V verifies that P has used the Thue system T.
2. V verifies that the dummy boxes contain n 0's and n 1's. Equality boxes contain more than n 1's and the commute boxes contain a few 0's and more than n 1's.

In each round the boxes are reconstructed. Otherwise V learns everything in two rounds. If the boxes are always reconstructed then V learns nothing. P can cheat in two ways.

1. He does not use T. Then he gets caught if (1) in step 3 is followed.
2. P does not use a concurrency alphabet and a morphism g. Then he gets caught if (2) in step 3 is followed.

4   A PKC Based on Free Partially Commutative Groups

We now give the definitions needed [1,15].

Definition 2. *Given an alphabet Σ, let $\Sigma^{-1} = \{a^{-1}/a \in \Sigma\}$ be a set of formal inverses for the letters in Σ and $\Sigma^{\pm 1} = \Sigma \cup \Sigma^{-1}$.*

A group G, or more precisely, a presentation of the group G, denoted by $G = <\Sigma, R>$, is given by an alphabet Σ and a set R of pairs in $(\Sigma^{\pm 1})^* \times \{\lambda\}$ called defining relations of G. If Σ and R are finite we say that G is a finitely presented group.

Given a group $G = <\Sigma, R>$, we consider the binary relation on $(\Sigma^{\pm 1})^*$, denoted by $\Longleftrightarrow_G$ or simply $\Longleftrightarrow$ and defined as follows:

For any $x, y \in (\Sigma^{\pm 1})^*$, $x \Longleftrightarrow_G y$ iff one of the following cases holds.

1. $x = urv$, $y = uv$ with $(r, \lambda) \in R$ and $u, v \in (\Sigma^{\pm 1})^*$
2. $x = uv$, $y = urv$ with $(r, \lambda) \in R$ and $u, v \in (\Sigma^{\pm 1})^*$
3. $x = ua^\sigma a^{-\sigma} v$, $y = uv$ with $a \in \Sigma, \sigma = \pm 1$ and $u, v \in (\Sigma^{\pm 1})^*$
4. $x = uv$, $y = ua^\sigma a^{-\sigma} v$ with $a \in \Sigma, \sigma = \pm 1$ and $u, v \in (\Sigma^{\pm 1})^*$

We define $\Longleftrightarrow_G^*$ to be the reflexive, transitive closure of $\Longleftrightarrow$. It is easy to see that $\Longleftrightarrow_G^*$ is a congruence and the quotient $(\Sigma^{\pm 1})^* / \Longleftrightarrow_G^*$ is a group, also denoted by G. The congruence class of a word x is denoted by $[x]_G$ or simply $[x]$. Evidently $x \Longleftrightarrow_G^* y$ iff $[x]_G = [y]_G$. We usually write $x =_G y$ instead of $[x]_G = [y]_G$ and say that the words x and y are equal in G.

Definition 3. *The word problem for the group G consists in deciding for any two words x, y in $(\Sigma^{\pm 1})^*$, whether $x =_G y$, or equivalently in deciding, for any given word x, whether $x =_G \lambda$, where λ is the empty word.*

A very well-known result, due to Novikov [8] says that there exists a finitely presented group whose word problem is undecidable.

Definition 4. *[15] Let Σ be an alphabet and θ_0 be a partially commutative (concurrency) relation on Σ. Let $\theta \subseteq \Sigma^{\pm 1} \times \Sigma^{\pm 1}$ be the extension of θ_0 to $\Sigma^{\pm 1} : \theta = \{(a, b), (a^{-1}, b), (a, b^{-1}), (a^{-1}, b^{-1}) : (a, b) \in \theta_0\}$*
This develops the following Thue system T on $\Sigma^{\pm 1}$, where

$$T = \{(aa^{-1}, \lambda), (a^{-1}a, \lambda) : a \in \Sigma\} \cup \{(cd, dc) : (c, d) \in \theta\}$$

This Thue system T presents the free partially commutative group $G(\theta_0)$. This $G(\theta_0)$ is the finitely presented group $<\Sigma, R>$ where the set of defining relations $R = \{(cdc^{-1}d^{-1}, \lambda) : (c, d) \in \theta\}$. If θ_0 is empty, then $G(\theta_0)$ is just the free group on Σ and if θ_0 contains every pair of distinct letters, then $G(\theta_0)$ is the free abelian group on Σ.

It has been proved that the word problem for the finitely presented free partially commutative group is decidable in linear time [15].

Construction of the PKC

Let Σ be an alphabet and θ_0 be a partially commutative relation on Σ such that $G(\theta_0)$ is the finitely presented free partially commutative group. A word $x_1 \in (\Sigma^{\pm 1})^*$ is chosen such that $x_1 \neq_{G(\theta_0)} \lambda$.

Another alphabet Δ whose cardinality is much greater than that of Σ is considered. Let g be a morphism from $\Delta^{\pm 1}$ to $\Sigma^{\pm 1} \cup \{\lambda\}$ such that

1. $g(c^{\sigma_1}) = a^{\sigma_2}$ implies $g(c^{-\sigma_1}) = a^{-\sigma_2}, c \in \Delta, \sigma_1, \sigma_2 = \pm 1$.
2. $g(c^{\sigma}) = \lambda$ implies $g(c^{-\sigma}) = \lambda, c \in \Delta, \sigma = \pm 1$.
3. If $g(c) = a, g(d) = b$ and $(a, b) \in \theta$, then c and d commutes where $c, d \in \Delta$.

Two words u_0 and u_1 from $(\Delta^{\pm 1})^*$ are selected such that $g(u_0) =_{G(\theta_0)} \lambda$ and $g(u_1) =_{G(\theta_0)} x_1$.

Fix a finite subset $\bar{R}$ of $(\Delta^{\pm 1})^* \times \{\lambda\}$ such that if $(uv^{-1}, \lambda) \in \bar{R}$, then one of the following is true:

1. $(g(u)(g(v))^{-1}, \lambda) \in R$
2. $g(u) =_{G(\theta_0)} \lambda$ and $g(v) =_{G(\theta_0)} \lambda$

Then $\bar{G} = < \Delta, \bar{R} >$ is a finitely presented group. The triple $(\bar{G}, u_o, u_1)$ is the public encryption key.

The secret decryption key is (G, x_1, g).

The encryption of a bit 0 is done by choosing any word w from $[u_o]_{\bar{G}}$.

The decryption of the word w is done by first finding $g(w)$ and checking whether $g(w) =_{G(\theta_0)} \lambda$. This can be done in linear time [15].

Example 1.

Let $\Sigma = \{a, b, c\}; \theta_0 = \{(a, c)\}; \theta = \{(a, c), (a^{-1}, c), (a, c^{-1}), (a^{-1}, c^{-1})\};$
         $R = \{(aca^{-1}c^{-1}, \lambda), (a^{-1}cac^{-1}, \lambda), (ac^{-1}a^{-1}c, \lambda), (a^{-1}c^{-1}ac, \lambda)\};$
         $x_o = \lambda; x_1 = ab^{-1}a;$
         $G(\theta_0) = G = < \Sigma, R >.$
         $\Delta = \{c_1, c_2, c_3, c_4, c_5, c_6\}.$ Define $g : \Delta^{\pm 1} \to \Sigma^{\pm 1} \cup \{\lambda\}$ by
         $g(c_1) = g(c_2^{-1}) = a; g(c_1^{-1}) = g(c_2) = a^{-1}$
         $g(c_4^{-1}) = b; g(c_4) = b^{-1}; g(c_6) = c; g(c_6^{-1}) = c^{-1}$
         $g(c_3) = g(c_3^{-1}) = g(c_5) = g(c_5^{-1}) = \lambda.$
         $u_0 = c_1^{-1}c_6^{-1}c_4^{-1}c_3c_5^{-1}c_4c_2^{-1}c_6; u_1 = c_2^{-1}c_4c_1.$
Let $\bar{R} = \{(c_2^{-1}c_6c_3c_1^{-1}c_6^{-1}, \lambda), (c_1^{-1}c_5c_2^{-1}c_3^{-1}, \lambda), (c_4c_5c_4^{-1}c_5c_3^{-1}, \lambda),$
         $(c_3c_5, \lambda), (c_2c_3c_2^{-1}c_5, \lambda)\}$
         $\bar{G} = < \Delta, \bar{R} >$
To encrypt 0,
         $u_o = c_1^{-1}c_6^{-1}c_4^{-1}c_3c_5^{-1}c_4c_2^{-1}c_6$
         $=_{\bar{G}} c_1^{-1}c_6^{-1}c_4^{-1}c_1^{-1}c_5c_2^{-1}c_5^{-1}c_4c_2^{-1}c_6$

To decrypt,
$$g(c_1^{-1}c_6^{-1}c_4^{-1}c_1^{-1}c_5c_2^{-1}c_5^{-1}c_4c_2^{-1}c_6) = a^{-1}c^{-1}ba^{-1}ab^{-1}ac$$
$$=_G a^{-1}c^{-1}bb^{-1}ac =_G a^{-1}c^{-1}ac =_G \lambda.$$
Hence the plaintext bit is 0.

Remark 1. A zero-knowledge protocol to convince a Verifier that there is a morphism which maps 'the public' finitely presented group into a free partially commutative group can be done by using the protocol given in Section 3 with the following modifications.

1. As the alphabet is $\Delta \cup \Delta^{-1}$, we construct $4n$ letter boxes, $4n$ dummy boxes, $16n^2$ equality boxes, $16n^2$ commute boxes where cardinality of $\Delta = n$.
2. As $\bar{R}$ is a set of defining relations, instead of left and right boxes and rule boxes, it is enough to contruct m relation boxes where m is the number of defining relations in $\bar{R}$.

The rest of the procedure is as follows : V, the Verifier flips a coin.

1. If head, P opens all letter boxes and relation boxes.
2. If tail, P opens all other boxes and the Verifier can verify that the Prover indeed has a morphism.

Remark 2. Given a finitely presented group $(\Delta, \bar{R})$, the problem of constructing a nontrival interpretation morphism g so that g maps $(\Delta, \bar{R})$ into a free partially commutative group on Σ with a concurrency relation θ_0 is NP - hard.
    The proof is similar to Theorem 1

Remark 3. One common problem with language theoretic cryptosystems is that of message expansion during encryption. In the above systems, this can be controlled to a certain level as there are two ways of rewriting, allowing for decrease as well as increase in the length of the word rewritten.

References

1. R.V.Book and H.N.Liu, Rewriting systems and word problems in a free partially commutative monoid, Information Processing Letters 26 (1987), 29-32.
2. R.V.Book, Confluent and other types of Thue systems, Journal of the ACM 29 (1982), 171-182.
3. R.Cori and D.Perrin, Automates et commutations partielles, RAIRO Theoretical Informatics and Applications 19(1985), 21-32.
4. V.Diekert and Y.Metivier, Partial Commutation and Traces, in Hand Book of Formal Languages, Vol.3 (Eds. G.Rozenberg and A.Salomaa), Springer - Verlag (1997), 457-533.
5. J.Kari, Observations concerning a public-key cryptosystem based on iterated morphisms, Theo.Comp.Sci. 66 (1989), 45-53.

6. G.Lallement, The Word Problem for Thue Rewriting Systems, LNCS 909, (1995), 27-38.
7. V.Niemi, Cryptology : Language - Theoretic Aspects, in Hand Book of Formal Languages, Vol.2 (Eds. G.Rozenberg and A. Salomaa), Springer Verlag (1997), 507-524.
8. P.S.Novikov, On the algorithmic unsolvability of the word problem in group theory, Trudy Math. Inst. Stelkov 44 (1955), 143 pp.
9. A.Salomaa, Computation and Automata, Cambridge University Press, 1986.
10. A.Salomaa, Public-Key Cryptography, Springer-Verlag, 1990.
11. G.Siromoney and R.Siromoney, A public key cryptosystem that defies cryptanalysis, Bull. of EATCS 28 (1986), 37-43.
12. G.Siromoney, R.Siromoney, K.G.Subramanian, V.R.Dare and P.J.Abisha, "Generalized Parikh vector and public key cryptosystems" in A Perspective in Theoretical Computer Science-Commemorative Volume for Gift Siromoney, Ed.R.Narasimhan, World Scientific, 1989, 301-323.
13. K.G.Subramanian, P.J.Abisha and R.Siromoney, A DOL/TOL public key cryptosystem, Information Processing Letters 26 (1987/88), 95-97.
14. N.R.Wagner and M.R.Magyarik, A public key cryptosystem based on the word problem, Crypto'84, LNCS 209 (1984), 19-36.
15. C.Wrathall, The word problem for free partially commutative groups, J.Symbolic Computation 6 (1988), 99-104.

Prime Numbers of Diffie-Hellman Groups for IKE-MODP*

Ikkwon Yie[1], Seongan Lim[2], Seungjoo Kim[2], and Dongryeol Kim[2]

[1] Inha University, Inchon, Korea
ikyie@math.inha.ac.kr
[2] KISA, Garak-Dong, Seoul, Korea
{seongan, skim, drkim}@kisa.or.kr

Abstract. For Discrete Logarithm Problem(DLP) based public key cryptography, the most time consuming task is the mathematical operations in the underlying finite field. For computational efficiency, a predeterminate form of prime p has been proposed to be used in Diffie-Hellman Groups for Internet Key Exchange(IKE). In this paper, we analyze the effect of pre-fixed bits of the prime numbers related to the security and efficiency and we suggest some alternative choices for prime p's for More Modular Exponential (MODP) Diffie-Hellman groups as a substitute for Internet Key Exchange(IKE) which has been published as RFC of IETF recently.

Keywords. IKE, prime numbers

1 Introduction

Most cryptographic schemes use a few arithmetic operations such as addition, multiplication, or exponentiation in finite fields repeatedly. There have been two different approaches considered in the literature to achieve efficiency in arithmetic operations. One is to develop more efficient algorithms and the other is to select computational environment that gives better efficiency in the arithmetic operations. In this paper, we consider the second approach.

Special structures of the prime p have been considered to improve computational efficiency in $GF(p)$ for large prime p. One of the most common ways is to prescribe some bits of the prime p. For example, a recently published IETF RFC 3526 [1] in IETF recommends, for IKE of IPSEC, Diffie-Hellman groups with modulus of the form

$$p = 2^n - 2^{n-64} + r_0 2^{64} - 1,$$

for $n = 1536, 2048, 3072, 4096, 8192$. The MODP Oakley Groups for IKE proposed in [4] have the same form for $n = 768, 1024$. The binary representation of p is

---

* Yie's work was supported by Inha Research Fund.

T. Johansson and S. Maitra (Eds.): INDOCRYPT 2003, LNCS 2904, pp. 228–234, 2003.

$$p = \overbrace{11\cdots11}^{64} \; \overbrace{**\cdots**}^{n-128} \; \overbrace{11\cdots11}^{64}.$$

In this paper, we consider primes of more general form $p = 2^n - 2^m + r2^k - 1$, $k < m < n$, where r is a random number of bit length $m - k$. As a notation, we will refer the integers of such form as integers of the form $P(n, m, k)$. We shall study the aspects of security and efficiency in terms of m, k in $P(n, m, k)$ and give a formula to determine m, k's for appropriate security level.

2   Efficiency Considerations

Here we consider:

- The probability of a randomly chosen integer of the form $P(n, m, k)$ being a prime
- The efficiency of the field operations (multiplication and exponentiation) in $GF(p)$, where p is a prime of the form $P(n, m, k)$.

As long as we choose an n-bit integer at random, the probability of this integer being a prime does not depend much on whether we impose the condition $P(n, m, k)$ or not. On the other hand, we can improve the efficiency of the modular arithmetic by assigning a smaller portion to the random parts.

2.1   Number of Primes of the Form $P(n, m, k)$

In order to estimate the number of primes of the form $P(n, m, k)$, we make use of Dirichlet Density Theorem on arithmetic progression (see, for example, [5]), which can be thought of as an extension of the Prime Number Theorem.

Theorem 1. (Dirichlet Density Theorem) *Let $x > 1$ be an integer and let a be relatively prime to x. Let P_a be the set of primes such that $p = a \pmod{x}$. Then the set P_a has density $1/\varphi(x)$. Here, φ is the Euler totient function.*

An integer of the form $P(n, m, k)$ can be characterized by the following properties:

- $p \in [A - 2^m, A]$, where $A = 2^n - 1$;
- $p \equiv -1 \pmod{2^k}$.

From the Prime Number Theorem, we can roughly estimate the number $\mathcal{N}$ of primes in the interval $I = [A - 2^m, A]$ as

$$\mathcal{N} = \pi(A) - \pi(A - 2^m),$$

where $\pi(x) = \frac{x}{\ln x}$. By Mean Value Theorem and the downward concavity of the graph $y = \pi(x)$, it follows that

$$\frac{2^m(\ln A - 1)}{(\ln A)^2} \leq \mathcal{N} \leq \frac{2^m}{\ln A}.$$

Thus, if we are willing to tolerate 0.1% error, $\mathcal{N}$ is approximately $\frac{2^m}{n \log 2}$. Hence, Dirichlet Density Theorem implies that the number $\mathcal{N}_k$ of primes $P(n, m, k)$ is approximately

$$\mathcal{N}_k \approx \frac{1}{\phi(2^k)} \times \frac{2^m}{n \ln 2} = \frac{2^{m-k+1}}{n \ln 2}.$$

Therefore the probability of a randomly chosen integer $P(n, m, k)$ being a prime is

$$\frac{1}{2^{m-k}} \times \frac{2^{m-k+1}}{n \ln 2} = \frac{2}{n \ln 2}.$$

Note that this is also the theoretical probability of a randomly chosen n-bit odd integer being a prime.

2.2   Computational Complexity of Modular Arithmetic

In order to compute a modular exponentiation, one has to perform modular reduction repeatedly. The classical division algorithm does not depend much on the special shape of the prime. However, Montgomery algorithm, which is another popular reduction algorithm, does depend on the shape of the prime.

Montgomery Algorithm Compute $X \pmod{p}$ for $X < p^2$, where p is an α-bit prime.

Step 1: Compute $p' = p^{-1} \pmod{2^\alpha}$.

Step 2: Compute $t = Xp' \pmod{2^\alpha}$.

Step 3: Compute $Y = \frac{X-tp}{2^\alpha}$. If $Y > p$, then set Y as $\frac{X-tp}{2^\alpha} - p$.

Note that $X \equiv 2^\alpha Y \pmod{p}$. To compute a modulo p exponentiation, one needs repeated application of Montgomery algorithm and then one modulo p reduction. Since modulo 2^α reduction is almost for free, the operations relevant to the complexity are multiplication by p' and integer multiplication by p. Thus the complexity of modulo p exponentiation using Montgomery algorithm depends only on the Hamming weight of the signed binary representation of p and p'. Since p' has to be regarded as somewhat random, the only thing that matters is the Hamming weight of p. Hence we see that p's with fewer (signed) Hamming weight gives better efficiency in terms of Montgomery algorithm.

3   Security Considerations

3.1   Pohlig-Hellman Attack on Subgroup DLP

It is required that $p - 1$ have a large enough (at least 160 bits) prime factor for a selected p in DLP based scheme to resist Pohlig-Hellman attack. Following Maurer [3], we may assume that cryptographically good primes are uniformly distributed. Thus, we assume that the security level of the subgroup DLP of $GF(p)$ is not affected by specifying the form $P(n, m, k)$.

3.2   Number Field Sieve for Discrete Logarithm Problem on $GF(p)$

If a prime number p of a fixed bit length α is required, it is usually selected at random from the interval $[2^{\alpha-1}, 2^{\alpha}]$. Up to this point, the most efficient technique for solving DLP on $GF(p)$ is the Number Field Sieve. In this paper, we consider primes from the subinterval $[2^{\alpha} - 2^m - 1, 2^{\alpha} - 1]$. Besides, our primes have all tailing k bits fixed as 1's. Note that the required complexity to solve DLP using General Number Field Sieve(GNFS) depends only on the bit-size of p. However, the Special Number Field Sieve (SNFS) takes advantage of the special features of the prime number. Thus we have to consider SNFS to see whether the security level of the DLP in the prime field is weakened by pre-specifying the form of primes.

SNFS uses irreducible polynomials $f(x)$ of degree s with $Y^s f(X/Y) \equiv 0$ (mod p) for some integers X and Y near $p^{1/s}$. From [6], the best choice of the degree s of the polynomial $f(x)$ for SNFS is

$$s_b = (3/2)^{1/3} \left(\frac{\log p}{\log \log p} \right)^{1/3}.$$

Following Gordon [2], we call a prime p is *unsafe* with respect to SNFS if we can find

- s is between 3 and s_b,
- X and Y are less than $1000\, p^{1/s}$,
- the absolute values of the coefficients of f are less than 500.

It is obvious that the total number of integers of the form $Y^s f(X/Y)$ satisfying the above conditions is bounded by

$$A_{s_b} = \sum_{i=3}^{s_b} (1000 \cdot 2^{\alpha/i})^2 1000^{i+1}.$$

Suppose that p is an unsafe prime with respect to SNFS. Then there is $f(x)$ of degree $i \leq s_b$ with $Y^i f(X/Y) \equiv 0$ (mod p). That is,

$$c_0 Y^i + c_1 X Y^{i-1} + \cdots + c_i X^i$$

is a multiple of the prime p, where $f(x) = c_0 + c_1 x^1 + \cdots + c_i X^i$.

On the other hand, assuming that $2^{\alpha-1} < p < 2^\alpha$, note that

$$i \le s_b = \left(\frac{3\log p}{2\log\log p}\right)^{1/3} \le \left(\frac{3\alpha}{2\log\alpha}\right)^{1/3}.$$

Further assuming that α is at least 768, we have

$$|c_0Y^i + c_1XY^{i-1} + \cdots + c_iX^i| \le 500(i+1)(1000)^i p \le 2^{\alpha-2}p.$$

Hence $Y^i f(X/Y)$ has at most one prime factor of bit length α.

Thus we see that the number of unsafe primes in the interval $[2^{\alpha-1}, 2^\alpha]$ with respect to SNFS is at most

$$A_{s_b} = \sum_{i=3}^{i=s_b} (1000 \cdot 2^{\alpha/i})^2 1000^{i+1}.$$

We also see that the rate of unsafe primes among primes of the form $P(n, m, k)$ is at most

$$\frac{1}{\mathcal{N}_k} \sum_{i=3}^{i=s_b} (1000 \cdot 2^{n/i})^2 1000^{i+1}.$$

It is generally recommended that the fraction of unsafe primes to be smaller than 2^{-100}. Note that in Section 2.1 we estimated $\mathcal{N}_k$ roughly as $2^{m-k+1}/n\ln 2$. For a given bit-size n, one of the decision criteria of m, k for primes of the form $P(n, m, k)$ to be secure enough with respect to SNFS is to satisfy

$$n\,2^{-(m-k)-1} \sum_{i=3}^{i=s_b} (1000 \cdot 2^{n/i})^2 1000^{i+1} < 2^{-100}.$$

It can be guaranteed if

$$n\,2^{-(m-k)-1}(1000 \cdot 2^{n/3})^2 10^{s_b+2} < 2^{-100} \tag{1}$$

since we have

$$\sum_{i=3}^{i=s_b} (1000 \cdot 2^{n/i})^2 1000^{i+1} \le (1000 \cdot 2^{n/3})^2 10^{s_b+2}.$$

4   Choices for p

By the above decision criteria, we can determine an appropriate form of the primes, which gives better efficiency and sustains the same security as randomly chosen primes. As we have seen before, the selection of such primes does not depend on the choice of k, but depends only on the choice of $m - k$. From inequality (1) it follows that

$$2^{m-k} > \frac{n}{2} \cdot (2^{10} \cdot 2^{n/3})^2 \cdot 10^{s_b+2} \cdot 2^{100}, \tag{2}$$

where $s_b = (3/2)^{1/3} \left(\frac{\log p}{\log \log p} \right)^{1/3}$ is the best choice for the degree of the polynomial $f(x)$ for SNFS in $GF(p)$. For the given bit-sizes 1024, 2048, 4096, 8192 of p, the corresponding s_b's are $4.74, 5.78, 7.08, 8.68$, respectively. Hence we may roughly set $s_b = 10$ in the inequality (2) to obtain (for primes of bit length up to 8192)

$$2^{m-k} > \frac{n}{2} \cdot (2^{10} \cdot 2^{n/3})^2 \cdot 10^{12} \cdot 2^{100},$$

which implies that

$$m - k > 159 + \frac{2n}{3} + \log_2 n.$$

For example, if $n = 1024$, IKE-MODP recommends p as the form of $P(1024, 960, 64)$. According to our analysis, we need to have $m - k \geq 852$, which implies that p can be any form of $P(1024, k + 852, k)$ for $0 < k < 172$. One such example is

$$P(1024, 853, 1) = \overbrace{11\cdots 11}^{171}\ \overbrace{* *\cdots * *}^{852}\ 1.$$

Table 1 gives alternative forms of p compare to the p's in IKE-MODP when the bit size of the prime varies as $n = 1024, 2048, 3072, 4096, 8192$.

Table 1. Comparison of the form of primes

	Suggested	IKE-MODP
1024	$P(1024, k + 852, k)$ for $0 < k < 172$	$P(1024, 960, 64)$
2048	$P(2048, k + 1536, k)$ for $0 < k < 512$	$P(2048, 1984, 64)$
3072	$P(3072, k + 2219, k)$ for $0 < k < 853$	$P(3072, 3008, 64)$
4096	$P(4096, k + 2902, k)$ for $0 < k < 1194$	$P(4096, 4062, 64)$
8192	$P(8192, k + 5634, k)$ for $0 < k < 2558$	$P(8192, 8128, 64)$

We have seen before that more MSB's filled with 1's allows better efficiency in terms of Montgomery algorithm. The number of MSB's filled by 1's can be determined in terms of the best efficiency as the word size of the processors. For example, if the processor's word size is 64-bits, then we can set the number of 1's by a maximum multiple of 64 in the above range and the remaining bits can be included in the random portion.

5 Conclusion

As the required key sizes for public key cryptosystems have increased, the IETF RFC 3526 [1] of IETF working group proposed new MODP groups for larger modulus with bit sizes $1536, 2048, 3072, 4096, 8192$. In this paper, we generalize the form of primes recommended in IKE [1,4] into $p = 2^n - 2^m + r2^k - 1$, $k < m < n$ and investigate the security and efficiency for given m, k. As a result, we get a decision criteria to determine m, k which can improve the computational efficiency without loss of the security level associated to the modulus p by pre-assigning more bits as 1. We also suggest that m and k in the representation of the prime $p = 2^n - 2^m + r2^k - 1$, $k < m < n$ may be selected as Table 1.

References

1. T. Kivinen, M. Kojo *More Modular Exponential (MODP) Diffie-Hellman groups for Internet Key Exchange (IKE)*, IETF RFC 3526, May, 2003
2. Daniel M. Gordon, *Designing and Detecting Trapdoors for Discrete Log Cryptosystems*, Advances of Cryptology-CRYPTO'92, 1992, pp. 66–75
3. Ueli M. Maurer, *Fast Generation of Prime Numbers asd Secure Public-Key Cryptographic Parameters*, J. Cryptology 1994, pp. .
4. H. Orman, *The Oakley Key Determination Protocol*, IETF RFC 2412, Nov. 1998
5. J.-P. Serre, A course in arithmetic, Springer-Verlag, New York, 1973
6. A.K.Lenstra, H.W. Lenstra, M.S. Manasse, J.M.Pollard, *The number field sieve*, The developement of number field sieve, Lecture Notes in Mathematics 1554, Springer, 1991

Polynomial Equivalence Problems and Applications to Multivariate Cryptosystems

Françoise Levy-dit-Vehel and Ludovic Perret

ENSTA, 32 Boulevard Victor, 75739 Paris cedex 15.
{levy,lperret}@ensta.fr

Abstract. At Eurocrypt'96, J.Patarin proposed a signature and authentication scheme whose security relies on the difficulty of the Isomorphism of Polynomials problem [P]. In this paper, we study a variant of this problem, namely the Isomorphism of Polynomials with one secret problem and we propose new algorithms to solve it, which improve on all the previously known algorithms. As a consequence, we prove that, when the number of polynomials (u) is close to the number of variables (n), the instances considered in [P] and [P1] can be broken. We point out that the case $n - u$ small is the most relevant one for cryptographic applications. Besides, we show that a large class of instances that have been presumed difficult in [P] and [P1] can be solved in deterministic polynomial time. We also give numerical results to illustrate our methods.

Keywords: multivariate polynomial equations, Isomorphism of Polynomials, Gröbner Bases.

1 Introduction

Alternatively to public key cryptosystems based on integer factorization and discrete log problems, there exists cryptographic schemes whose security relies on the difficulty of finding a common zero of a set of non linear polynomials in several variables. This problem is known to be solvable by Gröbner bases calculations[1]. Up to now, apart from the cryptanalysis of HFE by J.C Faugère *et al.* [FJ], this tool has not been used to attack these systems. The main reason is probably that there was no really efficient method to compute them. These past years, significant progress has been made [Fa99],[Fa02], which carried out to the design of a new efficient software to compute Gröbner bases: *fgb*[2].

In this paper, we are interested in variants of the *Isomorphism of Polynomials* (IP) problem - as introduced by J. Patarin in [P]. Our idea is to link these variants to the above problem of finding zeroes of a system of polynomials. Our approach is not only of theoretical interest but also gives in some cases efficient methods to solve these IP variants, since we are able to solve instances that

---

[1] Note that Gröbner bases provide provide not only a tool for finding a zero of a system of polynomials, but permits in fact to find all the zeroes.

[2] http://calfor.lip6.fr/ jcf/Software/Fgb/index.html

T. Johansson and S. Maitra (Eds.): INDOCRYPT 2003, LNCS 2904, pp. 235–251, 2003.

are used in cryptographic applications. The variants we consider are the *Isomorphism of Polynomials with one secret* (IP1S) problem, and its linear counterpart. IP1S can be outlined as follows: given two sets of multivariate polynomials $\mathcal{A} = \{a_1(x_1, \cdots, x_n), \cdots, a_u(x_1, \cdots, x_n)\}$ and $\mathcal{B} = \{b_1(x_1, \cdots, x_n), \cdots, b_u(x_1, \cdots, x_n)\}$ over a finite field $\mathbb{F}_q$, find - if any - an invertible matrix S and a vector T such that $b_i(x_1, \cdots, x_n) = a_i((x_1, \cdots, x_n)S + T)$ for all i, $1 \leq i \leq u$. The linear variant of IP1S is the one where we only look for a matrix S (i.e. T is the null vector of $\mathbb{F}_q^n$).

In [P], it has been shown how to derive a signature and an authentication scheme from IP1S. It is believed more difficult [CGP] than the IP problem itself and as evidence of its hardness, it is shown in [CGP] that the IP1S problem is at least as difficult as the Graph Isomorphism (GI) problem or in other words, a deterministic polynomial time algorithm solving the IP1S problem would also solve the GI problem (a result which has not been achieved for the IP problem).

In this paper, for reasons we explain in the next section, we rename the IP1S problem into the *Polynomial Affine Equivalence* (PAE) problem. We here study the PAE problem and its linear variant, which we call the *Polynomial Linear Equivalence* (PLE) problem. Apart from [GMS], no algorithm has been designed for these problems. We present here new algorithms for solving them, based on the link we exhibit between them and that of finding zeroes of a system of polynomials. When one of the two sets of polynomials $\mathcal{A}$ and $\mathcal{B}$ is bijective, we propose an algorithm of complexity $\mathcal{O}((n+1)D^n)$, where D is the maximum degree of the polynomials involved for solving the PAE problem, and of complexity $\mathcal{O}(nD^n)$ for the PLE problem. For the general case, we present algorithms of complexity $\mathcal{O}(f(n)D^n + g(n))$, where $f(n) \leq 2n, \forall n$, and $g(n)$ depends on the cardinality of the varieties involved.

The paper is organized as follows. We begin in section 2 by introducing our notations. We also define more formally the PAE and PLE problems, which are the main concerns of this paper. In section 3, we present some new properties of these problems. This section is divided into two parts, the first one investigates structural properties, whereas the other presents a geometrical interpretation of these problems. In section 4, we present two new algorithms for solving the PLE problem which are based on the properties of section 3. We compare in this part our algorithms with the one proposed in [GMS]. Independently from these new algorithms, we exhibit instances of the PLE problem which are solvable in deterministic polynomial time. In section 5, we generalize the algorithms of section 4 to the affine case.

The last part is devoted to applications: we investigate the security of cryptosystems based on the PAE and PLE problems. We prove that when the number u of polynomials and the number n of variables are such that $n - u \geq 0$ is small, the parameter sizes of the instances considered in [P] and [P1] do not guarantee a reasonable level of security. We point out that this case is the most relevant one for cryptographic applications (indeed, when $n - u$ small, the considered systems of polynomials have only one common zero, or at worse very few zeroes). We

also show that a large class of instances that have been presumed difficult in
[P] and [P1] can be solved in deterministic polynomial time. We give evidences
of the efficiency of the methods we propose by presenting experimental results
of our algorithms on presumably intractable instances. Moreover, we present an
efficient general total break ciphertext attack on any encryption system whose
security relies on the difficulty of the PAE problem. This is the case for example
for restricted (to IP1S) versions of C^* [MI], HFE [P] and TTM [TTM]. We end
the paper by some comments on the IP problem.

2   Preliminaries

Throughout this paper we use the following notations. We denote by $\mathbb{F}_q$ a finite
field with q elements, by X the vector $(x_1, \cdots, x_n)$, by $\mathbb{F}_q[X] = \mathbb{F}_q[x_1, \ldots, x_n]$
the polynomial ring in the indeterminates $x_1, \cdots, x_n$ over $\mathbb{F}_q$, by $\mathcal{M}_{m,n}(\mathbb{F}_q)$ the
set of $m \times n$ matrices whose components lie in $\mathbb{F}_q$ and by $GL_n(\mathbb{F}_q)$ the invertible
matrices in $\mathcal{M}_{n,n}(\mathbb{F}_q)$. For a subset $V \subset \mathbb{F}_q^n$, we shall denote by $Span(V)$ the
$\mathbb{F}_q$-vector space over $\mathbb{F}_q^n$ generated by all the linear combinations of vectors of V
and by $dim_{\mathbb{F}_q}(Span(V))$ its dimension.

A *term* is a product of a field element by a product of the variables $x_1, \cdots, x_n$.
We shall define the *total degree* of a term $cx_1^{\mu_1} \cdots x_n^{\mu_n}, c \in \mathbb{F}_q$ and $(\mu_1, \cdots, \mu_m) \in$
$\mathbb{N}^n$ by the sum $\sum_{i=1}^n \mu_i$. As usual, the *head term* of a polynomial $p \in \mathbb{F}_q[X]$ is
the biggest term of the terms of p (with respect to some admissible ordering on
the terms) and the *degree* of this polynomial is the total degree of its head term.

Let $\mathcal{F} = \{f_1, \cdots, f_s\}$ be a set of polynomials in $\mathbb{F}_q[X]^s$, we shall say that
$\mathcal{F}$ is *bijective* if the function $X \mapsto (f_1(X), \cdots, f_s(X))$ is a bijection. We shall
denote by $V_{\mathcal{F}} = \{(z_1, \cdots, z_n) \in \bar{\mathbb{F}}_q : f_i(z_1, \cdots, z_n) = 0, \forall 1 \leq i \leq s\}$, where $\bar{\mathbb{F}}_q$ is
the algebraic closure of $\mathbb{F}_q$, the *variety* associated to the ideal $< f_1, \cdots, f_s >$. A
Gröbner basis of the ideal $< f_1, \cdots, f_s >$ describes the variety $V_{\mathcal{F}}$. For a detailed
description of Gröbner bases and varieties, we refer the reader to [BeWe] and
[COX].

Let $\mathcal{A} = \{a_1(X), \cdots, a_u(X)\} \in \mathbb{F}_q[X]^u$ and $\mathcal{B} = \{b_1(X), \cdots, b_u(X)\} \in$
$\mathbb{F}_q[X]^u$ be two sets of polynomials. We shall say that these two sets are *linear-
equivalent*, denoted $\mathcal{A} \equiv_L \mathcal{B}$, if there exists $S \in GL_n(\mathbb{F}_q)$ such that $b_i(X) =$
$a_i(XS)$ for all $i, 1 \leq i \leq u$. We call such a pair *a linear equivalence matrix
between* $\mathcal{A}$ *and* $\mathcal{B}$. In the sequel, for convenience, we shall denote the equations
$b_i(X) = a_i(XS)$ for all $i, 1 \leq i \leq u$ by $\mathcal{B}(X) = \mathcal{A}(XS)$. The *Polynomial Linear
Equivalence* (PLE) problem is then the problem of finding a linear equivalence
matrix between $\mathcal{A}$ and $\mathcal{B}$, if any.

A natural extension is to consider bijective affine mappings over the $\mathbb{F}_q$-
vector space $\mathbb{F}_q^n$. We shall say that two sets of polynomials $\mathcal{A}$ and $\mathcal{B}$ are *affine-
equivalent*, denoted $\mathcal{A} \equiv_A \mathcal{B}$, if there exists $(S, T) \in GL_n(\mathbb{F}_q) \times \mathbb{F}_q^n$ such that
$\mathcal{A}(X) = \mathcal{B}(XS + T)$. We call such a pair *an affine equivalence pair between
$\mathcal{A}$ and $\mathcal{B}$, S being the *linear part* of this pair and T being its *affine part*. The

Polynomial Affine Equivalence (PAE) problem is then the problem of finding an affine equivalence pair between $\mathcal{A}$ and $\mathcal{B}$, if any.

This last problem was first introduced in [P] under the name *Isomorphism of Polynomials with one secret* problem, in reference to the well known graph isomorphism problem. We believe that this name is not well suited. Remember that two graphs are said to be isomorphic if and only if they are identical after a permutation of the vertices of one of the graphs. In such a setting, isomorphism is defined by a permutation and permutations are a special kind of bijective mappings. The problems which are addressed in [P] and here are much more general than the one of finding a permutation between two sets of polynomials. For this reason, we think that the name PLE and PAE we chose are better suited. Moreover, PLE and PAE are equivalence relations, as can be seen easily.

As pointed out by Geiselmann *et al.* in [GMS], it makes a difference whether the relations $\equiv_L$ and $\equiv_A$ are checked over $\mathbb{F}_q[X]$ or over $\mathbb{F}_q[X]/ < x_i^q - x_i >_{1 \leq i \leq n}$. Indeed, if $\mathcal{A} \equiv_A \mathcal{B}$(or $\mathcal{A} \equiv_L \mathcal{B}$) over $\mathbb{F}_q[X]$ then $\mathcal{A} \equiv_A \mathcal{B}$(or $\mathcal{A} \equiv_L \mathcal{B}$) over $\mathbb{F}_q[X]/ < x_i^q - x_i >_{1 \leq i \leq n}$ but the converse is not always true. In this paper, we only work with polynomials over $\mathbb{F}_q[X]/ < x_i^q - x_i >_{1 \leq i \leq n}$, since it appears to us to be the most natural space where cryptographic applications can be designed.

3   General Properties of Polynomial Equivalence

We quote here properties of the polynomial equivalence. Proofs can be found in [LP].

3.1   Structural Properties

For a polynomial $p \in \mathbb{F}_q[X]$, we shall denote by $p^{(d)}$ the terms of total degree d of this polynomial and by $p^{(\tilde{d})}$ his terms of highest total degree $\tilde{d}$. By extension, we shall denote by $\mathcal{A}^{(d)} = \{a_1^{(d)}(X), \cdots, a_u^{(d)}(X)\}$ and by $\mathcal{B}^{(d)} = \{b_1^{(d)}(X), \cdots, b_u^{(d)}(X)\}$, the terms of total degree d of $\mathcal{A}$ and $\mathcal{B}$. Note that if there exists an index i, $1 \leq i \leq u$, for which a_i and b_i do not have the same degree, then $\mathcal{A} \not\equiv_L \mathcal{B}$.

Proposition 1. *Let $S \in GL_n(\mathbb{F}_q)$. Then $\mathcal{B}(X) = \mathcal{A}(XS) \iff \mathcal{B}^{(j)}(X) = \mathcal{A}^{(j)}(XS)$ for all $j, 0 \leq j \leq \tilde{D}$, where $\tilde{D}$ is the maximum total degree of the polynomials of $\mathcal{A}$ and $\mathcal{B}$.*

Proposition 2. *Let $\alpha \in \mathbb{F}_q$, $\mathcal{A}_\alpha = \{a_1(\alpha X), \cdots, a_u(\alpha X)\}$, $\mathcal{B}_\alpha = \{b_1(\alpha X), \cdots, b_u(\alpha X)\}$ be sets of polynomials and $S \in GL_n(\mathbb{F}_q)$. Then $\mathcal{B}(X) = \mathcal{A}(XS) \iff \mathcal{B}_\alpha(X) = \mathcal{A}_\alpha(XS)$, for all α in $\mathbb{F}_q$.*

3.2   Geometrical Properties

In the sequel, we denote by $V_{\mathcal{A}}$ and $V_{\mathcal{B}}$ the varieties associated to $\mathcal{A}$ and $\mathcal{B}$.

Property 1. Let $(S,T) \in GL_n(\mathbb{F}_q) \times \mathbb{F}_q^n$. If $\mathcal{B}(X) = \mathcal{A}(XS + T)$ then $V_{\mathcal{A}} = V_{\mathcal{B}}S + T$, with $V_{\mathcal{B}}S + T = \{v_{\mathcal{B}}S + T : v_{\mathcal{B}} \in V_{\mathcal{B}}\}$.

Corollary 1. *Let* $S \in GL_n(\mathbb{F}_q)$. *If* $\mathcal{B}(X) = \mathcal{A}(XS)$ *then* $V_{\mathcal{A}} = V_{\mathcal{B}}S$.

By property 1 and corollary 1, we have that $\mathcal{A} \equiv_A \mathcal{B}$ or $\mathcal{A} \equiv_L \mathcal{B}$ implies $|V_{\mathcal{A}}| = |V_{\mathcal{B}}|$.

Property 2. Let $\mathcal{A}, \mathcal{B}, \mathcal{C}, \mathcal{D}$ be sets of polynomials, $V_{\mathcal{A}}, V_{\mathcal{B}}, V_{\mathcal{C}}, V_{\mathcal{D}}$ be the varieties associated to these sets and $(S,T) \in GL_n(\mathbb{F}_q) \times \mathbb{F}_q^n$.

$$(\mathcal{B}(X) = \mathcal{A}(XS + T) \text{ and } \mathcal{D}(X) = \mathcal{C}(XS + T)) \Longrightarrow V_{\mathcal{A}} \cap V_{\mathcal{C}} = (V_{\mathcal{B}} \cap V_{\mathcal{D}})S + T.$$

By adding the field equations $\{x_1^q - x_1, \cdots, x_n^q - x_n\}$ to a set of polynomials $\mathcal{A}$, we change the geometry of the solutions. In particular, the variety associated to this new set is equal to $V_{\mathcal{A}} \cap \mathbb{F}_q^n$, a subset of $\mathbb{F}_q^n$ and not of $\bar{\mathbb{F}}_q^n$. Hence, we have a finite number of points in this variety.

Corollary 2. *Let* $(S,T) \in GL_n(\mathbb{F}_q) \times \mathbb{F}_q^n$. *If* $\mathcal{B}(X) = \mathcal{A}(XS+T)$ *then* $V_{\mathcal{A}} \cap \mathbb{F}_q^n = (V_{\mathcal{B}} \cap \mathbb{F}_q^n)S + T$.

By using structural properties of the affine equivalence relation, we get:

Proposition 3. *Let* $(S,T) \in GL_n(\mathbb{F}_q) \times \mathbb{F}_q^n$.
If $\mathcal{B}(X) = \mathcal{A}(XS+T)$ *then* $V_{\mathcal{A}_{1,p}} \cap \mathbb{F}_q^n = (V_{\mathcal{B}_{1,p}} \cap \mathbb{F}_q^n)S + T$ *for any fixed* $p \in \mathbb{F}_q^n$, $V_{\mathcal{A}_{1,p}}$ *being the variety associated to* $< a_1(X) - b_1(p), \cdots, a_u(X) - b_u(p) >$ *and* $V_{\mathcal{B}_{1,p}}$ *the variety associated to* $< b_1(X) - b_1(p), \cdots, b_u(X) - b_u(p) >$.

Remark 1. The results of property 2, corollary 2 and proposition 3 are also true for the linear equivalence relation.

For proposition 3, we have only used the properties of the affine equivalence relation. We get the next proposition by using particular properties of the linear equivalence relation.

Proposition 4. *Let* $(\alpha, p) \in \mathbb{F}_q \times \mathbb{F}_q^n, \mathcal{A}_{\alpha,p} = \{a_1(\alpha X) - b_1(\alpha p), \cdots, a_u(\alpha X) - b_u(\alpha p)\}$, $\mathcal{B}_{\alpha,p} = \{b_1(\alpha X) - b_1(\alpha p), \cdots, b_u(\alpha X) - b_u(\alpha p)\}, U \subseteq \mathbb{F}_q$ *and* $S \in GL_n(\mathbb{F}_q)$.

If $\mathcal{B}(X) = \mathcal{A}(XS)$ *then* $\bar{V}_{\mathcal{A}_{U,p}} = \bar{V}_{\mathcal{B}_{U,p}}S$, *for any fixed* $p \in \mathbb{F}_q^n$, *with:*

$$\bar{V}_{\mathcal{A}_{U,p}} = (\cap_{\alpha \in U} V_{\mathcal{A}_{\alpha,p}}) \cap (\cap_{1 \leq j \leq \tilde{D}} V_{\mathcal{A}^{(d)}}) \cap \mathbb{F}_q^n \text{ and}$$
$$\bar{V}_{\mathcal{B}_{U,p}} = (\cap_{\alpha \in U} V_{\mathcal{B}_{\alpha,p}}) \cap (\cap_{1 \leq j \leq \tilde{D}} V_{\mathcal{B}^{(d)}}) \cap \mathbb{F}_q^n,$$

$\tilde{D}$ *being the maximum total degree of the polynomials of* $\mathcal{A}$ *and* $\mathcal{B}$.

4   Polynomial Linear Equivalence Algorithms

In this section, we present two new algorithms for solving the PLE problem. The first uses a link between this problem and that of finding the common zeroes of a set of polynomials. The properties given in 3.2 are used to design the second algorithm. We conclude this section by exhibiting instances which can be solved in deterministic polynomial time.

4.1   Our First Algorithm

The basic idea follows [GMS]. We know that when two sets of polynomials $\mathcal{A} = \{a_1(X), \cdots, a_u(X)\}$ and $\mathcal{B} = \{b_1(X), \cdots, b_u(X)\}$ are linear-equivalent, then the evaluation of the b_is on some vector $p \in \mathbb{F}_q^n$ is equal to the evaluation of the a_is in $\widetilde{p}' = pS$, for some linear equivalence matrix S between $\mathcal{A}$ and $\mathcal{B}$. Knowledge of the pair $(p, \widetilde{p}')$ allows us to obtain n linear equations in the components of S. The main idea of this algorithm is to convert the search of these pairs into the solving of a non linear system of equations. For a vector $p \in \mathbb{F}_q^n$, we notice that the variety associated to the ideal $< \{a_i(X) - b_i(p)\}_{1 \leq i \leq u}, \{x_i^q - x_i\}_{1 \leq i \leq n} >$ gives all the vectors $p' \in \mathbb{F}_q^n$ such that $\mathcal{B}(p) = \mathcal{A}(p')$. We know that there exists a unique vector $\widetilde{p}'$ in this variety such that $\widetilde{p}' = pS$. When the polynomial mapping $\mathcal{A}$ is bijective, the variety above gives exactly one such vector. But it is not the case in general and in order to improve the effectiveness of the algorithm, we must construct varieties whose cardinalities are as small as possible. In order to do that, we will use the properties of 3.1. More precisely, according to proposition 1 and 2, if there exists a linear equivalence matrix S between $\mathcal{A}$ and $\mathcal{B}$, then we have the following:

$$\exists (p, \widetilde{p}') \in \mathbb{F}_q^n \times \mathbb{F}_q^n, \mathcal{B}(p) = \mathcal{A}(\widetilde{p}') \Longrightarrow \begin{cases} \mathcal{B}^{(j)}(p) = \mathcal{A}^{(j)}(\widetilde{p}'), \forall 0 \leq j \leq \widetilde{D}, \\ \mathcal{B}_\alpha(p) = \mathcal{A}_\alpha(\widetilde{p}'), \forall \alpha \in U \subseteq \mathbb{F}_q, \end{cases} (I)$$

where $\widetilde{D}$ is the maximum total degree of the polynomials of $\mathcal{A}$ and $\mathcal{B}$.

We now give an algebraic interpretation of these constraints. Let α be in $\mathbb{F}_q$. We shall denote by $I_{\alpha,p} =< \{a_i(\alpha X) - b_i(\alpha p)\}_{1 \leq i \leq u}, \{x_i^q - x_i\}_{1 \leq i \leq n} >$ and by $V_{\alpha,p}$ the variety associated to this ideal. We also set $I_p^{(j)} =< \{a_i^{(j)}(X) - b_i^{(j)}(p)\}_{1 \leq i \leq u}, \{x_i^q - x_i\}_{1 \leq i \leq n} >$ and call $V_p^{(j)}$ the variety associated to this ideal. Finally $\bar{V}_{U,p} = (\cap_{\alpha \in U \cup \{1\}} V_{\alpha,p}) \cap (\cap_{1 \leq j \leq \widetilde{D}} V_p^{(j)})$ denotes the vectors $p' \in \mathbb{F}_q^n$ such that $\mathcal{A}(p') = \mathcal{B}(p)$ and achieve the constraints (I). With these notations, we point out that if there exists a vector $p \in \mathbb{F}_q^n$ such that $\bar{V}_{U,p} = \emptyset$ for some $U \subseteq \mathbb{F}_q$ then, $\mathcal{A} \not\equiv_L \mathcal{B}$.

Idea of the algorithm

From the polynomials given in input, we construct sets $L_j, 1 \leq j \leq n$ such that each L_j contains the j-th row of candidates for the linear equivalence matrices between the two inputs. When such a matrix doesn't exist, the algorithm returns $\emptyset$. Let $\{e_j\}_{1 \leq j \leq n}$ be the n vectors of the canonical basis of $\mathbb{F}_q^n$. The algorithm is the following:

Algorithm A
Input: Two sets of polynomials $\mathcal{A}$ and $\mathcal{B}$.
Output: A linear equivalence matrix between $\mathcal{A}$ and $\mathcal{B}$, if any and $\emptyset$ otherwise.
Choose $U \subseteq \mathbb{F}_q$ randomly
For j from 1 to n **do**
Compute $\bar{V}_{U,e_j}$

    If $\bar{V}_{U,e_j} \neq \emptyset$ **then** $L_j \leftarrow \bar{V}_{U,e_j}$ **Else Return** $\emptyset$

EndFor
$S \leftarrow SeekRows(\mathcal{A}, \mathcal{B}, \{L_1, \cdots, L_n\})$
Return S

For all $j, 1 \leq j \leq n$ the elements of L_j are candidates for the j-th row of a linear equivalence matrix between $\mathcal{A}$ and $\mathcal{B}$.

Remark 2. If $\mathcal{A} \equiv_L \mathcal{B}$ and if $\mathcal{A}$ or $\mathcal{B}$ is a bijection[3] then there exists a unique linear equivalence matrix between these two sets. Indeed, for all $j, 1 \leq j \leq n$, L_j only contains the j-th row of this matrix.

The function *SeekRows* outputs a linear equivalence matrix between $\mathcal{A}$ and $\mathcal{B}$, if any, and $\emptyset$ otherwise. To recover this matrix, if such a matrix exists, it checks for all the invertible matrices than can be constructed from the sets $L_1, \cdots, L_n$. In [GMS], Geiselmann *et al.* propose a heuristic to improve the search of the good candidates which is based on the fact that if l_i and l_j are the i-th and j-th rows of some linear equivalence matrix S between $\mathcal{A}$ and $\mathcal{B}$, then $\mathcal{B}(c_i l_i + c_j l_j) = \mathcal{A}(c_i e_i + c_j e_j)$ for all $(c_i, c_j) \in U \times U$, where $U \subseteq \mathbb{F}_q$. We propose in [LP] a slightly different heuristic which has the advantage to be easily parallelized.

Remark 3. The bigger the size of the subset $U \subseteq \mathbb{F}_q$, the better algorithm A is. Indeed, the more equations you have, the faster are done the calculations of the Groëbner bases [Fa] and the smaller the number of candidates in the varieties computed are. But remember that the generation of the equations must be efficiently done.

Complexity

The complexity of calculation of a Gröbner basis depends on the maximum degree of the polynomials occurring during this computation [BeWe]. This parameter depends on the set of polynomials but for polynomials which have a finite number of zeroes (varieties are so-called 0-dimensional varieties), which always occurs in cryptographic applications, it can be bounded from above by $\mathcal{O}(D^n)$ (see [BeWe] *p.513*). At each step $j, 1 \leq j \leq n$ the varieties $\bar{V}_{U,e_j}$ computed have a finite number of points. Hence, when one of the inputs is a bijection, the complexity of this algorithm is $\mathcal{O}(nD^n)$.

In the general case, the function *SeekRows* checks the invertibility (by Gaussian elimination) of at most $\prod_{i=1}^n |L_i|$ matrices. Finally, for generic instances, this algorithm has a complexity of $\mathcal{O}(nD^n + n^3 \prod_{i=1}^n |L_i|)$. Our algorithm recovers in

---

[3] When $\mathcal{A} \equiv_L \mathcal{B}$ and if one of the inputs is bijective then the other is also bijective

fact all the linear equivalence matrices between the two sets of polynomials. Indeed, this complexity is exactly the one of finding them all.

Previous work

To our knowledge, the only work done on this subject is presented by Geiselmann *et al.* in [GMS]. For a detailed description of their algorithm, we refer the reader to this article. We briefly recall in this part the principle of the algorithm proposed by these authors. We point out that it was dedicated for the PAE problem, but it can be easily adapted to the PLE problem. In this setting, the main idea is to remark that if $l \in \mathbb{F}_q^n$ is the j-th row of a linear equivalence matrix between two sets of polynomials $\mathcal{A}$ and $\mathcal{B}$, then $\mathcal{B}_\alpha(e_j) = \mathcal{A}_\alpha(l)$ for all $\alpha \in U \subseteq \mathbb{F}_q$. An exhaustive search among the vectors $l \in \mathbb{F}_q^n$ is then performed to recover these candidates.

The set $\{l \in \mathbb{F}_q^n : \mathcal{B}_\alpha(e_j) = \mathcal{A}_\alpha(l), \forall \alpha \in U\}$ is equal to $\cap_{\alpha \in U} V_{\alpha, e_j}$. Hence, we have substituted in our algorithm the exhaustive search of the elements of $\{l \in \mathbb{F}_q^n : \mathcal{B}_\alpha(e_j) = \mathcal{A}_\alpha(l), \forall \alpha \in U\}$ by the computation of a variety. In the worst case, the theoretical complexity of computing V_{α, e_j} is $\mathcal{O}(D^n)$, where D is the maximum degree of the polynomials of $\mathcal{A}$. This must be compared with the complexity $\mathcal{O}(q^n)$ of the exhaustive search of [GMS]. But the complexity of computing Gröbner bases depends in practice very much on the algorithm used and an efficient software, such as *fgb*, behaves much better than in the worst case[Fa].

In addition, by investigating structural properties of the PLE problem, we have also added new constraints which permit to decrease the size of the set of candidates and so to increase the efficiency of our algorithm.

4.2   A Second Algorithm

This algorithm is more particularly dedicated to sets of polynomials which are not bijective. The main idea is to use geometrical properties of the linear equivalence relation. According to corollary 1, when $\mathcal{A} \equiv_L \mathcal{B}$ the varieties $V_\mathcal{A}$ and $V_\mathcal{B}$ are such that $V_\mathcal{A} = V_\mathcal{B} S$, for some linear equivalence matrix S between $\mathcal{A}$ and $\mathcal{B}$. Hence, for each $v_\mathcal{B} \in V_\mathcal{B}$ there exists a unique vector $\widetilde{v}_\mathcal{A} \in V_\mathcal{A}$ such that $\widetilde{v}_\mathcal{A} = v_\mathcal{B} S$. According to property 4 the pair $(\widetilde{v}_\mathcal{A}, v_\mathcal{B})$ lies in $\bar{V}_{\mathcal{A}_U} \times \bar{V}_{\mathcal{B}_U}$, with:

$$\bar{V}_{\mathcal{A}_U} = (\cap_{\alpha \in U \cup \{1\}} V_{\mathcal{A}_\alpha}) \cap (\cap_{1 \leq j \leq \widetilde{D}} V_{\mathcal{A}^{(d)}}) \cap \mathbb{F}_q^n,$$
$$\bar{V}_{\mathcal{B}_U} = (\cap_{\alpha \in U \cup \{1\}} V_{\mathcal{B}_\alpha}) \cap (\cap_{1 \leq j \leq \widetilde{D}} V_{\mathcal{B}^{(d)}}) \cap \mathbb{F}_q^n,$$

where $\widetilde{D}$ is the maximum degree of the polynomials of $\mathcal{A}$ and $\mathcal{B}$ and $U \subseteq \mathbb{F}_q$. This property permits to improve the search of the pair $(\widetilde{v}_\mathcal{A}, v_\mathcal{B})$ since it can be done in $\bar{V}_{\mathcal{A}_U} \times \bar{V}_{\mathcal{B}_U}$, a subset of $V_\mathcal{A} \times V_\mathcal{B}$. Knowledge of the pairs $\{(\widetilde{v}_\mathcal{A}, v_\mathcal{B}), v_\mathcal{B} \in \bar{V}_{\mathcal{B}_U}\}$ allows us to get $n * dim_{\mathbb{F}_q}(Span(\bar{V}_{\mathcal{B}_U}))$ linearly independent equations in the components of S.

It could be that the number of equations is not sufficient to recover S. Let p be a vector which is not in $Span(V_\mathcal{B})$. If $\mathcal{A} \equiv_L \mathcal{B}$, we have, according to proposition 3, $V_\mathcal{A} = V_\mathcal{B} S$ but also $V_{\mathcal{A}_{1,p}} = V_{\mathcal{B}_{1,p}} S$. For each vector $v \in V_{\mathcal{B}_{1,p}}$, we also have a unique vector $\widetilde{v} \in V_{\mathcal{A}_{1,p}}$ such that $\widetilde{v} = vS$. As explained in property 4, the search of the suitable pairs can be done on a subset of $V_{\mathcal{A}_{1,p}} \times V_{\mathcal{B}_{1,p}}$ and more precisely in $\bar{V}_{\mathcal{A}_{U,p}} \times \bar{V}_{\mathcal{B}_{U,p}}$, with:

$$\bar{V}_{\mathcal{A}_{U,p}} = (\cap_{\alpha \in U \cup \{1\}} V_{\mathcal{A}_{\alpha,p}}) \cap (\cap_{1 \leq j \leq \widetilde{D}} V_{\mathcal{A}^{(d)}}) \cap \mathbb{F}_q^n,$$
$$\bar{V}_{\mathcal{B}_{U,p}} = (\cap_{\alpha \in U \cup \{1\}} V_{\mathcal{B}_{\alpha,p}}) \cap (\cap_{1 \leq j \leq \widetilde{D}} V_{\mathcal{B}^{(d)}}) \cap \mathbb{F}_q^n,$$

where U is a subset of $\mathbb{F}_q$.

Hence, the pairs $\{(\widetilde{v}, v), v \in \bar{V}_{\mathcal{B}_{U,p}}\}$ allow us to get $n * dim_{\mathbb{F}_q}(Span(\bar{V}_{\mathcal{B}_{U,p}}))$ new linearly independent equations in the components of S, in addition to the equations always given by the pairs $\{(\widetilde{v}_{\mathcal{A}}, v_{\mathcal{B}}), v_{\mathcal{B}} \in \bar{V}_{\mathcal{B}_U}\}$. Since p is chosen not to lie in $Span(V_{\mathcal{B}})$, at least n of these news equations are linearly independent from the equations given by $\{(\widetilde{v}_{\mathcal{A}}, v_{\mathcal{B}}), v_{\mathcal{B}} \in \bar{V}_{\mathcal{B}_U}\}$. Note that, with these notations, if $\bar{V}_{\mathcal{A}_{U,p}} = \emptyset$ for some subset $U \subseteq \mathbb{F}_q$ and for some vector $p \in \mathbb{F}_q^n$ then $\mathcal{A} \neq_L \mathcal{B}$.

Idea of the algorithm

As long as the number of equations in the components of the matrix we try to determine is not sufficient, we will compute, from the polynomials given in input, varieties $V^{1,k}$ and $V^{2,k}$, where $V^{1,k}$ is the $k-th$ variety $\bar{V}_{\mathcal{A}_{U,p}}$, for different choices of U and p. Variety $V^{2,k}$ is defined analogously (with respect to the set $\mathcal{B}$). Those varieties verify $V^{1,k} = V^{2,k}S$, for some linear equivalence matrix S between the two inputs, if any. When such a matrix doesn't exist, the algorithm returns $\emptyset$. The algorithm is the following:

Algorithm B
Input: Two sets of polynomials $\mathcal{A}$ and $\mathcal{B}$.
Output: A linear equivalence matrix between $\mathcal{A}$ and $\mathcal{B}$, if any and $\emptyset$ otherwise.
Initialization: $V^{1,1} = V^{2,1} = \cdots = V^{1,n} = V^{2,n} = \emptyset, P = \emptyset, cpt = 0, l = 1$.
Choose $U \subseteq \mathbb{F}_q$ randomly
While $cpt < n$ **do**

    Choose $p \in \mathbb{F}_q^n \setminus P$
    Compute $\bar{V}_{\mathcal{A}_{U,p}}$
    If $\bar{V}_{\mathcal{A}_{U,p}} \neq \emptyset$ **then**
        Compute $\bar{V}_{\mathcal{B}_{U,p}}$
        $(V^{1,l}, V^{2,l}) \leftarrow (\bar{V}_{\mathcal{A}_{U,p}}, \bar{V}_{\mathcal{B}_{U,p}})$
        $cpt \leftarrow cpt + dim_{\mathbb{F}_q}(Span(V^{2,l}))$
        $l \leftarrow l + 1$
        $P \leftarrow Span(P \cup V^{2,l})$
    Else Return $\emptyset$

EndWhile
$S \leftarrow SeekMatrix(\{V^{1,k}, V^{2,k}\}_{1 \leq k \leq l})$
Return S

Remark 4. Let $d_k = dim_{\mathbb{F}_q}(Span(V^{2,k}))$. At the end of the algorithm $\sum_{k=1}^{l} d_k \geq n$.

The function $SeekMatrix$ outputs a linear equivalence matrix between the two inputs, if any, and $\emptyset$ otherwise. It computes for each $k, 1 \leq k \leq l$ a basis $B_{V^{2,k}} = \{v_1^{2,k}, \cdots, v_{d_k}^{2,k}\}$ of $Span(V^{2,k})$. For all the elements:

$$\{((v_1^{1,1}, v_1^{2,1}), \cdots, (v_{d_1}^{1,1}, v_{d_1}^{2,1})), \cdots, ((v_1^{1,l}, v_1^{2,l}), \cdots, (v_{d_l}^{1,l}, v_{d_l}^{2,l}))\}$$

in the set $(V^{1,1} \times B_{V^{2,1}})^{d_1} \times \cdots \times (V^{1,l} \times B_{V^{2,l}})^{d_l}$, it checks if the linear system in the unknowns the components of a matrix M:

$$v_1^{2,1} M = v_1^{1,1}, \cdots, v_{d_1}^{2,1} M = v_{d_1}^{1,1}, \cdots, v_1^{2,l} M = v_1^{1,l}, \cdots, v_{d_l}^{2,l} M = v_{d_l}^{1,l}$$

is invertible and recovers if so this matrix. Finally, it checks if M is a linear equivalence matrix between $\mathcal{A}$ and $\mathcal{B}$.

Complexity

Let $N_k = \prod_{r=1}^{d_k} |V^{1,k}|$ and D be the maximum degree of the polynomials of $\mathcal{A}$ and $\mathcal{B}$. At each step $k, 1 \leq k \leq l$ of the while loop the varieties $V^{1,k}$ and $V^{2,k}$ have a finite number of points. Hence, the complexity of constructing these $2l$ sets is $\mathcal{O}(2lD^n)$. The function $SeekMatrix$ computes l bases of vector-spaces, checks invertibility and solves at most $\prod_{k=1}^{l} N_k$ linear systems. Moreover, the maximum number of steps performed by algorithm B is given :

Proposition 5. *If $\mathcal{A} \equiv_L \mathcal{B}$, then algorithm B performs at most n steps.*

Proof. Let $S \in GL_n(\mathbb{F}_q)$, such that $\mathcal{B}(X) = \mathcal{A}(XS)$. At the k-th step of algorithm B, the vector p is chosen in $\mathbb{F}_q^n \setminus Span(\cup_{0 < r < k} V^{2,r})$. Therefore, the vector pS is linearly independent from the vectors $\{vS : v \in \cup_{0 < r < k} V^{2,r}\}$. Hence, at each step of the while loop, algorithm B gives at least n new equations which are linearly independent from the equations given by $\{(vS, v) : v \in \cup_{0 < r < k} V^{2,r}\}$. $\square$

Finally, the complexity of this algorithm is $\mathcal{O}(2nD^n + n^6 \prod_{k=1}^{n} N_k)$. Remark that this algorithm recovers in fact all the linear equivalence matrices between two sets of polynomials. This complexity is again exactly the one of finding them all.

Comparison with algorithm A

For each $j, 1 \leq j \leq n$ the variety $\bar{V}_{U,e_j}$ computed in algorithm A is equal to the variety $\bar{V}_{\mathcal{A}_{U,e_j}}$ in algorithm B. By construction, these two varieties contain a vector $l_j \in \mathbb{F}_q^n$ which is the j-th row of some linear equivalence matrix S between $\mathcal{A}$ and $\mathcal{B}$. In A, we focus only on recovering the rows of a linear equivalence matrix. Hence, we have only to compute for each j the variety $\bar{V}_{U,e_j}$, which contains the candidates for the j-th row of this matrix. In B, when we compute $\bar{V}_{\mathcal{A}_{U,e_j}}$ and $\bar{V}_{\mathcal{B}_{U,e_j}}$, we also find the pair (l_j, e_j) but in addition we try to recover other pairs $(\tilde{v}, v) \in \bar{V}_{\mathcal{A}_{U,e_j}} \times \bar{V}_{\mathcal{B}_{U,e_j}}$ such that $\tilde{v} = vS$.

Selection Strategy

In fact, these two algorithms are complementary and in order to minimize the number of varieties computed, you can use the following strategy.

Start with algorithm A and for each $j, 1 \leq j \leq n$ of the *for* loop, compute
$t = \sum_{k=1}^{j} dim_{\mathbb{F}_q}(Span(V_{U,e_j}))$.
If $t \geq n$ and $2j < n$ then stop the execution of A, compute for all k, $1 \leq k \leq j$
the varieties $V_{\mathcal{B}_{U,e_j}}$ and recover a linear equivalence matrix with the function
$SeekMatrix(\{V_{U,e_j}, V_{\mathcal{B}_{U,e_j}}\}_{1 \leq k \leq j})$.
Else continue the execution of algorithm A.
Notice that $2j$ is the number of varieties computed in algorithm B, in the
case when the set of vectors chosen during the *while* loop of algorithm B were
$\{e_1, \cdots, e_j\}$.

4.3   Weak Instances of PLE

From both a practical and theoretical point of view, it is revelant to be able to
identify the instances which can be solved by a deterministic polynomial time
algorithm. It is of major interest when this problem is used in cryptography. In
this part, we present instances of the PLE problem admitting a deterministic
polynomial time algorithm.
When restricting the inputs of the PLE problem to sets of polynomials of degree
one, the problem can be reformulated as follows:
Input: Two matrices A and B in $\mathcal{M}_{u,n}(\mathbb{F}_q)$.
Question: Find if there exists a matrix $S \in GL_n(\mathbb{F}_q)$ such that $B = SA$.
If one of the inputs matrices is invertible[4], BA^{-1} is the unique solution to
this problem. More generally, consider two sets of polynomials $\mathcal{A}$ and $\mathcal{B}$. According to proposition 1, we know that if there exists $S \in GL_n(\mathbb{F}_q)$ such that
$\mathcal{B}(X) = \mathcal{A}(XS)$ then $\mathcal{B}^{(1)}(X) = \mathcal{A}^{(1)}(XS)$. In the other direction, if we know
that $\mathcal{A}^{(1)} \equiv_L \mathcal{B}^{(1)}$ and the mapping $\mathcal{A}^{(1)}$ or $\mathcal{B}^{(1)}$ is bijective then the unique linear equivalence matrix between $\mathcal{A}^{(1)}$ and $\mathcal{B}^{(1)}$ is $S^{(1)} = B^{(1)}(A^{(1)})^{-1}$, where $A^{(1)}$
and $B^{(1)}$ are the matrices representing the linear mapping $\mathcal{A}^{(1)}$ and $\mathcal{B}^{(1)}$. Since S
is also a linear equivalence matrix between $\mathcal{A}^{(1)}$ and $\mathcal{B}^{(1)}$, we have $S^{(1)} = S$. The
unique linear equivalence matrix between $\mathcal{A}^{(1)}$ and $\mathcal{B}^{(1)}$ is a linear equivalence
matrix between $\mathcal{A}$ and $\mathcal{B}$. Consequently, when the linear part of one of the inputs
of the PLE problem is bijective then we can find a solution by performing very
basic linear algebra operations.

5   Polynomial Affine Equivalence Algorithms

5.1   General Method

Since the PAE problem is very similar to the PLE problem, it seems natural to
try to reuse the algorithms described in section 4. A straightforward way to do
this is:

---

[4] Remark that if $\mathcal{A} \equiv_L \mathcal{B}$, then if A(*resp.* B) is invertible then B(*resp.* A) is also
   invertible !

For a in $\mathbb{F}_q^n$,
Try to find by algorithm A or B an $S \in GL_n(\mathbb{F}_q)$ such that $\mathcal{B}(X) = \mathcal{A}(XS + a)$.
If so, return (S, a).

This approach - which we shall call *general method* in the sequel - adds a factor q^n to the complexity of the algorithms A or B of section 4. Another method can be derived from the linear case by using an idea of 4.1, as follows.

5.2   Generalization of Algorithm A

We present here the affine version of algorithm A presented in 4.1. When $\mathcal{A} \equiv_A \mathcal{B}$, then the evaluation of the b_is on some vector $p \in \mathbb{F}_q^n$ is equal to the evaluation of the a_is in $\widetilde{p'} = pS + T$, for some affine equivalence pair (S, T) between $\mathcal{A}$ and $\mathcal{B}$. Hence, knowledge of the pair $(p, \widetilde{p'})$ allows us to obtain n linear equations in the components of S and T. In order to convert the search of this pair into the resolution of a non linear system of equations, we set $I_p =< \{a_i(X) - b_i(p)\}_{1 \le i \le u}, \{x_i^q - x_i\}_{1 \le i \le n} >$ and denote V_p the variety associated to this ideal. Let e_0 be the null vector of $\mathbb{F}_q^n$ and $\{e_j\}_{1 \le j \le n}$ be the n vectors of the canonical basis of $\mathbb{F}_q^n$. The algorithm is the following:

Algorithm A'
Input: Two sets of polynomials $\mathcal{A}$ and $\mathcal{B}$.
Output: An affine equivalence pair between $\mathcal{A}$ and $\mathcal{B}$, if any and $\emptyset$ otherwise.
Compute V_{e_0}

> **If** $V_{e_0} \neq \emptyset$ **then**
> > $L_0 \leftarrow V_{e_0}$
> > **For** j from 1 to n **do**
> > > Compute V_{e_j}
> > > > **If** $V_{e_j} \neq \emptyset$ **then** $L_j \leftarrow \{v_{e_j} - l_0 : (v_{e_j}, l_0) \in V_{e_j} \times L_0\}$ **Else Return** $\emptyset$
> > **EndFor**
> **Else Return** $\emptyset$

$(S, T) \leftarrow Seek(\mathcal{A}, \mathcal{B}, \{L_0, \cdots, L_n\})$
Return (S, T)

L_0 is the set of candidates for the affine part of an affine equivalence pair between $\mathcal{A}$ and $\mathcal{B}$ and for all $j, 1 \le j \le n$ the elements of L_j are candidates for the j-th row of the linear part of an affine equivalence pair. To recover this pair, if such a pair exists, the function $Seek$ checks for all the vectors in L_0 and for all the matrices than can be constructed from the sets $L_1, \cdots, L_n$. The improvement proposed in [LP] can also be adapted to this function.

Remark 5. If $\mathcal{A} \equiv_L \mathcal{B}$ and if $\mathcal{A}$ or $\mathcal{B}$ is a bijection then there exists a unique affine equivalence pair between these two sets.

Complexity of algorithm A'
Let D be the maximum degree of the polynomials of $\mathcal{A}$. When one of the inputs is a bijection, the complexity A' is $\mathcal{O}((n+1)D^n)$. In the general case, the complexity is $\mathcal{O}((n+1)D^n + (n+1)^3 \prod_{i=0}^n |L_i|)$.

5.3   Generalization of Algorithm B

The adaptation of algorithm B, which we shall call algorithm B', contains no difficulty. We omit its description and refer the reader to [LP].

Selection strategy
When u is "small" compared to n, it is very important to be able to decrease the size of the varieties computed in A'or in B'. Unfortunately, contrary to the linear case, the structural properties of the affine equivalence relation are useless in this context. In this case, we think that the general method together with algorithm B is the best choice to do. Indeed, the general method transforms a PAE problem into a PLE problem and even if it adds a factor q^n to the complexity of algorithm B, it allows to decrease the search space of the linear part of an equivalence pair. In the other case, one can use the following strategy to minimize the number of varieties computed.
Start with algorithm A' and for each $j, 1 \leq j \leq n$ of the *for* loop, compute $t = \sum_{k=0}^{j} dim_{\mathbb{F}_q}(Span(V_{e_j}))$.
If $t \geq n+1$ and $2j < n+1$, stop the execution of A', compute for all $k, 0 \leq k \leq j$ the varieties $V_{\mathcal{B}_{e_j}}$. Given $\{V_{e_k}\}_{0 \leq k \leq j}$ and $\{V_{\mathcal{B}_{e_j}}\}_{0 \leq k \leq j}$, you can recover an affine equivalence pair with a simple extension of the function $SeekMatrix$ described in section 4.2.
Else continue the execution of B'.

6   Applications

6.1   Security of Cryptosystems Based on PAE

Remember that we call PAE problem in this paper, the problem which is called isomorphism of polynomials with one secret in [P]. In this article, it has been shown how the PAE problem can be used to derive a signature and authentication scheme. We do not recall these constructions here, as our algorithms focus on the underlying problem which guarantees the security of these schemes. Let us recall the parameters for which the PAE problem was supposed intractable[P],[P1]. The two sets $\mathcal{A}$ and $\mathcal{B}$ are composed of $u \geq 2$ polynomials in n indeterminates of degree two whose coefficients lie in $\mathbb{F}_q$. The author recommends to choose the number of variables n and the size q of the field such that $q^{\sqrt{2}n^{3/2}} \geq 2^{64}$. We will now show that, in some cases, these parameters are far from being sufficient to achieve a reasonable level of security. In particular, in order to improve the efficiency of these schemes, the author suggests to restrict the affine equivalence to the linear equivalence. This restriction strongly acts on the safety of these schemes, since as explained in 4.3 the PLE problem has more properties than the PAE problem. Moreover, this problem admits a lot of instances which can be solved in deterministic polynomial time. Hence, without adding structural constraints on the shape of the polynomials of $\mathcal{A}$ and $\mathcal{B}$, which have not been given in the original design, these schemes are insecure. In order to avoid these weaknesses, the sets of polynomials $\mathcal{A}$ and $\mathcal{B}$ must be chosen in such a way that the linear parts $\mathcal{A}^{(1)}$ and $\mathcal{B}^{(1)}$ are not bijective. Even with these additional constraints, these parameters don't really give rise to difficult instances of the PLE problem and are not adapted to the design of secure applications, as we shown in the appendix. Furthermore, the polynomial affine equivalence problem admits also instances for which the complexity of resolution is far from the cryptographically safe bounds. Experimental results are given in the appendix.

6.2   Chosen Ciphertext Attack

A large family of multivariate asymmetric encryption cryptosystems, like C^*[MI], HFE [P] and TTM [TTM] can be sketched as follows. Alice generates a set of polynomials $\mathcal{A} = \{a_1(x_1, \cdots, x_n), \cdots, a_u(x_1, \cdots, x_n)\} \subset \mathbb{F}_q[x_1, \cdots, x_n]^u$ in such a way that for all $c = (c_1, \cdots, c_u) \in \mathbb{F}_q^u$ there exists a unique solution [5] to the system $\{a_1(x_1, \cdots, x_n) - c_1 = 0, \cdots, a_u(x_1, \cdots, x_n) - c_u = 0\}$ and this solution can be efficiently computed. In order to hide $\mathcal{A}$, Alice chooses two pairs $(S, T) \in GL_n(\mathbb{F}_q) \times \mathbb{F}_q^n$ and $(U, V) \in GL_u(\mathbb{F}_q) \times \mathbb{F}_q^u$, computes $\mathcal{B}(X) = U \circ \mathcal{A}(XS + T) + V^t$, denoted by $\mathcal{B} = \{b_1(x_1, \cdots, x_n), \ldots, b_u(x_1, \cdots, x_n)\}$, and publishes $\mathcal{B}$. When Bob wants to encrypt a message $(m_1, \cdots, m_n) \in \mathbb{F}_q^n$, he computes $c = (b_1(m_1, \cdots, m_n), \cdots, b_u(m_1, \cdots, m_n))$ and sends it to Alice. After receiving $c = (c_1, \cdots, c_u)$, Alice computes the solution $c' = (c'_1, \cdots, c'_n)$ of the system $\mathcal{A}(X) - U^{-1}c^t - U^{-1}V^t = 0$ and recovers the message sent by computing $(c'_1, \cdots, c'_n)S^{-1} - T = (m_1, \cdots, m_n)$.

An open question is to know whether or not these schemes remain secure if the set of polynomials $\mathcal{A}$ is public. In this situation, the security of these schemes relies not only on the difficulty of finding a common zero of a system of non linear equations but also on the difficulty of the Isomorphism of Polynomials (IP) problem (when $\mathcal{A}$ and $\mathcal{B}$ are given, the problem is to recover the pairs (S, T) and (U, V)). Until now, this question remains open since the best algorithm known to solve the IP problem[6] has a complexity of $O(q^{\frac{3n}{2}})$[CGP]. As pointed out in this paper, the PAE problem seems to be more difficult than the IP problem. Moreover, they have also shown that unless the polynomial hierarchy collapses, the PAE problem and the IP problem are not NP-hard. But contrary to the PAE problem, which is at least as difficult as the graph isomorphism problem, we would like to emphasize that there exists no theoretical evidence of the hardness of the IP problem. Hence, it is a natural question to ask whether the security of these schemes could be increased if Alice would choose $(U, V) = (I_u, \mathbf{0}_u)$[7] and publish $\mathcal{A}(X)$ and $\mathcal{B}(X) = \mathcal{A}(XS + T)$.

We now show that if the public key is generated in this way, an adversary is able to recover the secret pair $(S, T) \in GL_n(\mathbb{F}_q) \times \mathbb{F}_q^n$ with only $n+1$ queries to a deciphering oracle. Remark that due to the symmetry of the relation $\equiv_A$, we have $\mathcal{A}(X) = \mathcal{B}(XS' + T')$, with $S' = S^{-1}$ and $T' = -TS^{-1}$. Let $\{e_i\}_{1 \leq i \leq n}$, be the

---

[5] or very few solutions

[6] This algorithm works only in the particular case when the sets of polynomials $\mathcal{A}$ and $\mathcal{B}$ are bijective.

[7] $\mathbf{0}_u$ is here the null vector of $\mathbb{F}_q^u$

n canonical vectors of $\mathbb{F}_q^n$. In order to recover the secret vector T', an adversary sends $c_0 = (a_1(e_0), \cdots, a_u(e_0))$. The unique cleartext $m_0 \in \mathbb{F}_q^n$ corresponding to this ciphertext is such that $b_i(m_0) = a_i(e_0)$ for all i, $1 \le i \le u$. Hence, $m_0 = e_0 S' + T' = T'$. To recover the j-th row of the matrix S', $1 \le j \le n$, an adversary sends $c_j = (a_1(e_j), \cdots, a_u(e_j))$. The cleartext $m_j \in \mathbb{F}_q^n$ corresponding to this ciphertext is such that $b_i(m_j) = a_i(e_j)$ for all i, $1 \le i \le u$. We then have $m_j = e_j S' + T'$ and therefore the j-th row of the matrix S' is equal to $m_j - m_0$. Finally, knowledge of the pair (S', T') allows to recover easily the secret key (S, T) of Alice. Remark that when the cleartext is not unique, we obtain with this method not exactly the rows of the secret pair (S, T) but a list of candidates. Since for each row the number of candidates is not too big (otherwise Alice herself would not be able to decrypt), the secret affine pair can be recovered efficiently with the method described in [GMS] or with an extension of the method described in [LP]. Hence, the security of encryption schemes like HFE, can not be related to the difficulty of the PAE problem since in this situation the problem can be easily solved with the help of a deciphering oracle.

Let $(\mathcal{A}, U \circ \mathcal{A}(XS + T) + V^t)$ be the public-key of a multivariate encryption scheme. It is straightforward to see that if an adversary is able to recover the secret pair (U, V) then he can use the method described above to find the other pair (S, T). Hence, our method can be used in addition to an attack specifically designed to recover (U, V). Moreover, one sees at once that when the secret pair (S, T) is given then the pair (U, V) can be easily recovered. That is, when one of the two secret pairs is known, the other can be easily recovered. This is not the case for the underlying IP problem and so the problem considered in the security analysis of these schemes is weaker than the generic problem. It is left as an open problem whether or not the IP problem can be solved in deterministic polynomial time if we have access to a deciphering oracle.

7 Conclusion

We have presented in this paper new approaches to the IP with one secret problem, which lead to the design of efficient algorithms. We studied the security of crypnosystems based on this problem and it appears that the usually suggested parameters often yield weak instances. Advises concerning parameters sizes are given at the end of the appendix.

References

[BeWe]   T. Becker and V. Weispfenning: *Gröbner Bases, A Computational Approach to Commutative Algebra. In cooperation with Heinz Kredel.* Graduate Texts in Mathematics, 141. Springer-Verlag, New York, 1993.

[CGP]   N. Courtois, L. Goubin, J. Patarin: Improved Algorithms for Isomorphism of Polynomials. Eurocrypt'98, Springer-Verlag, pp 184-200.

[COX]   D. A. Cox, D. O'Shea, J.B. Little: *Ideals, Varieties, and Algorithms: An Introduction to Computational Algebraic Geometry and Commutative Algebra.* Undergraduate Texts in Mathematics. Springer-Verlag. New York, 1992.

[Fa]   J.-C. Faugère: Algebraic cryptanalysis of HFE using Gröbner bases. INRIA report: RR-4738. Available from http://www.inria.fr/rrrt/rr-4738.html.

[Fa99]   J.-C. Faugère: A new efficient algorithm for computing Gröbner basis: F_4. Journal of pure and applied algebra, vol. 139, 1999, pp. 61–68.

[Fa02]   J.-C. Faugère: A new efficient algorithm for computing Gröbner basis without reduction to zero: F_5. Proceedings of ISSAC, pages 75-83. ACM press, July 2002.

[FJ]   J.-C. Faugère, Antoine Joux: Algebraic cryptanalysis of hidden field equation (HFE) cryptosystems using Gröbner bases. Crypto 2003, LNCS 2729, Springer-Verlag, August 2003.

[GMS]   W. Geiselmann and W. Meier and R. Steinwandt: An Attack on the Isomorphisms of Polynomials Problem with One Secret. Cryptology ePrint Archive: Report 2002/143. Available from http://eprint.iacr.org/2002/143.

[LP]   F. Levy-dit-Vehel and L. Perret: Polynomial equivalence problems and applications to multivariate cryptosystems. Rapport INRIA 2003, to appear.

[Magma]   http://magma.maths.usyd.edu.au/magma/

[MI]   T. Matsumoto, H. Imai: Public Quadratic Polynomial-tuples for efficient signature-verification and message-encryption. Eurocrypt'88, Springer-Verlag, pp. 419-453.

[P]   J. Patarin: Hidden Fields Equations (HFE) and Isomorphisms of Polynomials (IP): two new families of Asymmetric Algorithms. Eurocrypt'96, Springer-Verlag, pp. 33-48.

[P1]   J. Patarin: Hidden Fields Equations (HFE) and Isomorphisms of Polynomials (IP): two new families of Asymmetric Algorithms - Extended version. http://www.cp8.com/sct/uk/partners/page/publi/eurocryptb.ps.

[TTM]   T.-T.Moh: A Fast Public Key System With Signature And Master Key Functions. CrypTEC'99, Hong Kong City University Press, pages 63-69, July 1999.

Appendix

Experimental results on the PLE and PAE problems

Conditions of the tests

We have generated a set of polynomials $\mathcal{A} = \{a_1(X), \cdots, a_u(X)\}$ with respect to the constraints given in 6.1 and we have chosen q and n such that $q^{\sqrt{2}n^{3/2}} \geq 2^{64}$. We have randomly chosen a matrix S in $GL_n(\mathbb{F}_q)$(*resp.* a pair $(S, T) \in GL_n(\mathbb{F}_q) \times \mathbb{F}_q^n$), we have computed $\mathcal{B} = \{a_1(XS), \cdots, a_u(XS)\}$ (*resp.* $\mathcal{B}\{a_1(XS + T), \cdots, a_u(XS + T)\}$). We have tested the algorithms described in section 4(*resp.* 5) with the selection strategy described in 4.2 (*resp.* 5.3). The Gröbner bases have been computed through the web interface of *fgb*, the others computation have been done on a standard PC, using Magma software [Magma]. The results are quoted below:

n	u	$field$	$q^{\sqrt{2}n^{3/2}}$	PLE $Time$	PAE $Time$
15	15	$\mathbb{F}_2$	2^{82}	$\approx 2\ min.$	$\approx 2\ min.$
15	13	$\mathbb{F}_2$	2^{82}	$\approx 7\ min.$	$\approx 7\ min.$
15	11	$\mathbb{F}_2$	2^{82}	$\approx 15\ min.$	$\approx 15\ min.$
10	10	$\mathbb{F}_{11}$	$\approx 2^{152}$	$\approx 8\ min.$	$\approx 10\ min.$
10	9	$\mathbb{F}_{11}$	$\approx 2^{152}$	$\approx 30\ min.$	$\approx 35\ min.$

Interpretation

Since we have chosen $u \approx n$, the cost of our algorithms is approximately the cost of computing several Gröbner Bases. Hence, we can exhibit a large number of instances illustrating the weakness of the security parameters given in [P]. We believe that the parameters chosen are significative of the behaviour of our algorithms.

When the two sets of polynomials lie in $\mathbb{F}_2[X]$, the efficiency of the algorithms dedicated to the PAE problem are similar to the ones dedicated to the PLE problem. Indeed, the PLE algorithms use the particular properties of the linear equivalence relation. But when the field is reduced to two elements, proposition 2 gives no information about the linear equivalence matrix. Whereas when the field is bigger, one can see that the PLE algorithms find a solution more quickly than the PAE algorithms.

Some parameter propositions

For cryptographic applications, we strongly believe that u must be chosen approximately equal to n. In this setting, we know that there exists very few solutions to our problems. When u is small compared to n, there will probably be a large number of solution to these problems. Hence, an improved exhaustive search like local search method or test and trials method could work rather well.

For the PLE problem, we believe that the instances must be composed of homogeneous polynomials over $\mathbb{F}_2[X]$. In this setting, one can see at once that proposition 1 and proposition 2 give no information on the linear equivalence matrices. With such instances, we are in the same situation than for the PAE problem. For this problem, we have not found any particular weakness in the structure of the instances.

Security Analysis of
Several Group Signature Schemes

Guilin Wang

Infocomm Security Department
Institute for Infocomm Research
21 Heng Mui Keng Terrace, Singapore 119613
glwang@i2r.a-star.edu.sg
http://www.i2r.a-star.edu.sg/icsd/staff/guilin/

Abstract. At Eurocrypt'91, Chaum and van Heyst introduced the concept of group signature. In such a scheme, each group member is allowed to sign messages on behalf of a group anonymously. However, in case of later disputes, a designated group manager can open a group signature and identify the signer. In recent years, researchers have proposed a number of new group signature schemes and improvements with different levels of security. In this paper, we present a security analysis of several group signature schemes proposed in [25, 27, 18, 31]. By using the same method, we successfully identify several *universally forging attacks* on these schemes. In our attacks, anyone (not necessarily a group member) can forge valid group signatures on any messages such that the forged signatures cannot be opened by the group manager. We also discuss the linkability of these schemes, and further explain why and how we find the attacks.

Keywords: digital signature, group signature, forgery, cryptanalysis.

1 Introduction

A group signature scheme, first introduced by Chaum and van Heyst in [7], allows each group member to sign messages on behalf of a group anonymously and unlinkably. However, in case of later disputes, a designated group manager can open a group signature and then identify the true signer. A secure group signature scheme must satisfy the following properties [1, 2, 4, 7] [1]:

- *Unforgeability:* Only group members are able to sign messages on behalf of the group.
- *Anonymity:* Given a valid signature of some message, identifying the actual signer is computationally hard for everyone but the group manager.
- *Unlinkability:* Deciding whether two different valid signatures were computed by the same group member is computationally hard.

---

[1] Note that the property of *Coalition-resistance* is not listed here since in essence it is implied by *Traceability*.

T. Johansson and S. Maitra (Eds.): INDOCRYPT 2003, LNCS 2904, pp. 252–265, 2003.

- *Exculpability:* Neither a group member nor the group manager can sign on behalf of other group members.
- *Traceability:* The group manager is always able to open a valid signature and identify the actual signer.

In general, group signature schemes can be classified into two different types: The schemes based on *signatures of knowledge* [4] and the schemes designed with *straightforward* and *ad-hoc methods*. The schemes in [4, 5, 1, 22] belong to the first type, while the schemes proposed by [10, 11, 24, 25, 26, 27, 18, 31] belong to the second type. Some of the first type schemes are provably secure, but all those schemes are not very efficient. For example, as one of the most efficient schemes belonging this type, the scheme in [5] still needs about 13,000 RSA modular multiplications in generation and verification a group signature (see Section 5.6 of [5]). The second type schemes are very efficient since generation and verification of a signature only need to compute several standard signatures. However, no existing scheme of this type has provable security.

In 1998, Lee and Chang presented an efficient group signature scheme based on the discrete logarithm [11]. Their scheme is obviously linkable since two same pieces of information are included in all signatures generated by the same group member. To provide unlinkability, Tseng and Jan proposed an improved group signature scheme in [24]. But Sun pointed out that this improvement is still linkable [23]. At the same time, based on Shamir's idea of identity(ID)-based cryptosystems [20], Tseng and Jan proposed an ID-based group signature scheme in [26]. However, Joye et al. [8, 9] showed that the schemes proposed in [11, 24, 26] all are *universally forgeable*, i.e., anyone (not necessarily a group member) is able to generate a valid group signature on any message, which cannot be opened by the group manager. After that, in [25] and [27], Tseng and Jan revised their schemes, and Popescu presented a modification to the Tseng-Jan ID-based scheme [26] in [18]. In addition, Xian and You proposed a new group signature scheme with strong separability such that the group manager can be split into a membership manager and a revocation manager [31].

In this paper, we present a security analysis of several group signature schemes proposed in [25, 27, 18, 31]. By using the same method originated from [3, 8, 9], we successfully identify different *universally forging attacks* on these schemes. That is, anybody can easily forge valid group signatures on arbitrary messages. At the same time, we point out that the schemes proposed in [26, 27, 18, 31] all are *linkable*. In our paper, we not only describe how to attack these schemes, but also explain why and how we find the attacks. Our attacks also demonstrate that no more group signatures should be constructed with such ad-hoc methods used by the above mentioned insecure schemes. In other words, from the contrary side of the same problem, the formal design methodology employed in [1, 4, 5, 22] are further confirmed.

In addition, using our method, the existing attacks on Kim et al.'s convertible group signature scheme [10] can be unified in a family. Those existing attacks are pointed out by [12, 19, 28, 6] independently and accidentally. Furthermore, we find a new problem in Kim et al.'s scheme, that is, a valid group signature signed by one group member is also a possible valid signature of other group members for the same message. Therefore, their scheme is information-theoretically *anonymous* even for the group manager, and hence all valid group signatures are completely *untraceable* and *unlinkable*.

The rest of this paper is organized as follows. We review and analyze Tseng-Jan (DLP-based) scheme [25], Tseng-Jan ID-based scheme [27], Popsecu's improved scheme [18], and Xia-You scheme [31] in Sections 2, 3, 4, and 5, respectively. The security analysis of Kim et al.'s scheme [10] is presented in the full version of this paper [29]. Finally, Section 6 gives some concluding remarks.

2   Tseng-Jan Group Signature Scheme

2.1   Review of Tseng-Jan Scheme

Setup. Let p and q be two large primes such that $q|(p-1)$, and g a generator of order q in $\mathbb{Z}_p$. A user U_i selects his secret key $x_i \in_R \mathbb{Z}_q^*$, and sets his public key as $y_i := g^{x_i} \bmod p$. Similarly, the group manager (GM) selects his secret key $x \in_R \mathbb{Z}_q^*$, and computes his public key $y := g^x \bmod p$. Furthermore, GM selects a one-way hash function $h(\cdot)$. To join the group, U_i sends his public key y_i to GM. Then, GM chooses a random number $k_i \in_R \mathbb{Z}_q^*$, computes and sends the following pair (r_i, s_i) to U_i privately:

$$r_i := g^{-k_i} \cdot y_i^{k_i} \bmod p, \quad s_i := k_i - r_i x \bmod q. \tag{1}$$

U_i can check the validity of his certificate (x_i, r_i, s_i) by

$$g^{s_i} y^{r_i} r_i \equiv (g^{s_i} y^{r_i})^{x_i} \bmod p. \tag{2}$$

Signing. To sign a message M, U_i first selects four random numbers $a, b, d, t \in_R \mathbb{Z}_q^*$, then calculates a signature (R, S, A, B, C, D, E) as follows:

$$
\begin{aligned}
A &:= r_i^a \bmod p, & B &:= as_i - b \cdot h(A||C||D||E) \bmod q, \\
C &:= r_i a - d \bmod q, & D &:= g^b \bmod p, \\
E &:= y^d \bmod p, & \alpha_i &:= g^B y^C E D^{h(A||C||D||E)} \bmod p, \\
R &:= \alpha_i^t \bmod p, & S &:= t^{-1}(h(M||R) - Rx_i) \bmod q.
\end{aligned} \tag{3}
$$

Verification. On receiving a signature (R, S, A, B, C, D, E) on a message M, a verifier first computes α_i as above and check its validity by

$$\alpha_i^{h(M||R)} \equiv (\alpha_i \cdot A)^R \cdot R^S \bmod p. \tag{4}$$

Note that the above equality holds since we have the following equations:

$$g^{s_i} y^{r_i} = g^{k_i} \bmod p, \quad \alpha_i = g^{ak_i} \bmod p, \quad \text{and} \quad \alpha_i A = \alpha_i^{x_i} \bmod p. \tag{5}$$

Open. To identify the signer of a valid group signature (R, S, A, B, C, D, E) on message M, GM first computes the corresponding α_i and then find the signer by searching which pair (r_i, s_i, k_i) satisfies $\alpha_i \equiv (g^C \cdot E^{x^{-1}})^{r_i^{-1} \cdot k_i} \bmod p$, where x^{-1} and $r_i^{-1} \cdot k_i$ all are computed in $\mathbb{Z}_q$.

2.2   Security Analysis of Tseng-Jan Scheme

Forging Signatures. We want to forge a group signature on an arbitrary message M even though we do not know any certificate, i.e., we need to find a tuple (R, S, A, B, C, D, E) that satisfies the following two verification equations:

$$\begin{cases} \alpha_i = g^B y^C E D^{h(A||C||D||E)} \bmod p, \\ \alpha_i^{h(M||R)} = (\alpha_i \cdot A)^R \cdot R^S \bmod p. \end{cases} \tag{6}$$

Note that in the signature generation, A, D, E and R all are some powers to the bases g and y. At the same time, C is embedded in the hash value $h(A||C||D||E)$. Therefore, we can define A, D, E, R as some known powers of g and y, and choose a value for C. Then, we try to solve B and S from equation (6). Hence, we select nine random numbers $a_1, a_2, a_3, a_4, b_1, b_2, b_3, b_4, C \in_R \mathbb{Z}_q$ to define A, D, E and R as follows (all in $\mathbb{Z}_p$)

$$A := g^{a_1} y^{b_1}, \quad D := g^{a_2} y^{b_2}, \quad E := g^{a_3} y^{b_3}, \quad R := g^{a_4} y^{b_4}.$$

Then, we evaluate the two hash values $h := h(A||C||D||E)$, $h' := h(M||R)$, and replace the corresponding variables in equation (6) with the above expressions. So we get the following two equations for unknown variables of B and S:

$$\begin{cases} (B + a_3 + a_2 h) h' = (B + a_3 + a_2 h) R + a_1 R + a_4 S \bmod q, \\ (C + b_3 + b_2 h) h' = (C + b_3 + b_2 h) R + b_1 R + b_4 S \bmod q. \end{cases} \tag{7}$$

Therefore, if $b_4 \neq 0$ and $R \neq h' \bmod q$ (i.e., $R \neq h(M||R) \bmod q$.), we get the following solutions for S and B:

$$\begin{cases} S = b_4^{-1} [(C + b_3 + b_2 h)(h' - R) - b_1 R] \bmod q, \\ B = (a_1 R + a_4 S)(h' - R)^{-1} - (a_3 + a_2 h) \bmod q. \end{cases} \tag{8}$$

For summary, in the Tseng-Jan group signature scheme [25], an attacker can forge a valid group signature on any message M as follows:

(1) Select nine numbers $a_1, a_2, a_3, a_4, b_1, b_2, b_3, b_4, C \in_R \mathbb{Z}_q$, where $b_4 \neq 0$.
(2) Define $A := g^{a_1} y^{b_1}$, $D := g^{a_2} y^{b_2}$, $E := g^{a_3} y^{b_3}$, and $R := g^{a_4} y^{b_4}$ (all in $\mathbb{Z}_p$).
(3) Evaluate $h := h(A||C||D||E)$ and $h' := h(M||R)$.
(4) Determine if $R \equiv h' \bmod q$. If yes, go to step (1); otherwise, continue.
(5) Compute S and B according to equation (8).
(6) Output (R, S, A, B, C, D, E) as a group signature for message M.

The correctness of the above attack can be verified directly. When one such forged group signature is given, GM cannot find the signer. At the same time, note that in the above attack $R = h' \bmod q$ occurs only with a negligible probability since $h(\cdot)$ is a one-way hash function. Therefore, in general, our attack will succeed just by one try. Furthermore, for simplicity, some of those nine random numbers can be set as zeroes. For example, if we set $a_1 = b_2 = b_3 = a_4 = 0$, A, D, E and R can be computed simply: $A := y^{b_1} \bmod p$, $D := g^{a_2} \bmod p$,

$E := g^{a_3} \bmod p$, $R := y^{b_4} \bmod p$. In such case, S and B can be computed by $S = b_4^{-1}(Ch' - CR - b_1R) \bmod q$ and $B = -(a_3 + a_2h) \bmod q$.

Forging Certificates. The authors of [11, 24, 25] noted that for any group member U_i, (r_i, s_i) is a Nyberg-Rueppel signature [15] on message $y_i^{k_i}$. However, this *does not* imply that only GM can generate a valid certificate. Now, we demonstrate how to forge a certificate $(\bar{x}_i, \bar{r}_i, \bar{s}_i)$ that satisfies equation (2). We first choose $a_0, b_0 \in \mathbb{Z}_q^*$, and define $\bar{r}_i := g^{a_0}y^{b_0} \bmod p$. Then, from equation (2), we have the following equation for unknown $\bar{x}_i$ and $\bar{s}_i$:

$$g^{\bar{s}_i}y^{\bar{r}_i}g^{a_0}y^{b_0} = (g^{\bar{s}_i}y^{\bar{r}_i})^{\bar{x}_i} \bmod p.$$

From the above equation, we get the following two equations for $\bar{x}_i$ and $\bar{s}_i$:

$$\bar{s}_i + a_0 = \bar{s}_i \cdot \bar{x}_i \bmod q, \quad \text{and} \quad \bar{r}_i + b_0 = \bar{r}_i \cdot \bar{x}_i \bmod q.$$

Therefore, we obtain the solutions for $\bar{x}_i$ and $\bar{s}_i$: $\bar{x}_i = 1 + b_0\bar{r}_i^{-1} \bmod q$ and $\bar{s}_i = a_0b_0^{-1}\bar{r}_i \bmod q$. The forged certificate $(\bar{x}_i, \bar{r}_i, \bar{s}_i)$ satisfies equation (2) since $g^{\bar{s}_i}y^{\bar{r}_i}\bar{r}_i = g^{a_0b_0^{-1}\bar{r}_i}y^{\bar{r}_i}g^{a_0}y^{b_0} = g^{a_0b_0^{-1}\bar{r}_i(1+b_0\bar{r}_i^{-1})}y^{\bar{r}_i(1+b_0\bar{r}_i^{-1})} = (g^{\bar{s}_i}y^{\bar{r}_i})^{\bar{x}_i} \bmod p$. So an attacker can use it to generate valid group signatures on any messages as a group member does.

Remark 1. The schemes proposed in [11, 24, 21] all are subject to similar attacks due to their similar structures. Especially, the above forged certificate can be directly used to generate valid group signatures in those schemes since all those schemes use the same certificate. Compared with Joye's attacks [8] on the two schemes in [11, 24], our above attacks not only constitute a family, but also are very simple (especially for the forging certificate attack.). The attack on Shi's scheme [21] specified independently by [34] is weaker than ours, because it assumed that a valid signature is known. In addition, there is a design error in Shi's scheme. That is, all the following numbered equations in [21] should be modified from modulo p to modulo q: (5), (6), (11), (15), and (17). Otherwise, Shi's scheme does not work since the signatures generated by honest group members cannot be successfully validated by verifiers. If this modification is made, however, Shi's scheme will become the same scheme proposed in [24].

3   Tseng-Jan ID-based Group Signature Scheme

3.1   Review of Tseng-Jan ID-based Scheme

The Tseng-Jan ID-based group signature scheme [27] involves four parties: a trusted authority (TA), the group manager (GM), the group members, and the verifiers. TA acts as a third party to setup the system parameters. GM selects the group public/secret key pair. He (jointly with TA) issues certificates to new users who want to join the group. Then, group members can anonymously sign on behalf of the group by using their membership certificates. In case of disputes, GM opens the contentious signature to reveal the identity of the actual signer.

System Initialization. In order to set up the system, TA sets a modulus $n = p_1 p_2$ where p_1 and p_2 are two large prime numbers (about 120 decimal digits) such that $p_1 = 3 \mod 8$, $p_2 = 7 \mod 8$, and $(p_1 - 1)/2$ and $(p_2 - 1)/2$ are smooth, odd and co-prime. Furthermore, $(p_1 - 1)/2$ and $(p_2 - 1)/2$ should contain several prime factors of about 20 decimal digits but no large prime factors. In this case, it is easy for TA to find the discrete logarithms for p_1 and p_2 [13, 14, 16, 17]. TA also defines e, d, v, t satisfying $ed = 1 \mod \phi(n)$ and $vt = 1 \mod \phi(n)$. Then, he selects an element g of large order in $\mathbb{Z}_n^*$, and computes $F := g^v \mod n$. TA also chooses a hash function $h(\cdot)$. The public parameters of TA are $(n, e, g, F, h(\cdot))$, and the secret parameters of TA are (p_1, p_2, d, v, t). To create a group, GM selects a secret key x and computes the corresponding group public key $y := F^x \mod n$.

When a user U_i (with identity information D_i) wants to join the group, TA and GM computes and sends the following s_i and x_i to U_i, respectively.

$$s_i := et \cdot \log_g ID_i \mod \phi(n), \quad \text{and} \quad x_i := ID_i{}^x \mod n. \tag{9}$$

where

$$ID_i := \begin{cases} D_i, & \text{if Jacobi symbol } (D_i|n) = 1; \\ 2D_i, & \text{if Jacobi symbol } (D_i|n) = -1. \end{cases} \tag{10}$$

Equation (10) guarantees the existence of the discrete logarithm of ID_i to the base g [14]. The membership certificate of the user U_i is (s_i, x_i).

Signing and Verification. To sign a message M, U_i chooses two random integers $r_1, r_2 \in_R \mathbb{Z}_n^*$, and then computes his group signature (A, B, C, D) as

$$
\begin{aligned}
A &:= y^{r_1} \mod n, & B &:= y^{r_2 e} \mod n \\
C &:= s_i + r_1 \cdot h(M||A||B) + r_2 e, & D &:= x_i \cdot y^{r_2 \cdot h(M||A||B||C)} \mod n.
\end{aligned} \tag{11}
$$

Upon receiving an alleged signature (A, B, C, D) for message M, a verifier can validate its validity by checking the following equality:

$$D^e A^{h(M||A||B)} B \equiv y^C B^{h(M||A||B||C)} \mod n. \tag{12}$$

Open. With the secret key x, GM can identify the signer of a valid signature by finding the ID_i that satisfies the following equation:

$$(ID_i)^{xe} \equiv D^e \cdot B^{-h(M||A||B||C)} \mod n. \tag{13}$$

3.2 Security Analysis of Tseng-Jan ID-based Scheme

In [27], Tseng and Jan provide detailed security analysis to demonstrate that their scheme is secure against forgeries and that the anonymity of the signer in their scheme depends on computing the discrete logarithm modulo for the composite number n. However, our analysis in this subsection shows that Tseng-Jan ID-based scheme is *linkable* and *universally forgeable*. Similarly, the scheme in [26] is also linkable (and forgeable [9]).

Linkability. It is easy to see that the value in the left side of equation (13) is an invariant for user U_i, since ID_i is the related information derived from user

U_i's real identity, and x, e both are fixed values. Therefore, given two valid group signatures (A, B, C, D) and $(\bar{A}, \bar{B}, \bar{C}, \bar{D})$ on messages M and $\bar{M}$, respectively, anybody can determine whether they are signed by the same group member by checking:

$$D^e B^{-h(M||A||B||C)} \equiv \bar{D}^e \bar{B}^{-h(\bar{M}||\bar{A}||\bar{B}||\bar{C})} \bmod n.$$

Forging Signatures. Note that in [26], the value D in equation (11) is computed in a different way: $D := x_i \cdot y^{r_2 \cdot h(M||A||B)} \bmod n$. However, this modification does not improve the security. Similar to what we did in Section 2.2, we want to forge a group signature for an arbitrary message M without any membership certificate. Note that the verification equation (12) is about some powers of A, B, D and y. So we first define A, B, D as some known powers to the base y, and then try to solve C from equation (12). We pick three random number r_1, r_2, r_4 and define A, B, D as follows (A and B have the same forms as in equation (11)):

$$A := y^{r_1} \bmod n; \quad B := y^{r_2 e} \bmod n; \quad D := y^{r_4} \bmod n.$$

Then, from equation (12), we get the condition for the value C:

$$r_4 e + r_1 \cdot h(M||A||B) + r_2 e = C + r_2 e \cdot h(M||A||B||C) \bmod \phi(n). \qquad (14)$$

We have selected r_1, r_2 and r_4, so A, B, D and then hash value $h(M||A||B)$ all are fixed. Therefore, finding a solution for unknown value C from equation (14) seems difficult because we do not know the modulus $\phi(n)$ and the value of C is embedded in the hash value $h(M||A||B||C)$. However, we note that solving equation (14) seems really difficult only if r_1, r_2 and r_4 are truly selected as *random* numbers. But, we are attackers. So we have the freedom to choose some special values for r_1, r_2 and r_4. In other words, to get a solution for the value C, we can let those numbers satisfy some specific relationships. It is not difficult to find the following solution for equation (14):

$$C := r_1 \cdot h(M||A||B) + r_2 e \in \mathbb{Z}^+; \quad r_4 := r_2 \cdot h(M||A||B||C) \in \mathbb{Z}^+.$$

We summary our attack on the Tseng-Jan ID-based scheme [27] as follows:

(1) Select two random numbers $r_1, r_2 \in_R \mathbb{Z}_n^*$.
(2) Define $A := y^{r_1} \bmod n$, and $B := y^{r_2 e} \bmod n$.
(3) Compute $C := r_1 \cdot h(M||A||B) + r_2 e \in \mathbb{Z}^+$.
(4) Define $r_4 := r_2 \cdot h(M||A||B||C) \in \mathbb{Z}^+$, and then compute $D := y^{r_4} \bmod n$.
(5) Output (A, B, C, D) as a group signature on message M.

In fact, if we choose a new random number r_3, the values of C and D in the above attack can be randomized by defining C and r_4 as follows

$$C := r_1 \cdot h(M||A||B) + r_2 e + r_3 e \in \mathbb{Z}^+; \quad r_4 := r_2 \cdot h(M||A||B||C) + r_3 \in \mathbb{Z}^+.$$

Furthermore, we have another idea to solve equation (14): First define A, B and C, then calculate hash values of $h(M||A||B)$ and $h(M||A||B||C)$, and finally solve r_4 for D. However, it seems difficult to find the value of r_4 from equation (14) since we do not know the values of modulus $\phi(n)$ and $e^{-1} \bmod \phi(n)$. But we can find the value of r_4 if e can be eliminated from equation (14). Here is the trick. We use $r_1 e$ to replace r_1 (i.e., $A := y^{r_1 e} \bmod p$) and define $C := r_3 e$ (in $\mathbb{Z}$) for some random number r_3, then r_4 can be attained:

$$r_4 := r_3 + r_2 \cdot h(M||A||B||C) - r_1 \cdot h(M||A||B) - r_2 \in \mathbb{Z}.$$

Forging Certificates. Note that the membership certificates in [26] and [27] are the same. Therefore, according to equation (9), for any positive integer k there are two ways to forge valid membership certificates: (1) A group member U_i can generate a new certificate $(ks_i, x_i^k \bmod n)$ using his certificate (s_i, x_i); (2) Anybody can use $(\bar{s}_i = ke, \bar{x}_i = y^k \bmod n)$ as a valid certificate [9].

4 Popescu's Improved Group Signature Scheme

4.1 Review of Popescu's Improved Scheme

Key Generation. TA selects two large primes p_1, p_2 as in [27] (see §3.1) and sets $n := p_1 p_2$. Then, TA selects g of large order in $\mathbb{Z}_n^*$, a large integer e (160 bits) such that $\gcd(e, \phi(n)) = 1$, and then computes d satisfying $de = 1 \bmod \phi(n)$. GM chooses a secret key x and computes the corresponding public key $y := g^x \bmod n$. GM also chooses a collision-resistant hash function $h(\cdot)$. The public parameters are (n, e, g, y, h), TA's secret key is (p_1, p_2, d) and GM's secret key is x.

When a user U_i with identity information $ID_i \in \mathbb{Z}_n$ wants to join the group, TA and GM compute the following s_i and x_i, respectively

$$s_i := ID_i{}^d \bmod n, \quad x_i := (ID_i + eg)^x \bmod n.$$

Then, the membership certificate (s_i, x_i) is sent to the user U_i securely.

Signing. To sign a message M, user U_i chooses two random numbers r_1, r_2, and then computes his group signature (A, B, C, D) as follows

$$\begin{aligned} A :&= y^{r_2 e} \bmod n, & B :&= x_i y^{s_i + r_1} \bmod n \\ C :&= x_i y^{r_2} \bmod n, & D :&= s_i h(M||A) + r_1 h(M||A). \end{aligned} \tag{15}$$

Verification. (A, B, C, D) is a valid group signature for message M iff

$$C^{eh(M||A)} y^{eD} \equiv B^{eh(M||A)} A^{h(M||A)} \bmod n. \tag{16}$$

Open. GM can reveal the signer of a valid signature (A, B, C, D) for message M by searching which identity ID_i satisfies

$$(ID_i + eg)^{xe} \equiv C^e A^{-1} \bmod n. \tag{17}$$

4.2   Security Analysis of Popescu's Improved Scheme

In [18], Popescu claimed that his scheme is unforgeable and unlinkable, since a non-group member (including TA and GM) does not have a valid membership certificate (s_i, x_i) and deciding the linkability of two group signatures is computationally hard under decisional Diffie-Hellman assumption. However, these claims are not true. In this subsection, we will show that Popescu's scheme is *linkable, universally forgeable* and does not satisfy *exculpability*.

Linkability. First of all, similar to the Tseng-Jan schemes [26, 27], Popescu's improved scheme is still linkable since the left side of equation (17) is an invariant for each user U_i.

Forging Signatures. Now, we want to forge a group signature on an arbitrary given message by using similar method as we used in previous sections, even if we do not know any member certificate (s_i, x_i). Since the verification equation (16) is about some powers of A, B, C and y, we choose three random numbers r_1, r_2, r_3 and define A, B, C as follows (A has the same form as in equation (15)):

$$A := y^{r_2 e} \bmod n, \quad B := y^{r_1} \bmod n, \quad C := y^{r_3} \bmod n.$$

Let $h = h(M||A)$. Then, from the verification equation (16), we get the condition for the value D: $r_3 eh + De = r_1 eh + r_2 eh \bmod \phi(n)$, i.e:

$$r_3 h + D = r_1 h + r_2 h \bmod \phi(n). \tag{18}$$

Though we do not know the modulus $\phi(n)$, equation (18) has a trivial solution $D := (r_1 + r_2 - r_3)h \in \mathbb{Z}^+$ if we choose r_1, r_2, r_3 such that $r_1 + r_2 > r_3$. This shows that Popsecu's scheme is universally forgeable. That is, an attacker can forge a valid group signature for any message M as follows:

(1) Select three random numbers r_1, r_2, r_3 such that $r_1 + r_2 > r_3$.
(2) Define $A := y^{r_2 e} \bmod n$, $B := y^{r_1} \bmod n$, and $C := y^{r_3} \bmod n$.
(3) Compute $h := h(M||A)$, and $D := (r_1 + r_2 - r_3)h \in \mathbb{Z}^+$.
(4) Output (A, B, C, D) as a group signature for message M.

Forging Certificates. We now want to derive the determining equation for a valid membership certificate. Let $(\bar{s}_i, \bar{x}_i)$ be a pair of two random numbers. We select two random numbers r_1, r_2 and compute (A, B, C, D) according to equation (15), as if we have a valid member certificate. Let $h = h(M||A)$. Then, we calculate the both sides of the verification equation (16) as follows:

$$C^{eh} y^{eD} := (\bar{x}_i y^{r_2})^{eh} \cdot y^{eh(\bar{s}_i + r_1)} = (\bar{x}_i)^{eh} \cdot y^{(\bar{s}_i + r_1 + r_2)eh} \bmod n,$$
$$B^{eh} A^h := (\bar{x}_i y^{\bar{s}_i + r_1})^{eh} \cdot (y^{r_2 e})^h = (\bar{x}_i)^{eh} \cdot y^{(\bar{s}_i + r_1 + r_2)eh} \bmod n.$$

They are identical. So we reveal an unbelievable fact: In Popsecu's modified scheme [18], any random number pair $(\bar{s}_i, \bar{x}_i)$ is a valid membership certificate!

No Exculpablility. Above fact not only strengthens the conclusion that Popsecu's scheme is universally forgeable, but also reveals another fact that

Popsecu's scheme has no exculpablility: The group manager, who knows the secret value x_i for user U_i, can generate a valid group signature for any message on behalf of U_i by using $(x_i, \bar{s}_i)$ as a membership certificate, where $\bar{s}_i$ is a chosen random number. If such a valid group signature (A, B, C, D) is opened, user U_i will be identified as the signer because $x_i^e = (ID_i + eg)^{xe} = C^e A^{-1} \bmod n$.

5   Xia-You Group Signature Scheme

5.1   Review of Xia-You Scheme

Setup of Trusted Authority (TA). TA generates two prime numbers p_1 and p_2 satisfying the same conditions listed in the Setup of Tseng-Jan ID-based scheme and sets $m := p_1 p_2$. In this case, it is easy for TA to find the discrete logarithms modulo p_1 and p_2. An integer g is chosen such that $g < \min\{p_1, p_2\}$. Finally, TA publishes (m, g) but keeps the prime factors p_1 and p_2 as his secret.

Generating Private Keys. Since a signer U_i's identity information D_i (which is smaller than m) is not guaranteed to have a discrete logarithm modulo the composite number m, TA computes ID_i by equation (10) (respect to modulus m). Now TA computes the private key x_i for U_i as the discrete logarithm of ID_i to the base g:

$$ID_i = g^{x_i} \bmod m. \tag{19}$$

Finally, TA sends x_i to U_i in a secure way and U_i can check the validity of x_i by verifying equation (19). The reader can refer to [31] for details.

Setup of Group Manager (GM). GM chooses two large primes p_3 and p_4 such that $p_3 - 1$ and $p_4 - 1$ are not smooth, and sets $n = p_3 p_4$ such that $n > m$. Let e be an integer satisfying $\gcd(e, \phi(n)) = 1$, and computes d such that $ed = 1 \bmod \phi(n)$. Then, GM chooses two integers $x \in \mathbb{Z}_m, h \in \mathbb{Z}_m^*$, and then computes $y := h^x \bmod m$ as the group public key. Let $H(\cdot)$ be a collision-resistant hash function that maps $\{0, 1\}^*$ to $\mathbb{Z}_m$. The group public key is (n, e, h, y, H) and GM's secret key is (x, d, p_3, p_4).

Generating Membership Keys. When a signer U_i wants to join the group, GM computes the membership key z_i of U_i as follows

$$z_i = ID_i^d \bmod n. \tag{20}$$

z_i is then sent to U_i in a secure way and U_i checks its validity by $ID_i \equiv z_i^e \bmod n$.

Signing. To sign a message M, U_i first chooses five random numbers $\alpha, \beta, \theta, \omega \in \mathbb{Z}_m$ and $\delta \in \mathbb{Z}_n$, and then computes the signature (A, B, C, D, E, F, G) as follows:

$$\begin{aligned}
A &:= y^\alpha \cdot z_i \bmod n, \quad B := y^\omega \cdot ID_i, \quad C := h^\omega \bmod m, \\
D &:= H(y\|g\|h\|A\|B\|\hat{B}\|C\|v\|t_1\|t_2\|t_3\|M), \\
E &:= \delta - D(\alpha e - \omega), \quad F := \beta - D\omega, \quad G := \theta - Dx_i,
\end{aligned} \tag{21}$$

where

$$\begin{aligned}
\hat{B} &:= B \bmod m, \quad v := (A^e/B) \bmod n; \\
t_1 &:= y^\delta \bmod n, \quad t_2 := y^\beta \cdot g^\theta \bmod m, \quad t_3 := h^\beta \bmod m.
\end{aligned}$$

Verification. A verifier accepts a signature (A, B, C, D, E, F, G) on a message M if and only if

$$D \equiv H(y||g||h||A||B||\hat{B}||C||v||t_1'||t_2'||t_3'||M), \tag{22}$$

where $\hat{B}$ and v are computed as in signing equation, i.e., $\hat{B} = B \bmod m, v = (A^e/B) \bmod n$, but t_1', t_2' and t_3' are given by the following equations

$$t_1' := v^D y^E \bmod n, \quad t_2' := \hat{B}^D y^F g^G \bmod m, \quad t_3' := C^D h^F \bmod m. \tag{23}$$

Open. Given a valid group signature (A, B, C, D, E, F, G) for message M, GM can identify the signer by finding the ID_i such that $ID_i \equiv B \cdot C^{-x} \bmod m$.

5.2   Security Analysis of Xia-You Scheme

Xia and You claimed that their scheme [31] satisfies all the security properties listed in Section 1. However, this subsection presents two attacks to show that Xia-You scheme [31] is insecure.

Linkability. From signing equation (21), B is a non-negative integer. Since $B = y^w \cdot ID_i$, we know $ID_i | B$ if U_i is the signer of a valid group signature (A, B, C, D, E, F, G). Therefore, for anyone who knows the identities of group members, he can identify the signer with a high probability. Usually, ID_i, a large integer (e.g. 160 bits), is computed as a hash value of U_i's real-world identity (e.g., name, network address, etc.). So it seems unlikely that there are two identities ID_i and ID_j such that $ID_i | B$ and $ID_j | B$. Hence, Xia-You scheme only satisfies a *weak* anonymity and unlinkability. In addition, even without knowledge of the identities of group members, it is also possible to break the linkability by using the great common divisors of the values of B's in several valid signatures.

Forging Signatures. Using similar method used in the previous sections, we can forge a group signature on an arbitrary given message M even without any membership certificate (ID_i, x_i, z_i). Note that to satisfy the verification equations (22) and (23), we can first choose A, B, C and t_1, t_2, t_3, then we get D by evaluating the corresponding hash value, and finally try to solve the values of E, F and G from equation (23). Observing equations (21)-(23) carefully, we will know that a good strategy is to choose A, B, t_1 and t_2 as some known representations of bases y and g, but C and t_3 as powers of h. Therefore, we can choose ten random numbers $a_1, a_2, a_3, a_4, a_5, b_1, b_2, b_3, b_4, b_5$ to define A, B, C and t_1, t_2, t_3 as follows:

$$
\begin{aligned}
A &:= y^{a_1} \cdot g^{a_2} \bmod n, & t_1 &:= y^{b_1} \cdot g^{b_5} \bmod n, \\
B &:= y^{a_3} \cdot g^{a_4}, & \text{and} \quad t_2 &:= y^{b_2} \cdot g^{b_3} \bmod m, \\
C &:= h^{a_5} \bmod m. & t_3 &:= h^{b_4} \bmod m.
\end{aligned}
$$

Then, we compute $\hat{B} := B \bmod m, v := (A^e/B) \bmod n = y^{a_1 e - a_3} \cdot g^{a_2 e - a_4} \bmod n$ and evaluate the hash value $D = H(y||g||h||A||B||\hat{B}||C||v||t_1||t_2||t_3||M)$. At last, to get the values of E, F and G, we replace the occurrences of $t_1', t_2', t_3', \hat{B}$ and

v in equations (23) by t_1, t_2, t_3, B and $y^{a_1e-a_3} \cdot g^{a_2e-a_4} \bmod n$, respectively, and then we have

$$b_1 = (a_1e - a_3)D + E \bmod \phi(n),$$
$$b_5 = (a_2e - a_4)D \bmod \phi(n),$$
$$b_2 = a_3D + F \bmod \phi(m),$$
$$b_3 = a_4D + G \bmod \phi(m),$$
$$b_4 = a_5D + F \bmod \phi(m).$$

In general, we cannot find a solution for (E, F, G) from the above equation system. However, we can set the ten numbers, i.e., $a_1, \cdots, b_5$, satisfying specific relationships such that the above equation system has one solution. Firstly, we should set $b_5 = 0$. Because D is determined by those ten numbers, we cannot require $b_5 = (a_2e - a_4)D \bmod \phi(n)$ again. $b_5 = 0$ also implies that $a_2e - a_4 = 0$, i.e., $a_4 = a_2e$ (in $\mathbb{Z}$). Secondly, note that F has to satisfy the third and the fifth equations at the same time, so we should set these two equations as the same one. This means that $a_5 = a_3$ and $b_4 = b_2$. Therefore, under the conditions of $b_5 = 0$, $a_4 = a_2e$, $a_5 = a_3$ and $b_4 = b_2$, we get the following solution for (E, F, G) even though we do not know the values of $\phi(m)$ and $\phi(n)$:

$$E := b_1 + (a_3 - a_1e)D \in \mathbb{Z}, \quad F := b_2 - a_3D \in \mathbb{Z}, \quad G := b_3 - a_2eD \in \mathbb{Z}.$$

So an attacker can forge a group signature on any message M as follows:

(1) Select six random numbers: $a_1, a_2, a_3, b_1, b_2, b_3$.
(2) Define $A := y^{a_1} \cdot g^{a_2} \bmod n$, $B := y^{a_3} \cdot g^{a_2e}$, $C := h^{a_3} \bmod m$, $t_1 := y^{b_1} \bmod n$, $t_2 := y^{b_2} \cdot g^{b_3} \bmod m$, and $t_3 := h^{b_2} \bmod m$.
(3) Compute $\hat{B} := B \bmod m$ and $v := (A^e/B) \bmod n = y^{a_1e-a_3} \bmod n$, and then evaluate $D := H(y||g||h||A||B||\hat{B}||C||v||t_1||t_2||t_3||M)$.
(4) Compute $E := b_1 + (a_3 - a_1e)D \in \mathbb{Z}$, $F := b_2 - a_3D \in \mathbb{Z}$, and $G := b_3 - a_2eD \in \mathbb{Z}$.
(5) Output (A, B, C, D, E, F, G) as a group signature for message M.

Forging Certificates. Similarly, we can get the following conditions for a valid membership certificate $(\overline{ID}_i, \bar{x}_i, \bar{z}_i)$:

$$\bar{z}_i^e = \overline{ID}_i \bmod n, \quad \text{and} \quad \overline{ID}_i = g^{\bar{x}_i} \bmod m.$$

These two conditions are the exact equations (19) and (20). So, it seems that valid membership certificates can only be generated jointly by TA and GM. However, for any positive integer k, it is not difficult to see that (1) A group member U_i with certificate (ID_i, x_i, z_i) can generate a valid membership certificates $(ID_i^k, kx_i, z_i^k \bmod n)$, and (2) anyone can use $(\overline{ID} := g^{ke}, \bar{x} := ke, \bar{z} := g^k \bmod n)$ as a valid certificate to forge group signatures on any messages.

Remarks 2. Different attacks on Xia-You scheme are also identified independently by Zhang and Kim [32], and Zhang et al. [33]. The attack in [32] is a special case of our forging signatures, and the two attacks in [33] are weaker than our (universally) forging certificate attack since their attacks can only be mounted by colluding group members.

6   Concluding Remarks

In this paper, by using the same method, we successfully identified different universally forging attacks on several group signature schemes proposed in [25, 27, 18, 31]. That is, our attacks allow anybody (not necessarily a group member) can forge valid group signatures on any messages of his/her choice. Therefore, all these group signature schemes are insecure. Our attacks also implied that no more group signatures should be constructed with such ad-hoc methods used by these insecure schemes. From the contrary side of the same problem, the formal design methodology employed in [1, 4, 5, 22] are further confirmed. In addition, the attacking method described in this paper can be used to test or analyze the security of group signatures in future design, and other signature schemes, such as proxy signatures [30], etc.

References

[1] G. Ateniese, J. Camenisch, M. Joye, and G. Tsudik. A practical and provably secure coalition-resistant group signature scheme. In: *Crypto'2000, LNCS 1880*, pp. 255-270. Springer-Verlag, 2000.

[2] G. Ateniese and G. Tsudik. Some open issues and new directions in group signature schemes. In: *Financial Cryptography (FC'99), LNCS 1648*, pp. 196-211. Springer-Verlag, 1999.

[3] S. Brands. An efficient off-line electronic cash systems based on the representation problem. *Technical Report CS-R9323*, Centrum voor Wiskunde en Inforamatica, April 1993.

[4] J. Camenisch and M. Stadler. Effient group signature schemes for large groups. In: *Crypto'97, LNCS 1294*, pp. 410-424. Springer-Verlag, 1997.

[5] J. Camenisch and M. Michels. A group signature scheme with improved efficiency. In: *Asiacrypt'98, LNCS 1514*, pp. 160-174. Springer-Verlag, 1998.

[6] C.-C. Chang and K.-F. Hwang. Towards the forgery of a group signature without knowing the group center's secret. In: *Information and Communications Security (ICICS'01), LNCS 2229*, pp. 47-51. Springer-Verlag, 2001.

[7] D. Chaum and E. van Heyst. Group Signatures. In: *Eurocrypt'91, LNCS 950*, pp. 257-265. Springer-Verlag, 1992.

[8] M. Joye, N.-Y. Lee, and T. Hwang. On the security of the Lee-Chang group signature scheme and its derivatives. In: *Information Security (ISW'99), LNCS 1729*, pp. 47-51. Springer-Verlag 1999.

[9] M. Joye, S. Kim, and N.-L. Lee. Cryptanalysis of two group signature schemes. In: *Information Security (ISW'99), LNCS 1729*, pp. 271-275. Springer-Verlag 1999.

[10] S.J. Kim, S.J. Park, and D.H. Won. Convertible group signatures.In: *Asiacrypt'96, LNCS 1163*, pp. 311-321. Springer-Verlag, 1996.

[11] W. Lee and C. Chang. Efficient group signature scheme based on the discrete logarithm. *IEE Proc. Comput. Digital Techniques*, 1998, 145 (1): 15-18.

[12] C.H. Lim and P.J. Lee. Remarks on convertible signatures of Asiacrypt'96. *Electronics Letters*, 1997, 33(5): 383-384.

[13] U. Maurer and Y. Yacobi. Non-interactive public-key cryptography. In: *Eurocrypt'91, LNCS 547*, pp. 498-507. Springer-Verlag, 1991.

[14] U. M. Maurer and Y. Yacobi. A non-interactive public-key distribution system. *Designs, Codes and Cryptography*, 1996, 9: 305-316.

[15] K. Nyberg and R.A. Rueppel. Message recovery for signature schemes based on the discrete logarithm problem. *Designs, Codes and Cryptography*, 1996, 7(1-2): 61-81.

[16] S.C. Pohlig and M.E. Hellman. An improved algorithm for computing logarithms over $GF(P)$ and its cryptographic significance. *IEEE Trans. Inform. Theory*, 1978, 24: 106-110.

[17] J.M. Pollard. Theorems on factorization and primality testing. *Proc. Cambridge Philos. Soc.*, 1974, 76: 521-528.

[18] C. Popescu. A modification of the Tseng-Jan group signature scheme. *Studia Univ. Babes-Bolyai, Informatica*, 2000, XLV(2): 36-40. http://www.cs.ubbcluj.ro/ ~studia-i/2000-2/ or http://citeseer.nj.nec.com/504016.html.

[19] S. Saeednia. On the security of a convertible group signature schemes. *Information Processing Letters*, 2000, 73: 93-96.

[20] A. Shamir. Identity-based cryptosystem based on the discrete logarithm problem. In: *Crypto'84, LNCS 196*, pp. 47-53. Springer-Verlag, 1985.

[21] Shi Rong-Hua. An efficient secure group signature scheme. In: *Proc. of TEN-CON'02*, pp. 109-112. IEEE Computer Society, 2002.

[22] D.X. Song. Practical forward secure group signature schemes. In: *Proc. of CCS'01*, pp. 225-234. ACM press, 2001.

[23] H. Sun. Comment: improved group signature scheme based on discrete logarithm problem. *Electronics Letters*, 1999, 35(13): 1323-1324.

[24] Y.-M. Tseng, and J.-K. Jan. Improved group signature based on discrete logarithm problem. *Electronics Letters*, 1999, 35(1): 37-38.

[25] Y.-M. Tseng, and J.-K. Jan. Reply: improved group signature scheme based on discrete logarithm problem. *Electronics Letters*, 1999, 35(20): 1324.

[26] Y.-M. Tseng, and J.-K. Jan. A novel ID-based group signature. In: T.L. Hwang and A.K. Lenstra, editors, *1998 International Computer Symposium, Workshop on Cryptology and Information Security*, Tainan, 1998, pp. 159-164.

[27] Y.-M. Tseng, and J.-K. Jan. A novel ID-based group signature. *Information Sciences*, 1999, 120: 131-141. Elsevier Science.

[28] C.-H. Wang, T. Hwang, and N.-Y. Lee. Comments on two group signatures. *Information Processing Letters*, 1999, 69: 95-97. Elsevier Science.

[29] G. Wang. Security analysis of several group signature schemes (full version of this paper). *Cryptology ePrint Archive*, http://eprint.iacr.org/2003/194/.

[30] G. Wang, F. Bao, J. Zhou, and R.H. Deng. Security analysis of some proxy signatures. *Cryptology ePrint Archive*, http://eprint.iacr.org/2003/196/.

[31] S. Xia, and J. You. A group signature scheme with strong separability. *The Journal of Systems and Software*, 2002, 60(3): 177-182. Elsevier Science.

[32] F. Zhang and K. Kim. Cryptanalysis of two new signature schemes. *Cryptology ePrint Archive*, http://eprint.iacr.org/2002/167/.

[33] J. Zhang, J.-L. Wang, and Y. Wang. Two attacks on Xia-You group signature. *Cryptology ePrint Archive*, http://eprint.iacr.org/2002/177/.

[34] F. Zhang and K. Kim. Security of a new group signature scheme from IEEE TENCON'02. *Technical Reports 2003*, CAIS Lab, Korea.

Forking Lemmas for Ring Signature Schemes⋆

Javier Herranz and Germán Sáez

Dept. Matemàtica Aplicada IV, Universitat Politècnica de Catalunya
C. Jordi Girona, 1-3, Mòdul C3, Campus Nord, 08034-Barcelona, Spain
{jherranz,german}@mat.upc.es

Abstract. Pointcheval and Stern introduced in 1996 some forking lemmas useful to prove the security of a family of digital signature schemes. This family includes, for example, Schnorr's scheme and a modification of ElGamal signature scheme.

In this work we generalize these forking lemmas to the ring signatures' scenario. In a ring signature scheme, a signer in a subset (or *ring*) of potential signers produces a signature of a message in such a way that the receiver can verify that the signature comes from a member of the ring, but cannot know which member has actually signed.

We propose a new ring signature scheme, based on Schnorr signature scheme, which provides unconditional anonymity. We use the generalized forking lemmas to prove that this scheme is existentially unforgeable under adaptive chosen-message attacks, in the random oracle model.

1 Introduction

Group-oriented cryptography deals with those situations in which a secret task (signing or decrypting) is performed by a group of entities or on behalf of such a group. Threshold cryptography is an approach to this situation. In a threshold scheme, some participants have shares of the unique secret key of the group. Participation of some determined subset of players is required to perform the corresponding secret task in a correct way.

Two related but different approaches are *ring signatures* and *group signatures*. In a ring signature scheme, an entity signs a message on behalf of a set (or ring) of members that includes himself. The verifier of the signature is convinced that it was produced by some member of the ring, but he does not obtain any information about which member of the ring actually signed. The real signer includes in the signature the identities of the members of the ring that he chooses, depending on his purposes, and probably without their consent.

The idea behind group signature schemes is very similar to that of ring signatures, but with some variations. First of all, there exists a group manager in charge of the join and revocation of the members in the group. Therefore, a user cannot modify the composition of the group. And second, some mechanisms

---

⋆ This work was partially supported by Spanish *Ministerio de Ciencia y Tecnología* under project TIC 2000-1044.

T. Johansson and S. Maitra (Eds.): INDOCRYPT 2003, LNCS 2904, pp. 266–279, 2003.

are added in order to allow (only) the group manager to recover the real identity of the signer of a message, for example in the case of a legal dispute.

Although the formalization and the name of ring signatures schemes have been recently given in [17], first proposals of such schemes can be found in [8,9,5]. Furthermore, some efficient proposals of group signature schemes [7,6,2] are constructed using as a basis the ring signature scheme that appears in Definition 2 of [5], by adding the necessary elements to achieve revocability of the anonymity on the part of the group manager.

Analogously to individual signature schemes, the highest proposed level of security exigible to ring signature schemes as well as group signature schemes is existential unforgeability under adaptive chosen-message attacks. The recent proposals of ring signature schemes in [17,4] reach this level of computational security, based on the hardness of the RSA problem, in the random oracle model [3].

In this paper, we extend to the ring signatures' scenario the forking lemmas introduced in [15] to prove the security of the Schnorr signature scheme. We propose a new ring signature scheme based on Schnorr signature scheme [18] which provides unconditional anonymity. We use the extended forking lemmas to prove that this scheme is existentially unforgeable under adaptive chosen-message attacks, in the random oracle model, assuming the hardness of the discrete logarithm problem in subgroups of prime order.

A different but related approach to ring signature schemes for discrete-log settings can be found in [1]. They consider a scenario in which the discrete-log parameters of each participant are different. The resulting scheme is less efficient than ours. However, they also propose a scheme for the particular case where the public parameters of all the participants are equal, which is as efficient as our scheme. The security of this last scheme is not explicitly proved, although the authors asserts that this can be done using reduction techniques similar to those that we use in this work. We do a formal and detailed study of the exact security of our proposed scheme.

The paper is organized as follows. In Section 2, we explain the general characteristics of a ring signature scheme, and the security properties that such a scheme must satisfy. In Section 3 we extend to the ring signatures' scenario some techniques introduced by Pointcheval and Stern in [15]. In Section 4 we propose our new ring signature scheme and prove its security (unconditional anonymity and existential unforgeability under adaptive chosen-message attacks). Finally, we sum up the work in Section 5.

2   Ring Signatures

Following the formalization about ring signatures proposed in [17], we explain in this section the basic definitions and the properties exigible to ring signature schemes, although not rigorously.

Each potential user A_i generates his pair of secret/public keys (sk_i, pk_i) by using a key generation protocol that takes as input a security parameter. The public keys of all the users are certified via a public key infrastructure.

A regular operation of a ring signature scheme consists of the execution of the two following algorithms:

Ring-sign: if a user A_s wants to compute a ring signature on behalf of a ring $A_1, \ldots, A_n$ that contains himself, he executes this probabilistic algorithm with input a message m, the public keys $pk_1, \ldots, pk_n$ of the ring and his secret key sk_s. The output of this algorithm is a ring signature σ for the message m.

Ring-verify: this is a deterministic algorithm that takes as input a message m and a ring signature σ, that includes the public keys of all the members of the corresponding ring, and outputs "True" if the ring signature is valid, or "False" otherwise.

The resulting ring signature scheme must satisfy the following properties:

Correctness: a ring signature generated in a correct way must be accepted by any verifier with overwhelming probability.

Anonymity: any verifier should not have probability greater than $1/n$ to guess the identity of the real signer who has computed a ring signature on behalf of a ring of n members. If the verifier is a member of the ring distinct from the actual signer, then his probability to guess the identity of the real signer should not have greater than $1/(n-1)$.

Unforgeability: among all the proposed definitions of unforgeability (see [13]), we consider the strongest one: any attacker must not have non-negligible probability of success in forging a valid ring signature for some message m on behalf of a ring that does not contain him, even if he knows valid ring signatures for messages, different from m, that he can adaptively choose.

The first proposals of ring signature schemes are previous to the formal definition of this concept. They can be found in [8,5] and they are used as a tool to construct group signature schemes. They use zero-knowledge proofs and witness indistinguishable proofs of knowledge for disjunctive statements (introduced in [9,10]).

In [17], Rivest, Shamir and Tauman formalize the concept of ring signature schemes, and propose a scheme which they prove existentially unforgeable under adaptive chosen-message attacks, in the ideal cipher model, assuming the hardness of the RSA problem [16]. This scheme also uses a symmetric encryption scheme and the notion of combining functions.

Bresson, Stern and Szydlo show in [4] that the scheme in [17] can be modified in such a way that the new scheme can be proved to achieve the same level of security, but under the strictly weaker assumption of the random oracle model. Furthermore, they propose a threshold ring signature scheme, in which a set of t users sign a message on behalf of a ring that contains themselves, in such a way that the verifier is convinced of the participation of t users in the generation of the signature, but he does not obtain any information about which t users have in fact signed the message.

Finally, in [1], Abe, Ohkubo and Suzuki give general constructions of ring signature schemes for a variety of scenarios, including those where signature schemes are based on one-way functions, and those where signature schemes are of the three-move type (for example, Schnorr's signature scheme).

3   Forking Lemmas for Generic Ring Signatures

In this section we prove some lemmas that we will use later to demonstrate the security of our proposal for a Schnorr ring signature scheme.

In [15], Pointcheval and Stern prove the security of a class of signature schemes, that they call *generic*, which includes Schnorr [18] and a modification of ElGamal [11] schemes. They introduce the *forking lemmas*, which are based on a reduction technique that they call *oracle replay attack*.

Our goal is to extend all these results to the ring signatures' scenario.

3.1   Generic Ring Signature Schemes

We define a class of ring signature schemes that we call also *generic*, and for which the results in this section are valid. Consider a security parameter k, a hash function which outputs k-bit long elements, and a ring $A_1, \ldots, A_n$ of n members. Given the input message m, a generic ring signature scheme produces a tuple $(m, R_1, \ldots, R_n, h_1, \ldots, h_n, \sigma)$, where $R_1, \ldots, R_n$ (randomness) take their values randomly in a large set G in such a way that $R_i \neq R_j$ for all $i \neq j$, h_i is the hash value of (m, R_i), for $1 \leq i \leq n$, and the value σ is fully determined by $R_1, \ldots, R_r, h_1, \ldots, h_n$ and the message m.

Another required condition is that no R_i can appear with probability greater than $2/2^k$, where k is the security parameter. This condition can be achieved by choosing the set G as large as necessary.

The security proofs of this paper are valid in the *random oracle model* [3], in which a cryptographic hash function is supposed to behave as a random and hidden function that outputs values independently of the input (the only restriction is that equal inputs must produce equal outputs). In the framework that we consider, the outputs of the random oracle will be k-bit long elements.

The basic idea of the forking lemmas in [15] and in the ring forking lemmas that we are going to introduce is the following: assuming that an attacker can forge a generic ring signature, another attacker could obtain, by replaying enough times the first attacker with randomly chosen hash functions (i.e. random oracles), two forged ring signatures of the same message and with the same randomness. Then, these two forged signatures could be used to solve some computational problem which is assumed to be intractable. In this way, the corresponding ring signature scheme is proved to be existentially unforgeable under no-message attacks. Some precautions must be taken in order to achieve unforgeability under chosen-message attacks, which is the standard level of security that a signature scheme can achieve (see [13]).

3.2   No-Message Attacks

Lemma 1 *Let Σ_{ring} be a generic ring signature scheme with security parameter k, and let n be the number of members of the corresponding ring. Let the forger $\mathcal{A}$ be a probabilistic polynomial time Turing machine whose input only consists of public data and which can ask Q queries to the random oracle, with $Q \geq n$. We denote as $V_{Q,n}$ the number of n-permutations of Q elements, that is, $V_{Q,n} = Q(Q-1) \cdot \ldots \cdot (Q-n+1)$. We assume that, within time bound T, $\mathcal{A}$ produces, with probability of success $\varepsilon \geq \frac{7 \, V_{Q,n}}{2^k}$, a valid ring signature $(m, R_1, \ldots, R_n, h_1, \ldots, h_n, \sigma)$. Then, within time $T' \leq \frac{16 V_{Q,n} T}{\varepsilon}$, and with probability $\varepsilon' \geq \frac{1}{9}$, a replay of this machine outputs two valid ring signatures $(m, R_1, \ldots, R_n, h_1, \ldots, h_n, \sigma)$ and $(m, R_1, \ldots, R_n, h'_1, \ldots, h'_n, \sigma')$ such that $h_j \neq h'_j$, for some $j \in \{1, \ldots, n\}$ and $h_i = h'_i$ for all $i = 1, \ldots, n$ such that $i \neq j$.*

Proof. The Turing machine $\mathcal{A}$, with random tape ω, can ask Q queries to the random oracle f. We denote by $\mathcal{Q}_1, \ldots, \mathcal{Q}_Q$ the Q distinct questions and by $\rho = (\rho_1, \ldots, \rho_Q)$ the list of the Q answers of the random oracle f. So we can see a random choice of the random oracle f as a random choice of such a vector ρ.

Now, for a random choice of (ω, f) and with probability ε, the machine $\mathcal{A}$ outputs a valid ring signature $(m, R_1, \ldots, R_n, h_1, \ldots, h_n, \sigma)$. Since f is a random oracle and its outputs are k-bit long elements, the probability that there exists some index i such that $\mathcal{A}$ has not asked the query (m, R_i) to the random oracle is less than $n/2^k$. Therefore, with probability at least $1 - n/2^k$, $\mathcal{A}$ has asked all the queries (m, R_i) to the oracle, for $1 \leq i \leq n$, and so we have that $Q \geq n$ is necessary.

With probability at least $\varepsilon - n/2^k$, the machine $\mathcal{A}$ is successful in forging a ring signature $(m, R_1, \ldots, R_n, h_1, \ldots, h_n, \sigma)$ and besides it has asked to the random oracle all the queries (m, R_i), for $i = 1, \ldots, n$. In this case, for all index i there exists an integer $\ell_i \in \{1, 2, \ldots, Q\}$ such that the query $\mathcal{Q}_{\ell_i}$ is precisely (m, R_i). Then, we define $L(\omega, f) = (\ell_1, \ell_2, \ldots, \ell_n)$ and $\beta(\omega, f) = \max\{\ell_i \mid (\ell_1, \ell_2, \ldots, \ell_n) = L(\omega, f)\}$. Note that, since the forged ring signature is a valid generic one, we have that all the R_i's are different, and so the integers ℓ_i are also all different.

Note that the fact that we must deal with sets of n indexes instead of a unique index is the main difference (and difficulty) with respect to the forking lemmas in [15] for individual signatures.

In the unlikely case where $\mathcal{A}$ has not asked some of the pairs (m, R_i) to the random oracle, then we say that $\ell_i = \infty$, and so $\beta(\omega, f) = \infty$. Now we define the sets

$$\mathcal{S} = \{(\omega, f) \mid \mathcal{A}(\omega, f) \text{ succeeds and } \beta(\omega, f) \neq \infty\},$$

$$\mathcal{S}_{\boldsymbol{\ell}} = \{(\omega, f) \mid \mathcal{A}(\omega, f) \text{ succeeds and } L(\omega, f) = \boldsymbol{\ell}\},$$

for all the vectors $\boldsymbol{\ell} \in L_n = \{(\ell_1, \ell_2, \ldots, \ell_n) \mid 1 \leq \ell_i \leq Q \text{ and } \ell_i \neq \ell_j \; \forall i \neq j\}$. Note that the cardinality of this set L_n is the number of n-permutations of Q elements, that is, $V_{Q,n} = Q(Q-1) \cdot \ldots \cdot (Q-n+1)$. Furthermore, the set

$\{S_\ell \mid \ell \in L_n\}$ is a partition of S. The pairs (ω, f) in the set S are called the successful pairs. We can find the lower bound $\nu = \Pr[S] \geq \varepsilon - \frac{n}{2^k} \geq \frac{6\varepsilon}{7}$.

Let I be the set formed by the most likely vectors, $I = \{\ell \in L_n \mid \Pr[S_\ell \mid S] \geq \frac{1}{2V_{Q,n}}\}$. Probabilities are taken over the random choice of (ω, f). The following lemma asserts that, in case of success, the corresponding vector of indexes lies in I with probability at least $\frac{1}{2}$.

Lemma 2 $\Pr[L(\omega, f) \in I \mid S] \geq \frac{1}{2}$.

Proof. Since the sets S_ℓ are disjoint, we have that $\Pr[L(\omega, f) \in I \mid S] = \sum_{\ell \in I} \Pr[S_\ell \mid S]$. This probability is equal to $1 - \sum_{\ell \notin I} \Pr[S_\ell \mid S]$. Since the complement of I contains fewer than $V_{Q,n}$ vectors, this probability is at least $1 - V_{Q,n} \cdot \frac{1}{2V_{Q,n}} = \frac{1}{2}$. □

The following lemma will be used in the same way as Pointcheval and Stern do (see [15] for the proof):

Lemma 3 *(The Splitting Lemma) Let $A \subset X \times Y$ such that $\Pr[(x, y) \in A] \geq \epsilon$. For any $\alpha < \epsilon$, define*

$$B = \{(x, y) \in X \times Y \mid \Pr_{y' \in Y}[(x, y') \in A] \geq \epsilon - \alpha\}$$

then the following statements hold:

1. $\Pr[B] \geq \alpha$.
2. *for any* $(x, y) \in B$, $\Pr_{y' \in Y}[(x, y') \in A] \geq \epsilon - \alpha$.
3. $\Pr[B \mid A] \geq \alpha/\epsilon$.

Now we run $2/\varepsilon$ times the attacker $\mathcal{A}$ with random ω and random f. Since $\nu = \Pr[S] \geq \frac{6\varepsilon}{7}$, with probability greater than $1 - (1 - 6\varepsilon/7)^{2/\varepsilon} \geq 1 - e^{-12/7} \geq \frac{4}{5}$, we get at least one pair (ω, f) in S.

For each vector $\ell \in I$, if we denote by β_ℓ the maximum of the coordinates of ℓ, we can apply the Splitting Lemma (Lemma 3). Following the notation of this lemma, and if we see the oracle f as a random vector $(\rho_1, \ldots, \rho_Q)$, then $A = S_\ell$, $X = \{(\omega, \rho_1, \ldots, \rho_{(\beta_\ell - 1)})\}_{\omega, f}$ and $Y = \{(\rho_{\beta_\ell}, \ldots, \rho_Q)\}_f$. We also refer to $(\rho_1, \ldots, \rho_{(\beta_\ell - 1)})$ as f_{β_ℓ}, the restriction of f to queries of index strictly less than β_ℓ. Since $\Pr[S_\ell] = \Pr[S] \cdot \Pr[S_\ell \mid S] \geq \frac{\nu}{2V_{Q,n}}$, we take $\epsilon = \frac{\nu}{2V_{Q,n}}$ and $\alpha = \frac{\nu}{4V_{Q,n}}$, and the Splitting Lemma proves that there exists a subset Ω_ℓ of executions (ω, f) such that, for any $(\omega, f) \in \Omega_\ell$,

$$\Pr_{f'}[(\omega, f') \in S_\ell \mid f_{\beta_\ell} = f'_{\beta_\ell}] \geq \frac{\nu}{4V_{Q,n}} \tag{1}$$

$$\text{and} \quad \Pr[\Omega_\ell \mid S_\ell] \geq \frac{1}{2}$$

Using again that the sets S_ℓ are all disjoint, we have that

$$\Pr_{\omega, f}[\exists \ell \in I \text{ s.t. } (\omega, f) \in \Omega_\ell \cap S_\ell \mid S] = \Pr\left[\bigcup_{\ell \in I}(\Omega_\ell \cap S_\ell) \mid S\right] =$$

$$= \sum_{\ell \in I} \Pr[\Omega_\ell \cap S_\ell \mid S] = \sum_{\ell \in I} \Pr[\Omega_\ell \mid S_\ell] \cdot \Pr[S_\ell \mid S] \geq \left(\sum_{\ell \in I} \Pr[S_\ell \mid S]\right)/2 \geq \frac{1}{4}.$$

For simplicity, we denote by ℓ the vector $L(\omega, f)$ corresponding to the successful pair (ω, f) obtained in the first $2/\varepsilon$ repetitions of the attack $\mathcal{A}$ with probability at least $4/5$, and by β the corresponding index $\beta(\omega, f)$. As we have seen, with probability at least $1/4$, we have that $\ell \in I$ and $(\omega, f) \in \Omega_\ell \cap \mathcal{S}_\ell$. Therefore, with probability greater than $1/5$, the $2/\varepsilon$ repetitions of the attack have provided a successful pair (ω, f), with $\ell = L(\omega, f) \in I$ and $(\omega, f) \in \Omega_\ell \cap \mathcal{S}_\ell$.

If now we replay the attack, with fixed random tape ω but randomly chosen oracle f' such that $f'_\beta = f_\beta$, we can use inequality (1) and thus we obtain that

$$\Pr_{f'}[(\omega, f') \in \mathcal{S}_\ell \text{ and } \rho_\beta \neq \rho'_\beta \mid f'_\beta = f_\beta] \geq \Pr_{f'}[(\omega, f') \in \mathcal{S}_\ell \mid f'_\beta = f_\beta] - \Pr[\rho_\beta = \rho'_\beta] \geq$$

$$\geq \frac{\nu}{4V_{Q,n}} - \frac{1}{2^k} \geq \frac{\varepsilon}{14V_{Q,n}} ,$$

where $\rho_\beta = f(\mathcal{Q}_\beta)$ and $\rho'_\beta = f'(\mathcal{Q}_\beta)$. If we now replay the attack $14V_{Q,n}/\varepsilon$ times with fixed ω and random oracle f' such that $f'_\beta = f_\beta$, we will get another success (or forking) $(\omega, f') \in \mathcal{S}_\ell$ with probability greater than $3/5$.

Summing up, after less than $\frac{2}{\varepsilon} + \frac{14V_{Q,n}}{\varepsilon} \leq \frac{16V_{Q,n}}{\varepsilon}$ executions of the machine $\mathcal{A}$, and with probability greater than $\frac{1}{5} \cdot \frac{3}{5} \geq \frac{1}{9}$, we obtain two valid ring signatures $(m, R_1, \ldots, R_n, h_1, \ldots, h_n, \sigma)$ and $(m, R'_1, \ldots, R'_n, h'_1, \ldots, h'_n, \sigma')$ from two executions of $\mathcal{A}$ with the same random tape ω (that is, with the same randomness $R_1 = R'_1, \ldots, R_n = R'_n$), but with two different random oracles f and f', that we can see as two different vectors $\rho = (\rho_1, \ldots, \rho_Q) = (f(\mathcal{Q}_1), \ldots, f(\mathcal{Q}_Q))$ and $\rho' = (\rho'_1, \ldots, \rho'_Q) = (f'(\mathcal{Q}_1), \ldots, f'(\mathcal{Q}_Q))$. These two oracles verify, furthermore, that $\rho_t = \rho'_t$, for all $t = 1, \ldots, \beta - 1$, and $\rho_\beta \neq \rho'_\beta$, where β is the index $\beta(\omega, f)$ corresponding to the first successful forgery performed by $\mathcal{A}$.

Therefore, if we denote as j the index such that (m, R_j) was the query $\mathcal{Q}_\beta$, then we have that $h_j = f(\mathcal{Q}_\beta) \neq f'(\mathcal{Q}_\beta) = h'_j$. However, the rest of pairs (m, R_i), for $i = 1, \ldots, n$, $i \neq j$, have been asked before the query $\mathcal{Q}_\beta$, and so the values obtained from the oracles f and f' have been the same. That is, $h_i = h'_i$, for all $i = 1, \ldots, n$ with $i \neq j$. $\qquad \square$

Theorem 1 *(The No-Message Ring Forking Lemma). Let Σ_{ring} be a generic ring signature scheme with security parameter k, and let n be the number of members of the corresponding ring. Let the forger $\mathcal{A}$ be a probabilistic polynomial time Turing machine whose input only consists of public data. We denote by Q the number of queries that $\mathcal{A}$ can ask to the random oracle. Assume that, within time bound T, $\mathcal{A}$ produces, with probability of success $\varepsilon \geq \frac{7\,V_{Q,n}}{2^k}$, a valid ring signature $(m, R_1, \ldots, R_n, h_1, \ldots, h_n, \sigma)$. Then there is another probabilistic polynomial time Turing machine which uses $\mathcal{A}$ and produces two valid ring signatures $(m, R_1, \ldots, R_n, h_1, \ldots, h_n, \sigma)$ and $(m, R_1, \ldots, R_n, h'_1, \ldots, h'_n, \sigma')$ such that $h_j \neq h'_j$, for some $j \in \{1, \ldots, n\}$ and $h_i = h'_i$ for all $i = 1, \ldots, n$ such that $i \neq j$, in expected time $T' \leq \frac{84480\,T V_{Q,n}}{\varepsilon}$.*

Proof. The idea is to construct a specific expected polynomial time Turing machine $\mathcal{B}$ that uses the Turing machine $\mathcal{A}$ as a sub-routine in order to obtain two valid ring signatures of the same message and with the desired properties (same randomness and all the h_i's but one equal). And then we must calculate the expectation of the random variable that counts the number of times that the machine $\mathcal{A}$ is invoked by $\mathcal{B}$. The design of $\mathcal{B}$ and the computation of this expectation can be performed exactly in the same way as in [15], changing their value Q by our value $V_{Q,n}$. The resulting expectation is less than $84480 V_{Q,n}/\varepsilon$. If T is a time bound for the machine $\mathcal{A}$, then a time bound for the machine $\mathcal{B}$ will be $\frac{84480 T V_{Q,n}}{\varepsilon}$. $\qquad\qquad\square$

3.3   Chosen-Message Attacks

Now we consider another kind of attacks against a ring signature scheme, the chosen-message attacks. They are the strongest ones usually considered, so if a ring signature scheme is proved to be unforgeable against them, then we can say that the scheme achieves the standard level of security.

In a chosen-message attack, an adversary is given the public data of the scheme (including the members of the ring and their public keys). Then he can ask to some real signers of the ring for valid ring signatures of a polynomial number of messages of his choice. A same message can be asked more than once, and the choice of the messages is adaptive, in the sense that the attacker can adapt his queries according to previous message-signature pairs.

If the attacker obtains, after this interaction with the signers, with non-negligible probability and in polynomial time a valid ring signature on a message that has not been previously signed by the real signers, then we say that the attack is successful, or that the ring signature scheme is *existentially forgeable under chosen-message attacks*.

If we want to prove the security of a ring signature scheme in the random oracle model, then the considered attacker will be able to ask a polynomial number of queries to the random oracle model, too.

The following theorem is a variation of Theorem 1 considering chosen-message attacks.

Theorem 2 *(The Chosen-Message Ring Forking Lemma). Let Σ_{ring} be a generic ring signature scheme with security parameter k, and let n be the number of members of the corresponding ring. Let $\mathcal{A}$ be a probabilistic polynomial time Turing machine whose input only consists of public data. We denote by Q and W the number of queries that $\mathcal{A}$ can ask to the random oracle and to some real signers of the ring, respectively. Assume that, within time bound T, $\mathcal{A}$ produces, with probability of success $\varepsilon \geq \frac{12\ V_{Q,n}+6(Q+Wn)^2}{2^k}$, a valid ring signature $(m, R_1, \dots, R_n, h_1, \dots, h_n, \sigma)$. Suppose that valid ring signatures can be simulated with a polinomially indistinguishable distribution of probability, with time bound T_s and without knowing any of the secret keys of the ring. Then there is another probabilistic polynomial time Turing machine which has control over the machine obtained from $\mathcal{A}$ by replacing interactions with the real signers by simulation, and which produces two valid ring signatures $(m, R_1, \dots, R_n, h_1, \dots, h_n, \sigma)$ and $(m, R_1, \dots, R_n, h'_1, \dots, h'_n, \sigma')$ such that $h_j \neq h'_j$, for some $j \in \{1, \dots, n\}$*

and $h_i = h_i'$ for all $i = 1, \ldots, n$ such that $i \neq j$, in expected time $T' \leq \frac{144823 V_{Q,n}(T + W T_s)}{\varepsilon}$.

Proof. We consider a machine $\mathcal{B}$ that executes the machine $\mathcal{A}$, in such a way that $\mathcal{B}$ simulates all the environment of $\mathcal{A}$. Therefore, $\mathcal{B}$ must simulate the interactions of $\mathcal{A}$ with the random oracle and with real signers. Then we could see $\mathcal{B}$ as a machine performing a no-message attack against the ring signature scheme.

We denote by $\mathcal{Q}_1, \ldots, \mathcal{Q}_Q$ the Q distinct queries of $\mathcal{A}$ to the random oracle, and by $m^{(1)}, \ldots, m^{(W)}$ the W queries (possibly repeated) to the real signers. Using the simulator, $\mathcal{B}$ can perfectly simulate the answers of the real ring of signers without knowing any of the secret keys of the ring. For a message $m^{(j)}$, the simulator answers a tuple $(m^{(j)}, R_1^{(j)}, \ldots, R_n^{(j)}, h_1^{(j)}, \ldots, h_n^{(j)}, \sigma^{(j)})$. Then $\mathcal{B}$ constructs a random oracle f by storing in a "random oracle list" the relations $f(m^{(j)}, R_i^{(j)}) = h_i^{(j)}$. The attacker $\mathcal{A}$ receives this signature, assumes that $f(m^{(j)}, R_i^{(j)}) = h_i^{(j)}$, where f is the random oracle, for all $1 \leq i \leq n$ and $1 \leq j \leq W$, and stores all these relations. When $\mathcal{A}$ makes a query $\mathcal{Q}_t = (m, R)$ to the random oracle, $\mathcal{B}$ looks for the value (m, R) in the random oracle list. If the value is already in the list, then $\mathcal{B}$ returns to $\mathcal{A}$ the corresponding $f(m, R)$. Otherwise, $\mathcal{B}$ chooses a random value h, sends it to $\mathcal{A}$ and stores the relation $f(m, R) = h$ in the list.

There is some risk of "collisions" of queries to the random oracle. In the definition of generic ring signature schemes, we made the assumption that no R_i can appear with probability greater than $2/2^k$ in a ring signature. If the simulator outputs ring signatures which are indistinguishable of the ones produced by a real signer of the ring, then we have that no $R_i^{(j)}$ can appear with probability greater than $2/2^k$ in a simulated ring signature, too. Since the values $h_i^{(j)}$ are the outputs of the random oracle, then we have that a determined $h_i^{(j)}$ appears in a ring signature (real or simulated) with probability less than $1/2^k$.

Then, three kinds of collisions can occur:

- A pair $(m^{(j)}, R_i^{(j)})$ that the simulator outputs, as part of a simulated ring signature, has been asked before to the random oracle by the attacker. In this case, it is quite unlikely that the relation $f(m^{(j)}, R_i^{(j)}) = h_i^{(j)}$ corresponding to the values output by the simulator coincides with the relation previously stored in the random oracle list. The probability of such a collision is, however, less than $Q \cdot nW \cdot \frac{2}{2^k} \leq \frac{\varepsilon}{6}$.

- A pair $(m^{(j_1)}, R_{i_1}^{(j_1)})$ that the simulator outputs, as part of a simulated ring signature, is exactly equal to another pair $(m^{(j_2)}, R_{i_2}^{(j_2)})$ also output by the simulator. The probability of this collision is less than $\frac{(nW)^2}{2} \cdot \frac{2}{2^k} \leq \frac{\varepsilon}{6}$.

- Two answers h_1 and h_2 of the random oracle chosen at random by $\mathcal{B}$ are exactly equal, while the two corresponding inputs $(m^{(1)}, R_1)$ and $(m^{(2)}, R_2)$ are different. The probability of such an event is less than $\frac{(Q+nW)^2}{2} \cdot \frac{1}{2^k} \leq \frac{\varepsilon}{12}$.

Altogether, the probability of collisions is less than $5\varepsilon/12$. Now we can compute:

$$\Pr_{(\omega,f)}[\mathcal{B} \text{ succeeds}] = \Pr_{(\omega,f)}[\text{no-collisions in the simulations and } \mathcal{A} \text{ succeeds}] \geq$$

$$\geq \Pr_{(\omega,f)}[\mathcal{A} \text{ succeeds} \mid \text{no-collisions in the simulations }] -$$

$$- \Pr_{(\omega,f)}[\text{collisions in the simulations}] \geq \varepsilon - \frac{5\varepsilon}{12} = \frac{7\varepsilon}{12} \ .$$

Summing up, we have a machine $\mathcal{B}$ that performs a no-message attack against the ring signature scheme with time bound $T + WT_s$ and with probability of success greater than $\frac{7\varepsilon}{12} \geq \frac{7V_{Q,n}}{2^k}$. So we can use Theorem 1 applied to the machine $\mathcal{B}$, and we will obtain two valid ring signatures in expected time bounded by $\frac{84480(T+WT_s)V_{Q,n}}{7\varepsilon/12} \leq \frac{144823V_{Q,n}(T+WT_s)}{\varepsilon}$.   □

4   Schnorr Ring Signature Scheme

In this section we present a ring signature scheme based on Schnorr signature scheme [18]. We first remind how this individual signature scheme works.

Let p and q be large primes such that $q|p-1$ and $q \geq 2^k$, where k is the security parameter of the scheme. Let g be an element of $\mathbb{Z}_p^*$ with order q, and let $H()$ be a collision resistant hash function which outputs elements in $\mathbb{Z}_q$.

A signer has a private key $x \in \mathbb{Z}_q^*$ and the corresponding public key $y = g^x \bmod p$. To sign a message m, the signer:

1. Chooses a random $a \in \mathbb{Z}_q^*$.
2. Computes $R = g^a \bmod p$ and $\sigma = a + xH(m, R) \bmod q$.
3. Defines the signature on m to be the pair (R, σ).

The validity of the signature is verified by the recipient by checking that $g^\sigma = Ry^{H(m,R)} \bmod p$.

In [15], Pointcheval and Stern proved that, in the random oracle model, an existential forgery under an adaptive chosen-message attack of Schnorr's scheme is equivalent to solving the discrete logarithm problem in the subgroup $< g >$ generated by the element g. The input of this problem is a tuple (p, q, g, y) such that $y \in < g >$, where g is an element of order q in $\mathbb{Z}_p$, and q is a prime that divides $p - 1$ (where p is also a prime number). The solution of the problem is the only element $x \in \mathbb{Z}_q$ such that $y = g^x \bmod p$. The discrete logarithm problem in subgroups of prime order is supposed to be computationally intractable.

4.1   The Proposed Scheme

The security parameter k, the public parameters p, q and g, and the hash function $H()$ are defined as in the Schnorr signature scheme.

Consider a set, or ring, of potential signers $A_1, \ldots, A_n$. Every potential signer A_i has a private key $x_i \in \mathbb{Z}_q^*$ and the corresponding public key $y_i = g^{x_i} \bmod p$.

Ring-sign: to sign a message m on behalf of the ring $A_1, \ldots, A_n$, a signer A_s, where $s \in \{1, \ldots, n\}$, acts as follows:

1. For all $i \in \{1, \ldots, n\}$, $i \neq s$, choose a_i at random in $\mathbb{Z}_q^*$, pairwise different. Compute $R_i = g^{a_i} \bmod p$, for all $i \neq s$.
2. Choose a random $a \in \mathbb{Z}_q$.
3. Compute $R_s = g^a \prod_{i \neq s} y_i^{-H(m, R_i)} \bmod p$. If $R_s = 1$ or $R_s = R_i$ for some $i \neq s$, then go to step 2.
4. Compute $\sigma = a + \sum_{i \neq s} a_i + x_s H(m, R_s) \bmod q$.
5. Define the signature of the message m made by the ring $A_1, \ldots, A_n$ to be $(m, R_1, \ldots, R_n, h_1, \ldots, h_n, \sigma)$, where $h_i = H(m, R_i)$, for all $1 \leq i \leq n$. Note that the values $h_1, \ldots, h_n$ can be derived from the rest of values in the signature, and so they do not need to be part of the final signature. We include them for clarity in the security proofs.

Ring-verify: the validity of the signature is verified by the recipient of the message by checking that $h_i = H(m, R_i)$ and that

$$g^\sigma = R_1 \cdot \ldots \cdot R_n \cdot y_1^{h_1} \cdot \ldots \cdot y_n^{h_n} \bmod p.$$

The property of correctness is satisfied. In effect, if the ring signature has been correctly generated, then the verification result is always "True":

$$R_1 \cdot \ldots \cdot R_n \cdot y_1^{h_1} \cdot \ldots \cdot y_n^{h_n} = g^a y_s^{h_s} \prod_{i \neq s} R_i = g^{a + x_s h_s + \Sigma_{i \neq s} a_i} = g^\sigma \bmod p.$$

4.2   Anonymity of the Scheme

In order to prove that our ring signature scheme is unconditionally anonymous, we must show that any attacker outside a ring of n possible users has probability $1/n$ to guess which member of the ring has actually computed a given signature on behalf of this ring.

Let $Sig = (m, R_1, \ldots, R_n, h_1, \ldots, h_n, \sigma)$ be a valid ring signature of a message m. That is, $h_i = H(m, R_i)$ and $g^\sigma = R_1 \cdot \ldots \cdot R_n \cdot y_1^{h_1} \cdot \ldots \cdot y_n^{h_n}$. Let A_s be a member of the ring. We now find the probability that A_s computes exactly the ring signature Sig, when he produces a ring signature of message m by following the method explained in Section 4.1.

The probability that A_s computes the correct $R_i \neq 1$ of Sig, pairwise different for $1 \leq i \leq n$, $i \neq s$, is $\frac{1}{q-1} \cdot \frac{1}{q-2} \cdot \ldots \cdot \frac{1}{q-n+1}$. Then, the probability that A_s chooses exactly the only value $a \in \mathbb{Z}_q$ that leads to the value R_s of Sig, among all possible values for R_s different to 1 and different to all R_i with $i \neq s$, is $\frac{1}{q-n}$.

Summing up, the probability that A_s generates exactly the ring signature Sig is

$$\frac{1}{q-1} \cdot \frac{1}{q-2} \cdot \ldots \cdot \frac{1}{q-n+1} \cdot \frac{1}{q-n} = \frac{1}{V_{q-1,n}}$$

and this probability does not depend on A_s, so it is the same for all the members of the ring. This fact proves the unconditional anonymity of the scheme.

4.3   Unforgeability of the Scheme

Proposition 1 *The ring signatures produced by the scheme proposed in Section 4.1 can be simulated in polynomial time, without knowing any of the secret keys of the ring, and with distribution of probability polinomially indistinguishable of ring signatures produced by a legitimate signer, in the random oracle model.*

Proof. The simulation of a Schnorr ring signature for a message m goes as follows:

1. Choose at random an index $s \in \{1, \ldots, n\}$.
2. For all $i \in \{1, \ldots, n\}$, $i \neq s$, choose a_i at random in $\mathbb{Z}_q^*$, pairwise different. Compute $R_i = g^{a_i} \bmod p$, for all $i \neq s$.
3. Choose independently and at random $h_1, h_2, \ldots, h_n$ in $\mathbb{Z}_q$.
4. Choose at random $\sigma \in \mathbb{Z}_q$.
5. Compute $R_s = g^{\sigma - \sum_{i \neq s} a_i} y_1^{-h_1} \ldots y_n^{-h_n}$. If $R_s = 1$ or $R_s = R_i$ for some $i \neq s$, then go to step 4.
6. Return the tuple $(m, R_1, \ldots, R_n, h_1, \ldots, h_n, \sigma)$.

It is easy to see that this simulation runs in polynomial time. We denote by T_s the time bound for an execution of each simulation. Note that, if we assume $H(m, R_i) = h_i$ (we are in the random oracle model), for all $i \in \{1, \ldots, n\}$, then the returned tuple is a valid Schnorr ring signature of the message m.

On the other hand, the statistic distance between the distribution of ring signatures generated by using the protocol explained in Section 4.1, and the distribution of ring signatures simulated as above is not only negligible, but exactly zero. Therefore, these two distributions are statistically indistinguishable, and so polynomially indistinguishable, as desired.   □

Theorem 3 *Let $\mathcal{A}$ be a probabilistic polynomial time Turing machine which obtains an existential forgery of the Schnorr ring signature scheme presented in Section 4.1, within a time bound T and under an adaptive chosen-message attack, with probability of success ε. Let n be the number of members of the ring. We denote respectively by Q and W the number of queries that $\mathcal{A}$ can ask to the random oracle and to the signing oracle. Assuming that $\varepsilon \geq \frac{12\ V_{Q,n} + 6(Q+Wn)^2}{2^k}$ (otherwise, ε would be negligible), then the discrete logarithm problem in subgroups of prime order can be solved within expected time less than $\frac{144823V_{Q,n}(T+WT_s)}{\varepsilon}$.*

Proof. Let (p, q, g, y) the input of an instance of the discrete logarithm problem in the subgroup $< g >$ of $\mathbb{Z}_p$ of order q, where q is a prime that divides $p - 1$.

We choose at random $\alpha_i \in \mathbb{Z}_q^*$ pairwise different, for $1 \leq i \leq n$, and define $y_i = y^{\alpha_i} \bmod p$. Then we initialize the attacker $\mathcal{A}$ with a ring of members $A_1, \ldots, A_n$ and corresponding public keys $y_1, \ldots, y_n$. Since our Schnorr ring signature scheme can be simulated, we can apply Theorem 2, and so we can obtain from a replay of attacker $\mathcal{A}$ two valid ring signatures $(m, R_1, \ldots, R_n, h_1,$

$\ldots, h_n, \sigma)$ and $(m, R_1, \ldots, R_n, h'_1, \ldots, h'_n, \sigma')$ such that $h_j \neq h'_j$, for some $j \in \{1, \ldots, n\}$ and $h_i = h'_i$ for all $i = 1, \ldots, n$ such that $i \neq j$. Then we have that

$$g^{\sigma} = R_1 \cdot \ldots \cdot R_n \cdot y_1^{h_1} \cdot \ldots \cdot y_j^{h_j} \cdot \ldots \cdot y_n^{h_n}$$

$$g^{\sigma'} = R_1 \cdot \ldots \cdot R_n \cdot y_1^{h'_1} \cdot \ldots \cdot y_j^{h'_j} \cdot \ldots \cdot y_n^{h'_n}$$

Dividing these two equations, we obtain $g^{\sigma - \sigma'} = y_j^{h_j - h'_j} = y^{\alpha_j(h_j - h'_j)}$, and so we have that

$$y = g^{\frac{\sigma - \sigma'}{\alpha_j(h_j - h'_j)}} \bmod p.$$

Therefore, we have found the discrete logarithm of y in base g.

$\square$

We note that this theorem gives the exact security of our ring signature scheme, i.e. the exact relation between the successful probabilities and the execution times of both attacks against our scheme and the discrete logarithm problem. A negative point is the presence of the factor $V_{Q,n}$ in the reduction coefficients, because this factor is exponential in the number n of members of the ring. Therefore, the size of the ring should be limited in some way. However, this does not seem to be a problem in realistic situations, where the number of possible signers is assumed to be bounded.

5   Conclusions

We have proposed a new ring signature scheme for the discrete-log setting, which we have proved to be unconditionally anonymous and existentially unforgeable, in the random oracle model, under adaptive chosen-message attacks, assuming the hardness of the discrete logarithm problem in subgroups of prime order. For proving these results, we have extended to the ring signatures' scenario some security lemmas introduced in [15] to prove the security of some generic signature schemes.

The forking lemmas in [15] can be applied in any signature scheme obtained from a honest-verifier zero-knowledge identification protocol, for example the ones by Schnorr [18], Fiat-Shamir [12], or Guillou-Quisquater [14]. Analogously, our extension of the forking lemmas to the ring signatures' scenario, that we have applied to a particular Schnorr ring signature scheme, could be used to prove the security of future ring signature schemes constructed from these three-move signature schemes.

Acknowledgments

The authors wish to thank Jacques Stern and the anonymous referees for their helpful comments.

References

1. M. Abe, M. Ohkubo and K. Suzuki. 1−out−of−n signatures from a variety of keys. *Advances in Cryptology-Asiacrypt'02*, LNCS **2501**, Springer-Verlag, pp. 415–432 (2002).

2. G. Ateniese, J. Camenisch, M. Joye and G. Tsudik. A practical and provably secure coalition-resistant group signature scheme. *Advances in Cryptology-Crypto'00*, LNCS **1880**, Springer-Verlag, pp. 255–270 (2000).

3. M. Bellare and P. Rogaway. Random oracles are practical: a paradigm for designing efficient protocols. *First ACM Conference on Computer and Communications Security*, pp. 62–73 (1993).

4. E. Bresson, J. Stern and M. Szydlo. Threshold Ring Signatures for Ad-hoc Groups. *Advances in Cryptology-Crypto'02*, LNCS **2442**, Springer-Verlag, pp. 465–480 (2002).

5. J. Camenisch. Efficient and generalized group signatures. *Advances in Cryptology-Eurocrypt'97*, LNCS **1233**, Springer-Verlag, pp. 465–479 (1997).

6. J. Camenisch and M. Michels. A group signature scheme with improved efficiency. *Advances in Cryptology-Asiacrypt'98*, LNCS **1514**, Springer-Verlag, pp. 160–174 (1998).

7. J. Camenisch and M. Stadler. Efficient group signature schemes for large groups. *Advances in Cryptology-Crypto'97*, LNCS **1294**, Springer-Verlag, pp. 410–424 (1997).

8. D. Chaum and E. van Heyst. Group signatures. *Advances in Cryptology-Eurocrypt'91*, LNCS **547**, Springer-Verlag, pp. 257–265 (1991).

9. R. Cramer, I. Damgård and B. Schoenmakers. Proofs of partial knowledge and simplified design of witness hiding protocols. *Advances in Cryptology-Crypto'94*, LNCS **839**, Springer-Verlag, pp. 174–187 (1994).

10. A. De Santis, G. Di Crescenzo, G. Persiano and M. Yung. On monotone formula closure of SZK. *Proceedings of FOCS'94*, IEEE Press, pp. 454–465 (1994).

11. T. ElGamal. A public key cryptosystem and a signature scheme based on discrete logarithms. *IEEE Transactions on Information Theory* **31**, pp. 469–472 (1985).

12. A. Fiat and A. Shamir. How to prove yourself: practical solutions of identification and signature problems. *Advances in Cryptology-Crypto'86*, LNCS **263**, Springer-Verlag, pp. 186–194 (1986).

13. S. Goldwasser, S. Micali and R. Rivest. A digital signature scheme secure against adaptative chosen-message attacks. *SIAM Journal of Computing*, **17 (2)**, pp. 281–308 (1988).

14. L.C. Guillou and J.-J. Quisquater. A practical zero-knowledge protocol fitted to security microprocessor minimizing both transmission and memory. *Advances in Cryptology-Eurocrypt'88*, LNCS **330**, Springer-Verlag, pp. 123–128 (1988).

15. D. Pointcheval and J. Stern. Security arguments for digital signatures and blind signatures. *Journal of Cryptology*, Vol. **13** (3), pp. 361–396 (2000).

16. R.L. Rivest, A. Shamir and L. Adleman. A method for obtaining digital signatures and public key cryptosystems. *Communications of the ACM*, **21**, pp. 120–126 (1978).

17. R. Rivest, A. Shamir and Y. Tauman. How to leak a secret. *Advances in Cryptology-Asiacrypt'01*, LNCS **2248**, Springer-Verlag, pp. 552–565 (2001).

18. C.P. Schnorr. Efficient signature generation by smart cards. *Journal of Cryptology*, Vol. **4**, pp. 161–174 (1991).

Practical Mental Poker Without a TTP Based on Homomorphic Encryption

Jordi Castellà-Roca[1], Josep Domingo-Ferrer[2], Andreu Riera[1], and
Joan Borrell[3]

[1] Scytl Online World Security S.A., Entença 95 4-1, E-08015 Barcelona, Catalonia.
{jordi.castella,andreu.riera}@scytl.com
http://www.scytl.com
[2] Universitat Rovira i Virgili, Dept. of Computer Engineering and Maths,
Av. Països Catalans 26, E-43007 Tarragona, Catalonia.
jdomingo@etse.urv.es
[3] Universitat Autònoma de Barcelona, Dept. of Computer Science,
E-08193 Bellaterra, Catalonia.
joan.borrell@uab.es

Abstract. A solution for obtaining impartial random values in on-line gambling is presented in this paper. Unlike most previous proposals, our method does not require any TTP and allows e-gambling to reach standards of fairness, security an auditability similar to those common in physical gambling.

Although our solution is detailed here for the particular case of games with reversed cards (*e.g.* poker), it can be easily adapted for games with open cards (*e.g.* blackjack) and for random draw games (*e.g.* keno). Thanks to the use of permutations of homomorphically encrypted cards, the protocols described have moderate computational requirements.

Keywords: Mental poker, E-gambling, Privacy homomorphisms.

Categories: Applications of cryptography (e-gambling), Multi-party computation.

1 Introduction

Computer networks and especially the Internet have allowed some common activities such as shopping, information search o gambling to become remote. This paper is about gambling over the Internet, also called e-gambling, and more specifically about mental poker or e-poker. E-gambling has a number of advantages for players, because it is space-independent (there is no need to physically go to the casino) and time-independent (an Internet casino can be available 24 hours a day or at least longer than physical casinos). The drawback of e-gambling is the difficulty of guaranteeing the same standards of security, fairness and auditability offered by physical gambling.

In a physical casino, each player sees the actions performed by other players, so that most unfair actions can immediately be detected and reported. In the digital world, things are more complex, as discussed below:

T. Johansson and S. Maitra (Eds.): INDOCRYPT 2003, LNCS 2904, pp. 280–294, 2003.

Card draw In a physical casino, any player can draw a card without the other players seeing the card she gets. In an Internet casino, however, unless some kind of encryption is used, a third party can see the card gotten by any player.

Randomness In a physical casino, physical devices can be used to obtain true randomness (roulettes, card shuffling, dice, etc.). The fairness, the unbiasedness and the outcomes of such devices can be verified by all players (at least in theory). In an on-line casino, the best that can be achieved is pseudorandomness and verifying fairness is much less trivial. Pseudo-random values should be generated in a way that they cannot be manipulated. A common way to achieve this goal is to use a Trusted Third Party (TTP). The usual setting is for the on-line casino to act as a TTP while, at the same time, playing an active role in the game. In this case, the on-line casino is in a privileged position, whereas players are helpless and cannot verify actions by other players or by the casino. As reported in [13], unlawful manipulation in remote gambling is quite common. A better option is for random values to be jointly generated by all participants using a cryptographic protocol. This is the approach proposed in this paper; our cryptographic protocols guarantee that, if the random value has been manipulated by some participant, such manipulation is detected.

Auditability Physical casinos are equipped with CCTV or other recording systems which allow dispute resolution when there is disagreement between players and the casino or between players themselves as to the development or outcome of the game. Auditing misbehaviors in on-line casinos is less obvious. For example, imagine the casino refuses payment to a player after the latter wins a game; unless special IT security measures are built in the on-line casino, the player might end up without any valid proof that might enable her to prove in court that she has been abused. In general, enough information on the development of the game should be logged so that posterior analysis of that information allows resolution of any dispute that may arise. Integrity and time-stamping of logfiles should be guaranteed, *i.e.* addition, deletion or modification of any log entry should be detectable; otherwise, the logfile owner could alter log entries at will. On the other hand, each log entry should be signed by the party having generated it, to prevent later repudiation.

1.1   Our Contribution

In this paper, we propose a method to shuffle and deal a deck of cards to the players that does not require any TTP, and allows e-gambling to reach standards of fairness, security and auditability similar to those common in physical gambling. Unlike for other TTP-free proposals, eliminating the TTP does not dramatically increase the computational complexity of our solution.

Casino games fall into three groups:

- Random draw games, with a single draw (*e.g.* dice, roulette) or with multiple draws (*e.g.* bingo, keno).

- Games where a value or a set of values are obtained in a non-secret way. Games where cards are visible (*e.g.* blackjack) fall into this category.
- Games where a value or a set of values are obtained in a secret way. Games where cards are reversed (*e.g.* poker) fall into this category.

There are several good solutions for the first and second categories. Two significant examples are [17] and [23]. Our contribution relates on the third category, which is the most complex one. Specifically, our method will be illustrated here on the poker game. Further detail and applications to other game categories can be found in patent [3].

Section 2 recalls previous work on secure e-gambling; special emphasis is made on mental poker, which is the name used in the literature for secure e-poker. Our solution is described in detail in Section 3. A security analysis considering various possible attacks is reported in Section 4. Section 5 is a conclusion.

2   Previous Work

Difficult problems have a special appeal. Mental poker is one such problem and has received substantial attention over the past two decades. In 1981, Shamir *et al.* [24] presented the first mental poker protocol, which is restricted to two players. In [21] and [5], it was shown that this protocol uses a cryptosystem allowing cards to be marked, which can leak information to the opponent and thus be fatal. In 1982, a new protocol was proposed in [14] which uses probabilistic encryption and prevents card marking, but is still limited to two players. Barany *et al.* proposed in 1983 a protocol for several players [1]. This protocol does not use cryptographic primitives and relies exclusively on permutations; its main drawback is that player confabulations are possible. Two years later, the problem of player confabulation was solved by Fortune and Merritt [12] using a TTP. In fact, most recent proposals [15,22,16,28,4] also use a TTP, because it is argued that using a TTP is the only way to obtain mental poker protocols which are both fair and reasonably efficient.

On the theoretical side, several solutions have been proposed, some of which do not require a TTP. However, those solutions share the drawback of requiring too much computation to be usable in practice. Along this line, Crépeau [6] presented in 1985 a mental poker protocol that minimizes the probability of successful player coalition, and one year later in [7] the same author presented an improved protocol that offers player confidentiality. This latter protocol uses a Zero-Knowledge-Proof (ZKP) subprotocol, which needs a substantial computing time. Later Kurosawa *et al.* in [18] and [19] and Schindelhauer in [25] also propose solutions that use ZKP. As an illustration of the amount of computing time needed by this kind of protocols, in [11] an implementation of the protocol [7] on three Sparc workstations is reported to have taken eight hours to shuffle a deck. Even though all the aforementioned proposals are very time consuming, [7,18,19] at least have the theoretical interest of not requiring disclosure of players' strategy after the game, a feature not offered by the protocol presented in this paper.

As mentioned above, we present a ZKP-free and TTP-free mental poker protocol which can be used in practical gaming scenarios, because it requires an amount of computation which is not much higher than the computation required by proposals using a TTP.

3   A Protocol Suite for E-gambling with Reversed Cards

In our protocol suite for e-games with reversed cards, deck shuffling is carried out in an efficient and fair way by the players themselves.

All players co-operate in shuffling, so that no player or player coalition can force a particular outcome, *i.e.* determine the card that will be obtained after shuffling. Every player generates a random permutation of the card deck and keeps it secret; the player then commits to her permutation using a bit commitment protocol. The shuffled deck is formed by the composition of all player permutations. The use of permutations to shuffle the deck was introduced in [1], and also was used in [6,7].

Note that reversing cards in the physical world translates to encrypting cards in e-gambling. Now, permuting (*i.e.* shuffling) encrypted cards requires encryption to be homomorphic, so that the outcome of permuting and decrypting (*i.e.* opening) a card is the same that would be obtained if the card had been permuted without prior encryption (*i.e.* reversal). The use of an additive homomorphic cryptosystem to shuffle the deck of cards and maintain the privacy of the cards was initially proposed in [18].

3.1   Notation

The following notation will be used in what follows:

- PL_i: i-th player
- $m_1|m_2$: Concatenation of messages m_1 and m_2.
- P_{entity}, S_{entity}: Asymmetric key pair of *entity*, where P_{entity} is the public key and S_{entity} is the private key.
- $S_{entity}\{m\}$: Digital signature of message m by *entity*, where digital signature means computing the hash value of message m using a collision-free one-way hash function and encrypting this hash value under S_{entity}.
- K_{entity}: Secret symmetric key of *entity*.
- $E_{K_{entity}}\{m\}$: Encryption of message m under K_{entity}.
- $D_{K_{entity}}\{c\}$: Decryption of message c under K_{entity}.

3.2   Card Representation and Permutation

In most e-gambling approaches, a prescribed ordering of cards in the deck is assumed, so that a card is represented by a scalar corresponding to its rank. In our protocol, a card representation is needed which allows card operations and permutations. Thus, we will map the usual scalar representation to a vector representation in the way described below.

Definition 1 (Card vector representation). *Let t be the number of cards in the deck. Let z be a prime number chosen by a player. A card can be represented as a vector*

$$v = (a_1, \cdots, a_t) \tag{1}$$

where there exists a unique $i \in \{1, \cdots, t\}$ for which a_i mod $z \neq 0$, whereas $\forall j \neq i$ it holds that a_j mod $z = 0$. The value of the card is i; assuming a prescribed ordering of cards, i is interpreted as a rank identifying a particular card.

The above vector representation for cards allows card permutations to be represented as matrices in a natural way.

Definition 2 (Card permutation matrix). *A permutation π over a deck of t cards is a bijective mapping that can be represented as a square matrix Π with t rows called* card permutation matrix, *where rows and columns are vectors of the form described by Expression (1):*

$$\Pi = \begin{pmatrix} \pi_{11} & \pi_{12} & \cdots & \pi_{1t} \\ \vdots & \ddots & & \vdots \\ \vdots & & \ddots & \vdots \\ \pi_{t1} & \pi_{t2} & \cdots & \pi_{tt} \end{pmatrix} \tag{2}$$

The i-th row of matrix Π is card $\pi(i)$, i.e. the card resulting from applying permutation π to the card having rank i. Thus, all elements in the i-th row of Π are 0 mod z except π_{ij}, where $j = \pi(i)$.

See Example 1 in Appendix A for an illustration of the above definition. The result below is straightforward from the above construction:

Proposition 1 (Permutation algebra). *With the above representation for cards and permutations, the result $w = \pi(v)$ of permuting a card v using a permutation π can be computed in vector representation as $w = v \cdot \Pi \pmod{z}$, where $\cdot$ denotes vector product. For this computation to work properly, the same value z must be used to represent v and Π.*

Example 2 in Appendix A illustrates the above proposition. Every player PL_i will use her own prime modulus z_i for representing and permuting cards. In order for PL_i to be able to operate her card permutation matrix with a card coming from another player PL_j, player PL_i must represent her permutation matrix using the modulus z_j corresponding to PL_j. This means transforming her permutation matrix based on z_i into an equivalent permutation matrix based on z_j, as defined below:

Definition 3 (Equivalent card permutation matrices). *Let Π and Π' be two $t \times t$ card permutation matrices using moduli z and z', respectively. Then Π and Π' are said to be equivalent if they represent the same permutation π of a* *set of t cards. Specifically, $\Pi = \{\pi_{ij}\}$ and $\Pi' = \{\pi'_{ij}\}$ are equivalent if and only if, $\forall i, j \in \{1, \cdots, t\}$, one has*

$$\pi_{ij} \bmod z \neq 0 \Leftrightarrow \pi'_{ij} \bmod z' \neq 0$$

$$\pi_{ij} \bmod z = 0 \Leftrightarrow \pi'_{ij} \bmod z' = 0$$

3.3   Protocol Description

In order to meet the game auditability requirement, we introduce in this paper a new tool called distributed notarization chains (DNC). DNCs have a philosophy similar to the one of Lamport's hash chains [20] and their construction is described in Appendix B. Whenever a player builds a link of a DNC, she sends it the other players, so that all of them see the same DNC. Furthermore, each link is digitally signed by the player who built it, which guarantees authentication, integrity and non-repudiability for that link. A link can only be appended at the end of the DNC and no participant can modify or delete any link without being detected.

The initialization protocol is as follows:

Protocol 1 (Initialization)

1. *Each player PL_i is assumed to have an asymmetric key pair (P_i, S_i) whose public key has been certified by a recognized certification authority. Assume the card deck consists of t cards.*
2. *PL_i does:*
   (a) *Generate a permutation π_i of the card deck and keep it secret.*
   (b) *Generate a symmetric secret key K_i corresponding to a homomorphic cryptosystem allowing algebraic operations (additions and multiplications) to be carried out directly on encrypted data. Security requirements on this homomorphism are limited to resistance against ciphertext-only attacks; possible choices are [9,10].*
   (c) *Choose a prime value z_i which falls within the range of the cleartext space of the homomorphic cryptosystem used.*
   (d) *Build a link of the DNC which contains the value z_i used by PL_i. Link building can be done in parallel by all players, which expands the DNC at this point.*
   (e) *Build the card permutation matrix Π_i corresponding to π_i, using z_i.*
   (f) *Commit to this permutation Π_i using a bit commitment protocol [26]. Denote the resulting commitment by Cp_i.*
   (g) *Build the next link of the DNC following the previous expanded link. The new link contains the commitment Cp_i.*
   (h) *Choose s values $\{\delta_1, \cdots, \delta_s\}$ such that $\delta_j \bmod z_i = 0$, $\forall j \in \{1, \cdots, s\}$ and $s > t$.*
   (i) *Choose s values $\{\epsilon_1, \cdots, \epsilon_s\}$ such that $\epsilon_j \bmod z_i \neq 0$, $\forall j \in \{1, \cdots, s\}$ and $s > t$.*
   (j) *Homomorphically encrypt the previous values under the symmetric key K_i to get $d_j = E_{K_i}(\delta_j)$ and $e_j = E_{K_i}(\epsilon_j)$, $\forall j \in \{1, \cdots, s\}$.*

(k) *Build the next link of the DNC which contains the set* $\mathbf{D} = \{d_1, \cdots, d_s\}$.
(l) *Build another link of the DNC which contains the set* $\mathbf{E} = \{e_1, \cdots, e_s\}$.
(m) *Generate the vector representation for the t cards in the deck* $\{w_1, \cdots, w_t\}$ *and encrypt them under* K_i *using the aforementioned homomorphic cryptosystem to obtain* $w'_j = E_{K_i}(w_j)$.
(n) *Randomly permute the encrypted cards* $\{w'_1, \cdots, w'_t\}$.
(o) *Build the next link of the DNC which contains the card deck encrypted and permuted by* PL_i.

Once initialization is over, the player playing the croupier role performs chain contraction by building a link with a chaining value mixing together the last link created by every player PL_i. This link indicates that initialization is over and that the game can start.

Note that, even though each player encrypts the deck under her key during initialization, the deck of the game is a single one, namely a shuffled deck resulting from the composition of all players' secret permutations. Also, the initialization protocol results in each player PL_i publishing in the DNC her z_i and a commitment to her *secret* permutation Π. When PL_i wants a card, the following protocol is started:

Protocol 2 (Card draw)

1. PL_i *does:*
   (a) *Pick an integer value* v_0 *such that it falls within the range of cards in the deck, i.e.* $1 \leq v_0 \leq t$, *and which has not previously been requested. This operation is simple because it is public. All participants know the initial values chosen in previous steps.*
   (b) *Build the next link of the DNC which contains the vector representation* w_0 *of card value* v_0 *chosen by* PL_i.
2. PL_1 *does:*
   (a) *Check the validity of the link sent by* PL_i, *compute her equivalent card permutation* Π'_1 *for the modulus* z_i *published by* PL_i *in the DNC and permute* w_0 *to obtain* $w_1 = w_0 \cdot \Pi'_1$.
   (b) *Build the next link of the DNC, which contains* w_1 *and the name of the next player* PL_2 *in the computation.*
3. *For* $j = 2$ *to* $i - 1$, *player* PL_j *does:*
   (a) *Check the validity of the link sent by* PL_{j-1}, *compute her equivalent card permutation matrix* Π'_j *for the modulus* z_i *published by* PL_i *and permute* w_{j-1} *to obtain* $w_j = w_{j-1} \cdot \Pi'_j$.
   (b) *Build the next link of the DNC, which contains* w_j *and the name of the next player* PL_{j+1} *in the computation.*
4. *Player* PL_i *does:*
   (a) *Check the validity of the link sent by* PL_{i-1} *and permute* w_{i-1} *with her permutation matrix* Π_i *to obtain* $w_i = w_{i-1} \cdot \Pi_i$.
   (b) *Modify the m-th row of* Π_i *where* $m \in \{1, \cdots, t\}$ *is the value of card* w_{i-1}. *All values in the m-th row are changed to values that are nonzero modulo* z_i

(c) *Pick the encrypted card w_i' corresponding to clear card w_i. Note that the encrypted deck has been published in the last step of Protocol 1.*

(d) *Build the next link of the DNC, which contains w_i' and the name of the next player PL_{i+1} in the computation.*

5. *For $j = i + 1$ to n, player PL_j does:*

(a) *Check the validity of the link sent by PL_{j-1} and compute her equivalent card permutation matrix Π_j' for the modulus z_i published by PL_i. Use Protocol 3 below to encrypt Π_j' as Π_j^c under the key K_i corresponding to PL_i.*

(b) *Permute the encrypted card w_{j-1}' using her encrypted matrix Π_j^c to obtain $w_j' = w_{j-1}' \cdot \Pi_j^c$.*

(c) *Build the next link of the DNC, which contains w_j'. If $j < n$, the link also indicates the name of the next player PL_{j+1} in the computation.*

6. *When PL_i sees the link computed by PL_n, she does:*

(a) *Check the validity of the link computed by PL_n.*

(b) *Decrypt the card w_n' contained in the link computed by PL_n under her private key K_i to obtain the drawn card $w_n = D_{K_i}(w_n')$. This finishes the card draw protocol.*

An extension of Protocol 2 for drawing several cards simultaneously (multi-card) is described in Appendix C. A loose end that remains is how does PL_j encrypt her permutation matrix Π_j' under PL_i's secret key K_i in Step (5a) of Protocol 2. To do that, player PL_j can only use the sets of encrypted values $\mathbf{D}$ and $\mathbf{E}$ published by PL_i in Step (2k) of Protocol 1. The specific procedure is described below:

Protocol 3 (Permutation matrix homomorphic encryption)

Let the homomorphic encryption of the cleartext matrix $\Pi_j = \{\pi_{kl}\}$ under K_i be $\Pi_j^c = \{\pi_{kl}^c\}$. To compute π_{kl}^c, for $1 \leq k, l \leq t$, player PL_j does the following:

1. *Generate a pseudorandom value g, such that $1 \leq g \leq s$, where s is the size of the sets $\mathbf{D}, \mathbf{E}$ of encrypted values published by PL_i in Protocol 1.*

2. *Randomly pick g values $\{d_1, \cdots, d_g\}$ of the set $\mathbf{D}$ and add them to obtain $h = d_1 + \cdots + d_g$. Remember that values in $\mathbf{D}$ are 0 modulo z_i, because they are homomorphically encrypted versions of values which are 0 modulo z_i, so by the homomorphic properties, h is also 0 modulo z_i.*

3. *Generate a pseudorandom value c such that $c \bmod z_i \neq 0$ and compute $h' = c \cdot h$.*

4. *If $\pi_{kl} \bmod z_j = 0$, then $\pi_{kl}^c := h'$.*

5. *If $\pi_{kl} \bmod z_j \neq 0$, then*

(a) *Generate a pseudorandom value g' such that $1 \leq g' \leq s$.*

(b) *$\pi_{kl}^c := h' + e_{g'}$, where $e_{g'}$ is the g'-th element of the set E.*

A player can also discard a card w by building a link of the DNC which says that the player is discarding a card and contains the encrypted version of the discarded card.

3.4   Game Validation

When a hand of the game is over, players should reveal their encryption keys and their permutations. The validation process is specified next:

Protocol 4 (Game validation)
Each player does the following:

1. *Check that the permutation revealed and used by each other player PL_i is the same permutation π_i to which she committed when publishing the commitment Cp_i in Protocol 1 (initialization). This check implies verifying the bit commitment for player PL_i.*
2. *Decrypt cards $\{w'_1, \ldots, w'_t\}$ published by each other player PL_i in the last step of Protocol 1 and check that the card deck is correct.*
3. *Use the private key K_i of each other player PL_i to decrypt the result of permuting encrypted cards at Step (5b) of Protocol 2. Check that permutations were correctly performed.*
4. *Check that cards discarded by other players have not been used during the game.*
5. *If necessary, use the DNC to prove any detected misbehaviors by any other player to a third-party (casino, court, etc.).*

4   Security Analysis

Let us examine a collection of possible attacks and check that they fail:

A coalition wants to get the cards drawn by a player Player PL_i draws one or more cards in an instance of Protocol 2. In subsequent instances belonging to the same hand, a coalition of players attempt to determine the cards drawn by PL_i. To do that, the coalition must construct a (multi)card (see Appendix C) with the card value(s) drawn by PL_i and encrypt that (multi)card. However, the coalition will not get PL_i's card(s) because, if a card is requested which had been previously requested, what is obtained is a vector with all components different from 0 modulo z_i. This is due to the modification of matrix Π_i at Step (4b) of Protocol 2. The resulting multicard has all components different from 0 modulo z_i and is thus a non-valid multicard (by definition, see Appendix C).

A player uses a wrong modulus Assume that, to permute a card coming from player PL_{k-1}, player PL_k computes her equivalent permutation matrix Π'_k using $z_j \neq z_{k-1}$. This is detected by PL_i in the last step of Protocol 2, because w'_n decrypts into a non-valid card (with too many nonzero components modulo z_i). Player PL_i has no option other than reporting the wrong decryption; otherwise PL_i would not be able to show her cards during game validation. Verification performed during Protocol 4 discloses the identity of the cheater PL_k.

A player does not use the permutation she committed to   A player may choose to use a permutation Π different from the one she committed to during initialization.

If the player changes her permutation for the whole game, this is detected during game validation. If the player changes her permutation only during some parts of the game, two things may happen:

- Some of the other players get duplicated cards. A player getting a duplicated card is forced to report it (otherwise she will not be able to show her cards during game validation). Upon such report, the game is stopped and game validation started.
- The change is not detected during the game. In this case, it is detected during game validation and the dishonest player is identified.

A player supplies wrong sets D or E If, during initialization, a player does not supply correct sets $\mathbf{D} = \{d_1, \cdots, d_s\}$ or $\mathbf{E} = \{e_1, \cdots, e_s\}$, then permutations of encrypted cards cannot be correctly computed. Two things can happen:

- The card obtained by PL_i at the end of Protocol 2 is not valid. In this case, PL_i is forced to report the problem.
- The card obtained by PL_i is valid. In this case, the problem is detected during game validation.

A player supplies a wrong encrypted deck The game validation protocol verifies that all players have supplied correct encrypted decks during initialization.

A player builds an incorrect DNC link During execution of Protocol 2, each player checks that the previous link of the DNC has correctly been built. Any wrong link is reported (all players see all links so it is risky not to report a wrong link when discovered).

A player requests a card which had already been requested Cards requested at the beginning of Protocol 2 are recorded in the DNC. If a player requests a card that had already been requested, this is detected by the rest of players (all of them see the DNC links).

A player does not correctly encrypt her permutation An incorrectly encrypted card permutation matrix can yield non-valid or duplicated cards, which is detected during the game. Even if all cards are valid, a wrongly encrypted permutation is detected during game validation.

Player withdrawal Depending on when a player withdraws from the game, different things happen:

- If a player withdraws during initialization, the game simply proceeds without her.
- If a player withdraws immediately after initialization, the game proceeds without her. In this case, the composition of permutations must be computed without using the permutation of the withdrawn player.
- If a player withdraws during the game (*i.e.* during Protocol 2), the game stops at the moment of withdrawal. Players show their cards and game validation is started. If the player has withdrawn without justification, she may be fined.

5 Conclusion

A solution for obtaining impartial random values in on-line gambling has been presented in this paper. Unlike most previous proposals, our method does not require any TTP and allows e-gambling to reach levels of fairness, security and auditability similar to those common in physical gambling. In addition, eliminating the TTP does not result in a dramatical increase of computation.

The solution has been specified for the particular case of games with reversed cards (*e.g.* poker), but it can be easily adapted for games with open cards (*e.g.* blackjack) and for random draw games (*e.g.* keno).

Appendix A. Examples of Card Representation and Permutation

Example 1. Let $t = 5$ and consider the following permutation of five cards

$$\pi = (3\ 5\ 4\ 1\ 2) \tag{3}$$

If $z = 5$ is taken, a possible matrix representation of π is as follows

$$\Pi = \begin{pmatrix} 5 & 10 & 3 & 5 & 15 \\ 25 & 10 & 35 & 5 & 6 \\ 50 & 10 & 10 & 8 & 5 \\ 4 & 60 & 5 & 50 & 25 \\ 15 & 7 & 35 & 60 & 10 \end{pmatrix} \tag{4}$$

Note that, in the first row, the only element which is nonzero modulo 5 is the third one, that is, the one in the position corresponding to $\pi(1) = 3$. Similarly, in the second row, the only nonzero element modulo 5 is in the 5-th position, because $\pi(2) = 5$. And so on for the other rows.

Example 2. Using $t = 5$, $z = 5$ and the permutation matrix of Example 1, the card $v = (10, 15, 8, 20, 20)$ (vector representation modulo 5 of a card with value 3) is permuted as

$$w = (10, 15, 8, 20, 20) \cdot \Pi = (1205, 1670, 1435, 2389, 980) \tag{5}$$

Since the only component of the permuted card w that is nonzero modulo z is the fourth one, w is the vector representation of a card with value 4 (the fourth card of the ranked deck). In this way, $\pi(3) = 4$, which is consistent with the fact that the only component that is nonzero modulo z in the third row of the permutation matrix is the fourth one.

Appendix B. Distributed Notarization Chains (DNCs)

Each operation performed during the e-game will be notarized as a link of a distributed notarization chain (DNC). DNCs are efficiently computable and consist of links which are constructed and chained as follows:

Link structure Every link m_k of the DNC is formed by two fields: a data field D_k and a chaining value X_k. The data field D_k consists of three subfields:
- Timestamp T_k, which contains the link generation time according to the clock of the participant who generated the link (synchronizing the clocks of all participants is not needed).
- Link subject or concept C_k, which describes the information contained in the link, *e.g.* a step in the game, a commitment or an outcome.
- Additional attributes V_k, which depend on the subject C_k. For a link corresponding to a commitment, V_k will contain the encrypted commitment.

Link chaining Chaining is guaranteed by chaining values X_k included in each link. First, the chaining value X_{k-1} of the previous link is concatenated with the data field D_k of the current link; then the hash value of the concatenation is computed and signed with the private key of the author of the k-th link, *i.e.*

$$X_k = S_{author}\{D_k|X_{k-1}\} \tag{6}$$

There are moments during the game at which operations do not need to be sequential, but can be carried out in parallel by the participants. Parallel execution can be accomodated in the distributed notarization system by introducing two operations:

Chain expansion Parallel execution can be notarized by *expanding the DNC*, whereby participants $\{PL_1, \cdots, PL_n\}$ independently compute their chaining values $X_k^{PL_1}, \cdots, X_k^{PL_n}$ using the chaining value X_{k-1} of the previous link:

$$X_k^{PL_1} = S_{PL_1}\{D_k^{PL_1}|X_{k-1}\}$$

$$\vdots$$

$$X_k^{PL_n} = S_{PL_n}\{D_k^{PL_n}|X_{k-1}\} \tag{7}$$

Chain contraction It happens when the protocol requires sequential execution after a stage of expansion where n participants have computed links in parallel. A single link is obtained which is chained to previous links. The participant initiating the sequential execution concatenates its data field D_k with all chaining values of previous links $\{X_{k-1}^{PL_1}|\ldots|X_{k-1}^{PL_n}\}$, computes the hash of the concatenation and signs it to obtain the chaining value X_k of the first sequential link:

$$X_k = S_{PL_i}\{D_k|X_{k-1}^{PL_1}|\cdots|X_{k-1}^{PL_n}\} \tag{8}$$

The following properties of a DNC make it a good tool for distributed notarization:

- The DNC is not possessed by a single participant, but by all of them. Whenever a participant builds a link of a DNC, she sends it to the other participants, so that all of them see the same DNC. If a participant recomputes the chain to add false links to it, the manipulated chain will not match the copies held by the rest of participants, and manipulation will be detected.
- If someone deletes or modifies one or more links, the chain will show an inconsistency at the point of deletion.
- The chaining and the structure of links allow the exact sequence and time of link construction to be securely determined.
- As links are signed by the participant who built them, link authorship can be securely determined.
- Link computation is performed in parallel whenever the protocol allows it (using the expansion operation). This results in improved performance without degrading security.

Appendix C. Extensions

In each run of Protocol 2, player PL_i gets only one card. The protocol can be extended so that the player draws several cards in a single protocol run. To do this, let us define a way to pack several cards together:

Definition 4 (Multicard). *Given a deck with t cards and a prime value z, a multicard ξ is a vector of t elements*

$$\xi = (a_1, \cdots, a_t) \tag{9}$$

where there are up to $M < t$ components a_i, such that $a_i \bmod z \neq 0$. The index i of each component a_i that is nonzero modulo z represents one of the cards contained in the multicard. By convention, a multicard with more than M components that are nonzero modulo z is not valid.

Protocol 2 can be adapted to multicards as explained below:

- At Step (1a) of Protocol 2, PL_i should choose the cards she wishes. Let the vector representation of these be $w_0^1, \cdots, w_0^x$, with $x < M$.
- A multicard $\xi_0 = \sum_{j=1}^{x} w_0^j$ is obtained by adding the chosen cards.
- At Step (1b), the DNC link would be computed using the multicard ξ_0 rather than a single card w_0.
- The protocol carries on until Step (4b), where all rows corresponding to values of cards in the multicard are modified.
- The protocol then proceeds as described above. In the final Step (6b), PL_i decrypts the card computed by PL_n and obtains a multicard.

Acknowledgments

The second author is partly supported by the Spanish Ministry of Science and Technology and the European FEDER fund under project TIC2001-0633-C03-01 "STREAMOBILE". Thanks go to Jordi Herrera for useful comments on some parts of this work.

References

1. I. Barany and Z. Furedi, "Mental poker with three or more players", Technical report, Mathematical Institute of the Hungarian Academy of Sciences, 1983.
2. M. Blum, "Coin flipping by telephone: A protocol for solving impossible problems", in *Proceedings of the 24th IEEE Computer Conference (CompCon.)*, pp. 175-193, 1982.
3. Jordi Castellà-Roca, Andreu Riera-Jorba, Joan Borrell-Viader and Josep Domingo-Ferrer, "A method for obtaining an impartial result in a game over a communications network and related protocols and programs", international patent PCT ES02/00485, Oct. 14, 2002.
4. J. S. Chou and Y. S. Yeh, "Mental poker game based on a bit commitment scheme through network", *Computer Networks*, vol. 38, pp. 247-255, 2002.
5. D. Coppersmith, "Cheating at mental poker", in *Advances in Cryptology - Crypto '85* (ed. H. C. Williams), LNCS 218, Berlin: Springer Verlag, pp. 104-107, 1986.
6. C. Crépeau, "A secure poker protocol that minimizes the effect of player coalitions", in *Advances in Cryptology - Crypto '85* (ed. H. C. Williams), LNCS 218, Berlin: Springer Verlag, pp. 73-86, 1986.
7. C. Crépeau, "A zero-knowledge poker protocol that achieves confidentiality of the players' strategy or how to achieve an electronic poker face", in *Advances in Cryptology - Crypto'86* (ed. A. M. Odlyzko), LNCS 263, Berlin: Springer-Verlag, pp. 239-250, 1986.
8. R. DeMillo and M. Merritt, "Protocols for data security", *Computer*, vol. 16, pp. 39-50, 1983.
9. J. Domingo-Ferrer, "A new privacy homomorphism and applications", *Information Processing Letters*, vol. 60, pp. 277-282, 1996.
10. J. Domingo-Ferrer, "A provably secure additive and multiplicative privacy homomorphism", in *Information Security* (eds. A. Chan and V. Gligor), LNCS 2433, pp. 471-483, 2002.
11. J. Edwards, *Implementing Electronic Poker: A Practical Exercise in Zero-Knowledge Interactive Proofs*. Master's thesis, Department of Computer Science, University of Kentucky, 1994.
12. S. Fortune and M. Merritt, "Poker protocols", in *Advances in Cryptology: Proceedings of Crypto'84* (eds. G. R. Blakley and D. Chaum), LNCS 196, Berlin: Springer-Verlag, pp. 454-466, 1985.
13. Gambling Review Body, Department for Culture Media and Sport of Great Britain, chapter 13, page 167, July 17, 2001. http://www.culture.gov.uk/role/gambling_review.html,
14. S. Goldwasser and S. Micali, "Probabilistic encryption & how to play mental poker keeping secret all partial information", in *Proceedings of the 18th ACM Symposium on the Theory of Computing*, pp. 270-299, 1982.

294   J. Castellà-Roca et al.

15. C. Hall and B. Schneier, "Remote electronic gambling", in *13th ACM Annual Computer Security Applications Conference*, pp. 227-230, 1997.
16. L. Harn, H. Y. Lin and G. Gong, "Bounded-to-unbounded poker game", *Electronics Letters*, vol. 36, pp. 214-215, 2000.
17. M. Jakobsson, D. Pointcheval and A. Young, "Secure mobile gambling", in *Topics in Cryptology - CT-RSA 2001*, (ed. D. Naccache), LNCS 2020, Springer-Verlag, pp. 100-109, San Francisco, CA, USA, April 8-12, 2001.
18. K. Kurosawa, Y. Katayama, W. Ogata and S. Tsujii, "General public key residue cryptosystems and mental poker protocols", in *Advances in Cryptology - EuroCrypt'90* (ed. I. B. Damgaard) LNCS 473, Berlin: Springer-Verlag, pp. 374-388, 1990.
19. K. Kurosawa, Y. Katayama and W. Ogata, "Reshufflable and laziness tolerant mental card game protocol", *TIEICE: IEICE Transactions on Communications/Electronics/Information and Systems*, vol. E00-A, 1997.
20. L. Lamport, "Password authentication with insecure communications", *Communications of the ACM*, vol. 24, pp. 770-771, 1981.
21. R. Lipton, "How to cheat at mental poker", in *Proc. AMS Short Course on Cryptography*, 1981.
22. R. Oppliger and J. Nottaris, "Online casinos", in *Kommunikation in verteilten Systemen*, pp. 2-16, 1997.
23. E. Kushilevitz and T. Rabin, "Fair e-lotteries and e-casinos", in *Topics in Cryptology CT-RSA 2001*, (ed. D. Naccache), LNCS 2020, Springer-Verlag pp. 110-119, San Francisco, CA, USA, April 8-12, 2001.
24. A. Shamir, R. Rivest and L. Adleman, "Mental poker", *Mathematical Gardner*, pp. 37-43, 1981.
25. C. Schindelhauer, "A toolbox for mental card games", Medizinische Universität Lübeck, 1998. http://citeseer.nj.nec.com/schindelhauer98toolbox.html
26. B. Schneier, *Applied Cryptography: Protocols, Algorithms and Source Code in C, 2nd Edition*, New York: Wiley, 1996.
27. M. Yung, "Cryptoprotocols: Subscription to a public key, the secret blocking and the multi-player mental poker game", in *Advances in Cryptology: Proceedings of Crypto'84* (eds. G. R. Blakley and D. Chaum), LNCS 196, Berlin: Springer-Verlag pp. 439-453, 1985.
28. W. Zhao, V. Varadharajan and Y. Mu, "Fair on-line gambling", in *16th IEEE Annual Computer Security Applications Conference (ACSAC'00)*, New Orleans, Louisiana, pp. 394-400, 2000.

Lightweight Mobile Credit-Card
Payment Protocol

Supakorn Kungpisdan[1], Bala Srinivasan[2], and Phu Dung Le[1]

[1] School of Network Computing, Monash University
McMahons Road, Frankston, Victoria 3199 Australia
hotkeng@mail1.monash.edu.au
pdle@monash.edu.au
[2] School of Computer Science and Software Engineering, Monash University
900 Dandenong Road, Caulfield East, Victoria 3145 Australia
srini@monash.edu.au

Abstract. Recently, making Internet credit-card payment is widely accepted. Several payment protocols have been proposed to secure the credit-card payment on fixed networks. However, these protocols do not apply well to wireless networks due to the limitations of wireless devices and wireless networks themselves. In this paper, we propose a simple and powerful credit-card payment protocol for wireless networks. We implement a secure cryptographic technique that works well under this protocol. We show that our proposed protocol is more suitable for applying to wireless networks than SET and iKP in that client's computation is reduced. The protocol also satisfies all security properties provided by both SET and iKP. Moreover, it offers the ability to resolve disputes and recover from failures which are normally occurred in wireless environment. Furthermore, client's credit-card information is not required to be sent in the protocol. It results in the security enhancement of the system.
Keywords: Electronic commerce and payment, credit-card payment protocol, mobile commerce

1 Introduction

Credit-card payment has been accepted as a worldwide payment method for several years. Traditionally, it has been done physically by presenting credit card to merchant and signing on the payment slip as an evidence of payment. With the evolution of the Internet, it can be done online. Recently, with wireless communications technology, we can access the Internet to make the payment via wireless devices e.g. cellular phones or PDAs (Personal Digital Assistants).

Ideally, Internet applications should be able to be applied to wireless environment. However, some limitations cause inefficient performance [12, 16]. Firstly, it comes from mobile devices which are assumed to have low-powered and computational capability [17]. Even though some advanced technologies such as secure coprocessor [18] are proposed to enhance its performance, but it is still high costs for users. Also, the bandwidth and reliability of wireless networks are lower compared to that of the fixed one, hence resulting in longer processing time and more

T. Johansson and S. Maitra (Eds.): INDOCRYPT 2003, LNCS 2904, pp. 295–308, 2003.

failures. Furthermore, the connection cost of wireless networks is higher. Thus, making payment on wireless networks seems to be unacceptable by users.

These limitations also affect the credit-card payment protocols which have been successfully implemented for fixed networks e.g. SET [14] and iKP [3]. This is because they are based on public-key infrastructure (PKI) that every party, especially client, is required to perform asymmetric operations. This results in high client's computation. Moreover, the structures of those protocols are complex. It results in long message length which affects wireless users.

To overcome these limitations, we propose a simple and powerful credit-card payment protocol (so called KSL protocol) which is suitable for wireless networks. This protocol can be implemented using a secure cryptographic technique which satisfies all transaction securities stated in [2]. In KSL protocol, client is not required to trust payment gateway as required in SET and iKP. Therefore, client can ensure that her sensitive information will not be leaked to adversary. In addition, client is not required to send the credit-card information in the protocol. This results in the security enhancement of the system.

We compare our KSL protocol with SET and iKP in terms of performance and security properties. We show that the client's computation of our protocol is lower than that of the other two protocols since no public-key operation is required. Thus, mobile users can have efficient e-commerce.

According to the security aspects, KSL protocol also offers the ability to resolve disputes among parties. Furthermore, it provides an efficient method to deal with failures.

This paper is organized as follows: section 2 describes the overviews of SET and iKP. In section 3, we introduce KSL protocol. Section 4 discusses about the proposed protocol in terms of security and performance. In section 5, we compare our KSL protocol with SET and iKP. Section 6 concludes our work.

2   Existing Credit-Card Payment Protocols

2.1   Payment Model

From the general payment model proposed in [1], it is composed of 4 involved parties: client, merchant, issuer (client's financial institution), and acquirer (merchant's financial institution). Both issuer and acquirer are represented by payment gateway which acts as a medium between them and both client and merchant for the clearing purpose.

From [1], there are 3 primitive transactions (as shown in Fig.1): *Payment*, *Value Subtraction*, and *Value Claim*: *Payment* is made by client about the payment to merchant, *Value Subtraction* is made by client in order to request payment gateway (on behalf of issuer) to deduct the money from the client's account, and *Value Claim* is made by merchant in order to request payment gateway (on behalf of acquirer) to transfer money to the merchant's account.

Note that both SET and iKP transactions are based on the above payment model [7]. The high-level protocol steps of both protocols are shown as follows:

$C \rightarrow$M: *Payment(Request), Value-Subtraction(Request)*
$M \rightarrow$PG: *Value-Subtraction(Request), Value-Claim(Request)*
$PG \rightarrow$M: *Value-Claim(Response), Value-Subtraction(Response)*
$M \rightarrow$C: *Payment(Response), Value-Subtraction(Response)*

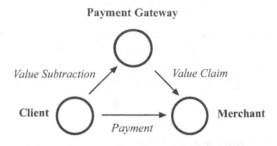

Fig. 1. Primitive Transactions

2.2   Notations

We introduce general notations which are used throughout this paper as follows:

- $\{C, M, PG, I, A\}$: the set of client, merchant, payment gateway, issuer, and acquirer, respectively.
- $\{K_A, K_A^{-1}\}$: the set of public/private key pair of a party A.
- $\{ID_C, ID_M, ID_I\}$: the set of identities of client, merchant, and issuer, respectively. It contains the contact information of each party.
- $Cert_A$: the certificate of a party A which contains $\{ID_A, K_A\}$.
- OI : order information.
- PI : payment information which contains credit-card information (CCI).
- OD : order descriptions which contains goods descriptions.
- $Price$: the amount and currency.
- TID contains the identity of transaction and the date of transaction $Date$.
- Yes/No : the status of transaction *approved/rejected*.
- $TIDReq$: the request for TID.
- $MIDReq$: the request for ID_M.
- $\{M\}_X$: the message M symmetrically encrypted with the shared key X.
- $\{M\}_{K_X}$: the message M encrypted with the public key of the party X.
- $\{M\}_{K_X^{-1}}$: the message M signed with the private key of the party X.
- $h(M)$: the one-way hash function of the message M.
- $MAC(M, K)$: the message authentication code (MAC) of the message M with the key K.

2.3   SET and iKP Overviews

SET (Secure Electronic Transaction) [14] and its ancestor, iKP (Internet Key Protocol), are the most well-known credit-card payment protocols. Most of the structures of SET are similar to iKP. In SET, all parties are required to possess public-key certificates. In iKP, it consists of three versions depending on the number of certificates of engaging parties, in that, in 1KP, only payment gateway is required to possess its own certificate. Both client and merchant can authenticate themselves to payment gateway and each other using PIN (Private Identification Number). In 2KP, only client is required to possess certificate. Finally, the 3KP protocol requires all engaging parties to possess their own certificates. The details of SET and 3KP protocols are shown as follows.

SET Protocol

PinitReq: C→M: *Initial Request*
PinitRes: M→C: $\{TID\}_{K_A^{-1}}, Cert_M, Cert_{PG}$
PReq: C→M: *OI, h(PI),* $\{h(OI), h(PI)\}_{k_1}, \{k_1\}_{K_C^{-1}}, \{h(OI), PI\}_{k_2},$
$\{k_2\}_{K_{PG}}, Cert_C$
AuthReq: M→PG: $\{\{TID, Price, Date, h(OI), (OI), \{h(OI), h(PI)\}_{k_1},$
$\{k_1\}_{K_C^{-1}}, \{h(OI), PI\}_{k_2}, \{k_2\}_{K_{PG}}\}_{K_M^{-1}}\}_{K_{PG}}$
AuthRes: PG→M: $\{\{TID, Price, Date, Yes/No\}_{K_{PG}^{-1}}\}_{K_M}$
PRes: M→C: $\{TID, Date, Yes/No\}_{K_M^{-1}}$

Where $\{k_1, k_2\}$ stands for the set of session keys generated by client, *OI =* *{ TID, h(OD, Price)}*, and *PI = { TID, h(OD, Price), ID_M, Price, CCI}*.

iKP Protocol

Initiate: C→M: ID_C
Invoice: M→C: ID_M, *h(OI)*, *Date*
Payment: C→M: *PI,* $\{PI, h(OI)\}_{K_C^{-1}}$
AuthReq: M→PG: ID_M, *h(OI)*, *Date*, *h(OD)*, *PI,* $\{h(OI)\}_{K_M^{-1}},$
$\{PI, h(OI)\}_{K_C^{-1}}$
AuthRes: PG→M: *AI,* $\{AI, h(OI)\}_{K_{PG}^{-1}}$
Confirm: M→C: *AI,* $\{h(OI)\}_{K_M^{-1}}, \{AI, h(OI)\}_{K_{PG}^{-1}}$

Where *PI = {Price, h(OI), CCI}*$_{K_{PG}}$, *OI = { TID, Price, ID_C, ID_M, Date, h(OD)}*, and *AI* stands for authorization information which contains *Yes/No*.

Referred to the payment model described in section 2.1, in **PReq** (or **Payment**), client sends both *Payment Request* to merchant and *Value-Subtraction Request* to PG. Merchant then retrieves *OI* and sends **AuthReq** which is *Value-Claim Request* that contains the forwarded *Value-Subtraction Request* to PG. PG consults issuer and acquirer to ask for the approval of transaction. After that, PG sends **AuthRes** which contains *Value-Claim Response* to merchant. Merchant retrieves the result of transaction and sends **PRes** (or **Confirm**) which contains *Payment Response* to client.

Although SET and iKP are successfully implemented for fixed networks, they do not apply well to wireless ones because of the following limitations:

- SET and iKP are heavy-loaded protocols. Client who has low-computational capability wireless device is required to perform high computational operations such as public-key encryptions.
- Since SET and iKP are PKI-based, each party's certificate sent to client has to be verified by CA. It results in additional communication passes at client.

3   KSL Protocol

3.1   Initial Assumptions

We state initial assumptions of our proposed KSL protocol as follows:

1. Client is considered to have any wireless device which is able to access the Internet.
2. All involved parties except client are required to have their own certificates.
3. Client's credit-card information (*CCI*) is shared between client and issuer.
4. Issuer issues the shared secret Y to client. Both client and issuer then generate a set of secrets Y_i, where $i = 1, \ldots, n$, by using the same generation technique (will be described in section 3.2) and store them in their terminals.
5. Issuer is trusted by client since the issuer issues credit card to client.
6. Client is not required to trust payment gateway.
7. Client and merchant have agreed on the price and description of the goods or services.
8. It is easy to compute the hash function *h(x)* from the given *x*, and it is computational infeasible to compute *x* which *y = h(x)* from the given *y*. Moreover, the MAC algorithm is assumed to be a fast and secure version.

3.2   Key Generation Technique

In the proposed protocol, two sets of shared keys are generated: X_i (shared between client and merchant) and Y_i (shared between client and issuer), where $i = 1, \ldots, n$. We present two possible efficient key generation techniques: one used for generating a set of X_i from the given secret X and the other used for generating a set of Y_i from the given secret Y. The main concept of both techniques is to apply one hash algorithm with one-bit shift (either left shift or right shift) of a master secret at each time of generating a session key. The details of both techniques are shown as the following:

Generating X_i
$X_1 = h(1\text{-}bit\text{-}shift\text{-}of\text{-}X)$, $X_2 = h(2\text{-}bit\text{-}shift\text{-}of\text{-}X)$,..., $X_n = h(n\text{-}bit\text{-}shift\text{-}of\text{-}X)$

Generating Y_i
$Y_1 = h(1\text{-}bit\text{-}shift\text{-}of\text{-}(CCI,\ Y))$, $Y_2 = h(2\text{-}bit\text{-}shift\text{-}of\text{-}(CCI,\ Y))$,...,
$Y_n = h(n\text{-}bit\text{-}shift\text{-}of\text{-}(CCI,\ Y))$

Note that it is not restricted to the above two algorithms in order to generate the sets of X_i and Y_i. They are just presented as examples of possible key generation techniques. Before making payment, client needs to register herself to issuer and merchant. Client can register to issuer either by phone or via the issuer's website. After the registration is successful, client receives client's wallet software by mail or downloading from the issuer's site. The wallet contains both key generation and payment software. After the wallet is successfully installed, a set of Y_i is generated and stored in the client's device. To generate Y_i, issuer sends client the secret Y using authenticated key exchange protocol (will be discussed in section 4.3.7). Client then uses Y, together with her CCI, to generate a set of Y_i by using the above key generation technique. To generate a set of X_i, client runs *merchant registration protocol* to register herself to merchant in order to share the secret X with merchant. The details of the merchant registration protocol will be shown in section 3.3.

3.3   KSL Protocol

KSL protocol is composed of two sub-protocols: *merchant registration protocol* and *payment protocol*. In merchant registration protocol, client shares the master secret X with merchant when she newly registers herself to merchant or she wants to update a set of X_i. After the sets of X_i and Y_i are successfully generated, client can start the payment protocol. The details of both protocols are shown as follows:

Merchant Registration Protocol
Before making payment, client needs to register herself to merchant to share the master secret X with merchant. The details of the protocol are shown as follows:

C→M: $\{ID_C, X, n\}_k$
M→C: $\{n\}_k$

Client generates the secret X which is to be shared with merchant and then sends merchant ID_C, a nonce n, and X encrypted with the session key k generated from running the authenticated key exchange protocol (its detail will be discussed in section 4.3.7) with merchant. Merchant then confirms the client registration by sending n encrypted with k to client. After the completion of the protocol, both of them can generate a new set of X_i by using the same key generation technique (as described in section 3.2). Note that the objective of the merchant registration protocol is to distribute and update the master secret X. Thus, it is not required to run in every transaction, but only when either client or merchant wants to update the key X.

Payment Protocol

After generating a set of secret X_i and Y_i, client is able to perform a payment transaction by running payment protocol as follows:

Step1
C→M: ID_C, i, *TIDReq*, *MIDReq*
M→C: $\{TID, ID_M\}_{X_i}$
Step2
C→M: $\{OI, Price, ID_C, ID_I, MAC[(Price, h(OI), ID_M), Y_i]\}_{X_i}$,
$MAC[(OI, Price, ID_C, ID_I), X_{i+1}]$
Step3
M→PG: $\{\{MAC[(Price, h(OI), ID_M), Y_i], Price\}_{K_{PG}}$,
$h(OI), i, TID, ID_C, ID_I\}_{K_M^{-1}}$
Step4 Under private network,
PG→I: $MAC[(Price, h(OI), ID_M), Y_i], h(OI), i, TID, Price, ID_C, ID_M$
PG→A: $Price, ID_M$
I,A→PG: $Yes/No, \{h(OI), Yes/No\}_{Y_i}$
Step5
PG→M: $\{\{h(OI), Yes/No\}_{Y_i}, \{h(OI), Yes/No\}_{K_M}\}_{K_{PG}^{-1}}$
Step6
M→C: $\{\{h(OI), Yes/No\}_{Y_i}\}_{X_{i+1}}$

Note that $OI = \{TID, h(OD, Price)\}$. The details of the payment protocol are shown as follows:

Step 1: Client and merchant exchange necessary information to start the protocol.

Step 2: Client sends *Payment Request* (referred to the payment model described in section 2.1) to merchant. *Payment Request* contains OI used to inform merchant about the goods and price requested. It also contains $MAC[(Price, h(OI), ID_M), Y_i]$ which represents *Value-Subtraction Request* that is to be forwarded to issuer. It can be noted that, although merchant has X_i, she cannot generate this message since she does not have Y_i used for constructing $MAC[(Price, h(OI), ID_M), Y_i]$. Consequently, we can ensure that the message is really originated by client.

Step 3: Merchant decrypts the message to retrieve OI. Merchant then sends *Value-Claim Request* signed with the merchant's private key to PG. *Value-Claim Request* contains the forwarded *Value-Subtraction Request*. We can see that *Value-Subtraction Request*, together with *Price*, is encrypted with PG's public key. This is to ensure that only PG is intended recipient of the message. Note that ID_C and ID_I are used to identify client and issuer, respectively, and the index i is used to identify the current session key of Y_i. In addition, PG cannot generate *Value-Claim Request* by itself since it does not have Y_i.

Step 4: After receiving the message and verifying the merchant's signature, PG recognizes the issuer from ID_I. PG then passes *Value-Subtraction Request* together with relevant information, including the index i, to issuer. PG can recognize merchant from the merchant's signature. PG then sends ID_M and the requested price (*Price*) to claim to acquirer that she is the party whom the money will be transferred to. After checking the validity of the client's account, either issuer or acquirer sends the approval result *Yes/No* to PG. Note that this step is done under banking private network, hence we do not concern about its security issue.

Step 5: After receiving an approval, PG sends both *Value-Claim Response* to merchant and *Value-Subtraction Response* to client via merchant. Note that $\{h(OI), Yes/No\}_{K_M}$ represents *Value-Claim Response* and $\{h(OI), Yes/No\}_{Y_i}$ represents *Value-Subtraction Response*. Merchant verifies the PG's signature and retrieves the response of her request. Merchant can check whether or not the message is the response of her request by comparing the received *h(OI)* with her own *OI*. If they are not matched, merchant rejects the transaction. Merchant then encrypts *Value-Subtraction Response* with X_{i+1}, and forwards to client as *Payment Response*.

Step 6: Client retrieves the result of her request and compares between the received *h(OI)* and her own *OI*. As well as that in **Step 5**, if they are not matched, client can reject the transaction.

In next purchases, client does not have to run the merchant registration protocol again since she can use other value in the set of X_i to perform transactions until being notified to update the secret X. Then client runs the merchant registration protocol to get a new secret X.

We can see that the payment protocol is the main sub-protocol of KSL protocol which needs to be performed in every transaction whereas the merchant registration protocol is used only for updating the master secret X. Note that issuer can update the master secret Y and send to client by using any authenticated key exchange protocol for wireless networks. Note also that after each X_i and Y_i has been used, they are put into all parties' revocation lists in order to prevent the replay of the secrets from both client and merchant.

4   Discussions

4.1   The Protocol Goals

At the completion of SET protocol [14], the following goals [15] must be satisfied:

1. Client is ensured that merchant has committed to deliver the goods she has ordered.
2. Merchant is ensured that PG has committed to transfer the money at the amount requested to her account.
3. PG has completed its tasks: sending the commitments for both deducting money at the amount requested by client from the client's account and transferring the money at the amount requested by merchant to the merchant's account.

Compared to KSL protocol, the goal *(1)* is satisfied by the protocol **Step 5** that merchant sends to client as a receipt of payment. It is encrypted with X_{i+1} which can be used to authenticate the merchant. It also contains the result of transaction *Yes/No*. Moreover, *h(OI)* contains the goods descriptions which will be sent to client. The goal *(2)* is satisfied by the protocol **Step 4**. PG is identified by the message signed with its private key, and *Yes/No* notifies the merchant about the result of transaction, and *h(OI)* contains the approved price. As well as the goal *(3)* is satisfied by the protocol **Step 4** and **Step 5**. After merchant receives the protocol **Step 4** and client receives the protocol **Step 5**, it means that PG has completed its tasks. As a result, KSL protocol achieves all of the goals relevant to all involved parties.

4.2   The Cryptographic Technique

We realize that applying asymmetric cryptographic operations on wireless device is not applicable [16, 17]. However, symmetric cryptography still has disadvantages compared to asymmetric one. As described in [10], one of the main advantages of asymmetric cryptography over symmetric cryptography is the ability to identify the originator of message which can be done from its digital signature, whereas it cannot be done in symmetric cryptography since the secret key is shared between two parties.

We apply the technique proposed by [6] and [13] to mobile payment scenario by using symmetric cryptography to secure the communications between client and involved parties. The concept is that a party can authenticate herself to others by adding a secret which cannot be generated by those participants into the message. The recipient can recognize the originator of the message from the attached secret. Furthermore, we provide message integrity by using MAC. The technique can be formalized as the following:

$$\{Message, Y, MAC[(Message, Y), X_2]\}_{X_1}$$

The following example can describe how our technique works and its properties. Let $\{X_1, X_2\}$, where $X_1 \neq X_2$, is the set of secrets shared between client and merchant, and issuer has given a secret Y to client. Note that Y can be any kind of message (hashed or MAC). It can be seen that, if the keys are not compromised, this message has been sent from client since she holds all secrets ($X_1, X_2,$ and Y), whereas merchant does not have Y. Moreover, it cannot be generated by issuer since the issuer does not have X_1 and X_2. In the case that a dispute occurs, with the assistance of issuer, client is able to prove to a verifier that she is the originator of the message. Note that this technique requires no conspiracy between issuer and merchant. However, in the payment scenario, issuer would never give the secret Y to merchant. Our technique offers the ability to resolve disputes among parties in that all parties can provide non-repudiable evidence to prove to a third party that they are responsible for transactions [11].

4.3   Security Issues

4.3.1 Transaction Security

The most concerned security aspect of any payment system is transaction securities in that each transaction must be performed in a secure manner. In this section, we consider whether or not KSL protocol satisfies the following transaction securities for any payment systems [2]: (i) *party authentication*, (ii) *transaction privacy*, (iii) *transaction integrity*, and (iv) n*on-repudiation of transactions*. From the message format discussed in section 4.2, all transaction security properties are satisfied. The details are shown as follows:

(i) Party authentication is ensured by symmetric encryption and Y shared between client and issuer. The encryption ensures that either client or merchant generated the message and Y ensures that the client is the originator of message.

(ii) Transaction privacy is ensured by symmetric encryption.

(iii) Transaction integrity is ensured by MAC.

(iv) Non-repudiation of transactions is ensured by Y (or $h(Y)$) in that merchant is able to provide a non-repudiable evidence to prove to other parties that client has sent a message or requested merchant to perform transaction. This is because Y cannot be generated by merchant, but by client or issuer, and she is ensured that client is the party who has sent the message to her because the message is encrypted with X_1 which is shared between client and merchant.

4.3.2 Security of Secret Keys

We realize that it is easier to break the symmetric-key cryptosystems, which each of the two parties shares the same secret, than asymmetric-key cryptosystems, which use pairs of public and private keys. With our symmetric-cryptographic technique discussed in section 4.2, the secrets are shared between each of the two parties. We employ two different keys to provide two levels of security; one used for encryption and the others used for MAC. Even though the encrypting key is compromised, we can ensure the integrity of the message by MAC with the other key. Each transaction exploits different session keys based on the same master secrets until being updated by the system.

According to the key distribution issue, each time when a new key is distributed, it is possible to be intercepted by an attacker. In KSL protocol, client is not required to send her credit-card information in the protocol. The credit-card information is used only for generating a set of Y_i which can be done offline. Although an attacker may able to intercept both X and Y during the key distribution Y_i since he does not have client's credit-card information.

4.3.3 The Client's Credit-Card Information

In general, credit-card information can be used to identify client's identity during transactions. Physically, to purchase goods or services, client has to present credit card to merchant. However, for electronic payment transactions, it is not necessary to do that because the goal of sending credit-card information is to authenticate client to issuer. We realize that normally credit-card information is shared between client and issuer. Thus, it can be changed into another form which is difficult to be attacked.

In KSL protocol, client is not required to send credit-card information. The credit-card information is used only as a component for generating a set of Y_i shared between client and issuer. With efficient key generation technique, it is hard to retrieve credit-card information although an attacker can intercept the message and successfully retrieve the MAC key Y_i. It is hard to run reverse operation of hash function. In addition, as Y_i is used for two purposes: symmetric encryption and hashing, if the MAC algorithm used in the protocol requires shorter key length than the encryption algorithm, the MAC key length has to be shortened. Thus, it is hard to retrieve the pre-image of Y_i from the intercepted message.

4.3.4 Trust Relationships
In any payment system, a party should not trust others unless a valid proof of trustworthiness can be provided. However, in SET and iKP, client and merchant need to trust payment gateway since some confidential information, particularly client's credit-card information, has to be revealed to payment gateway in order to be forwarded to issuer/acquirer for the clearing purpose. Unfortunately, payment gateway may be a company that is monitoring the system. It may possibly have a conspiracy with an attacker, or even merchant, so that the attacker can get credit-card information without any attempt to decrypt messages.

In KSL protocol, we state the trust relationship between client and issuer instead of client and payment gateway since the issuer issues credit card to client. Therefore, we do not need to concern about the honesty of payment gateway as that in SET and iKP.

4.3.5 Failure Recovery
The failure recovery is relevant to key synchronization process, in that, after recovered from any failure, the protocol should have an efficient key synchronization mechanism. Consider when the failure occurs in our proposed protocol, especially when performing several sessions concurrently. For example, five sessions are performing concurrently. The session keys $X_1 - X_5$ and $Y_1 - Y_5$ are being used. If session 3 fails, after its recovery, client can restart the session 3 with new X_i and Y_i where i can be any value in the sets of X_i and Y_i, not restricted to the their next values. Merchant and issuer can recognize the new session keys X_i and Y_i from the index i attached in the message sent from client. As each session key is generated independently, client can select any value in the index.

4.3.6 Dispute Resolution
KSL protocol offers the ability to resolve disputes among parties. From [10, 11], each party should be able to provide accountable evidence to prove the third party in case of disputes. In the protocol, the secret X_i and Y_i contained in each message can be used to identify the originator of message, and OI (or its hash)

and the party's identity in each message can be used as the details of the payment which has already been occurred. Such information can be used as evidence to resolve disputes, and it is also enough to prove the security requirements of the iKP protocol.

4.3.7 Authenticated Key Exchanged Protocols for Wireless Networks

As mentioned in section 3.3, we initially establish the communications between client and merchant using authenticated key exchange (AKE) protocol for wireless networks. In this paper, we leave the AKE protocol as a block of operation since it is out of our scope. Instead, we provide an overview of existing AKE protocols for wireless networks in order that the readers can apply an appropriate one to their applications.

Many AKE protocols for wireless networks [8, 4, 19, 17] have been proposed including their analyses [4, 8]. [8] and [5] employ the elliptic-curve cryptosystems to reduce the computation and resource consumption of involved parties, especially the client. The difference between these protocols is that [8] is PKI-based whereas [5] is password-based. [17]'s protocol is based on the challenge-response. Recently, [19] proposed a password-based AKE protocol using RSA algorithm. Horn et al [9] proposed the analysis of the existing AKE protocols for wireless networks including [4] and [8]. They argued that both protocols are suitable for wireless communications even though [4] requires more communication passes.

5   Comparing KSL Protocol with SET and iKP

In this section, we show that our proposed KSL protocol has advantages over SET and iKP in terms of the transaction security and efficiency when applied to wireless networks.

Based on the payment model described in section 2.1, in KSL protocol, the message in each protocol step can deliver the same information and purposes as that of SET and iKP. The important security properties of both SET and iKP e.g. non-repudiation are also satisfied since every party is capable to deliver non-repudiable evidence to prove to the third party in case of resolving disputes.

According to the transaction efficiency, we mainly focus on the computational load of involved parties, particularly the number of cryptographic operations applied to the protocol. Table 1 demonstrates the number of cryptographic operations applied to SET, iKP, and our KSL protocols, respectively.

It is not hard to see that, in KSL protocol, client is required to perform only symmetric-cryptographic operations and hash functions which do not require high computation. Compared with SET and iKP, the client is required to perform both public key encryptions and signature verifications which are high computational tasks. As a result, our KSL protocol has the advantage over SET and iKP on the reduction of computational tasks at the client's wireless device. Note that we compare KSL protocol and SET with 3KP because we concern about the comparison at the same security level as provided by SET whereas 1KP and 2KP cannot provide that security level.

Table 1. The number of cryptographic operations of SET, iKP, and KSL protocol at the client, merchant, and payment gateway, respectively

Cryptographic operations	Numbers of cryptographic operations								
	Client			Merchant			Payment Gateway		
	SET	iKP	KSL	SET	iKP	KSL	SET	iKP	KSL
1. Public-key encryption	1	1	-	1	-	1	1	-	1
2. Public-key decryption	-	-	-	1	-	1	2	1	1
3. Digital signature	1	1	-	3	1	1	1	1	1
4. Signature verification	2	3	-	2	2	1	1	2	1
5. Symmetric-key encryption/decryption	1	-	3	-	-	4	1	-	-
6. Hash function	3	2	3	2	4	3	-	1	-
7. Keyed-hash function	-	-	2	-	1	-	-	-	-
8. Key generation	-	-	2	-	-	1	-	-	-

6 Conclusion

In this paper, we have propose a new credit-card payment protocol which is suitable for wireless networks. We overcome the limitations of wireless devices by using simple cryptographic operations at the client side. Due to the poor reliability of wireless networks, our protocol structure, composed of short and simple message passes, offers the ability to secure transactions after the network is recovered from failure. KSL protocol offers the practical usefulness and advantages over SET [14] and iKP [3] as the following:

1. The structure and cryptographic technique applied to KSL protocol are simpler, but have powerful capability, than those applied in SET and iKP.
2. KSL protocol satisfies important security properties as provided by SET and iKP. Especially at client, even though no public-key operation is exploited, the security properties at client are still provided at the same level as that of SET and iKP. The party authentication property and the necessary information contained in each protocol step offer the ability to resolve disputes among parties.
3. KSL protocol has lower client's computation since it does not exploit any asymmetric-key operation. Thus, no wireless device with advanced features is required.
4. Client is not required to trust payment gateway as required in SET and iKP. Thus, client can ensure that her confidential information will not be compromised by payment gateway.
5. Client is not required to send the credit-card information to her issuer. The issuer can identify the client and infer the credit-card information from the

session keys generated from the master secret. Therefore, the credit-card information is not susceptible to attacks.

As a result, with KSL protocol, mobile users can have efficient and secure payment transactions which result in gaining more acceptability than existing protocols.

References

1. Abad-Peiro, J. L., Asokan, N., Steiner, M., Waidner, M.: Designing a generic payment service. IBM Systems Journal, Vol. 37(1). (1998) 72-88
2. Ahuja, V.: Secure Commerce on the Internet. Academic Press (1996)
3. Bellare, M., Garay, J. A., Hauser, R., Herzberg, A., Krawczyk, H., Steiner, M., Tsudik, G., Herreweghen, E. V., Waidner, M.: Design, Implementation, and Deployment of the iKP Secure Electronic Payment System. IEEE Journal of Selected Areas in Communications (2000)
4. Boyd, C., Park, D. G.: Public Key Protocols for Wireless Communications. Proceedings of the ICISC 1998, Seoul, Korea (1998) 47-57
5. Boyd, C., Montague, P., Nguyen, K.: Elliptic Curve Based Password Authenticated Key Exchange Protocols. Proceedings of ACISP, LNCS Vol. 2119 (2001) 487-501
6. Cimato, S.: Design of an Authentication Protocol for GSM Javacards. Proceedings of ICICS 2001, LNCS Vol. 2288 (2002) 355-368
7. Herreweghen, E. V.: Non-Repudiation in SET: Open Issues. Proceedings of the Financial Cryptography 1999, LNCS, Vol. 1962 (1999) 140-156
8. Horn, G., Preneel, B.: Authentication and Payment in Future Mobile Systems. Proceedings of 5th European Symposium on Research in Computer Security, Belgium (1998) 277-293
9. Horn, G., Martin, K. M., Mitchell, C. J.: Authentication Protocols for Mobile Network Environment Value-Added Services. IEEE Transactions on Vehicular Technology, Vol. 51(2) (2002) 383-392
10. Kailar, R.: Accountability in Electronic Commerce Protocols. IEEE Transactions on Software Engineering, Vol. 22(5) (1996)
11. Kungpisdan, S., Permpoontanalarp, Y.: Practical Reasoning about Accountability in Electronic Commerce Protocols. LNCS Vol. 2288 (2002) 268-284
12. Kungpisdan, S., Srinivasan, B., Le, P. D.: A Practical Framework for Mobile SET Payment. Proceedings of the IADIS International E-Society Conference (2003) 321-328
13. Marvel L. M.: Authentication for Low Power Systems. Proceedings of IEEE MILCOM (2001)
14. Mastercard and Visa. SET Protocol Specifications. (1997)
    http://www.setco.org/set_specifications.html
15. Meadows, C., Syverson, P.: A Formal Specification of Requirements for Payment Transactions in the SET Protocol. Proceedings of FC 1998 (1998) 15p.
16. Romao A., da Silva, M.: An Agent-based Secure Internet Payment Systems. Proceedings of TREC 1998. LNCS, Vol. 1402 (1998) 80-93
17. Wong, D. S., Chan, A. H.: Efficient and Mutually Authentication Key Exchange for Low Power Computing Devices. LNCS, Vol. 2248 (2001) 272-289
18. Yee, B. S.: Using Secure Coprocessor. PhD thesis. Carnegie Mellon University. (1994)
19. Zhu, F., Wong, D. S., Chan, A. H., Ye, R.: Password Authenticated Key Exchange Based on RSA for Imbalanced Wireless Networks. LNCS, Vol. 2433 (2002) 150-161

On the Construction of Prime Order Elliptic Curves*

Elisavet Konstantinou[1,2], Yannis C. Stamatiou[3,4], and Christos Zaroliagis[1,2]

[1] Computer Technology Institute, P.O. Box 1122, 26110 Patras, Greece
[2] Dept of Computer Eng. & Informatics, University of Patras, 26500 Patras, Greece
{konstane,zaro}@ceid.upatras.gr
[3] Dept of Mathematics, University of the Aegean, 83200 Samos, Greece
[4] Joint Research Group (JRG) on Communications and Information Systems
Security (University of the Aegean and Athens University of Economics and Business)
stamatiu@aegean.gr

Abstract. We consider a variant of the Complex Multiplication (CM) method for constructing elliptic curves (ECs) of prime order with additional security properties. Our variant uses Weber polynomials whose discriminant D is congruent to 3 (mod 8), and is based on a new transformation for converting roots of Weber polynomials to their Hilbert counterparts. We also present a new theoretical estimate of the bit precision required for the construction of the Weber polynomials for these values of D. We conduct a comparative experimental study investigating the time and bit precision of using Weber polynomials against the (typical) use of Hilbert polynomials. We further investigate the time efficiency of the new CM variant under four different implementations of a crucial step of the variant and demonstrate the superiority of two of them.

1 Introduction

Elliptic Curve (EC) cryptography has proven to be an attractive alternative for building fast and secure public key cryptosystems. One of the fundamental problems in EC cryptography is the generation of cryptographically secure ECs over prime fields, suitable for use in various cryptographic applications. A typical requirement of all such applications is that the *order* of the EC (number of elements in the algebraic structure induced by the EC) possesses certain properties (e.g., robustness against known attacks [5], small prime factors [1], etc), which gives rise to the problem of how such ECs can be generated.

A specific application domain that is our main concern in this work involves implementations of EC-based cryptosystems in computing devices with limited resources, or in systems operating under strict timing response constraints. Two specific scenarios in this framework involve: (i) The development of a proactive

---

* This work was partially supported by the IST Programme of EU under contracts no. IST-1999-14186 (ALCOM-FT) and no. IST-1999-12554 (ASPIS), and by the Human Potential Programme of EU under contract no. HPRN-CT-1999-00104 (AMORE).

T. Johansson and S. Maitra (Eds.): INDOCRYPT 2003, LNCS 2904, pp. 309–322, 2003.
© Springer-Verlag Berlin Heidelberg 2003

cryptosystem approach (e.g., in the sense of [12]) in networks of resource limited hardware devices (e.g., microcontroller chips, smart dust clouds) working for some highly critical – with respect to security – task and which for that reason are frequently requested to refresh their security parameters. (ii) A wireless and web-based environment, as it is described in [11], in which millions of (resource-limited) client devices connect to secure servers. Clients may be frequently requested to choose different key sizes and EC parameters depending on vendor preferences, security requirements, and processor capabilities.

A frequently employed approach for generating ECs whose order satisfies certain desirable properties is the so-called *Complex Multiplication* (CM) method. This method was used by Atkin and Morain [1] for the construction of elliptic curves with good properties in the context of primality proving, and since then has been adapted to give rise to ECs with good security properties by Spallek [24], and independently by Lay and Zimmer [18]. Furthermore, a number of works appeared that compare variants of the CM method and also present experimental results concerning the construction efficiency, such as the recent works of Müller and Paulus [20], as well as the theses of Weng [26] and Baier [4].

In the case of prime fields, the CM method takes as input a given prime (the field's order) and determines a specific parameter, called the *CM discriminant* D of the EC. The EC of the desirable order is generated by constructing certain class field polynomials based on D and finding their roots. The construction and the location of the roots (modulo the finite field's order) is one of the most crucial steps in the whole process. The most commonly used class field polynomials are the Hilbert (original version of the CM method) and the Weber polynomials. Their main differences are: (i) the coefficients of Hilbert polynomials grow unboundedly large as D increases, while for the same D the Weber polynomials have much smaller coefficients (although their coefficients also grow with D) and thus are easier and faster to construct; (ii) the roots of the Hilbert polynomial construct directly the EC, while the roots of the Weber polynomial have to be transformed to the roots of its corresponding Hilbert polynomial to construct the EC. For a general discussion and comparison on class field polynomials, see [9].

The use of Hilbert polynomials in the CM method requires high precision in the arithmetic operations involved in their construction, resulting in considerable increase in computing resource requirements. This makes them rather inappropriate for fast and frequent generation of ECs. To overcome these shortcomings of Hilbert polynomials, two alternatives have been recently proposed: either to compute them off-line in powerful machines, and store them for subsequent use (see e.g., [22]), or to use Weber polynomials for certain values of D (see e.g., [2,4,15,17,18,25]) and produce the required Hilbert roots from them. The former approach [22] tackles adequately the efficient construction of ECs, setting as a sole requirement for cryptographic strength that the order of the EC is prime which in turn implies that $D \equiv 3 \pmod{8}$ (a prime order is necessary in certain situations – see e.g., [6]). However, there may still be problems with storing and handling several Hilbert polynomials with huge coefficients on hardware devices with limited resources. These problems are addressed by the second approach.

Despite the space and time efficiency though, the known studies do not treat the case of $D \equiv 3 \pmod 8$ as these values of D give Weber polynomials with a degree three times larger than that of their corresponding Hilbert polynomial. For example, the case of $D \equiv 7 \pmod 8$ and not divisible by 3 is treated in [2,4,15,18], while the cases of $D \not\equiv 3 \pmod 8$ and $D \not\equiv 0 \pmod 3$ were treated in [17,25]. In addition, there are works that consider the generation of prime order ECs over extension fields, but either they use the CM method with Hilbert polynomials [3], or they generate the EC parameters at random and use a point counting algorithm to compute the order of the curve [21]. To the best of our knowledge, the use of Weber polynomials within the CM method for the generation of prime order ECs along with the necessary transformation of the Weber roots to their Hilbert counterparts for the case $D \equiv 3 \pmod 8$ has not been studied before.

The first contribution of this paper is a new transformation for converting roots of Hilbert polynomials to roots of their corresponding Weber polynomials (Section 5) for the case $D \equiv 3 \pmod 8$, resulting in a new CM variant for generating ECs of prime order (Section 3) and which also satisfies the three conditions for cryptographic strength posed in [5, Sec. V.7]. We also investigate the theoretical (Section 4) and experimental (Section 6) bit-precision for the construction of Hilbert and Weber polynomials and present a new approximation bound of the precision required for the construction of Weber polynomials in the case $D \equiv 3 \pmod 8$. Our experiments showed that the new approximate bound is very close to the actual precision needed.

Another important step of the CM method is the determination of the order p of the underlying prime field and the construction of the order m of the EC. This step is independent of the computation of Hilbert or Weber polynomials (a computation that can be performed off-line as we remarked above for various values of the discriminant D). We consider four different ways for implementing this step in the new CM variant (Section 3). The first method is similar to that in [17] and uses the modified Cornacchia's algorithm [8]. The second method generates p and m at random as it is described in [22]. The third method is the very efficient algorithm given in Baier's PhD thesis [4, p. 68]. The fourth method, which we introduce here, resembles the third one and constitutes a simpler and more space-efficient alternative to it.

The second contribution of this paper is a comparative experimental study (Section 6) regarding the four methods mentioned above for the computation of p and m in the construction of an EC, as well as an investigation of the time and bit precision requirements when constructing Weber polynomials on-line, in comparison with their Hilbert counterparts. Such an investigation is considerably important in resource-limited hardware systems which are requested to frequently change their security parameters. Our experiments revealed that despite the fact that Weber polynomials have a degree which is three times larger, the new CM variant using any of the four methods for computing p and m is considerably more space and time efficient than its counterpart which uses Hilbert polynomials. Regarding now the comparison of the four methods for computing

p and m, Baier's method turns out to be the most time-efficient, followed very closely by the new method we present here. Hence, the latter could be used as a simpler, space-efficient, and easy-to-use alternative.

2   A Brief Overview of Elliptic Curve Theory

In this section we review some basic concepts and results of elliptic curve theory. It is assumed that the reader has some familiarity with elementary number theory. A detailed presentation of EC theory and its cryptographic significance can be found in [5,23].

An *elliptic curve* over a finite field F_p, p a prime larger than 3, is denoted by $E(F_p)$ and it is comprised of all the points $(x, y) \in F_p$ (in affine coordinates) such that

$$y^2 = x^3 + ax + b, \tag{1}$$

with $a, b \in F_p$ satisfying $4a^3 + 27b^2 \neq 0$. These points, together with a special point denoted by $\mathcal{O}$ (the *point at infinity*) and a properly defined addition operation form an Abelian group. This is the *Elliptic Curve group* and the point $\mathcal{O}$ is its identity element (see [5,23] for more details on this group). Finally, let m be the *order* of $E(F_p)$, that is, the number of points in the group.

The difference between m and p is measured by the so-called *Frobenius trace* $t = p + 1 - m$ for which Hasse's theorem (see e.g., [5,23]) states that $|t| \leq 2\sqrt{p}$, implying that

$$p + 1 - 2\sqrt{p} \leq m \leq p + 1 + 2\sqrt{p}. \tag{2}$$

This is an important inequality that provides lower and upper bounds on the number of points in an EC group.

If $P \in E(F_p)$, then the *order* of the point P is the smallest positive integer n for which $nP = \mathcal{O}$. According to Langrange's theorem the order of a point $P \in E(F_p)$ must divide the order of the EC group and, thus, $mP = \mathcal{O}$ for any $P \in E(F_p)$. This also implies that the order of a point is never larger than the order of the EC.

Among the most important quantities defined for an EC $E(F_p)$ given by Eq. (1) are the *curve discriminant* Δ and the *j-invariant*. These two quantities are given by the equations $\Delta = -16(4a^3 + 27b^2)$ and $j = -1728(4a)^3/\Delta$.

For a specific j-invariant $j_0 \in F_p$ (where $j_0 \neq 0, 1728$) two ECs can be readily constructed. Let $k = j_0/(1728 - j_0) \bmod p$. One EC is given by Eq. (1) with $a = 3k \bmod p$ and $b = 2k \bmod p$. The other, which is called the *twist* of the first, is given by

$$y^2 = x^3 + ac^2x + bc^3 \tag{3}$$

where c is any quadratic non-residue in F_p. If m_1 and m_2 are the orders of an EC and its twist respectively, then $m_1 + m_2 = 2p + 2$. This implies that if one of the curves has order $p + 1 - t$, then its twist has order $p + 1 + t$, or vice versa (see [5, Lemma VIII.3]).

EC cryptosystems base their security on the difficulty of solving efficiently the discrete logarithm problem (DLP) on the EC group. To increase the difficulty of the solution of DLP (and hence the security of the EC cryptosystem), the order m of the EC should obey certain conditions which guarantee resistance to all known attacks [5, Sec. V.7]. An order m that satisfies these conditions is called *suitable*. We would like to note that sometimes there exists a fourth security requirement regarding the degree h of the class field polynomial. To the best of our knowledge, such requirement is only posed by the German Information Security Agency, which requires that h should be greater than 200. The reason is that there are few ECs produced from class field polynomials with smaller degrees and which may be amenable to specific attacks. However, no such attacks are known to date and this requirement is not part of the security requirements in any international security standard [2]. Despite this fact, we have taken into consideration such large values of h in our experimental study.

3   CM Method and Variants

The main idea behind the CM method is as follows (see [5,13] for a detailed discussion). According to Hasse's theorem the quantity $Z = 4p - (p + 1 - m)^2$ is positive. Thus, there is a unique factorization of Z of the form Dv^2, with D a square free positive integer. Consequently,

$$4p = u^2 + Dv^2 \qquad (4)$$

for some integer u satisfying

$$m = p + 1 \pm u. \qquad (5)$$

The number D is called a *CM discriminant* for the prime p and the EC has a *CM by D*. The CM method uses D to determine the j-invariant, and then constructs an EC of order $p + 1 - u$ or $p + 1 + u$.

The CM method requires as input a prime p. Then the smallest D is chosen that along with integers u, v satisfy Eq. (4). The next step is to check whether $p + 1 - u$ and/or $p + 1 + u$ is a suitable order. If none of them is suitable, then the whole process is repeated with another prime p as input. If one, however, is found to be suitable, then the Hilbert polynomial (see Section 4) is constructed and its roots (modulo p) are computed. A root of the Hilbert polynomial is the j-invariant we are seeking. Then, the EC and its twist are constructed as explained in Section 2. Since only one of these ECs has the required suitable order, it can be found using Langrange's theorem by picking random points P in each EC until a point is found in some curve for which $mP \neq \mathcal{O}$. Then, the other curve is the one we are seeking.

The most time consuming part of the CM method is the construction of the Hilbert polynomial, as it requires high precision floating point and complex arithmetic. In order to overcome the high computational requirements of this construction, a variant of the CM method was proposed in [22]. In contrast with the CM method described above, this variant does not start with a specific p but with a CM discriminant $D \equiv 3 \pmod 8$, since it requires that the EC order m

is prime (it is not hard to verify this justification for D). It then computes p and the EC order m (the primality of m is the only requirement for cryptographic strength set in [22]). The prime p is found by first picking randomly u and v of appropriate sizes, and then checking if $(u^2 + Dv^2)/4$ is prime. An important aspect of the variant concerns the computation of the Hilbert polynomials: since they depend only on D (and not on p), they can be constructed in a preprocessing phase and stored for later use. Hence, the burden of their construction can be excluded from the generation of the EC.

In [17], another variant to the CM method was given which uses Weber polynomials. This variant starts with a discriminant $D \not\equiv 3 \pmod 8$ and a specific prime p chosen at random, or from a set of prescribed primes. It then computes u and v using Cornacchia's algorithm [8] to solve Eq. (4), and requires that the resulting EC order m is suitable (cf. Section 2) but not necessarily prime. Moreover, like in [22], the Weber polynomials can be constructed in a preprocessing phase as they also depend only on D.

In the rest of the section, we shall describe yet another variant of the CM method which shares similarities with those in [22,17], but also differentiates from them in several aspects. The new variant generates ECs of *prime and suitable* order, hence taking as input values of D which are congruent to 3 (mod 8), and determines the pair (u, v) that specifies p using four alternative implementations. Moreover, since Weber polynomials are used, which for these values of D have a degree that is three times the degree of their corresponding Hilbert polynomials, a new transformation is presented for transforming Weber roots to Hilbert roots for this case (Section 5).

We are now ready to present the main steps of our variant. It starts with a CM discriminant $D \equiv 3 \pmod 8$ for the computation of the Weber polynomial[5], and then generates at random, or selects from a pool of precomputed *good* primes (e.g., Mersenne primes), a prime p and computes odd integers u, v such that $4p = u^2 + Dv^2$. Those odd integers u, v can be computed with four different ways, which we will outline below. If no such numbers u and v can be found, then take another prime p and repeat. Otherwise, proceed with the next steps, which are similar to those of the original CM method.

We now turn to the four different methods for computing u and v. The first is to use the modified Cornacchia's algorithm [7] as in [17]. The second is to generate them at random as it is done in [22]. The third method was proposed in [4, p. 68] and uses some clever heuristic in order to speed up the discovery of a suitable prime p. Despite its efficiency, this approach is quite complicated and uses an auxiliary table and two sieving arrays. Motivated by this approach we have developed a fourth method which is simpler and does not use any auxiliary tables or sieving arrays. The method is outlined in the following paragraph.

From Eqs. (4) and (5) we know that if we compute u and v such that $4p = u^2 + Dv^2$, then the order m of the EC is given either by $p + 1 - u$ or $p + 1 + u$ (recall that m is prime). We will denote the former by m^- and the latter with

---

[5] Although the variant defaults to the use of Weber polynomials, Hilbert polynomials can be used as well.

m^+. Since m is prime, u and v must be odd. In addition, u and v should not have common divisors because then p would not be a prime. With this observation in mind, we start our method by randomly picking odd u and v of appropriate sizes such that $u = 210x + 1$ and $v = 210y + 105$, where x, y are random numbers. In this way, u and v do not have common divisors the numbers 3, 5 and 7 ($3 \cdot 5 \cdot 7 = 105$). Then, we check whether $(u^2 + Dv^2)/4$ is prime. If it is, then we check for primality the quantities m^- and m^+. If $(u^2 + Dv^2)/4$ is not prime, then we add to u an integer keeping the same value for v, we calculate a new value for p, and repeat the whole process. An issue arises here as to what integer we add to u. Note, that when $u = 210x + 1 \equiv 1 \pmod{3}$, then $p \equiv 1 \pmod{3}$, $m^- \equiv 1 \pmod{3}$ and $m^+ \equiv 0 \pmod{3}$. Thus, only m^- can be a prime. If u were equal to $u = 210x + 107 \equiv 2 \pmod{3}$, then again $p \equiv 1 \pmod{3}$, but $m^- \equiv 0 \pmod{3}$ and $m^+ \equiv 1 \pmod{3}$. Therefore, at the first iteration of our method we select $u = 210x + 1$, at the second $u = 210x + 107$, and so on, in order to check for primality m^- and m^+ in tandem. In particular, if the choice $u = 210x + 1$ does not give primes p and m, then we add to u the number 106, in the next iteration we add 104, and so on. In this way, u is at one step congruent to 1 (mod 3) and at the next step congruent to 2 (mod 3).

As mentioned earlier, the other most complicated part of the CM method is the construction of the polynomials (Weber or Hilbert), which is addressed in the next Section.

4   Hilbert and Weber Polynomials

The only input for the construction of the Hilbert or the Weber polynomials, denoted by $H_D(x)$ and $W_D(x)$ respectively, is the CM discriminant D. They both require complex floating point arithmetic. The Hilbert polynomial $H_D(x)$, for a given positive value of D, is defined as

$$H_D(x) = \prod_\tau (x - j(\tau)) \tag{6}$$

for a set of values of τ obtained by the expression $\tau = (-\beta + \sqrt{-D})/2\alpha$, for all integers α, β, and γ that satisfy the following conditions: (i) $\beta^2 - 4\alpha\gamma = -D$, (ii) $|\beta| \le \alpha \le \sqrt{D/3}$, (iii) $\alpha \le \gamma$, (iv) $\gcd(\alpha, \beta, \gamma) = 1$, and (v) if $|\beta| = \alpha$ or $\alpha = \gamma$, then $\beta \ge 0$. The 3-tuple of integers $[\alpha, \beta, \gamma]$ satisfying these conditions, is a *primitive, reduced quadratic form* of $-D$, and τ is a root of the quadratic equation $\alpha z^2 + \beta z + \gamma = 0$. It can be proved that the set of primitive reduced quadratic forms of discriminant $-D$, denoted by $\mathcal{H}(-D)$, is finite. Moreover, it is possible to define an operation that gives to $\mathcal{H}(-D)$ the structure of an Abelian group whose neutral element is called the *principal form*. The principal form is equal to $[1, 0, D/4]$ if $D \equiv 0 \pmod{4}$ and $[1, -1, (D+1)/4]$ if $D \equiv 3 \pmod{4}$. This means that $\tau = \sqrt{-D}/2$ for the first principal form and $\tau = (1 + \sqrt{-D})/2$ for the second. The quantity $j(\tau)$ in Eq. (6) is called *class invariant* and is defined as follows. Let $z = e^{2\pi\sqrt{-1}\tau}$ and $h(\tau) = \frac{\Delta(2\tau)}{\Delta(\tau)}$, where $\Delta(\tau) = z\left(1 + \sum_{n \ge 1} (-1)^n \left(z^{n(3n-1)/2} + z^{n(3n+1)/2}\right)\right)^{24}$. Then, $j(\tau) = \frac{(256h(\tau)+1)^3}{h(\tau)}$.

If h is the *degree* or *class number* of $H_D(x)$, the bit precision required for the generation of $H_D(x)$ according to [18] is

$$\text{H-Prec}(D) \approx \frac{\ln 10}{\ln 2}(h/4 + 5) + \frac{\pi\sqrt{D}}{\ln 2}\sum_{\tau}\frac{1}{\alpha}$$

where the sum runs over the same values of τ as the product in Eq. (6). Note that this is much smaller than the precision given in [1,5].

The Weber polynomials are defined using the Weber functions (see [1,13]):

$$f(y) = q^{-1/48}\prod_{m=1}^{\infty}(1 + q^{(m-1)/2}) \qquad f_1(y) = q^{-1/48}\prod_{m=1}^{\infty}(1 - q^{(m-1)/2})$$

$$f_2(y) = \sqrt{2}\; q^{1/24}\prod_{m=1}^{\infty}(1 + q^m) \qquad \text{where } q = e^{2\pi y\sqrt{-1}}.$$

Then, the Weber polynomial $W_D(x)$, which has degree $3h$ as $D \equiv 3 \pmod 8$, is defined as

$$W_D(x) = \prod_{\ell}(x - g(\ell)) \tag{7}$$

where $\ell = \frac{-b+\sqrt{-D}}{a}$ satisfies the equation $ay^2 + 2by + c = 0$ for which $4b^2 - 4ac = -4d$, where $d = D/4$ if $D \equiv 0 \pmod 4$, and $d = D$ if $D \equiv 3 \pmod 4$. Let $\zeta = e^{\pi\sqrt{-1}/24}$. The class invariant $g(\ell)$ for $W_D(x)$ is defined by

$$g(\ell) = \begin{cases} \zeta^{b(c-a-a^2c)} \cdot f(\ell) & \text{if } 2 \nmid a \text{ and } 2 \nmid c \\ -(-1)^{\frac{a^2-1}{8}} \cdot \zeta^{b(ac^2-a-2c)} \cdot f_1(\ell) & \text{if } 2 \nmid a \text{ and } 2 \mid c \\ -(-1)^{\frac{c^2-1}{8}} \cdot \zeta^{b(c-a-5ac^2)} \cdot f_2(\ell) & \text{if } 2 \mid a \text{ and } 2 \nmid c \end{cases} \tag{8}$$

if $D \not\equiv 0 \pmod 3$, and

$$g(\ell) = \begin{cases} \frac{1}{2}\zeta^{3b(c-a-a^2c)} \cdot f^3(\ell) & \text{if } 2 \nmid a \text{ and } 2 \nmid c \\ -\frac{1}{2}(-1)^{\frac{3(a^2-1)}{8}} \cdot \zeta^{3b(ac^2-a-2c)} \cdot f_1^3(\ell) & \text{if } 2 \nmid a \text{ and } 2 \mid c \\ -\frac{1}{2}(-1)^{\frac{3(c^2-1)}{8}} \cdot \zeta^{3b(c-a-5ac^2)} \cdot f_2^3(\ell) & \text{if } 2 \mid a \text{ and } 2 \nmid c \end{cases} \tag{9}$$

if $D \equiv 0 \pmod 3$. In [25], an upper bound of $v_0 + \frac{\pi\sqrt{D}}{\ln 2}\sum_{\ell}\frac{1}{a}$ is given for the precision required for the construction of $W_D(x)$, where the sum runs over the same values of ℓ as the product in Eq. (7) and v_0 is a positive constant that handles round-off errors (typically $v_0 = 33$). In [18] a more accurate precision estimate is given for the computation of Weber polynomials with discriminants $D \equiv 7 \pmod 8$. In particular, the bit precision in this case is given by

$$\text{W-Prec}(D) \approx \frac{\ln 10}{\ln 2}\left(\frac{h/4 + 5 + \frac{\pi\sqrt{D}}{\ln 10}\sum_{\tau}\frac{1}{\alpha}}{47} + 1\right)$$

where τ takes the same values as in the product in Eq. (6) for $H_D(x)$. This precision estimate however, can not be used in the case $D \equiv 3 \pmod 8$ which is of our concern. For this reason, we provide in the following lemma a new precision estimate specifically for this case.

Lemma 1. *The bit precision required for the construction of Weber polynomials with discriminant* $D \equiv 3 \pmod 8$ *and* $D \not\equiv 0 \pmod 3$ *is approximately* $3h + \frac{\pi\sqrt{D}}{24\ln 2}\sum_{\ell}\frac{1}{a}$, *where the sum runs over the same values of* ℓ *as the product* $W_D(x) = \prod_{\ell}(x - g(\ell))$. *For the case of* $D \equiv 3 \pmod 8$ *and* $D \equiv 0 \pmod 3$ *the approximate precision becomes* $3h + \frac{\pi\sqrt{D}}{8\ln 2}\sum_{\ell}\frac{1}{a}$.

Proof. From the proof of Proposition (B4.4) in [16], if the Weber polynomial is written in the form $W_D(x) = x^{3h} + w_{3h-1}x^{3h-1} + \ldots + w_1 x + w_0$, then $|w_i| \leq 2^{3h}M$, where $M = \prod_{\ell}\max(1, |g(\ell)|)$. This means that the bit precision required for the coefficient w_i of the polynomial is $\log_2(|w_i|) \leq 3h + \log_2 M \leq 3h + \sum_{\ell}\log_2(|g(\ell)|)$. Therefore, the bit precision required for the construction of the whole polynomial (i.e., the construction of its coefficients) is at most $3h + \sum_{\ell}\log_2(|g(\ell)|)$.

For the case $D \equiv 3 \pmod 8$ and $D \not\equiv 0 \pmod 3$, the precision required by each $g(\ell)$ is the same with the precision required by $f(\ell)$, $f_1(\ell)$ or $f_2(\ell)$ as it is evident from Eq. (8). In addition, it is known that $j(z) = \frac{(f^{24}(z)-16)^3}{f^{24}(z)} = \frac{(f_1^{24}(z)+16)^3}{f_1^{24}(z)} = \frac{(f_2^{24}(z)+16)^3}{f_2^{24}(z)}$. These equalities imply that the precision needed for $j(\ell)$ is approximately 48 times the precision needed for $f(\ell)$, $f_1(\ell)$ or $f_2(\ell)$. Using the expansion of j in terms of its Fourier series [5], we obtain that $|j(\ell)| \approx |e^{-2\pi\sqrt{-1}\ell}| = e^{2\pi\sqrt{D}/a}$. Therefore, the bit precision that is required for the computation of $j(\ell)$ is $\log_2|j(\ell)| \approx \frac{2\pi\sqrt{D}}{a\ln 2}$ and, consequently, the precision required for $g(\ell)$ is given by $\log_2|g(\ell)| \approx \frac{2\pi\sqrt{D}}{48a\ln 2} = \frac{\pi\sqrt{D}}{24a\ln 2}$. This, in turn, results in the total bit precision requirements for the computation of the Weber polynomial: $3h + \frac{\pi\sqrt{D}}{24\ln 2}\sum_{\ell}\frac{1}{a}$.

In the case $D \equiv 3 \pmod 8$ and $D \equiv 0 \pmod 3$, the precision required by $g(\ell)$ is three times the precision required by $f(\ell)$, $f_1(\ell)$ or $f_2(\ell)$ as it is evident from Eq. (9). Using an analysis similar to the analysis used in the previous case, we obtain that the bit precision requirements in this case is given by $3h + \frac{\pi\sqrt{D}}{8\ln 2}\sum_{\ell}\frac{1}{a}$ which completes the proof of the lemma. □

5   Transforming Weber Roots to Hilbert Roots

In this section we elaborate on the transformation of roots of Weber polynomials to roots of the corresponding (generated from the same discriminant value D) Hilbert polynomials. Note that for the particular case we consider ($D \equiv 3 \pmod 8$), the degree of the Weber polynomial is *three* times larger that the degree of its Hilbert counterpart, and this introduces an additional difficulty. We start with some basic relationships between the Weber functions and $j(z)$ (defined in the previous section). In particular,

$$f(z)f_2\left(\frac{1+z}{2}\right) = e^{\pi\sqrt{-1}/24}\sqrt{2} \tag{10}$$

$$j(z) = \frac{(f^{24}(z) - 16)^3}{f^{24}(z)} = \frac{(f_1^{24}(z) + 16)^3}{f_1^{24}(z)} = \frac{(f_2^{24}(z) + 16)^3}{f_2^{24}(z)}. \tag{11}$$

Hence, $f^{24}(z)$, $-f_1^{24}(z)$, and $-f_2^{24}(z)$ are the roots of the cubic equation $(x - 16)^3 - xj(z) = 0$.

It can be proved that any transformation of a real root of a weber polynomial to a real root of the corresponding Hilbert polynomial holds also for the roots of the polynomials when taken (mod p). Suppose R_W is a real root of $W_D(x)$ to be transformed to the corresponding real root R_H of $H_D(x)$. In addition, $R_H = j(\tau)$, where τ corresponds to the principal form. First, R_W is transformed into one of the quantities $f^{24}(\tau)$, $-f_1^{24}(\tau)$ or $-f_2^{24}(\tau)$ (we will denote either of these quantities by A) and we set $R_H = (A - 16)^3/A$. The most complex part of the transformations is the first, which depends on the discriminant D. For different values of D, different class invariants are used, which in turn, define the relationship between R_W and the Weber functions.

The class invariant for $D \equiv 3 \pmod{8}$ and $D \not\equiv 0 \pmod 3$ is $f(\sqrt{-D})$. That is, $R_W = f(\sqrt{-D})$. The principal form for such discriminants is $[1, 1, (D+1)/4]$, and hence one of the roots of Eq. (11) is $f^{24}(\tau)$, $-f_1^{24}(\tau)$, or $-f_2^{24}(\tau)$, where $\tau = (1 + \sqrt{-D})/2$. According to Eq. (10)

$$f_2(\tau) = f_2\left(\frac{1 + \sqrt{-D}}{2}\right) = e^{\frac{\pi\sqrt{-1}}{24}}\sqrt{2}f^{-1}(\sqrt{-D}) = e^{\frac{\pi\sqrt{-1}}{24}}\sqrt{2}R_W^{-1}.$$

Consequently, $f_2^{24}(\tau) = -2^{12}R_W^{-24}$ and since $A = -f_2^{24}(\tau)$, we obtain

$$R_H = \frac{(A - 16)^3}{A} = \frac{(2^{12}R_W^{-24} - 16)^3}{2^{12}R_W^{-24}}.$$

The class invariant for $D \equiv 3 \pmod 8$ and $D \equiv 0 \pmod 3$ is $f^3(\sqrt{-D})/2$. That is, $R_W = f^3(\sqrt{-D})/2$. The principal form is, again, $[1, 1, (D+1)/4]$ and one of the roots of Eq. (11) is $f^{24}(\tau)$, $-f_1^{24}(\tau)$, or $-f_2^{24}(\tau)$, where $\tau = (1+\sqrt{-D})/2$. Following the same procedure as before we obtain that $f_2^{24}(\tau) = -2^{12}f^{-24}(\sqrt{-D}) = -2^4R_W^{-8}$. Since $A = -f_2^{24}(\tau)$, then

$$R_H = \frac{(A - 16)^3}{A} = \frac{(2^4R_W^{-8} - 16)^3}{2^4R_W^{-8}}.$$

6   Implementation and Experimental Results

As mentioned in the introduction, one of our main concerns was to investigate the efficiency of implementing CM variants in resource-limited hardware devices (e.g., embedded systems). For that reason and for reasons of proper comparison, we have made all of our implementations in a unified framework using the same language and software libraries. Since the vast majority of language tools developed for such devices are based on ANSI C, we have made all of our implementations in this language using the (ANSI C) GNUMP [10] library for high precision floating point arithmetic and also for the generation and manipulation of integers of unlimited precision. Our goal was to boost portability as well as adaptability to the development tools for resource-limited hardware devices. Note that there are highly efficient and optimized C++ libraries (e.g., LiDIA [14]) which however result in executables of a few MB, since they call dynamically linked libraries at run time. In contrast, our code does not call any such libraries at runtime. In particular, we have carried out our implementations and experiments on a Pentium III (933 MHz) running Linux and equipped with 256 MB of main memory. The Weber (resp. Hilbert) version of our code had size 53KB (resp. 49KB) including the code for the generation of the polynomials; exclusion of the latter (i.e., when polynomials are computed off-line) reduces the code size to 29KB if the modified Cornacchia's algorithm is used, to 25KB if p and m are selected at random as it is done in [22], to 28KB if Baier's algorithm [4, p. 68] is used, and to 26KB if the new method (described in Section 3) is used. All reported experimental values are averages over 3000 ECs for each value of the discriminant D. We considered two prime field sizes, 192 and 224 bits, which are typically used in such experiments.

Our experiments first focused on the bit precision and the time requirements needed for the construction of Hilbert and Weber class field polynomials. We have considered various values of D and h and made several experiments. We observed a big difference in favor of Weber polynomials both w.r.t. precision and time. This was evident even for small values of D and h. Figure 1(left) illustrates the actual and the approximate estimate of the bit precision for both Weber and Hilbert polynomials.

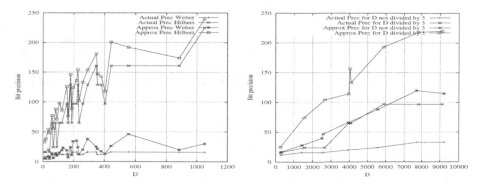

Fig. 1. Bit precision for the construction of Hilbert and Weber polynomials (left), and for the construction of Weber polynomials only (right).

As it is evident from the figure, there is a large difference in the required precision between the two types of polynomials. The difference grows considerably larger for bigger values of D and h. We also observe (see Fig. 1(left)) that the approximate precision estimates are very close to the actual precision used in the implementation. For Hilbert polynomials the approximation from

Eq. (4) was used, while for Weber polynomials that of Lemma 1. Regarding the precision requirements of Weber polynomials and their theoretical estimates, illustrative results are reported in Figure 1(right). It is clear that the precision required for the case of $D \equiv 0 \pmod 3$ is bigger than the precision required for $D \not\equiv 0 \pmod 3$ for similar values of D and h. The approximate precision is larger than the actual precision for all values of D. The difference in the precision requirements of Weber polynomials for the two cases of D (divisible or not divisible by 3) is also reflected in the time requirements for their construction, shown in Figure 2(left). The degree h of the polynomials ranges from 50 to 150, while D ranges from 11299 to 69315 (for $D = 69211$ and $h = 150$ the time for the construction of the polynomial is only 7.57 seconds). The difference between these two cases can be readily explained from the EC theory: the class invariants for such values of D are raised to the power of three, and since they increase in magnitude the time requirements are expected to be much larger than the requirements for Weber polynomials corresponding to other values of the discriminant. This fact implies that values of D divisible by 3 should be avoided.

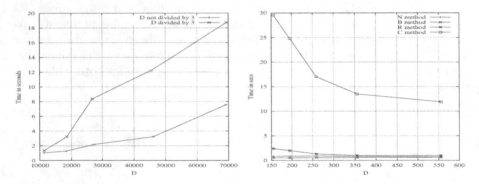

Fig. 2. Time in seconds for the construction of Weber polynomials (left) and for the computation of p, and m in the 224-bit field (right).

We next turn to the four methods for the calculation of the prime order p of the underlying field and the prime (and suitable) order m of the EC. We shall refer to these methods as R (random choice used in [22]), C (modified Cornacchia's algorithm), B (Baier's algorithm in [4, p. 68]), and N (new method). We have made several experiments both in the 192-bit and in the 224-bit fields with various values of D and h. We report on the most representative results in Figures 2(right), and 3. Figure 2(right) presents the time requirements of the four methods for various discriminants D in the 224-bit field. Clearly, C is by far the slowest, even for small values of h ($h \leq 10$ in Fig. 2(right)); this is due to its time complexity which is $O(\log^4 p)$. Hence, we do not consider C when reporting results with larger values of D and h, and concentrate on the comparison among methods R, B, and N. The difference in efficiency among these three methods can be seen in Figure 3. Figure 3(left) involves values of D ranging from 163 to 2099, and values of h ranging from 10 to 20, while Figure 3(right) involves values of (D, h) in $\{(125579, 200), (184091, 250), (223739, 300), (294971, 350), (428819, 400), (539579, 450)\}$.

In either case, we observe a similar behavior in the relative efficiency among the three methods: R is the most time consuming, while the most efficient is B. The new method (N) is slightly slower than B, but it is simpler and uses less memory. The difference between R, and B or N becomes more apparent as D and h increase (cf. Fig. 3(right)). We would also like to note that the timings obtained by our implementation of B using GNUMP are very close to those reported in [4], which were based on a C++ implementation of B using the advanced C++ library LiDIA [14] and carried out on a similar machine.

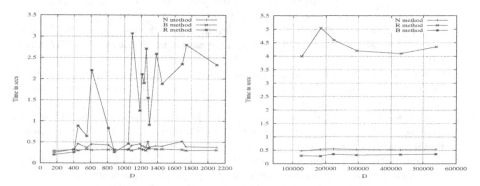

Fig. 3. Time requirements for the computation of p and m for various degrees $h \in [10, 20]$ (left) and $h \in [200, 400]$ (right).

References

1. A.O.L. Atkin and F. Morain, Elliptic curves and primality proving, *Mathematics of Computation* 61(1993), pp. 29-67.
2. H. Baier and J. Buchmann, Efficient construction of cryptographically strong elliptic curves, in *Progress in Cryptology* – INDOCRYPT 2000, LNCS Vol. 1977 (Springer-Verlag, 2000), pp. 191-202.
3. H. Baier, Elliptic Curves of Prime Order over Optimal Extension Fields for Use in Cryptography, in *Progress in Cryptology* – INDOCRYPT 2001, LNCS Vol. 2247 (Springer-Verlag, 2001), pp. 99-107.
4. H. Baier, Efficient Algorithms for Generating Elliptic Curves over Finite Fields Suitable for Use in Cryptography, PhD Thesis, Dept. of Computer Science, Technical Univ. of Darmstadt, May 2002.
5. I. Blake, G. Seroussi, and N. Smart, *Elliptic curves in cryptography*, London Mathematical Society Lecture Note Series 265, Cambridge Univ. Press, 1999.
6. D. Boneh, B. Lynn, and H. Shacham, Short signatures from the Weil pairing, in *ASIACRYPT 2001*, LNCS 2248, pp. 514-532, Springer-Verlag, 2001.

7. H. Cohen, *A Course in Computational Algebraic Number Theory*, Graduate Texts in Mathematics, **138**, Springer-Verlag, Berlin, 1993.
8. G. Cornacchia, Su di un metodo per la risoluzione in numeri interi dell' equazione $\sum_{h=0}^{n} C_h x^{n-h} y^h = P$, *Giornale di Matematiche di Battaglini* 46 (1908), pp. 33-90.
9. A. Enge and F. Morain, Comparing Invariants for Class Fields of Imaginary Quadratic Fields, in *ANTS V*, LNCS Vol. 2369, pp. 252-266, 2002.
10. GNU multiple precision library, edition 3.1.1, September 2000. Available at: http://www.swox.com/gmp.
11. N. Gura, H. Eberle, and S.C. Shantz, Generic Implementations of Elliptic Curve Cryptography using Partial Reduction, in *Proc. 9th ACM Conf. on Computer and Communications Security* – CCS'02, pp. 108-116.
12. A. Herzberg, M. Jakobsson, S. Jarecki, H. Krawczyk, and M. Yung, Proactive Public Key and Signature Systems, in *Proc. 4th ACM Conf. on Computer and Communications Security* – CCS'97, pp. 100-110.
13. IEEE P1363/D13, *Standard Specifications for Public-Key Cryptography*, 1999. http://grouper.ieee.org/groups/1363/tradPK/draft.html.
14. LiDIA. *A library for computational number theory*, Technical University of Darmstadt. Available from http://www.informatik.tu-darmstadt.de/TI/LiDIA/Welcome.html.
15. E. Kaltofen, T. Valente, and N. Yui, An Improved Las Vegas Primality Test, in *Proc. ACM-SIGSAM 1989 International Symposium on Symbolic and Algebraic Computation*, pp. 26-33, 1989.
16. E. Kaltofen and N. Yui, Explicit construction of the Hilbert class fields of imaginary quadratic fields by integer lattice reduction. Research Report 89-13, Renseelaer Polytechnic Institute, May 1989.
17. E. Konstantinou, Y. Stamatiou, and C. Zaroliagis, On the Efficient Generation of Elliptic Curves over Prime Fields, in *Cryptographic Hardware and Embedded Systems* – CHES 2002, LNCS Vol. 2523 (Springer-Verlag, 2002), pp. 333-348.
18. G.J. Lay and H. Zimmer, Constructing Elliptic Curves with Given Group Order over Large Finite Fields, in *Algorithmic Number Theory* – ANTS-I, Lecture Notes in Computer Science Vol. 877, Springer-Verlag, pp. 250-263, 1994.
19. F. Morain, Building Cyclic Elliptic Curves Modulo Large Primes, in *Advances in Cryptology* – *Eurocrypt '91*, LNCS 547 (Springer Verlag), pp. 328-336, 1991.
20. V. Müller and S. Paulus, On the Generation of Cryptographically Strong Elliptic Curves, preprint, 1997.
21. Y. Nogami and Y. Morikawa, Fast generation of elliptic curves with prime order over $F_{p^{2c}}$, in *Proc. of the International workshop on Coding and Cryptography*, March 2003.
22. E. Savaş, T.A. Schmidt, and Ç.K. Koç, Generating Elliptic Curves of Prime Order, in *Cryptographic Hardware and Embedded Systems* – CHES 2001, LNCS Vol. 2162 (Springer-Verlag, 2001), pp. 145-161.
23. J. H. Silverman, *The Arithmetic of Elliptic Curves*, Springer, GTM 106, 1986.
24. A.-M. Spallek, *Konstruktion einer elliptischen Kurve über einem endlichen Körper zu gegebener Punktgruppe*, Master Thesis, Universität GH Essen, 1992.
25. T. Valente, *A distributed approach to proving large numbers prime*, Rensselaer Polytechnic Institute Troy, New York, PhD Thesis, August 1992.
26. A. Weng, *Konstruktion kryptographisch geeigneter Kurven mit komplexer Multiplikation*, PhD thesis, Institut für Experimentelle Mathematik, Universität GH Essen, 2001.

Counting Points on an Abelian Variety over a Finite Field

Farzali A. Izadi[1] and V. Kumar Murty[1,2]

[1] GANITA Lab, Department of Mathematics and Computational Sciences,
University of Toronto at Mississauga, 3359 Mississauga Road North, Mississauga,
ON L5L 1C6, CANADA
[2] Department of Mathematics, University of Toronto, 100 St. George Street, Toronto,
ON M5S 3G3, CANADA,
murty@math.toronto.edu

Abstract. Matsuo, Chao and Tsujii [16] have proposed an algorithm
for counting the number of points on the Jacobian variety of a hyperelliptic curve over a finite field. The Matsuo-Chao-Tsujii algorithm is
an improvement of the 'baby-step-giant-step' part of the Gaudry-Harley
scheme. This scheme consists of two parts: firstly to compute the residue
modulo a positive integer m of the order of a given Jacobian variety,
and then to search for the actual order by a square-root algorithm. In
this paper, following the Matsuo-Chao-Tsujii algorithm, we propose an
improvement of the square-root algorithm part in the Gaudry-Harley
scheme by optimizing the use of the residue modulo m of the characteristic polynomial of the Frobenius endomorphism of an Abelian variety. It
turns out that the computational complexity is $O\left(q^{\frac{4g-2+i^2-i}{8}}/m^{\frac{i+1}{2}}\right)$,
where i is an integer in the range $1 \leq i \leq g$. We will show that for each
g and each finite field $\mathbb{F}_q$ of $q = p^n$ elements, there exists an i which
gives rise to the optimum complexity among all three corresponding algorithms.

1 Introduction

The use of hyperelliptic curves (or more precisely, their Jacobians) in public-key
cryptography was first proposed by Koblitz [12] in 1989, following the publication
of elliptic curve cryptography, introduced independently by both Koblitz [11] and
Miller [17] in 1987. One feature of Jacobians of hyperelliptic curves of genus > 1
(and more generally, of Abelian varieties of dimension > 1) is that it is possible
to work over a smaller base field to achieve the same level of security as elliptic
curves. The security of such systems depends in an essential way on the orders
of the group of points, either of the Jacobian varieties in the case of hyperelliptic
curves or Abelian varieties in general. For a secure system one should select an
Abelian variety over a finite field $\mathbb{F}_q$ such that the order, $\#A(\mathbb{F}_q)$, of the group
of points has a large prime divisor. However, if a random variety is chosen, then
it is necessary to have an efficient algorithm for computing its order $\#A(\mathbb{F}_q)$.

T. Johansson and S. Maitra (Eds.): INDOCRYPT 2003, LNCS 2904, pp. 323–333, 2003.
© Springer-Verlag Berlin Heidelberg 2003

Therefore, computing the order of a random variety is one of the most important problems in the construction of secure cryptosystems.

A significant amount of research has been done for computing the number of points on different classes of Abelian varieties. Among other things, extensive work has been done to compute the order of the Jacobian varieties of hyperelliptic curves over finite fields of arbitrary characteristic. For small characteristic efficient algorithms can be found in [2], [4],[5],[7],[10],[13]. These algorithms make it possible to construct cryptosystems with key sizes appropriate for practical usage (e.g., 160 bits) over finite fields of small characteristic. For large characteristic, there are some theoretical results [1], [8],[9],[18], but a practical algorithm was first proposed and implemented by Gaudry-Harley [3], [6].

The Gaudry-Harley algorithm consists of two parts: firstly to compute the residue modulo a positive integer m of the order of a given Jacobian variety, and then to search for the order by a square-root algorithm. Matsuo, Chao and Tsujii had the idea that if the entire characteristic polynomial of Frobenius is known modulo m, then the Gaudry-Harley algorithm could be improved. In this paper, following the Matsuo-Chao-Tsujii algorithm, we propose a further improvement of the square-root algorithm part in the Gaudry-Harley scheme by optimizing the use of the residue modulo m of the characteristic polynomial of the Frobenius endomorphism of an Abelian variety.

We cast our work in the context of Abelian varieties rather than restricting ourselves to hyperelliptic Jacobians as this seems to be the natural environment for the theoretical discussion. From a practical point of view, computing on an Abelian variety that is not a Jacobian is still in the early stages of research. Results of the kind presented here might serve as a guide in that work.

2   Abelian Varieties over Finite Fields and Their Characteristic Polynomials

Let p be an odd prime, $\mathbb{F}_q$ a finite field of order q with $\mathrm{char}(\mathbb{F}_q) = p$. Let g be a positive integer and A be a Abelian variety of dimension g over $\mathbb{F}_q$. Then the characteristic polynomial $P_A(t)$ of the q-th power Frobenius endomorphism of A is a monic polynomial of degree $2g$ given by

$$\begin{aligned}
P_A(t) = t^{2g} - s_1 t^{2g-1} + \cdots + (-1)^g s_g t^g \\
+ (-1)^{g-1} s_{g-1} q t^{g-1} + \cdots - s_1 q^{g-1} t + q^g,
\end{aligned} \tag{1}$$

where $s_i \in \mathbb{Z}$ are numbers satisfying

$$|s_i| \le \binom{2g}{i} q^{\frac{i}{2}}. \tag{2}$$

By the Riemann hypothesis for Abelian varieties [19] the roots of $P_A(t)$ in (1) have modulus $\sqrt{q}$. Furthermore, the number $\#A(\mathbb{F}_q)$ is completely determined by $P_A(t)$ according to the formula

$$\#A(\mathbb{F}_q) = P_A(1).$$

It follows that

$$\#A(\mathbb{F}_q) = P_A(1) = q^g + 1 - s_1(q^{g-1} + 1) + s_2(q^{g-2} + 1) + \\ \cdots + (-1)^{g-1} s_{g-1}(q+1) + (-1)^g s_g. \tag{3}$$

An important consequence of this is that the number $\#A(\mathbb{F}_q)$ is constrained to a rather small interval, the Hasse-Weil interval:

$$[(\sqrt{q} - 1)^{2g}] \leq \#A(\mathbb{F}_q) \leq [(\sqrt{q} + 1)^{2g}]. \tag{4}$$

This interval has width w bounded by $2gq^{g-\frac{1}{2}} + 4^g q^{g-1}$.
Consequently we have

$$|\#A(\mathbb{F}_q) - (q^g + 1)| \leq 2gq^{g-\frac{1}{2}} + 4^g q^{g-1}. \tag{5}$$

3 Standard Baby-Step-Giant-Step Algorithm and Its Refinement

In this section, as a preliminary to the upcoming discussions, we describe the standard baby-step-giant-step algorithm and its refinement applied by Gaudry-Harley for point counting of the Jacobian varieties of the hyperelliptic curves [3], [6]. Set $N = \#A(\mathbb{F}_q) - (q^g + 1)$. From (5) we have $|N| \leq 2gq^{g-\frac{1}{2}} + 4^g q^{g-1}$. Let L be a positive integer (to be specified). Let us write

$$N = a + bL, \quad 0 \leq a < L, \quad b < (2gq^{g-\frac{1}{2}} + 4^g q^{g-1})/L.$$

Now suppose that $\#A(\mathbb{F}_q)$ is a prime number. By taking a random point $Q \in A(\mathbb{F}_q)$, we get

$$\#A(\mathbb{F}_q)Q = (N + q^g + 1)Q = O.$$

where O is the neutral element in the Jacobian group. This is equivalent to

$$(bL + q^g + 1)Q = -aQ.$$

Hence, to determine N, we need

$$\ll L + (2gq^{g-\frac{1}{2}} + 4^g q^{g-1})/L$$

computations. Now, if we choose

$$L = \left[\sqrt{2gq^{g-\frac{1}{2}} + 4^g q^{g-1}} \right],$$

then we need $\mathbf{O}(L)$ computations.

Next, suppose one is given $N_1 \in \mathbb{Z}$ such that for some $0 \leq t \in \mathbb{Z}$,

$$N = N_1 + mt, 0 \leq N_1 < m$$

for some small m. Then, the order $\#A(\mathbb{F}_q)$ can be obtained by searching for t among

$$0 \leq t < (2gq^{g-\frac{1}{2}} + 4^g q^{g-1})/m.$$

Now, to determine t, one can use a baby-step-giant-step algorithm as follows. For some L (to be specified), let

$$t = u + vL \quad 0 \leq u < L \quad 0 \leq v < (2gq^{g-\frac{1}{2}} + 4^g q^{g-1})/mL. \tag{6}$$

Then the values of $t = u + vL$ can be obtained by finding u and v as (6) such that

$$(N_1 + mu)Q = -mLvQ \tag{7}$$

for all $Q \in A(\mathbb{F}_q)$ by searching for a collision between RHS and LHS of (7). Assuming that $\#A(\mathbb{F}_q)$ is a prime and q is large enough, one can compute $\#A(\mathbb{F}_q)$ from the pair u and v obtained by the above algorithm as follows:

$$\#A(\mathbb{F}_q) - (q^g + 1) = N_1 + m(u + Lv).$$

As there are L choices for u and $2gq^{g-\frac{1}{2}} + 4^g q^{g-1}/mL$ choices for v, if we choose

$$L = q^{(2g-1)/4}(2g + 4^g q^{-1/2})^{1/2}/m^{1/2}. \tag{8}$$

then the above algorithm requires $\mathbf{O}(L)$ computations.

4   An Improved Baby-Step-Giant-Step Algorithm - Part I (M-C-T)

In this case, instead of $N \pmod{m}$ one supposes that $P_A(t) \pmod{m}$ is known. More precisely, we assume that for $i = 1, 2, \ldots, g$ we are given integers $r_i \in \mathbb{Z}$ such that

$$s_i = r_i + mt_i, \quad 0 \leq r_i < m. \tag{9}$$

According to (2) each t_i is bounded by

$$|t_i| \leq \binom{2g}{i} q^{\frac{i}{2}}/m. \tag{10}$$

Moreover, for some L (to be specified), let u_g, v_g be integers such that

$$t_g = u_g + v_g L, \quad 0 \leq u_g < L. \tag{11}$$

According to (10) for v_g we have

$$|v_g| < \binom{2g}{g} q^{\frac{g}{2}}/mL.$$

Now by substituting (9) and (11) into (3) we find that

$$\#A(\mathbb{F}_q) = q^g + 1 - r_1(q^{g-1} + 1) - m(q^{g-1} + 1)t_1 + r_2(q^{g-2} + 1)$$
$$+ m(q^{g-2} + 1)t_2 + \cdots + (-1)^{g-1}r_{g-1}(q + 1)$$
$$+ (-1)^{g-1}m(q + 1)t_{g-1} + (-1)^g r_g + (-1)^g m(u_g + v_g L).$$

Hence $\#A(\mathbb{F}_q)$ can be computed by finding the $(g + 1)$-tuple

$$(t_1, t_2, \ldots, t_{g-1}, v_g, u_g)$$

such that

$$\left(q^g + 1 - r_1(q^{g-1} + 1) - m(q^{g-1} + 1)t_1 + \right.$$
$$\left. \cdots + (-1)^g r_g + (-1)^g m v_g L\right) Q = (-1)^{g+1} m u_g Q \qquad (12)$$

for all $Q \in A(\mathbb{F}_q)$ in the corresponding ranges for the above $(g + 1)$-tuple. We search for a collision between the LHS and RHS of (12) among different candidates of the above $(g + 1)$-tuple. It turns out that there are $B(g)q^{\frac{g(g+1)}{4}}/m^g L$ choices for the g-tuple $(t_1, t_2, \ldots, t_{g-1}, v_g)$ and L choices for u_g, where

$$B(g) = 2^g \prod_{k=1}^{g} \binom{2g}{k}.$$

In order to make the algorithm most efficient, we choose

$$L = (B(g))^{1/2} q^{\frac{g(g+1)}{8}} / m^{\frac{g}{2}}.$$

Then, the algorithm requires the computation of $\mathbf{O}(L)$ point multiples.

5   An Improved Baby-Step-Giant-Step Algorithm - Part II

In this section, we propose an improved baby-step-giant-step algorithm for $g > 2$ which optimizes the use of s_i's (mod m) in the characteristic polynomial. To this end, we chop $P_A(1)$ in two parts in the following way:

(*) $P_A(1) = q^g + 1 - s_1 \left(q^{g-1} + 1\right) + \cdots + (-1)^i s_i(q^{g-i} + 1) + b,$

where $i \leq g$, and

$$|b| < 2(g - i) \binom{2g}{g} q^{g-(\frac{i+1}{2})}. \qquad (13)$$

Note that this bound is the result of the following approximations:

(a) for each j, $\binom{2g}{j} \leq \binom{2g}{g}$.

(b) for each $j \geq i$, $q^{g-(j+1)} \leq q^{g-(i+1)}$.

Now, suppose that for some m, one knows $b_1 \in \mathbb{Z}$, and $r_j \in \mathbb{Z}$ $(1 \leq j \leq i \leq g)$ such that

$$s_j = r_j + mt_j, \quad 0 \leq r_j < m, \quad \text{and} \tag{14}$$

$$b = b_1 + tm, \quad 0 \leq b_1 < m. \tag{15}$$

According to (2) and (13) each t_j and t are bounded by

$$|t_j| < \binom{2g}{j} q^{\frac{j}{2}}/m, \tag{16}$$

$$|t| < 2(g - i) \binom{2g}{g} q^{g - \left(\frac{i+1}{2}\right)}/m. \tag{17}$$

Furthermore, for some $0 < L \in \mathbb{Z}$ to be specified, let the integers u, v be such that

$$t = u + vL, \quad 0 \leq u < L. \tag{18}$$

According to (17), for v we have the bound

$$|v| < 2(g - i) \binom{2g}{g} q^{g - \left(\frac{i+1}{2}\right)}/mL. \tag{19}$$

Now by substituting (18) into (15) we get

$$b = b_1 + m(u + vL) = b_1 + mu + mvL. \tag{20}$$

Substituting this last relation (20) along with (14) into (*) we find that

$$\#A(\mathbb{F}_q) = q^g + 1 - (r_1 + mt_1)(q^{g-1} + 1) +$$
$$\cdots + (-1)^i(r_i + mt_i)(q^{g-i} + 1) + b_1 + mu + mvL.$$

Hence, the order $\#A(\mathbb{F}_q)$ can be computed by finding the $(i + 2)$-tuple

$$(t_1, t_2, \ldots, t_i, v, u)$$

such that

$$\left(q^g + 1 - (r_1 + mt_1)(q^{g-1} + 1) + \right.$$
$$\left. \cdots + (-1)^i(r_i + mt_i)(q^{g-i} + 1) + (b_1 + mvL)\right) Q = -muQ \tag{21}$$

for all $Q \in A(\mathbb{F}_q)$ in the corresponding ranges for the above $(i + 2)$-tuple. We search for a collision between the LHS and RHS of (21) among different candidates of the above $(i + 2)$-tuple. It turns out that there are

$$C(g, i) q^{\frac{4g - 2 + i^2 - i}{4}}/m^{i+1} L$$

choices for the $(i+1)$-tuple $(t_1, t_2, \ldots, t_i, v)$ and L choices for u, where

$$C(g, i) = 2^{i+2}(g - i) \binom{2g}{g} \prod_{j=1}^{i} \binom{2g}{j}.$$

To make the algorithm most efficient, we set

$$L = \left[(C(g, i))^{1/2} q^{\frac{4g-2+i^2-i}{8}} / m^{\frac{i+1}{2}} \right].$$

With this choice, the computational complexity of the algorithm is

$$\ll L \ll q^{\frac{4g-2+i^2-i}{8}} / m^{\frac{i+1}{2}}.$$

6   Theoretical Comparison and Conclusion

First of all, for $g = 2$, the Gaudry-Harley algorithm costs $O(q^{\frac{3}{4}}/\sqrt{m})$ time, whereas the Matsuo-Chao-Tsujii algorithm costs $O(q^{\frac{3}{4}}/m)$. Taking $i = 1$ in our optimized case we get $O(q^{\frac{3}{4}}/m)$ which is the same as the M-C-T bound. For $g = 3$, and $i = 1$, our bound is better than the Gaudry-Harley bound comparing $O(q^{\frac{5}{4}}/m)$ to $O(q^{\frac{5}{4}}/m^{\frac{1}{2}})$. For $g = 3$, $i = 2$, it coincides with the M-C-T bound, i.e., $O(q^{\frac{3}{2}}/m^{\frac{3}{2}})$. Thus, we see that if $m < q^{\frac{1}{2}}$, using $g = 3$ and $i = 1$ gives an algorithm that is faster than that of Gaudry-Harley and Matsuo-Chao-Tsujii.

In general, the value of i in each case should be selected so that the proposed algorithm achieves better complexity than the two previous ones. To meet this condition, we select a curve with a fixed genus g and try to find a best possible i for the different extensions $\mathbb{F}_{p^n}$ of the finite field $\mathbb{F}_p$. To be more precise, let us state the following proposition.

Proposition. *Let $\mathbb{F}_q$ be the finite field of characteristic p with the extension of degree n over the prime field $\mathbb{F}_p$. Then for either $(g \geq 2, n \geq 2)$, or $(n = 1, g \neq 3, 4, 5)$, and using $m = 2p$, the proposed algorithm is optimized by choosing $i = 1$, while for $n = 1$ and $g = 3, 4, 5$ it is optimized by choosing $i = 2$.*

Remark 1. In fact for the case of genus $g = 5$, we have the following options.

(a) For $i = 1$, we have

$$\log_p O(L_{PROP}) = \log_p O(L_{MCT}) < \log_p O(L_{GH})$$

where here, and below, we denote by L_{PROP} (respectively, L_{MCT}, L_{GH}) the choice of L in the proposed algorithm (respectively in the MCT and GH algorithms).

(b) For $i = 2$, we have

$$\log_p O(L_{PROP}) < \log_p O(L_{MCT}) < \log_p O(L_{GH}).$$

Proof. As we have already shown, the above three algorithms give rise to the following complexities:

$$L_1 = L_{GH} = q^{2g-1/4}/m^{1/2},$$

$$L_2 = L_{MCT} = q^{g(g+1)/8}/m^{g/2},$$

$$L_3 = L_{PROP} = q^{(4g-2+i^2-i)/8}/m^{(i+1)/2}.$$

By assuming $m = 2p$ and $q = p^n$, we get

$$L_1 = 2^{-1/2}p^{(4ng-2n-4)/8}$$

$$L_2 = 2^{-g/2}p^{(ng^2+ng-4g)/8}$$

$$L_3 = 2^{-(i+1)/2}p^{(4ng-2n+ni^2-ni-4i-4)/8}.$$

Up to some constants for the functions $M_k = 8 \log_p L_i (i = 1, 2, 3)$, we have
  $M_1(g, n) = n(4g - 2) - 4$,
  $M_2(g, n) = n(g^2 + g) - 4g$,
  $M_3(g, n, i) = n(4g - 2 + i^2 - i) - 4i - 4$.
In terms of the different curves of genus g we have the following different cases.
  Case(1). Let $g = 2$. Then for $n = 1$ we have
  $M_1(2, 1) = 2$, $M_2(2, 1) = -2$, $M_3(2, 1, i) = i^2 - 5i + 2$. Now for $i = 1$, we get
$M_3(2, 1, 1) = -2$. This shows that $M_3 = M_2 < M_1$.
Similarly for $n = 2$, we have
  $M_1(2, 2) = 8$, $M_2(2, 2) = 4$, $M_3(2, 2, i) = 2i^2 - 6i + 8$. Consequently, for $i = 1$,
we get $M_3(2, 2, 1) = 4$. Thus $M_3 = M_2 < M_1$. We summarize the above different
cases in the TABLE 1 for $g = 2$:

Extension	Optimum index	G-H	M-C-T	PROPOSED
n	i	$\log_p O(L)$	$\log_p O(L)$	$\log_p O(L)$
1	1	2	−2	−2
2	1	8	4	4
3	1	14	10	10
4	1	20	16	16
5	1	26	22	22
6	1	32	28	28

Table 1. Complexities of 3 different algorithms for a hyperelliptic curve of genus
2 over the different extensions $\mathbb{F}_{p^n}$ of the finite field $\mathbb{F}_p$

Likewise TABLE 2 - TABLE 4 illustrate the same analysis for $g = 3, 4, 5$, respectively.

Finally for all $g \geq 6$, and all n, the corresponding i is equal to 1, in which
we have

$$\log_p O(L_{PROP}) < \log_p O(L_{GH}) < \log_p O(L_{MCT})$$

It turns out that, except for a few cases, namely for $n = 1$, and $g = 3, 4, 5$,
which enforce the value $i = 2$, the choice for all other cases is $i = 1$. TABLE 5
illustrates these facts in a schematic way.

Extension	Optimum index	G-H	M-C-T	PROPOSED
n	i	$\log_p O(L)$	$\log_p O(L)$	$\log_p O(L)$
1	2	6	0	0
2	1	16	12	12
3	1	26	24	22
4	1	36	36	32
5	1	46	48	42
6	1	56	60	52

Table 2. Complexities of 3 different algorithms for a hyperelliptic curve of genus 3 over the different extensions $\mathbb{F}_{p^n}$ of the finite field $\mathbb{F}_p$

Extension	Optimum index	G-H	M-C-T	PROPOSED
n	i	$\log_p O(L)$	$\log_p O(L)$	$\log_p O(L)$
1	2	10	4	4
2	1	24	24	20
3	1	38	44	34
4	1	52	64	48
5	1	66	84	62
6	1	80	104	76

Table 3. Complexities of 3 different algorithms for a hyperelliptic curve of genus 4 over the different extensions $\mathbb{F}_{p^n}$ of the finite field $\mathbb{F}_p$

Remark 2. As we see from the different values of g, there are at most two different options for selecting the optimal index i to obtain a best possible complexity. These facts not only give rise to find the smallest value of M_3 comparing to M_1, and M_2, but also make it possible to achieve the smallest coefficient $C(g,i)$ as well. To see how this works, we need to compare the corresponding coefficients, i.e., $C_2 = (B(g))^{1/2}$ and $C_3 = (C(g,i))^{1/2}$. Without loss of generality, let us assume that $i = 1$ (this is the optimum value of i for $n \geq 2$). Then comparing the value of C_3 to C_2, we find that

$$(C_2/C_3)^2 = 2^g((2g^2 - g)/(g^2 - g)) \prod_{k=3}^{g-1} \binom{2g}{k}$$

which grows fast as g increases. This fact along with the previous one about $M_3 \leq M_2$ tells us how much time and space we can save in the baby-step part of the proposed algorithm comparing those factors to the M-C-T counterparts. Another advantage of the smallness of the index i lies in the fact that the number of cases in the giant-step part of the proposed algorithm is far fewer than that of M-C-T algorithm especially as g ($g \geq 3$) increases. This is because the *case-space* of the giant-step in the M-C-T algorithm i.e., $(t_1, t_2, .., t_{g-1}, v)$ is g dimensional while this space i.e., $(t_1, .., t_i, v)$ has dimension $(i+1)$ in the proposed algorithm, where i is equal to 1 except for a very few cases over the prime field $\mathbb{F}_p$. It follows

Extension	Optimum index	G-H	M-C-T	PROPOSED
n	i	$\log_p O(L)$	$\log_p O(L)$	$\log_p O(L)$
1	1, 2	14, 14	10, 10	10, 8
2	1	32	40	28
3	1	50	70	46
4	1	68	100	64
5	1	86	130	82
6	1	104	160	100

Table 4. Complexities of 3 different algorithms for a hyperelliptic curve of genus 5 over the different extensions $\mathbb{F}_{p^n}$ of the finite field $\mathbb{F}_p$

Genus of the curve	Degree of the extension	The optimum index
g	n	i
2	1	1
3	1	2
4	1	2
5	1	2
≥ 6	1	1
≥ 2	≥ 2	1

Table 5. The optimum i for different genera and different extensions

that for $g = 2$, the two algorithms coincide while for all $g \geq 3$, the algorithm proposed here is significantly faster.

7   Implementation Comparison and Conclusion

We have implemented our algorithm as well as the M-C-T counterpart by using the Magma algebra system [14]. For the time being, our main objective was merely to compare the practical performance of the proposed algorithm with the M-C-T analog. Because of this, less time was spent to optimize code. In particular, for the computation of $s_i \bmod p$ using the Cartier-Manin operator [15], we simply used the basic DFT which is slightly slower than the asymptotically fast FFT multiplication. For example, for a hyperelliptic curve of genus 2 over a field of $q = p^3$ elements with $p = 1342181$, (M-C-T. Example 1) the latter is approximately 5 times faster than the first technique. Though this can be a drawback from the computational viewpoint, it is not a key issue for the sake of comparison, as we copied every single bit of the coding from one algorithm to the other, except for the dimension of the case-space and its ranges.

To see the difference between the two algorithms practically, we have worked out numerous case of genus 3 hyperelliptic curves with prime order Jacobian over $\mathbb{F}_{p^2}$ for different primes p. Working on an Athlon XP 1400+ with 1GB of memory, the results of such computations yielded a speedup by a factor between 3-5 from the MCT algorithm. In almost all cases we also found that the MCT algorithm required space of of about 1.92 times that required for our proposed algorithm.

References

1. L. M. Adleman, and M. D. Huang, Counting rational points on curves and Abelian varieties over finite fields, ANTS-II, ed.H. Cohen, Lecture Notes in Computer Science, no.1122, pp.1-16, Springer-Verlag, 1996.
2. J. Denef and F. Vercauteren, An extension of Kedlaya's algorithm to Artin-Schreier curves in characteristic 2, ANTS-V, ed. C. Fieker and D. Kohel, Lecture Notes in Computer Science, no.2369, pp.308-323, Springer-verlag, 2002.
3. P. Gaudry, Algorithmique des courbes hyperellitiques et applications á la cryptologie, Ph.D. thesis, École polytechnique, 2000.
4. P. Gaudry, Algorithms for counting points on curves. Talk at ECC 2001, The fifth Workshop on elliptic Curve Cryptography, U. Waterloo, 2001.
5. P. Gaudry and N. Gürel, An extension of Kedlaya's point counting algorithm to superelliptic curves, Advances in Cryptology - ASIACRYPT2001, ed. C. Boyd, Lecture Notes in Computer Science, no.2248, pp.480-494, Springer-Verlag, 2001.
6. P. Gaudry and R. Harley, Counting points on hyperelliptic curves over finite fields, ANTS-IV, ed. W. Bosma, Lecture Notes in Computer Science, no.1838, pp.297-312, Springer-Velag,2000.
7. R. Harley, Counting points with the arithmetic-geometric mean, Rump talk at EUROCRYPT 2001. (joint work with J.F. Mestre and P. Gaudry).
8. M.D. Huang and D. Ierardi, Counting rational point on curves over finite fields, J. Symbolic Computation, 25(1998), 1-21.
9. W. Kampkötter, Explizite Gleichungen für Jacobische Varietäten hyperelliptischer Kurven, Ph.D. thesis, GH Essen, 1991.
10. K.S. Kedlaya, Counting points on hyperelliptic curves using Monsky-Washinitzer cohomology, J. Ramanujan Math. Soc., 16(2001), 323-338.
11. N. Koblitz, Elliptic curve cryptosystems, Math. Comp., 48(1987), 203-209.
12. N. Koblitz, Hyperelliptic curve cryptosystems. J of Cryptology, 1(1989), 139-150.
13. A. Lauder and D. Wan, Computing zeta functions of Artin-Schreier curves over finite fields, LMS J. Comput. Math., 5(2002), 33-55.
14. The magma algebraic system, http://www.maths.usyd.edu.au: 8000/u/magma/.
15. J. I. Manin, The Hasse-Witt matrix of an algebraic curve. Transl. Amer. Math. Soc., 45(1965), 245-264.
16. K. Matsuo, J. Chao and S. Tsujii, Baby step giant step algorithms in point counting of hyperelliptic curves, IEICE Trans. Fundamentals, vol.E86-A, no.4,April 2003.
17. V. Miller, Uses of elliptic curves in cryptography, Advances in Cryptology-Proc. Crypto,85, Lecture Notes in Compu. Sci., vol.218, Springer-Verlag, Berlin, 1986, pp. 417-426.
18. J. Pila, Frobenius maps of Abelian varieties and finding roots of unity in finite fields, Math. Comp., 55(1990), 745-763.
19. A. Weil, Variétés Abéliennes et courbes algébriques, Hermann, Paris, 1948.

Side Channel Attack on Ha-Moon's Countermeasure of Randomized Signed Scalar Multiplication

Katsuyuki Okeya[1] and Dong-Guk Han[2]

[1] Hitachi, Ltd., Systems Development Laboratory,
292, Yoshida-cho, Totsuka-ku, Yokohama, 244-0817, Japan
`ka-okeya@sdl.hitachi.co.jp`
[2] Center for Information and Security Technologies (CIST),
Korea University, Seoul, KOREA
`christa@cist.korea.ac.kr`

Abstract. Side channel attacks (SCA) are serious attacks on mobile devices. In SCA, the attacker can observe the side channel information while the device performs the cryptographic operations, and he/she can detect the secret stored in the device using such side channel information. Ha-Moon proposed a novel countermeasure against side channel attacks in elliptic curve cryptosystems (ECC). The countermeasure is based on the signed scalar multiplication with randomized concept, and does not pay the penalty of speed. Ha-Moon proved that the countermeasure is secure against side channel attack theoretically, and confirmed its immunity experimentally. Thus Ha-Moon's countermeasure seems to be very attractive. In this paper we propose a novel attack against Ha-Moon's countermeasure, and show that the countermeasure is vulnerable to the proposed attack. The proposed attack utilizes a Markov chain for detecting the secret. The attacker determines the transitions in the Markov chain using side channel information, then detects the relation between consecutive two bits of the secret key, instead of bits of the secret key as they are. The use of such relations drastically reduces the search space for the secret key, and the attacker can easily reveal the secret. In fact, around *twenty* observations of execution of the countermeasure are sufficient to detect the secret in the case of the standard sizes of ECC. Therefore, Ha-Moon's countermeasure is not recommended for cryptographic use.

Keywords: *Elliptic Curve Cryptosystem, Side Channel Attacks, SPA, DPA, Ha-Moon's Countermeasure, Finite Markov Chain*

1 Introduction

Mobile devices such as smart cards are penetrating in our daily life in order for us to be comfortable. Since mobile devices are equipped with scarce resources only, the cryptographic algorithms on them should be optimized. Above all, elliptic

T. Johansson and S. Maitra (Eds.): INDOCRYPT 2003, LNCS 2904, pp. 334–348, 2003.

curve cryptosystems (ECC) [Kob87, Mil86] are suitable for them because of their short key size. On the other hand, side channel attacks are serious attacks on the mobile devices [Koc96, KJJ99, Cor99]. If the implementation is careless, the attacker can reveal the secret information stored in the device, by observing the side channel information such as power consumption of the device. In order to resist against side channel attacks, several countermeasures were proposed. Ha-Moon's countermeasure of randomized signed scalar multiplication [HaM02] is one of them. It is a novel countermeasure since its security against side channel attacks was proved in a theoretical way and confirmed in an experimental way. In addition, the countermeasure does not pay the penalty of speed. Thus, Ha-Moon's countermeasure seems to be very attractive.

In this paper we propose a novel attack against Ha-Moon's countermeasure, and show that it breaks the countermeasure completely. Therefore, Ha-Moon's countermeasure is not recommended for cryptographic use. The proposed attack utilizes a finite Markov chain, which is converted from the original Markov chain proposed by Ha-Moon, in order to break the countermeasure. Note that the converted Markov chain and the original one are same in the sense that their input, output and intermediate values are same. While Ha-Moon observed that the transitions between the states in the original Markov chain do not provide the attacker with bits of the secret key as they are, we show that using transitions in the converted Markov chain, the attacker can reveal the *relation of consecutive two bits* of the secret key, that is, whether k_i is equal to k_{i+1} or not, where k_i and k_{i+1} respectively denote the i-th and the $(i + 1)$-th bit of the secret key k. The use of the relations drastically reduces the search space for the secret key. In fact, around *twenty* observations of the execution of Ha-Moon's countermeasure are sufficient to detect the secret key k in the case of the standard sizes of ECC.

This paper is organized as follows: Section 2 reviews side channel attacks and Ha-Moon's countermeasure. Section 3 proposes the proposed attack against Ha-Moon's countermeasure, and shows that it breaks the countermeasure completely. Finally, Section 4 concludes the paper.

2   Side Channel Attacks and Their Countermeasures

In this section we review side channel attacks and their countermeasures. First, we mention side channel attacks and classify their countermeasures into four categories. The randomized addition chains type is one of them. Then we introduce Ha-Moon's countermeasure, which is a countermeasure of the randomized addition chains type.

2.1   Side Channel Attacks

Side channel attacks (SCA) are allowed to access the additional information linked to the operations using the secret key, e.g., timings, power consumptions, etc [Koc96, KJJ99]. The attack aims at guessing the secret key (or some related information). For example, Binary_Method can be broken by the SCA. It calculates the elliptic addition (ECADD) if and only if the i-th bit is not zero. The standard implementation of ECADD is different from that of the elliptic doubling (ECDBL) [CMO98], and thus the ECADD in the scalar multiplication can be detected using SCA.

Binary_Method

INPUT A point P, and $k = \sum_{i=0}^{n-1} k_i 2^i$, $k_i \in \{0, 1\}$
OUTPUT $Q = kP$

---

1. $Q \leftarrow \mathcal{O}$
2. for $i = n - 1$ downto 0
   2.1. $Q \leftarrow \text{ECDBL}(Q)$
   2.2. if $k_i = 1$ then $Q \leftarrow \text{ECADD}(Q, P)$
3. Return Q

Here, $\mathcal{O}$ denotes the identity element of the elliptic addition, namely the point at infinity.

If the attacker is allowed to observe the side channel information only a few times, it is called the simple power analysis (SPA). The above SCA is a typical example of SPA. In SPA, the sequence of operations such as ECADD and ECDBL is detected by using side channel information, and the secret is revealed from the sequence. If the attacker can analyze several side channel information using a statistical tool, it is called the differential power analysis (DPA). The standard DPA utilizes the correlation function that can distinguish whether a specific bit is related to the observed calculation. In order to resist DPA, we need to randomize the parameters of elliptic curves.

2.2   Classification of the Known Countermeasures

According to Okeya-Takagi [OT03], known countermeasures against side channel attacks are classified into four categories; fixed procedure type, randomized addition chains type, indistinguishable operations type, and data randomization type.

The fixed procedure type computes scalar multiplication using a predetermined fixed procedure of operations, which helps to prevent against timing and SPA attacks. This type of countermeasures includes Coron's dummy method [Cor99], Montgomery ladder methods [OS00, BJ02, FGKS02, IT02], and fixed window methods [Möl01a, Möl01b, OT03].

The randomized addition chains type computes scalar multiplication using randomized addition chains, that is, each execution uses a different addition chain, which helps to prevent against timing, SPA and DPA attacks. This type of countermeasures includes randomized exponent methods [Cor99], randomized window method [LS01, IYTT02], and randomized addition-subtraction chains method [OA01].

The indistinguishable operations type blinds the attacker to the distinguishability of addition and doubling using same addition formulae, which helps to prevent against SPA attacks. This type of countermeasures includes indistinguishable addition formulae [JQ01, LS01, BJ02], which are constructed on several forms of elliptic curves.

The data randomization type randomizes computing objects, which helps to prevent against DPA attacks. This type can be combined with SPA countermeasures such as the fixed procedure type, which prevents against both SPA and DPA. This type of countermeasures includes randomized projective coordinates method [Cor99, OMS01], and random isomorphic curves method [JT01, IT02].

2.3   Randomized Addition Chains Type

While countermeasures of the randomized addition chains type aim to resist against side channel attacks, some of them were broken.

Oswald-Aigner proposed the randomized addition-subtraction chains countermeasure [OA01], which is a countermeasure of the randomized addition chains type. The countermeasure utilizes random decisions in automata. However, Okeya-Sakurai [OS02a] broke the basic version of Oswald's countermeasure. In addition, the advanced version of Oswald's countermeasure was also broken by Okeya-Sakurai [OS03] and Han et al. [HCJ+03]. Furthermore, a generalized version of Oswald's countermeasure was broken by Walter [Wal03a].

Liardet-Smart [LS01] proposed a countermeasure of the randomized addition chains type, namely the randomized window method. An enhanced version of the randomized window method was proposed by Itoh et al. [IYTT02]. However, Walter [Wal02c] broke the Liardet-Smart version of the randomized window method.

Some other countermeasure in this type is MIST [Wal02a]. However, Walter proved that MIST is less secure than that it was expected [Wal02b, Wal03b]. Therefore, the security of this type is controversial.

2.4   Ha-Moon's Countermeasure

Ha-Moon [HaM02] proposed a countermeasure of the randomized addition chains type; randomized signed scalar multiplication. The countermeasure utilizes non-adjacent form (NAF), and uses a randomization concept. The countermeasure first encodes a secret integer into another representation, then computes scalar multiplication using the encoded representation. To generate a different representation for each execution provides the security against side channel attacks.

Let k be an integer, and express it as $k = \sum_{i=0}^{n-1} k_i 2^i$, $k_i \in \{0,1\}$ for some n. The randomized signed scalar representation encodes k into $d = \sum_{i=0}^{n} d_i 2^i$ with $d_i \in \{-1, 0, 1\}$. Note that $k = d$ as an integer (i.e. $\sum_{i=0}^{n-1} k_i 2^i = \sum_{i=0}^{n} d_i 2^i$). For each i, d_i is determined by k_i, k_{i+1}, c_i, and r_i using Table 1, where c_i is the auxiliary carry and r_i is a random bit. The auxiliary carry c_{i+1} is updated according to Table 1. For the encoded n, scalar multiplication $Q = kP$ is computed by using Addition-Subtraction_Method. Because of $k = d$ as an integer, the output dP is equal to kP. Note that ECSUB(Q, P) denotes the elliptic subtraction, which is the operation ECADD$(Q, -P)$.

Table 1. Random signed scalar recording method

Input				Output	
k_{i+1}	k_i	c_i	r_i	c_{i+1}	d_i
0	0	0	0	0	0
0	0	0	1	0	0
0	0	1	0	0	1
0	0	1	1	1	-1
0	1	0	0	0	1
0	1	0	1	1	-1
0	1	1	0	1	0
0	1	1	1	1	0
1	0	0	0	0	0
1	0	0	1	0	0
1	0	1	0	1	-1
1	0	1	1	0	1
1	1	0	0	1	-1
1	1	0	1	0	1
1	1	1	0	1	0
1	1	1	1	1	0

Addition-Subtraction_Method

INPUT A point P, and $d = \sum_{i=0}^{n} d_i 2^i$, $d_i \in \{-1, 0, 1\}$
OUTPUT $Q = dP$

1. $Q \leftarrow \mathcal{O}$
2. for $i = n$ downto 0
    2.1. $Q \leftarrow$ ECDBL(Q)
    2.2. if $d_i = 1$ then $Q \leftarrow$ ECADD(Q, P)
    2.3. if $d_i = -1$ then $Q \leftarrow$ ECSUB(Q, P)
3. Return Q

Table 2 shows an example that how to compute a random recorded number d with a random number r when $k = (101010001011)_2$ is given. Note that this example is used in Section 3.3.

Ha-Moon [HaM02] analyzed the security of this scheme in two ways; theoretical viewpoint and experimental one. As a result, they concluded that their proposed scheme is secure against side channel attacks.

First, they analyzed their countermeasure from the theoretical point of view. They utilized the theory of finite Markov chain for analyzing the security. Since the encoded bit d_i is determined by (k_i, k_{i+1}, c_i, r_i), the number of possible states is sixteen. That is, the finite Markov chain associated with the countermeasure has sixteen states and its transition matrix is determined by Table 1. It is easy to show that this Markov chain is irreducible and aperiodic. Thus, the Markov chain has the stationary distribution. They calculated the distribution, and showed that the output d_i is independent on (k_i, k_{i+1}). In other words, the randomized signed scalar multiplication is secure against side channel attacks.

Table 2. Example of finding the random recorded number $d^{(j)}$ when $k = (101010001011)_2$

Index	12	11	10	9	8	7	6	5	4	3	2	1	0
key k		1	0	1	0	1	0	0	0	1	0	1	1
carry c	1	1	1	0	0	0	0	1	1	0	0	0	0
random number r		0	0	1	1	0	0	0	1	1	0	0	1
$d^{(1)}$	1	0	-1	-1	0	1	0	1	-1	-1	0	1	1
carry c	0	0	0	0	0	0	0	0	1	1	1	1	0
random number r		0	1	0	1	0	0	1	0	1	0	1	0
$d^{(2)}$	0	1	0	1	0	1	0	0	1	0	-1	0	-1
carry c	1	1	1	1	1	0	0	0	0	0	0	0	0
random number r		1	0	1	0	1	1	0	1	0	1	0	1
$d^{(3)}$	1	0	-1	0	-1	-1	0	0	0	1	0	1	1

Second, they implemented their countermeasure, and experimentally analyzed the immunity against side channel attacks. According to their experiment, no peaks appeared in their countermeasure. Note that the peak implies that the attacker can correctly estimate the corresponding bit of the secret integer k. Thus, no peaks prove that the attacker cannot determine the secret using side channel information. Therefore, the randomized signed scalar multiplication is secure against side channel attacks from experimental viewpoint.

Furthermore, they analyzed the efficiency of their countermeasure. The computational cost of the countermeasure is nECDBLs$+\frac{n}{2}$ECADDs [HaM02]. This means that the countermeasure is as fast as Binary_Method. In other words, the countermeasure does not pay the penalty of speed. Therefore, the randomized signed scalar multiplication is fast, and secure against side channel attacks. Consequently, the countermeasure is very attractive. However, our proposed attack completely breaks this countermeasure, therefore the countermeasure is not recommended for cryptographic use.

3   Proposed Attack

In this section we propose a side channel attack against Ha-Moon's countermeasure, and show that the countermeasure is vulnerable to the proposed attack. First, we introduce the main idea of the proposed attack, and describe the attack algorithm. Then, we display an example of the proposed attack, and show an experimental result.

3.1   Main Idea

We describe the main idea of our proposed attack. The proposed attack utilizes a finite Markov chain, and reveals the transition between states, which is related to the secret key.

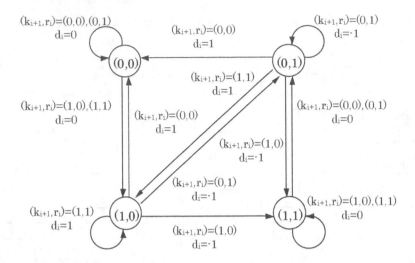

Fig. 1. The finite Markov chain associated with Ha-Moon's countermeasure

Notations : Let $k = \sum_{i=0}^{n-1} k_i 2^i$ with $k_i \in \{0,1\}$ be the n-bit secret binary value and $d = \sum_{i=0}^{n} d_i 2^i$ with $d_i \in \{-1,0,1\}$ be the $(n+1)$-bit random recorded number generated from k by Ha-Moon's random recording method. Let the j-th random recorded number of k be denoted as $d^{(j)} := \sum_{i=0}^{n} d_i^{(j)} 2^i$, where $d_i^{(j)} \in \{-1,0,1\}$.

While Ha-Moon utilized the finite Markov chain with sixteen states as we mentioned in the previous section, we convert it into another finite Markov chain whose input, output and intermediate values are the same to the original. Figure 1 shows the converted Markov chain.[3] It has four states; $(k_i, c_i) = (0,0), (0,1), (1,0),$ and $(1,1)$. Its transition matrix $\mathcal{T}$ is as follows:

$$\mathcal{T} = \begin{pmatrix} (0,0): & 1/2 & 0 & 1/2 & 0 \\ (0,1): & 1/4 & 1/4 & 1/4 & 1/4 \\ (1,0): & 1/4 & 1/4 & 1/4 & 1/4 \\ (1,1): & 0 & 1/2 & 0 & 1/2 \end{pmatrix}.$$

If the current state is $(k_i, c_i) = (0,0)$ and $(k_{i+1}, r_i) = (1,0)$ or $(1,1)$, then the next state is $(k_i, c_i) = (1,0)$ and the flow outputs $d_i = 0$. This event happens with the probability $1/2$. If $(k_{i+1}, r_i) = (0,0)$ or $(0,1)$, then the flow remains at the state $(0,0)$ and outputs $d_i = 0$. This event happens with the probability $1/2$. If the current state is $(k_i, c_i) = (0,1)$ and $(k_{i+1}, r_i) = (0,0)$, then the next state is $(k_i, c_i) = (0,0)$ and the flow outputs $d_i = 1$. This event happens with the probability $1/4$. Other states and transitions are given in the same way.

---

[3] Using the same Markov chain, Ebeid and Hasan [EH03] presented the average computational cost of Ha-Moon's method.

The converted Markov chain has the following properties.

Property 1. The state $(0, 1)$ and $(1, 0)$ always output $d_i \neq 0$, and the state $(0, 0)$ and $(1, 1)$ always output $d_i = 0$.

Property 2. In the case that the current state is $(k_i, c_i) = (0, 0)$ or $(1, 1)$, then

- if $k_i = k_{i+1}$, the flow remains at the same state,
- if $k_i \neq k_{i+1}$, the flow goes to the other states.

Property 3. The least significant bit (LSB) of the secret key is always revealed. Namely, if $d_0 = 0$ then $k_0 = 0$, and if $d_0 \neq 0$ then $k_0 = 1$.

Proposition 1. *If $d_i = d_{i+1} = 0$ then $k_i = k_{i+1}$. Also, if $d_i = 0$ and $d_{i+1} \neq 0$ then $k_i \neq k_{i+1}$.*

Note that Proposition 1 can be easily derived from Property 1 and 2.

Proposition 1 reveals the relation between k_i and k_{i+1} from the output d_i and d_{i+1}, if $d_i = 0$. Thus each event such as $d_i = d_{i+1} = 0$, or $d_i = 0$ and $d_{i+1} \neq 0$ provides 1-bit information about the secret key k. To the contrary, if $d_i \neq 0$, it does not provides any relation between k_i and k_{i+1}. In order to detect the relation, we need to use another random recorded number $d^{(j)}$ with $d_i^{(j)} \neq 0$, which was generated from the same k. Repeating this process, we obtain chains which consist of the relations between consecutive two bits of k. For example, the obtained chains are $k_0 \neq k_1 = \ldots \neq k_l$ and $k_{l+1} = k_{l+2} \neq \ldots = k_{n-1}$. Once the chains are revealed, to detect the secret k is easy task for the attacker. The more detail description of attack algorithm is covered in the next subsection.

3.2 Description of the Proposed Attack

We presents the proposed attack against Ha-Moon's countermeasure. First, we assume that the attacker has the following capability:

Assumption 1 : *ECADD and ECDBL are distinguishable by a single measurement of power consumption, whereas ECADD and ECSUB are indistinguishable.*

On the other hand, we should note that Addition-Subtraction_Method in Section 2.4 has the following property.

Property 4. Suppose the variable Q is not the point at infinity. Then the recorded digit d_i is not zero if and only if ECADD or ECSUB is performed. That is to say, the recorded digit d_i is zero if and only if ECADD and ECSUB are not performed, only ECDBL is performed.

But we should not overlook the fact that such special cases of ECDBL and ECADD as $Q = 2 * \mathcal{O}$ or $Q = P + \mathcal{O}$ can be avoided in the implementation of scalar multiplication dP. In the ordinary implementation, instead of ECDBL or ECADD operation, the point duplication or assignment can be used.

For instance, when $d_{12}^{(1)} = 1$ in Table 2, ECDBL and ECADD are not operated but just point duplication or assignment is performed in Addition-Subtraction_Method because the variable Q is the point at infinity. But when $d_{11}^{(1)} = 0$, ECDBL is fully performed, since the variable Q is not the point at infinity and Property 4. Thus we can deduce the following property.

Property 5. If ECDBL appears firstly in the AD sequence, then the previous bit is one.

For simplicity, ECADD and ECDBL are referred to as **A** and **D**, respectively. **A** and **D** are written with time-increasing from left to right.

Attack Algorithm. The concrete attack works as follows.

1. **Gathering AD sequences :** The attacker inputs an elliptic curve point into a cryptographic device with Ha-Moon's random recording method, and obtains a sequence of **A** and **D** (AD sequence). He/she repeats this procedure m times and gathers m AD sequences. Let $S^{(j)}$ be the j-th AD sequence $(1 \leq j \leq m)$.

2. **Data conversion :** As **A** and **D** are written with time-increasing from left to right, the attacker converts the obtained AD sequence $S^{(j)}$ into the signed-scalar number $d^{(j)}$ as follows.

   2.1. Split the obtained AD sequence by symbol | between **D** and **DA** from right to left.
   2.2. Match **D** $\Leftrightarrow$ the random recorded digit $d_i^{(j)} = 0$.
   2.3. Match **DA** $\Leftrightarrow$ the random recorded digit $d_i^{(j)} \neq 0$.
   2.4. If the last **D** (or **DA**) appears in the obtained AD sequence then the random recorded digits $d_i^{(j)} = 0$ (or $d_i^{(j)} \neq 0$) and $d_{i+1}^{(j)} = 1$.

3. **Determination of the relation between consecutive two bits (k_i, k_{i+1}) of the secret key k :** The attacker determines the relations using Proposition 1.

   - For $i = 0$ to $n - 2$ do

     * If $d_i^{(j)} = d_{i+1}^{(j)} = 0$ for some $1 \leq j \leq m$, then $k_i = k_{i+1}$.
     * If $d_i^{(j)} = 0$ and $d_{i+1}^{(j)} \neq 0$ for some $1 \leq j \leq m$, then $k_i \neq k_{i+1}$.
     * If $d_i^{(j)} \neq 0$ for all $1 \leq j \leq m$, the attacker obtains no relations between k_i and k_{i+1}.

4. **Key testing :**

 4.1. If the attacker obtains all relations between k_i and k_{i+1} for $0 \leq i \leq n-2$, he/she can determine the secret key exactly as the most significant bit (MSB) is 1.

 4.2. Using the known pair of plaintext and ciphertext, the attacker checks all combinations of bit-pattern which are obtained from Step 3. Then he/she finds the secret key. Note that the LSB is always revealed from Property 3.

3.3 Example

We now illustrate the attack against the Ha-Moon's countermeasure on a toy example.

Step 1 Gathering the AD sequences: Assume that the attacker obtains the following AD sequences for a 12-bit secret key k:

$$S^{(1)} = DDADADDADDADADADDADA$$
$$S^{(2)} = DDADDADDDADDADDA$$
$$S^{(3)} = DDADDADADDDDADDADA$$

Table 3. The correspondence between d and AD sequences

Index	12	11	10	9	8	7	6	5	4	3	2	1	0
key k		1	0	1	0	1	0	0	0	1	0	1	1
$d^{(1)}$	1	0	-1	-1	0	1	0	1	-1	-1	0	1	1
AD sequence $S^{(1)}$		D	DA	DA	D	DA	D	DA	DA	DA	D	DA	DA
$d^{(2)}$	0	1	0	1	0	1	0	0	1	0	-1	0	-1
AD sequence $S^{(2)}$			D	DA	D	DA	D	D	DA	D	DA	D	DA
$d^{(3)}$	1	0	-1	0	-1	-1	0	0	0	1	0	1	1
AD sequence $S^{(3)}$		D	DA	D	DA	DA	D	D	D	DA	D	DA	DA

Note that these come from Table 2. For example, when $d^{(1)}_{12} = 1$ in Table 3, AD sequence dose not appear as ECDBL and ECADD, since they are not operated but just point duplications are performed. In the same manner, when $d^{(2)}_{12} = 0$ and $d^{(2)}_{11} = 1$, AD sequence also dose not appear. Note that we use k and d in Table 3 for elucidation of the example, the attacker does not know these values in Step 1.

Step 2 Data conversion: The attacker converts the obtained AD sequences $S^{(j)}$ into the signed-scalar number $d^{(j)}$ in accordance with Step 2 of the proposed attack algorithm:

$$d^{(1)} = (d_{12}^{(1)}, d_{11}^{(1)}, ..., d_0^{(1)}) = (1, 0, *, *, 0, *, 0, *, *, *, 0, *, *)$$
$$d^{(2)} = (d_{11}^{(2)}, d_{10}^{(2)}, ..., d_0^{(2)}) = (1, 0, *, 0, *, 0, 0, *, 0, *, 0, *)$$
$$d^{(3)} = (d_{12}^{(3)}, d_{11}^{(3)}, ..., d_0^{(3)}) = (1, 0, *, 0, *, *, 0, 0, 0, *, 0, *, *),$$

where $*$ denotes a non-zero digit, namely 1 or -1.

Step 3 Determination of the relation of (k_i, k_{i+1}): The attacker determines the relation between consecutive two bits (k_i, k_{i+1}):

- $k_0 = 1$ as $d_0^{(1)} \neq 0$.
- $k_1 \neq k_2$ as $d_1^{(2)} = 0$ and $d_2^{(2)} \neq 0$.
- $k_2 \neq k_3$ as $d_2^{(1)} = 0$ and $d_3^{(1)} \neq 0$.
- $k_3 \neq k_4$ as $d_3^{(2)} = 0$ and $d_4^{(2)} \neq 0$.
- $k_4 = k_5$ as $d_4^{(3)} = d_5^{(3)} = 0$.
- $k_5 = k_6$ as $d_5^{(2)} = d_6^{(2)} = 0$.
- $k_6 \neq k_7$ as $d_6^{(1)} = 0$ and $d_7^{(1)} \neq 0$.
- $k_8 \neq k_9$ as $d_8^{(1)} = 0$ and $d_9^{(1)} \neq 0$.
- $k_9 \neq k_{10}$ as $d_9^{(3)} = 0$ and $d_{10}^{(3)} \neq 0$.
- $k_{10} \neq k_{11}$ as $d_{10}^{(2)} = 0$ and $d_{11}^{(2)} \neq 0$.

- Thus the attacker obtain $k_0 = 1$, $k_1 \neq k_2 \neq k_3 \neq k_4 = k_5 = k_6 \neq k_7$, and $k_8 \neq k_9 \neq k_{10} \neq k_{11} = 1$, i.e., $(k_8, k_9, k_{10}, k_{11}) = (0, 1, 0, 1)$.

Step 4 Key testing: As a consequence, the secret key k is $(101010001011)_2$ or $(101001110101)_2$. The true secret key k can be easily checked by using the known pair of plaintext and ciphertext. In fact, the secret key k was $(101010001011)_2$ in this case.

3.4 Experimental Result

We have implemented our proposed attack algorithm according to the number of AD sequences for standard 163, 193, and 233-bit secret keys. Table 4 shows the average testing number and the efficiency improvement.

For instance, the average testing numbers in the case of standard 163-bit key with 20 AD sequences in Table 4 is obtained as follows:

- For $l = 1$ to 100000 do
  - Select a 163-bit string randomly.
  - Obtain 20 AD sequences using another program that outputs characters A and D depending on the elliptic curve operations it executes while computing a scalar multiplication using the Ha-Moon's randomized addition chains method.
  - Convert AD sequence $S^{(j)}$ into the signed-scalar number $d^{(j)}$ where $1 \leq j \leq 20$.

- Compute the testing number $TestNum_l$ which is needed to recover the secret key:

$$TestNum_l = 2^{\#\{\ i\ \mid\ d_i^{(j)} \neq 0\ for\ all\ 1 \leq j \leq 20,\ where\ 0 \leq i \leq 163\} - 1}$$

- The average testing number of standard 163-bit key with 20 AD sequences $AveTestNum(163, 20)$ is computed as follows:

$$AveTestNum(163, 20) = \frac{\sum_{l=1}^{100000} TestNum_l}{100000} \approx 2^8$$

Table 4. Implementation results for standard $163, 193$, and 233-bit keys with m AD sequences

#(Obtained AD sequences) m	The length of secret value					
	163-bit		193-bit		233-bit	
	#(Ave. Testing)	Imp.	#(Ave. Testing)	Imp.	#(Ave. Testing)	Imp.
2	2^{54}	2^{28}	2^{64}	2^{33}	2^{77}	2^{40}
3	2^{40}	2^{42}	2^{48}	2^{49}	2^{58}	2^{59}
4	2^{32}	2^{50}	2^{38}	2^{59}	2^{46}	2^{71}
5	2^{27}	2^{55}	2^{32}	2^{65}	2^{38}	2^{79}
6	2^{23}	2^{59}	2^{27}	2^{70}	2^{33}	2^{84}
7	2^{20}	2^{62}	2^{24}	2^{73}	2^{29}	2^{88}
8	2^{18}	2^{64}	2^{21}	2^{76}	2^{26}	2^{91}
9	2^{16}	2^{66}	2^{19}	2^{78}	2^{23}	2^{94}
10	2^{15}	2^{67}	2^{18}	2^{79}	2^{21}	2^{96}
20	2^{8}	2^{74}	2^{9}	2^{88}	2^{11}	2^{106}
30	2^{6}	2^{76}	2^{7}	2^{90}	2^{8}	2^{109}
40	2^{4}	2^{78}	2^{5}	2^{92}	2^{6}	2^{111}
50	2^{4}	2^{78}	2^{4}	2^{93}	2^{5}	2^{112}
60	2^{3}	2^{79}	2^{4}	2^{93}	2^{4}	2^{113}
70	2^{3}	2^{79}	2^{3}	2^{94}	2^{4}	2^{113}
80	2^{2}	2^{80}	2^{3}	2^{94}	2^{3}	2^{114}
90	2^{2}	2^{80}	2^{3}	2^{94}	2^{3}	2^{114}
100	2^{2}	2^{80}	2^{2}	2^{95}	2^{3}	2^{114}

When an attacker obtained 20 AD sequences, for instance, in the case of standard 163-bit key, we can deduce that we only have to test about 2^8 possible keys on average. In the case of standard 193-bit key, we have to test about 2^9 possible keys on average. Also, in the case of standard 233-bit key, we have to test about 2^{11} possible keys on average.

Note that the values in the *Imp.* columns imply the efficiency improvement as compared to the testing number $2^{n/2}$ obtained from SPA when key size is n bit. Our results in Table 4 do not depend on the elliptic curve and the underlying field.

4   Conclusion

We have proposed the attack against Ha-Moon's countermeasure, and shown that the countermeasure is vulnerable to the proposed attack. The proposed attack utilizes the Markov chain for detecting the secret. First, the attacker determines the transitions in the Markov chain using side channel information, then detects the relation between consecutive two bits of the secret key, instead of bits of the secret key as they are. The use of such relations drastically reduces the search space for the secret key, and the attacker can easily reveal the secret. Our experiment has shown that around *twenty* observations of execution of the countermeasure are sufficient to detect the secret in the case of the standard sizes of ECC. Therefore, Ha-Moon's countermeasure is not recommended for cryptographic use.

Acknowledgments

We would like to thank TaeHyun Kim at the CIST for helping the implementation of the attack algorithm.

References

[BJ02]   Brier, É., Joye, M., *Weierstrass Elliptic Curves and Side-Channel Attacks*, Public Key Cryptography (PKC 2002), LNCS2274, (2002), 335-345.

[CMO98]   Cohen, H., Miyaji, A., Ono, T., *Efficient Elliptic Curve Exponentiation Using Mixed Coordinates*, Advances in Cryptology - ASIACRYPT '98, LNCS1514, (1998), 51-65.

[Cor99]   Coron, J.S., *Resistance against Differential Power Analysis for Elliptic Curve Cryptosystems*, Cryptographic Hardware and Embedded Systems (CHES '99), LNCS1717, (1999), 292-302.

[EH03]   Ebeid, N., Hasan, A., *Analysis of DPA Countermeasures Based on Randomizing the Binary Algorithm*, Technical Report of the University of Waterloo, No. CORR 2003-14. http://www.cacr.math.uwaterloo.ca/techreports/2003/corr2003-14.ps

[FGKS02]   Fischer, W., Giraud, C., Knudsen, E.W., Seifert, J.P., *Parallel scalar multiplication on general elliptic curves over $\mathbf{F}_p$ hedged against Non-Differential Side-Channel Attacks*, International Association for Cryptologic Research (IACR), Cryptology ePrint Archive 2002/007, (2002). http://eprint.iacr.org/2002/007/

[HaM02]   Ha, J., and Moon, S., *Randomized Signed-Scalar Multiplication of ECC to Resist Power Attacks*, Workshop on Cryptographic Hardware and Embedded Systems 2002 (CHES 2002), LNCS 2523, (2002), 551-563.

[HCJ+03]   Han, D.-G., Chang, N.S., Jung, S.W., Park, Y.-H., Kim, C.H., Ryu, H., *Cryptanalysis of the Full version Randomized Addition-Subtraction Chains*, The 8th Australasian Conference in Information Security and Privacy (ACISP 2003), LNCS2727, (2003), 67-78.

[IYTT02]   Itoh, K., Yajima, J., Takenaka, M., and Torii, N., *DPA Countermeasures by improving the Window Method*, Workshop on Cryptographic Hardware and Embedded Systems 2002 (CHES 2002), LNCS 2523, (2002), 318-332.

[IT02]   Izu, T., Takagi, T., *A Fast Parallel Elliptic Curve Multiplication Resistant against Side Channel Attacks*, Public Key Cryptography (PKC 2002), LNCS2274, (2002), 280-296.

[JQ01]   Joye, M., Quisquater, J.J., *Hessian elliptic curves and side-channel attacks*, Cryptographic Hardware and Embedded Systems (CHES 2001), LNCS2162, (2001), 402-410.

[JT01]   Joye, M., Tymen, C., *Protections against differential analysis for elliptic curve cryptography: An algebraic approach*, Cryptographic Hardware and Embedded Systems (CHES 2001), LNCS2162, (2001), 377-390.

[Kob87]   Koblitz, N., *Elliptic curve cryptosystems*, Math. Comp. 48, (1987), 203-209.

[Koc96]   Kocher, C., *Timing Attacks on Implementations of Diffie-Hellman, RSA, DSS, and Other Systems*, Advances in Cryptology - CRYPTO '96, LNCS 1109, (1996), 104-113.

[KJJ99]   Kocher, C., Jaffe, J., Jun, B., *Differential Power Analysis*, Advances in Cryptology - CRYPTO '99, LNCS1666, (1999), 388-397.

[LS01]   Liardet, P.Y., Smart, N.P., *Preventing SPA/DPA in ECC systems using the Jacobi form*, Cryptographic Hardware and Embedded System (CHES 2001), LNCS2162, (2001), 391-401.

[Mil86]   Miller, V.S., *Use of elliptic curves in cryptography*, Advances in Cryptology - CRYPTO '85, LNCS218, (1986), 417-426.

[Möl01a]   Möller, B., *Securing Elliptic Curve Point Multiplication against Side-Channel Attacks*, Information Security (ISC 2001), LNCS2200, (2001), 324-334.

[Möl01b]   Möller, B., *Securing elliptic curve point multiplication against side-channel attacks, addendum: Efficiency improvement.* http://www.informatik.tu-darmstadt.de/TI/Mitarbeiter/moeller/ecc-scaisc01.pdf, (2001).

[OA01]   Oswald, E., Aigner, M., *Randomized Addition-Subtraction Chains as a Countermeasure against Power Attacks*, Cryptographic Hardware and Embedded Systems (CHES 2001), LNCS2162, (2001), 39-50.

[OMS01]   Okeya, K., Miyazaki, K., Sakurai, K., *A Fast Scalar Multiplication Method with Randomized Projective Coordinates on a Montgomery-form Elliptic Curve Secure against Side Channel Attacks*, The 4th International Conference on Information Security and Cryptology (ICISC 2001), LNCS2288, (2002), 428-439.

[OS00]   Okeya, K., Sakurai, K., *Power Analysis Breaks Elliptic Curve Cryptosystems even Secure against the Timing Attack*, Progress in Cryptology - INDOCRYPT 2000, LNCS1977, (2000), 178-190.

[OS02a]   Okeya, K., Sakurai, K., *On Insecurity of the Side Channel Attack Countermeasure using Addition-Subtraction Chains under Distinguishability between Addition and Doubling*, The 7th Australasian Conference in Information Security and Privacy, (ACISP 2002), LNCS2384, (2002), 420-435.

[OS03]   Okeya, K., Sakurai, K., *A Multiple Power Analysis Breaks the Advanced Version of the Randomized Addition-Subtraction Chains Countermeasure against Side Channel Attacks*, in the proceedings of 2003 IEEE Information Theory Workshop (ITW 2003), (2003), 175-178.

[OT03]   Okeya, K., Takagi, T., *The Width-w NAF Method Provides Small Memory and Fast Elliptic Scalar Multiplications Secure against Side Channel Attacks*, Topics in Cryptology, The Cryptographers' Track at the RSA Conference 2003 (CT-RSA 2003), LNCS2612, (2003), 328-342.

[Wal02a]   Walter, C.D., *MIST: An Efficient, Randomized Exponentiation Algorithm for Resisting Power Analysis*, Cryptographers' Track RSA conference (CT-RSA 2002), LNCS2271, (2002), 53-66.

[Wal02b]   Walter, C.D., *Some Security Aspects of the Mist Randomized Exponentiation Algorithm*, Workshop on Cryptographic Hardware and Embedded Systems 2002 (CHES 2002), LNCS 2523, (2002), 564-578.

[Wal02c]   Walter, C.D., *Breaking the Liardet-Smart Randomized Exponentiation Algorithm*, Proceedings of CARDIS'02, USENIX Assoc, (2002), 59-68.

[Wal03a]   Walter, C.D., *Security Constraints on the Oswald-Aigner Exponentiation Algorithm*, International Association for Cryptologic Research (IACR), Cryptology ePrint Archive 2003/013, (2003).
http://eprint.iacr.org/2003/013/

[Wal03b]   Walter, C.D., *Seeing through Mist Given a Small Fraction of an RSA Private Key, Topics in Cryptology*, Topics in Cryptology, The Cryptographers' Track at the RSA Conference 2003 (CT-RSA 2003), LNCS2612, (2003), 391-402.

Systolic and Scalable Architectures for Digit-Serial Multiplication in Fields $GF(p^m)$

Guido Bertoni[1], Jorge Guajardo[2*], and Gerardo Orlando[3]

[1] Politecnico di Milano, P.za L. Da Vinci 32, 20133 Milano, Italy
bertoni@elet.polimi.it
[2] Infineon Technologies AG, Secure Mobile Solutions Division, Munich, Germany
Jorge.Guajardo@infineon.com
[3] General Dynamics C4 Systems, 77 A St. Needham MA 02494, USA
gerardo.orlando@gdc4s.com

Abstract. This contribution defines systolic digit-serial architectures for fields $G(p^m)$. These architectures are scalable in the sense that their instantiations support multiplication in different fields $GF(p^m)$ for which p is fixed and m is variable. These features make the multiplier architectures suitable for ASIC as well as FPGA implementations. In addition, the same architectures are easily applicable to tower fields $GF(q^m)$ for a given ground field $GF(q)$, where q itself is a prime power. We simulated the basic cell of a systolic LSDE multiplier on 0.18 μm CMOS technology to verify the functionality of the architectures. Finally, we provide specific values for $GF(2^m)$ and $GF(3^m)$ fields which are of particular interest in recent cryptographic applications, for example, the implementation of short signature schemes based on the Tate pairing.

1 Introduction

Galois field arithmetic has received considerable attention in recent years due to their application in public-key cryptography schemes and error correcting codes. In particular, public-key cryptosystems based on finite fields stand out: systems based on the difficulty to solve the discrete logarithm (DL) problem in finite fields such as DSS, XTR, LUC, and other variants; and elliptic curve (EC) and hyperelliptic (HE) cryptosystems. Both, prime fields and extension fields, have been proposed for use in such cryptographic systems. For applications based on elliptic curves and hyperelliptic curves, however, the focus has been mainly on fields of characteristic 2 due to the straight forward manner in which elements of $GF(2)$ can be represented. For these types of fields, both software implementations and hardware architectures have been extensively studied. In recent years, $GF(p^m)$ fields, p odd, have gained interest in the research community. Mihălescu [11] and independently Bailey and Paar [1] introduced the concept of Optimal Extension Fields (OEFs) in the context of elliptic curve cryptography.

---

* Part of this work was performed while the author was at the COSY group, Ruhr-Universität Bochum, Germany.

T. Johansson and S. Maitra (Eds.): INDOCRYPT 2003, LNCS 2904, pp. 349–362, 2003.

References [1, 11] and other works based on OEFs has only been concerned with efficient software implementations. Other applications of $GF(p^m)$ fields, p small, include: [8], where an ECDSA implementation is described over fields of characteristic 3 and 7, and [13], where a method to implement ECC over fields of *small odd characteristic* ($p < 24$) is described. More recently, Boneh and Franklin [5] introduced an identity-based encryption scheme based on the Weil and Tate pairings. The work in [5] and other applications based on ID-cryptography consider EC defined over fields of characteristic 2 and 3. Because characteristic 2 field arithmetic has already been extensively studied in the literature, authors have concentrated their efforts to improve the performance of systems based on characteristic 3 arithmetic[4]. For example, [2] describes algorithms to improve the efficiency of pairing computations. To the authors' knowledge, the only works that deal with the implementation of $GF(p^m)$ fields in hardware platforms are [12] and [3]. [12] treats the hardware implementation of fields of characteristic 3. Their design is only geared towards fields of characteristic 3. Reference [3] presents *general* $GF(p^m)$ multipliers architectures designs which trade area for performance based on the number of coefficients that the multiplier processes at one time. [3] also provide the first cubing architectures presented in the literature for fields $GF(3^m)$ by taking advantage of FPGA reconfigurability.

1.1   Our Contributions

The research community's interest on cryptographic systems based on fields of odd characteristic and the lack of hardware architectures for general odd characteristic fields is evident. Reference [3] has given a partial answer to this problem but their methods have the drawback of using global signals and long wires and they require reconfigurability to achieve their full potential, and thus, these solutions lack flexibility in other hardware platforms such as ASICs. Hence, in this work, we move a step forward towards the design of scalable and flexible hardware architectures for odd $GF(p^m)$ fields. In particular, we propose systolic architectures for arithmetic in $GF(p^m)$ fields. Systolic architectures solve the above mention problems in several ways: First, by using a systolic design we use localized routing, thus avoiding the need for global control signals and long wires. In addition, this methodology allows for ease of design and offers functional and layout modularity all of which are properties envisioned in good VLSI designs. Second, we modify the design of [3] to allow for scalability as introduced in [15]. In other words, for a fixed value of the digit-size D [14, 3] and parameter d, we can perform a multiplication for any value of m in $GF(p^m)$, with fixed p, i.e., we support multiple irreducible polynomials making unnecessary the use of reconfigurability in FPGAs. Thus, these architectures are well suited for very large multipliers and a large number of hardware platforms, including FPGAs and ASICs. At the end of this contribution, we also provide actual performance data from simulations (area and frequency) of a prototype implementation coded in VHDL and mapped to a 0.18 μm CMOS standard cell library.

---

[4] Use of characteristic 3 fields is preferred in some applications due to the improved bandwidth requirements implied by the security parameters

The remainder of this contribution is organized as follows. Section 2 surveys previously documented architectures relevant to this work. Section 3 reviews the most relevant mathematical concepts. Section 4 describes the LSE architecture introduced in [3] and studies its data dependencies, which allows us to develop a systolic version of the LSE multiplier. Section 5 describes how to transform the LSDE algorithm introduced in [3] into an algorithm that provides the basis for a systolic and scalable architecture. Section 6 describes the new LSDE architecture. Section 7 describes synthesis results and Section 8 summarizes the conclusions.

2   Related Work

We refer to [3] for a good overview of work done on $GF(p^m)$ architectures. Notice that this reference constitutes the basis for the systolic architectures presented in this contribution.

There has been significant work on systolization of modular multiplication (i.e. multiplication in $GF(p)$). The first attempt to provide systolic architectures for modular multiplication was presented in [9], however these architectures suffered from excessive latency and a slow clock as a result of the unsuitability of the regular multiply and then reduce algorithm to systolization. The first systolic architectures for Montgomery multiplication were introduced in [17]. Several systolic architecture designs of the binary type (i.e., using radix 2 to represent and process operands) exist, including both 1D-based arrays [10, 16] and 2D-based arrays [7]. Higher-radix systolic arrays have also been proposed, see [6, 4].

We also build on the concept of scalability presented in [15], which only treats Montgomery multiplication over $GF(2^k)$ and $GF(p)$. Scalability, as defined in [15], means that an arithmetic unit (AU) (in our case an AU to perform arithmetic operations in $GF(p^m)$) can be used or replicated in order to generate long-precision[5] results independently of the data path precision for which the unit was originally designed. A very important design choice in [15] is the use of a word-based algorithm. Rather than processing one of the inputs in a bitwise manner, [15] uses circuits that process multiple bits of the operands at a time. Based on the data dependencies of the Montgomery algorithm, the authors define processing units, which, when combined with pipelining and word-operand processing, result in scalable architectures for modular multiplication. The authors in [15] also examine trade-offs among desired performance, area, number of processing units, operand word-length, and precision.

3   Mathematical Background

Here, we briefly review the theory that we will need to develop the architectures of this paper. In the following, we will consider the field $GF(p^m)$ defined by an

---

[5] In our context long-precision refers to polynomials that require more than one word to be represented.

irreducible polynomial $p(x) = x^m + P(x) = x^m + \sum_{i=0}^{m-1} p_i x^i$ over $GF(p)$ of degree m. Let α be a root of $p(x)$, i.e. $p(\alpha) = 0$, then we can represent $A \in GF(p^m)$ in polynomial basis as $A(\alpha) = \sum_{i=0}^{m-1} a_i \alpha^i$, $a_i \in GF(p)$. Therefore,

$$\alpha^m = -P(\alpha) = \sum_{i=0}^{m-1}(-p_i)\alpha^i \qquad (1)$$

gives an easy way to perform modulo reduction whenever we encounter powers of α greater than $m - 1$. In what follows, it is assumed that $A, B, C \in GF(p^m)$, with $A = \sum_{i=0}^{m-1} a_i \alpha^i$, $B = \sum_{i=0}^{m-1} b_i \alpha^i$, $C = \sum_{i=0}^{m-1} c_i \alpha^i$, and $a_i, b_i, c_i \in GF(p)$. Addition in $GF(p^m)$ can be achieved as $C(\alpha) \equiv A(\alpha) + B(\alpha) = \sum_{i=0}^{m-1}(a_i + b_i)\alpha^i$, where the addition $a_i + b_i$ is done in $GF(p)$. We write the multiplication of two elements $A, B \in GF(p^m)$ as $C(\alpha) = \sum_{i=0}^{m-1} c_i \alpha^i \equiv A(\alpha) \cdot B(\alpha)$, where the multiplication is understood to happen in the finite field $GF(p^m)$ and all α^t, with $t \geq m$ can be reduced with (1). We abuse our notation and throughout the text we will write $A \bmod p(\alpha)$ to mean *explicitly* the reduction step described previously. Finally, we refer to A as the multiplicand and to B as the multiplier.

4 Systolic Least-Significant Element (LSE) First Architecture

The LSE scheme, introduced in [3] in the context of $GF(p^m)$ architectures, processes first coefficient b_0 of the multiplier and continues with the remaining coefficients one at a time in ascending order. This multiplier computes the result $C \equiv AB \bmod p(\alpha)$ according to

$$C \equiv b_0 A + b_1(A\alpha \bmod p(\alpha)) + \ldots + b_{m-1}(A\alpha^{m-1} \bmod p(\alpha)) \qquad (2)$$

Equation (2) can be implemented in a loop with m iterations. Each iteration requires the accumulation of a partial product, which is achieved with polynomial addition and the computation of the quantity $A\alpha \bmod p(\alpha)$. Using (1) and rewriting summation indices, $A\alpha \bmod p(\alpha)$ can be calculated as follows:

$$A\alpha \bmod p(\alpha) \equiv (-p_0 a_{m-1}) + \sum_{i=1}^{m-1}(a_{i-1} - p_i a_{m-1})\alpha^i \qquad (3)$$

where all coefficient arithmetic is done modulo p. Using (3) we can write expressions for A and C at iteration i as follows:

$$C^{(i)} = \sum_{j=0}^{m-1} c_j^{(i)} \alpha^j \equiv b_i A^{(i)} + C^{(i-1)} = \sum_{j=0}^{m-1}(b_i a_j^{(i)} + c_j^{(i-1)})\alpha^j, \qquad (4)$$

$$A^{(i)} = \sum_{j=0}^{m-1} a_j^{(i)} \alpha^j \equiv A^{(i-1)}\alpha \equiv (-p_0 a_{m-1}^{(i-1)}) + \sum_{j=1}^{m-1}(a_{j-1}^{(i-1)} - p_j a_{m-1}^{(i-1)})\alpha^j \qquad (5)$$

with $C^{(-1)} = 0$ and $A^{(-1)} = A$. Notice that if you initialize C to a value different from 0, say I, before you begin the multiplication, you will compute $C \equiv A \cdot B + I \bmod p(\alpha)$. This multiply-and-accumulate operation turns out to be very useful in elliptic curve systems and it is obtained at no extra cost. Using (4) and (5), one can define Algorithm 4.1.

Algorithm 4.1 Low Level LSE Multiplier

Input: $A = \sum_{i=0}^{m-1} a_i \alpha^i$, $B = \sum_{i=0}^{m-1} b_i \alpha^i$, $C = \sum_{i=0}^{m-1} c_i \alpha^i$, $p(\alpha) = \alpha^m + \sum_{i=0}^{m-1} p_i \alpha^i$ where $a_i, b_i, c_i, p_i \in GF(p)$

Output: $C \equiv A \cdot B + C \bmod p(\alpha)$

1: **for** $i = 0$ to $m - 1$ **do**
2:    **for** $j = m - 1$ to 0 **do**
3:       $c_j \leftarrow b_i a_j + c_j$
4:    **end for**
5:    $a^*_{m-1} = a_{m-1}$
6:    **for** $j = m - 1$ to 0 **do**
7:       $a_j \leftarrow a_{j-1} - p_j a^*_{m-1}$ {Note: $a_{-1} = 0$}
8:    **end for**
9: **end for**
10: Return (C)

4.1 Architecture

In this section, we analyze data dependencies of Algorithm 4.1. Steps 3 and 7 in Algorithm 4.1 are completely independent of each other. In other words, at step i, one can calculate coefficient c_j of C and, at the same time, compute a_j for iteration $i + 1$ of the outer loop. To best study the data dependencies in Algorithm 4.1, two dependency graphs (DG) are used. Figure 1 shows the DG for the computation of $A\alpha \bmod p(\alpha)$. Every square corresponds to the computation of Step 7 in Algorithm 4.1. Thus, each cell contains one $GF(p)$ multiplier and a $GF(p)$ adder. Each column corresponds to a new iteration of the outer loop (i-loop). Notice that because of the dependence of $a_j^{(i)}$ on $a_{j-1}^{(i-1)}$, there is a two cycle delay between the processing of a column at iteration i and $i + 1$ (this can be better seen on Figure 2, where the b_i's are labeled). Figure 2 shows the DG for Algorithm 4.1 as a whole. This figure is obtained by superimposing the computation of Step 3 on Figure 1. To make this clearer, we have used dotted lines to indicate the inputs and outputs corresponding to Step 3 while using solid lines for those that correspond to Step 7 of Algorithm 4.1. Figure 2, contains two types of cells. The white cells can compute the values of c_j and a_j for the next i iteration. The black cells can compute $c_{m-1}^{(i)}$ and $c_{m-2}^{(i)}$ given the values of $c_{m-1}^{(i-1)}$ and $c_{m-2}^{(i-1)}$, respectively. In other words, they compute one more c_j value than the white cells. Thus, the white cells contain two $GF(p)$ multipliers and two $GF(p)$ adders while the black cells contain three $GF(p)$ multipliers and three $GF(p)$ adders. The critical path for both types of cells is just the delay corresponding to one $GF(p)$ adder and one $GF(p)$ multiplier. As in Figure 1, there is a two cycle delay between columns, which is a result of the dependence

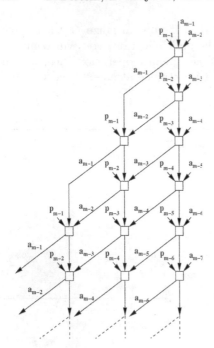

Fig. 1. DG for $A\alpha \bmod p(\alpha)$

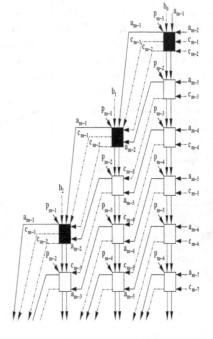

Fig. 2. DG for the LSE algorithm

of $a_j^{(i)}$ on $a_{j-1}^{(i-1)}$. Similarly to [15], each column in Figure 2 may be computed by a different processing element (PE) and the data generated by one PE may be passed to another PE in a pipeline manner.

5   Systolic Least-Significant Digit Element (LSDE) First Architecture

This section develops the theory for the scalable least-significant digit-element first multiplier (LSDE). Digit multipliers are a trade-off between speed, area, and power consumption. This is achieved by processing several of B's coefficients at the same time. The number of coefficients that are processed in parallel is defined to be the digit-size and we denote it by D. For a digit-size D, we can denote by $d = \lceil m/D \rceil$ the total number of digits in a polynomial of degree $m - 1$. Thus, we can re-write the multiplier as $B = \sum_{i=0}^{d-1} B_i \alpha^{Di}$, where $B_i = \sum_{j=0}^{D-1} b_{Di+j} \alpha^j$ with $0 \le i \le d - 1$ and we assume that B has been padded with zero coefficients such that $b_i = 0$ for $m - 1 < i < d \cdot D$ (i.e. B's size is $d \cdot D$ coefficients but $\deg(B) < m$). Hence,

$$C \equiv AB \bmod p(\alpha) = A \sum_{i=0}^{d-1} B_i \alpha^{Di} \bmod p(\alpha) \tag{6}$$

$$\equiv [B_0 A + B_1 (A\alpha^D \bmod p(\alpha)) + \ldots + B_{d-1} (A\alpha^{D(d-2)} \alpha^D \bmod p(\alpha))] \bmod p(\alpha)$$

Equation (6) looks very similar to (2), except that instead of multiplying by b_i, we now multiply by B_i (we process now D coefficients as opposed to just one coefficient of B) and instead of multiplying A by α, we multiply A by α^D. The authors in [3] define optimal irreducible polynomials of the form $p(\alpha) = \alpha^m + p_k\alpha^k + \sum_{i=0}^{k-1} p_i\alpha^i$ as those that satisfy the constraint $k \leq m - D$. These polynomials allow one to perform the reduction of $A\alpha^D$ modulo $p(\alpha)$ in one clock cycle. We illustrate the reduction $A\alpha^D \bmod p(\alpha)$ with a small example.

Example 1. Let $A = \sum_{i=0}^{d-1} A_i\alpha^{Di} \in GF(p^m)$ with $d = \lceil m/D \rceil$, A_i a digit, and $p(\alpha) = \alpha^m + p_k\alpha^k + \sum_{i=0}^{k-1} p_i\alpha^i$ an optimum irreducible polynomial in the sense of [3]. Then, $A\alpha^D \equiv \alpha^D \sum_{i=0}^{d-1} A_i\alpha^{Di} \bmod p(\alpha) = A_{d-1}\alpha^{Dd} + \sum_{i=1}^{d-1} A_{i-1}\alpha^{Di} \bmod p(\alpha)$. Further simplifying, we obtain $A\alpha^D \equiv A_{d-1}\alpha^{Dd-m}\left(-p_k\alpha^k - \sum_{i=0}^{k-1} p_i\alpha^i\right) + \sum_{i=1}^{d-1} A_{i-1}\alpha^{Di}$. Notice that the first term in the reduced result depends on the value of m, in other words on the field size. In fact, one needs to multiply by α^{Dd-m}, which can be instantiated as a *variable* shifter in hardware. This might be undesirable if scalability of the multiplier is desired.

Before we continue we prove a small proposition, the result of which we will use in the development of new architectures.

Proposition 1. *Let $A, B \in GF(p^m)$, $p(\alpha) = \alpha^m + \sum_{i=0}^{m-1} p_i\alpha^i$, be an irreducible polynomial over $GF(p)$, and $d = \lceil m/D \rceil$. Then, $A \cdot B \bmod p(\alpha) \equiv [A \cdot B \bmod \overline{p(\alpha)}] \bmod p(\alpha)$, where $\overline{p(\alpha)} = \alpha^{Dd-m}p(\alpha)$.*

Intuitively, Proposition 1 says that we can perform reductions modulo $\overline{p(\alpha)} = \alpha^{Dd-m}p(\alpha)$ and still obtain a result that when reduced modulo $p(\alpha)$ returns the correct value. We make this more formal by first introducing Algorithm 5.1.

Algorithm 5.1 Modified LSDE Multiplier

Input: $A = \sum_{i=0}^{d-1} A_i\alpha^{Di}$ with $A_i = \sum_{j=0}^{D-1} a_{Di+j}\alpha^j$, $B = \sum_{i=0}^{d-1} B_i\alpha^{Di}$ with $B_i = \sum_{j=0}^{D-1} b_{Di+j}\alpha^j$, $\overline{C} = \sum_{i=0}^{d} \overline{C}_i\alpha^{Di}$ with $\overline{C}_i = \sum_{j=0}^{D-1} c_{Di+j}\alpha^j$, $\overline{p(\alpha)} = \alpha^{Dd-m}p(\alpha)$, $a_i, b_i, c_i \in GF(p)$, and $d = \lceil \frac{m}{D} \rceil$

Output: $\overline{C} \equiv A \cdot B + \overline{C} \bmod \overline{p(\alpha)}$

1: **for** $i = 0$ to $d - 1$ **do**
2: $\quad \overline{C} \leftarrow B_i A + \overline{C}$
3: $\quad A \leftarrow A\alpha^D \bmod \overline{p(\alpha)}$
4: **end for**
5: Return $(\overline{C} \bmod \overline{p(\alpha)})$

Algorithm 5.1 suggests the following computation strategy. Given three inputs $A, B, C \in GF(p^m)$ one can compute $C = A \cdot B + C \bmod p(\alpha)$ by first computing $\overline{C} = A \cdot B + C \bmod \overline{p(\alpha)}$ using Algorithm 5.1, and then computing $C = \overline{C} \bmod p(\alpha)$. The second step follows as a consequence of Proposition 1. In practice, the second step can be performed at the end of a long range of computations, similar to the procedure done when using Montgomery multiplication. Step 3 in Algorithm 5.1 requires a modular multiplication. There, it would be desirable to reduce $A\alpha^D$ in just one iteration as in [3] and, at the same time, make the reduction process independent of the value of m and, thus, of the field $GF(p^m)$. Given $p(\alpha) = \alpha^m + \sum_{i=0}^{m-1} p_i \alpha^i$, then we define

$$
\begin{aligned}
\overline{p(\alpha)} = \alpha^{Dd-m} p(\alpha) &= \alpha^{Dd} + \alpha^{Dd-m} \sum_{i=0}^{m-1} p_i \alpha^i = \alpha^{Dd} + \sum_{i=0}^{m-1} p_i \alpha^{Dd+i-m} \\
&= \alpha^{Dd} + \sum_{i=0}^{Dd-1} \overline{p}_i \alpha^i = \alpha^{Dd} + \sum_{i=0}^{d-1} \overline{P}_i \alpha^{Di}
\end{aligned}
\tag{7}
$$

where $\overline{p}_i = 0$ for $0 \leq i < Dd - m$, $\overline{p}_i = p_{i+m-Dd}$ for $Dd - m \leq i < Dd$, and $\overline{P}_i = \sum_{j=0}^{D-1} \overline{p}_{Di+j} \alpha^j$. Then, we can compute $A\alpha^D \bmod \overline{p(\alpha)}$ as follows:

$$
\alpha^D A \bmod \overline{p(\alpha)} = \alpha^D \sum_{i=0}^{d-1} A_i \alpha^{Di} \bmod \overline{p(\alpha)} = A_{d-1} \alpha^{Dd} + \sum_{i=0}^{d-2} A_i \alpha^{D(i+1)} \bmod \overline{p(\alpha)}
\tag{8}
$$

Using (7), we obtain

$$
\begin{aligned}
\alpha^D A \bmod \overline{p(\alpha)} &= A_{d-1} \left(-\sum_{i=0}^{d-1} \overline{P}_i \alpha^{Di} \right) + \sum_{i=0}^{d-2} A_i \alpha^{D(i+1)} \bmod \overline{p(\alpha)} \\
&= \sum_{i=0}^{d-2} A_i \alpha^{D(i+1)} - \sum_{i=0}^{d-1} \left(A_{d-1} \overline{P}_i \right) \alpha^{Di} \bmod \overline{p(\alpha)}
\end{aligned}
\tag{9}
$$

Equation (9) contains $\bmod \ \overline{p(\alpha)}$ because it is possible that $\deg \left(A_{d-1} \overline{P}_{d-1} \alpha^{D(d-1)} \right) \geq Dd$, in which case it would require a further reduction. M-LSDE optimal polynomials, defined here as those for which $\overline{P}_{d-1} = p_{m-D} \in GF(p)$, solve this problem. Theorem 1 summarizes the above discussion.

Theorem 1. *Let* $A = \sum_{i=0}^{d-1} A_i \alpha^{Di}$ *be as defined in Algorithm 5.1 and* $\overline{p(\alpha)} = \alpha^{Dd-m} p(\alpha) = \alpha^{Dd} + \sum_{i=0}^{d-1} \overline{P}_i \alpha^{Di}$ *be as defined in (7), in particular* $p(\alpha)$ *is irreducible over* $GF(p)$ *of degree* m. *Then, if* $\overline{P}_{d-1} = p_{m-D} \in GF(p)$, $A\alpha^D \bmod \overline{p(\alpha)}$ *can be computed in one reduction step. Moreover,* $\overline{P}_{d-1} = p_{m-D}$ *implies that for* $p(\alpha) = \alpha^m + \sum_{i=0}^{m-1} p_i \alpha^i$, *coefficients* $p_{m-1} = p_{m-2} = \cdots = p_{m-D+1} = 0$.

Notice that Theorem 1 implies that if $p(\alpha) = \alpha^m + p_k \alpha^k + \sum_{i=0}^{k-1} p_i \alpha^i$ is to be an optimal M-LSDE polynomial, then $k \leq m - D$, which agrees with the findings in [14, 3]. Notice also that (9), which defines the way modular reduction is performed in Step 3 of Algorithm 5.1, is independent of the value of m and thus of the field. The price of this field independence is that now we do not obtain anymore the value of $A \cdot B \bmod p(\alpha)$ but rather $A \cdot B \bmod \overline{p(\alpha)}$ thus, requiring one more reduction at the end of the whole processing. In addition, we need to multiply *once* at initialization $p(\alpha)$ by α^{Dd-m}. This, however, can be thought of as analogous to the Montgomery initialization, and thus, can be neglected when considering the total costs of complex computations which is customary practice in cryptography. In the remainder of the paper, we only consider M-LSDE optimal polynomials.

In what follows, we re-write the steps in Algorithm 5.1 to make them suitable for systolic implementation. The beginning of our work is (9), assuming that $\overline{P}_{d-1} = p_{m-D}$, which we re-write in (10) as a recurrence. We also re-write the limits of the summation for convenience.

$$A^{(i)} = A^{(i-1)}\alpha^D \bmod \overline{p(\alpha)} = \sum_{j=1}^{d-1} A_{j-1}^{(i-1)}\alpha^{Dj} - \sum_{j=0}^{d-1}\left(A_{d-1}^{(i-1)}\overline{P}_j\right)\alpha^{Dj} \qquad (10)$$

Notice (10) is not entirely in terms of digits. Thus, we can write $A_{d-1}^{(i-1)}\overline{P}_j$ as

$$A_{d-1}^{(i-1)}\overline{P}_j = R_j^{(i-1)}\alpha^D + S_j'^{(i-1)} \qquad (11)$$

where $R_j^{(i-1)}$ is a polynomial of maximum degree $D-2$ and $S_j'^{(i-1)}$ is of maximum degree $D-1$. Plugging (11) into (10) and performing some algebra and index manipulation we obtain

$$A^{(i-1)}\alpha^D \bmod \overline{p(\alpha)} = \sum_{j=1}^{d-1} A_{j-1}^{(i-1)}\alpha^{Dj} - \sum_{j=0}^{d-1} S_j'^{(i-1)}\alpha^{Dj} - \sum_{j=1}^{d-1} R_{j-1}^{(i-1)}\alpha^{Dj}. \qquad (12)$$

Note that for optimal M-LSDE polynomials $R_{d-1}^{(i-1)}$ is always zero, and thus the high index is $d-1$ for the last summation in (12). Similarly, we can write an expression for Step 2 of Algorithm 5.1 at iteration i as follows:

$$\overline{C}^{(i)} = B_i A^{(i-1)} + \overline{C}^{(i-1)} = \sum_{j=0}^{d-1}\left(B_i A_j^{(i-1)}\right)\alpha^{Dj} + \sum_{j=0}^{d}\overline{C}_j^{(i-1)}\alpha^{Dj}. \qquad (13)$$

Notice that $B_i A_j^{(i-1)}$ is of the same form as (11), thus we can write $B_i A_j^{(i-1)} = R_j'^{(i-1)}\alpha^D + S_j''^{(i-1)}$, which when plugged back into (13) gives us

$$\overline{C}^{(i)} = \sum_{j=1}^{d} R_{j-1}'^{(i-1)}\alpha^{Dj} + \sum_{j=0}^{d-1} S_j''^{(i-1)}\alpha^{Dj} + \sum_{j=0}^{d}\overline{C}_j^{(i-1)}\alpha^{Dj}. \qquad (14)$$

In a similar manner, we can derive an expression for the last reduction ($\overline{C} \bmod \overline{p(\alpha)}$) of Algorithm 5.1. In particular, for optimal M-LSDE polynomials

$$\overline{C} \bmod \overline{p(\alpha)} = \sum_{j=0}^{d-1}\overline{C}_j\alpha^{Dj} - \sum_{j=0}^{d-1}\left(\overline{C}_d\overline{P}_j\right)\alpha^{Dj} \qquad (15)$$

$$= \sum_{j=0}^{d-1}\overline{C}_j\alpha^{Dj} - \sum_{j=1}^{d-1} R_{j-1}''\alpha^{Dj} - \sum_{j=0}^{d-1} S_j'''\alpha^{Dj}.$$

Using (12), (14), and (15) we readily obtain Algorithm 5.2.

Algorithm 5.2 Low Level Modified LSDE Multiplier

Input: $A = \sum_{i=0}^{d-1} A_i \alpha^{Di}$ with $A_i = \sum_{j=0}^{D-1} a_{Di+j} \alpha^j$, $B = \sum_{i=0}^{d-1} B_i \alpha^{Di}$ with $B_i = \sum_{j=0}^{D-1} b_{Di+j} \alpha^j$, $\overline{C} = \sum_{i=0}^{d} \overline{C}_i \alpha^{Di}$ with $\overline{C}_i = \sum_{j=0}^{D-1} c_{Di+j} \alpha^j$, $\overline{p(\alpha)} = \alpha^{Dd-m} p(\alpha) = \alpha^{Dd} + \sum_{i=0}^{d-1} \overline{P}_i \alpha^{Di}$ with $\overline{P}_{d-1} = p_{m-D}$, $a_i, b_i, c_i \in GF(p)$, and $d = \lceil \frac{m}{D} \rceil$

Output: $\overline{C} \equiv A \cdot B + \overline{C} \bmod \overline{p(\alpha)}$

1: **for** $i = 0$ to $d - 1$ **do**
2:   **for** $j = d$ to 0 **do**
3:     $R'_{j-1} \alpha^D + S''_{j-1} \leftarrow B_i A_{j-1}$ {Note: $A_{-1} = 0$}
4:     $\overline{C}_j \leftarrow R'_{j-1} + S''_j + \overline{C}_j$ {Note: $R'_{-1} = 0, S''_d = 0$}
5:   **end for**
6:   $A^*_{d-1} = A_{d-1}$
7:   **for** $j = d$ to 0 **do**
8:     $R_{j-1} \alpha^D + S'_{j-1} \leftarrow A^*_{d-1} \overline{P}_{j-1}$ {Note: $P_{-1} = 0$}
9:     $A_j \leftarrow A_{j-1} - S'_j - R_{j-1}$ {Note: $A_{-1} = 0, R_{-1} = R_{d-1} = 0, S'_d = A_{d-1}$}
10:   **end for**
11: **end for**
12: Return $(\overline{C} \bmod \overline{p(\alpha)})$ {The reduction is done according to (15)}

---

6 Architecture Description

Addition in fields $GF(p^m)$ are carry free operations, therefore digit addition is also carry free. Two digits result from the multiplication of two digits; for example, $A_j \cdot B_i = R\alpha^D + S'$. One can consider R as a carry. Due to carry free addition, carries are consumed in the next digit position without generating further carries. This operation can be transformed so that carries flow towards the least significant digit positions using the following primitive: $A_j \cdot B_i / \alpha^D = (R\alpha^D + S')/\alpha^D$. These principles are used by the architectures presented here to perform scalar multiplications in a most significant to least significant digit order. Scalar multiplications refer to the multiplication of a $GF(p^m)$ field element by a digit. Figure 3 shows a dependency graph for LSDE multiplication for $d = 4$. In this graph the flow of execution progresses horizontally in the i dimension from left to right and vertically in the j dimension from top to bottom. The cut set lines, shown with dotted lines, show the timing boundaries used here to develop a systolic architecture. Data propagation in the i dimension need to be registered once between cells, while data in the j direction needs to be registered twice. Data traveling diagonally need to be registered once. Dots at the top of the graphs represent the delays required to synchronize inputs with array execution. The dependency graph shows two types of cells. In the following discussion, the term type 1 refers to the cells with rounded corners and the term type 2 refers to the cells with right angle corners. Type 1 cells are used to compute Steps 1 through 11 of Algorithm 5.2 while type 2 cells are used to compute Step 12.

By folding the dependency graph along the i dimension one obtains a digit-serial multiplier where each processing unit processes a row of the dependency graph. If one then folds again along the j dimension, one obtains a scalable architecture, where processing unit x performs the functions in rows $j \equiv x \bmod e$, where e represents the number of processing units in the circuit. The folding just described requires that one of the processing units be able to perform the functions of both type 1 and type 2 cells. LSE multipliers do not need to compute

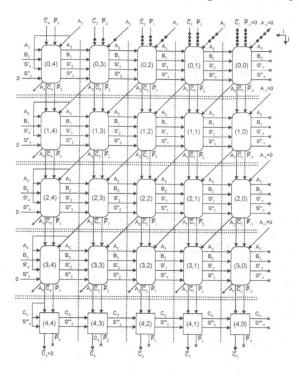

Fig. 3. Dependency graph for LSDE multiplication ($d = 4$)

the reduction in the last row, and thus require processing units to perform the functions of type 1 cells only. Figure 4 shows a block diagram of an LSE/LSDE multiplier that uses two processing units ($e = 2$). Processing unit 0 performs the function of type 1 cells. Processing unit 1 performs the functions of type 2 cells. For this work we developed a single processing unit that can perform the functions of cells of type 1 and 2.

For the LSDE multipliers, the data lines, represented with solid lines, transport D elements (width of data paths is D times the width of an element). For LSE multipliers the data lines carry one element. Dotted lines represent control signals. Note that control is sent from one cell to another in a way that allows data synchronization in the processing units. The figure shows the set of inputs to processing unit 0 with the subscript 0. The same scheme is used for processing unit 1. The scheduling of operands for the LSE/LSDE algorithms require that operands $\overline{P}$ be fed one digit or element at time in a cyclic manner. The rotator (Rot) in the block diagram performs this function. The digits from $\overline{P}$ are bypassed by processing unit 0. Shift registers load the operands B and A into the multiplier. The multiplicand A is loaded into the multiplier through a multiplexer during the computation of the top row in the dependency graph. Thereafter, processing unit 0 gets operands from processing unit 1. Even though it is not shown, the C_0 input can be enhanced with a circuit similar

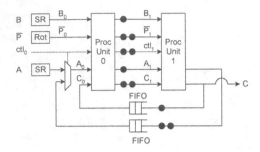

Fig. 4. LSDE scalable multiplier $(e = 2)$

to that used to multiplex A. This circuit would allow the multiplier to perform multiply-and-accumulate operations. The FIFOs are the main components that support scalability. These FIFOs allow processing unit 1 to buffer data destined for processing unit 0. For example, if processing unit 0 is working on row 0 and processing unit 1 is working on row 1 of the dependency graph, the partial results from processing unit 1 corresponding to row 1 are stored in the FIFOs. When processing unit 0 starts working on row 2, it starts consuming the partial results corresponding to row 1 of the dependency graph. Note that partial results from processing unit 0 to processing unit 1 require one or two register delays.

6.1 Complexity and Performance

Table 1 summarizes the most significant characteristics of LSE/LSDE multipliers. The results in the table assume that d is a multiple of e for LSE multipliers and that $d + 1$ is a multiple of e for LSDE multipliers. This arrangement aligns the data so that results can be gathered from the last processing unit in the pipeline. Note that the LSDE algorithm involves extra iterations in its processing loops. The table identifies two cases for d. When $d = e$ the multiplication

Table 1. Characteristics of LSE/LSDE multipliers

Parameter	Condition	LSE	LSDE
Throughput (# clocks/1 mult.)	$d \geq e$	d^2/e	$(d+1)^2/e$
Latency (# clocks)	$d \geq 2e$	$(d/e)d$	$((d+1)/e)(d+1)+1$
	$d = e$	$2e$	$2e+1$
FIFO storage (# digits)	$d \geq 2e$	$d - 2e$	$(d+1) - 2e$
	$d = e$	0	0
Critical path delay	$d \geq e$	$T_M + T_A$	$T_M + T_A$
Cell complexity	$d \geq e$	$2M + 4A + 11R + 5X$	$2M + 4A + 11R + 5X$

does not involve the feedback path. This is analogous to having a digit-serial systolic multiplier. When $d \geq 2e$ the feedback path is used. The timing metrics shown in the table can be deduced from the LSDE dependency graph. In Table 1, M, A, R, and X represent, $GF(p)$ digit/element multiplier, adder, register, and multiplexer, respectively. T_M and T_A represent, respectively, the delay of a $GF(p)$ digit/element multiplier and adder.

7   Prototypes Description

Since the multiplier is scalable, we report synthesis results for only one processing unit. The complexity of a scalar multiplier is a function of the number of processing units it contains. The prototyped processing units were synthesized with Synopsys Design Compiler and they were mapped on a 0.18 μm CMOS technology library from STMicroelectronics. For comparisons with other technology libraries, a gate density of 85 $Kgate/mm^2$ can be assumed for the STMicroelectronics library. The frequency of operation reported represents worst environment conditions (80 °C). Table 2 presents results for ground fields $GF(2)$ and $GF(3)$ and for digit sizes $D = 4, 8, 16$. One and two bits were used to represent elements of the fields $GF(2)$ and $GF(3)$, respectively. The results show that

Table 2. Area and Latency of the basic cell for an LSDE multiplier

Digit size	GF(3)		GF(2)	
	Frequency	Area	Frequency	Area
4	333	23900	454	6200
8	256	61600	357	14900
16	181	181000	344	43400

increases in digit size result in increases in the critical path delay and, thus, a reduction of the maximum frequency. In addition, the $GF(2^m)$ processing units exhibit superior time-area products, even when considering inputs to the processing units of the same width (for example, compare the results for $GF(2)$ with $D = 16$ and the results for $GF(3)$ with $D = 8$).

8   Conclusions

In this paper, we have generalized the LSE/LSDE architectures of [3] by making them suitable for systolic architectures. In addition, we have shown that by computing modulo a multiple of the irreducible field polynomial $p(\alpha)$, we can make the architectures scalable and independent of the value of m. The price to pay is that we do not obtain the result of the multiplication in least residue form. However, in many applications this does not constitute a major drawback since it is possible to perform several multiplications and additions before having to reduce the final result to its least residue form. The basic cell for the LSDE multiplier has been implemented in VHDL and synthesized on a commercial library for $p = 2, 3$ and for different digit sizes (4,8 and 16). An interesting property of our architectures is that the same architectures can be used for any field $GF(q^m)$, where $q = p^n$. This follows easily from our design.

References

[1] D. V. Bailey and C. Paar. Optimal Extension Fields for Fast Arithmetic in Public-Key Algorithms. In H. Krawczyk, editor, *Advances in Cryptology — CRYPTO '98*, volume LNCS 1462, pages 472–485, Berlin, Germany, 1998. Springer-Verlag.

[2] P. S. L. M. Barreto, H. Y. Kim, B. Lynn, and M. Scott. Efficient Algorithms for Pairing-Based Cryptosystems. In M. Yung, editor, *Advances in Cryptology — CRYPTO 2002*, volume LNCS 2442, pages 354–368. Springer-Verlag, 2002.

[3] G. Bertoni, J. Guajardo, S. Kumar, G. Orlando, C. Paar, and T. Wollinger. Efficient $GF(p^m)$ Arithmetic Architectures for Cryptographic Applications. In M. Joye, editor, *Topics in Cryptology — CT-RSA 2003*, volume LNCS 2612, pages 158–175. Springer-Verlag, April 13-17, 2003.

[4] T. Blum and C. Paar. High radix Montgomery modular exponentiation on reconfigurable hardware. *IEEE Transactions on Computers*, 50(7):759–764, July 2001.

[5] D. Boneh and M. Franklin. Identity-Based Encryption from the Weil Pairing. In J. Kilian, editor, *Advances in Cryptology — CRYPTO 2001*, volume LNCS 2139, pages 213–229. Springer-Verlag, 2001.

[6] W.L. Frecking and K. K. Parhi. Performance-Scalable Array Architectures for Modular Multiplication. In *IEEE International Conference on Application-Specific Systems, Architectures, and Processors — ASAP'00*, pages 149–162, July 10 - 12, 2000.

[7] Y.J. Jeong and W.P. Burleson. VLSI array algorithms and architectures for RSA modular multiplication. *IEEE Transactions on VLSI Systems*, 5(2):211–217, 1997.

[8] N. Koblitz. An elliptic curve implementation of the finite field digital signature algorithm. In H. Krawczyk, editor, *Advances in Cryptology — CRYPTO 98*, volume LNCS 1462, pages 327–337. Springer-Verlag, 1998.

[9] Ç. K. Koç and C. Y. Hung. Bit-level systolic arrays for modular multiplication. *Journal of VLSI Signal Processing*, 3(3):215–223, 1991.

[10] P. Kornerup. A systolic, linear-array multiplier for a class of right-shift algorithms. *IEEE Transactions on Computers*, 43(8):892–898, August 1994.

[11] P. Mihăilescu. Optimal Galois Field Bases which are not Normal. Recent Results Session — FSE '97, 1997.

[12] D. Page and N. P. Smart. Hardware implementation of finite fields of characteristic three. In B. S. Kaliski, Jr., Ç. K. Koç, and C. Paar, editors, *Workshop on Cryptographic Hardware and Embedded Systems — CHES 2002*, volume LNCS 2523, pages 529–539. Springer-Verlag, 2002.

[13] N. Smart. Elliptic Curve Cryptosystems over Small Fields of Odd Characteristic. *Journal of Cryptology*, 12(2):141–151, Spring 1999.

[14] L. Song and K. K. Parhi. Low energy digit-serial/parallel finite field multipliers. *Journal of VLSI Signal Processing*, 19(2):149–166, June 1998.

[15] A. F. Tenca and Ç K. Koç. A Scalable Architecture for Montgomery Multiplication. In Ç K. Koç and C. Paar, editors, *Workshop on Cryptographic Hardware and Embedded Systems — CHES'99*, volume LNCS 1717, pages 94–108, Berlin, Germany, August 12-13, 1999. Springer-Verlag.

[16] W.C. Tsai, C.B. Shung, and S.J. Wang. Two systolic architectures for modular multiplication. *IEEE Transactions on VLSI Systems*, 8(1):103–110, 2000.

[17] C. D. Walter. Systolic Modular Multiplication. *IEEE Transactions on Computers*, 42(3):376–378, March 1993.

Cryptanalysis of Block Based Spatial Domain Watermarking Schemes

Tanmoy Kanti Das

Infocomm Security Department
Institute for Infocomm Research
21 Heng Mui Keng Terrace
Singapore 119613
stukdt@i2r.a-star.edu.sg

Abstract. In this paper we propose a block based method of image reconstruction which acts as a generic attack against any block based watermarking scheme in spatial domain. Our reconstruction method relies on the geometric features of the image to discover and modify the inconsistencies in the pixel values introduced due to watermarking. To show its effectiveness we mount the attack on a few existing watermarking schemes. These schemes divide the image into some disjoint sets of pixels. Intensities of constituent pixels of each such set is modified to embed one bit of watermark. And the watermark is a secret binary string which comes from error correcting code. In the process of cryptanalysis we reconstruct an image from a single watermarked copy. Experimental result shows that from the attacked image, watermark extraction becomes impossible, making cryptanalysis successful.

Keywords: *Digital Watermarking, Image Reconstruction, Cryptanalysis, Single Copy Attacks.*

1 Introduction

The aim of digital watermarking is to embed invisible mark into digital objects for copyright protection. As the researchers taking active interest in designing new watermarking schemes, watermarking schemes are becoming resistant to standard signal processing attacks. But enough attention has never been paid to security of these schemes. Very few attempt have been made to analyze individual watermarking schemes and present attacks directed towards a specific scheme. Here, we concentrate on a class of watermarking scheme [9,1] and present a successful cryptanalysis.

Let us now very briefly discuss about watermarking schemes for digital images. An image I can be interpreted as a two dimensional matrix, where each cell contains the pixel value of a particular pixel. For a typical gray scale image, these values can be in the range of 0 to 255. This is called spatial domain representation of an image. Different transform domain representation [like Fast

T. Johansson and S. Maitra (Eds.): INDOCRYPT 2003, LNCS 2904, pp. 363–374, 2003.

Fourier Transform(FFT), Discrete Cosine Transform(DCT), Wavelet Transform, etc.] are also possible. Note that if we change values of the matrix by very small amount image quality may not degrade and remain visually indistinguishable from original image. Given an image I, neighborhood of the image $N(I)$ consists of those images which are visually indistinguishable from I. It is expected from any watermarking system that they meet some basic criteria which includes

1. **Imperceptibility**: Embedded watermark must be below the perceptible threshold i.e. while designing watermarking scheme embedding process must take into account some human visual model to quantify the noise and control the embedding process in such a manner that original & marked signal is perceptually indistinguishable.
2. **Redundancy:** Embedding process must embed the information redundantly to ensure robustness against several kind of attack.
3. **Keys:** Keys used to control the embedding process must be cryptographically secure to ensure robustness against manipulation and eraser.

A typically watermarking scheme adds a signal $s^{(i)}$ to the original image I in such a manner that $I^W = I + s^{(i)}$ remains in $N(I)$. This image is given to i^{th} buyer. Depending on whether original image is required for watermark detection or not, watermarking schemes are divided into two groups, namely non-oblivious and oblivious. In non-oblivious scheme (e.g. CKLS scheme [2]) original image is used at the time of retrieval. Generally in non-oblivious schemes original image I is subtracted from the watermarked image I^W to retrieve a signal $s^\# = I^W - I$. Similarly in oblivious scheme (e.g. [12,8]) signal $s^\#$ is recovered without using the original image but using some other information related to original image. Buyer i is suspected if correlation between recovered signal $s^\#$ and embedded signal $s^{(i)}$ is significant.

It is expected from any robust watermarking scheme that it will be able to identify the malicious buyer properly and never wrongly implicate an honest buyer. At least probability of wrongly implicating a buyer should be very small. As most of the existing watermarking schemes are correlation based, i.e. to identify the watermark it relies on correlation between embedded and recovered signal, it is natural that an attacker would try to break the correlation between recovered signal $s^\#$ and embedded signal $s^{(i)}$ to evade detection. To achieve his goal, given an watermarked image I^W, the attacker constructs an image $I^{W\#}$ in such a manner that $I^{W\#} \in N(I)$ and $s^\# = I^{(i)} - I$ is uncorrelated with $s^{(i)}$. Thus it is not possible to identify the malicious buyer i anymore.

An obvious requirement of any watermarking system is that, it is not possible for an attacker to construct an good approximation of host image from a single copy at his possession (known as single copy attack). This type of single copy attack is reported in [3,7,10,11]. Replacement attack proposed in [7] utilizes redundancy in multimedia content. Upon analyzing a multimedia content, one can find several segments which are exact replica of others. Each such segment is replaced by a perceptually similar segment generated by some combination of similar segments found within the same content or with in the library of media contents. Due to replacement, large part of the watermark got removed. It is well known that any digital watermarking scheme is vulnerable to collusion attack [4]. But that requires a large number of watermarked copies which may not be practical. On the other-hand, here we present a very strong cryptanalytic attack based on single watermarked copy. Before proceeding further let us highlight the importance of such an attack. *Existing watermarking schemes never been analyzed properly using cryptanalytic techniques as always done in any standard cryptographic schemes. We look into the watermarking scheme from the cryptographic point of view and present a very strong attack which is analogous to cipher text only attack. We successfully mount the attack on the schemes presented in [9,1]. It seems that digital watermarking schemes are not robust enough and susceptible to attack specifically designed for it.*

In this paper we present an image reconstruction method using a single watermarked copy. Reconstructed image is of very high visual quality. Unlike replacement attack which relies on finding redundancy in multimedia content (which may not work in many cases due to lack of redundancy) our method of image reconstruction uses the geometric features of the watermarked image and these features mostly remain intact even after watermark embedding. Image reconstruction acts as a generic attack against any spatial domain watermarking scheme, particularly block based watermarking schemes. It is also effective against few other block based watermarking scheme which embeds the watermark in other than spatial domain. We are mounting the attack against the scheme presented in [9,1] to prove it's effectiveness. We have also mounted this attack against some other block based schemes but due to space constraint we are not reporting it here.

The paper is organized as follows. In section 2 we present tools for cryptanalysis. In section 3 we present overview of schemes to be cryptanalized. In section 4 we present the exact algorithms used for cryptanalysis. In last section we conclude the paper.

2   Image Reconstruction

Spatial domain watermarking schemes change pixel values to embed watermark in the host image. Most of them do not take into account geometric features of the image at the time of embedding the watermark. They consider the pixel values only and change the pixel values by very small amount, so that watermarked image remain indistinguishable from the original image. In the process geometric features of the image remain undisturbed. Before discussing how to identify the geometric feature(s) of an image, let us discuss about some basic relationship among pixels [5, pages 40-45].

– **Neighbors of a pixel** : A pixel p at location (α, β) has four neighbors at $(\alpha+1, \beta)$, $(\alpha-1, \beta)$, $(\alpha, \beta+1)$, $(\alpha, \beta-1)$. These are horizontal and vertical neighbors of p and belong to set $N_4(p)$. Similarly there are four diagonal neighbors of p and represented by set $N_d(p)$. All the eight neighbors of p belongs to set $N_8(p)$.

- **8-connectivity** : Two pixels p and p' with similar pixel values are 8-connected if $p' \in N_8(p)$. Similarly we can define 4-connectivity. A *path* between pixel p_1 at location (α_0, β_0) and pixel p_2 at location (α_n, β_n) is a series of distinct pixels at location

$$(\alpha_0, \beta_0), \ (\alpha_1, \beta_1), \ldots, \ (\alpha_n, \beta_n)$$

  where $p_i \in N_8(p_{i-1})$, $1 \leq i \leq n$.
- **Connected component** : Let I be the set of pixels from an image. If there is path between pixel p_1 and p_2 consisting pixels from I, then p_1 and p_2 is connected. For any pixel p in I, the subset of pixels in I that are connected to p is termed as a *connected component* of I. Thus distinct connected components are disjoint.

It is easy to study image's geometric feature(s) if the image is a binary image. But one can always transform an (gray scale) image to binary by thresholding. However due to thresholding, many important geometric features will be lost if we set a single threshold for entire image. One way to solve the problem (to some extent) is to divide the image into several small blocks of size say 4×4 and binarize each of the available blocks separately. Threshold for individual blocks is set dynamically, depending on the constituent pixels of the block. Identification of connected components is done separately for each binary block. A connected component represents a geometric feature of the image. As the size of the block is not big enough, it is most unlikely that each connected component will be represented by pixels with different intensities in the original gray scale image. But if there is any inconsistency in pixel values within a connected component then that is rectified to generate a new block. Here we propose an algorithm to reconstruct the block.

Algorithm 1

1. *Read the input image I. Let the size of the image I be $2^a \times 2^b$.*
2. *Divide the image into number of blocks of size $2^\alpha \times 2^\beta$. Let the number of such blocks be γ and $\gamma = 2^{(a+b)-(\alpha+\beta)}$.*
3. *For $i = 0 \ldots \gamma - 1$ do.*
   (a) *Copy block b_i to b_{temp} i.e, $b_{temp} = b_i$.*
   (b) *Find minimum and maximum pixel values p_{min} & p_{max} from b_{temp}.*
   (c) *If $(p_{max} - p_{min} \leq diff)$ then replace all the pixels of block b_i by p_{avg}, where p_{avg} is the average pixel value of b_i and $diff$ is certain pre-defined threshold.*
   (d) *Else*
      i. *Set threshold $p_t = \frac{p_{min} + p_{max}}{2}$.*
      ii. *Binarize the block b_{temp} using p_t as the threshold i.e. where pixel value is less than p_t mark it as 1(black) otherwise mark it 0(white).*
      iii. *Following steps are performed twice. First for black pixels, second for white pixels.*
      iv. *Find out the number of distinct connected components. Let number of such connected component be n.*

     v. Label each distinct connected component with different label.

     vi. Let S_j be the set of pixel locations connected to each other(i.e. they have same label). Where $j = 1 \ldots n$.

    vii. For $j = 1 \ldots n$ do

        A. Find out average intensity P of pixels $b_i(k,l)$ where $(k,l) \in S_j$.

        B. Set $b_i(k,l) = P$ where $(k,l) \in S_j$.

4. Reconstruct the image using image blocks b_i (for $i = 0 \ldots \gamma - 1$).

Thus geometric feature based image processing would be able to discover/correct the inconsistency in pixel values introduced by watermark embedding. As described earlier, algorithm 1 groups image pixels(constituent pixels of a connected component termed as a group) using image's geometric features. Within a group there is very little variation in pixel values. Thus replacing these pixels with their average value degrades the image very little and simulation result show that image quality of reconstructed image is very good. And PSNR of reconstructed image w.r.t original image is as high as 35dB for peppers image. Also algorithm 1 acts as our tool for cryptanalysis.

3 Overview of Some Spatial Domain Watermarking Schemes

3.1 Overview of Mukherjee, Maitra, and Acton's Scheme

The scheme presented in [9] is a spatial domain watermarking scheme and it does not take into account any information regarding the host image except the pixel values. Thus it operates passively on the host image to generate the watermarked copy. Watermark, also known as **buyer key**, is basically a binary string which can uniquely identify a particular buyer and it comes from specific error correcting code. Likewise an image key is associated with each image. Thus for a particular image there is only one image key but multiple buyer keys. Let us consider an image I of size $2^a \times 2^b$. It is divided into m subgroups $G_0, \ldots, G_{m-2}, G_{m-1}$ in such a manner that each subgroup contains $\frac{2^a \times 2^b}{m}$ number of pixel location. It should be noted that $G_i \bigcap G_j = \emptyset$ for $0 \le i \ne j \le m-1$. This group information is known as **image key** K and a label matrix L^I is used to maintain the group information. Number of option to select such a group is prohibitively large and is grater then $2^{2^{n-1}(a+b-1)}$, where $m = 2^n$. Let us present the scheme for watermark embedding [9] next.

Algorithm 2

1. Let I be the host image of size $2^a \times 2^b$.

2. Spatially divide the image into several subgroup $G_0, G_1, \ldots, G_{m-2}, G_{m-1}$. Pixels for each subgroup are chosen in random fashion. Label matrix L^I maintains this group information.

3. For $i = 1 \ldots 2^a - 1$ do
  (a) For $j = 1 \ldots 2^b - 1$ do
    i. Let $I_{i,j}$ be the pixel value at location i, j.
    ii. Let $L_{i,j}^I = k$ i.e., pixel at location i, j belongs to group G_k.
    iii. If $B_k = 0$, then $I_{i,j}^W = I_{i,j}$.
    iv. Else if $B_k = 1$, then $I_{i,j}^W = I_{i,j} + \beta_{i,j}$

Here $\beta_{i,j}$ is either positive or negative for all the pixels belonging to a specific subgroup. Value of $\beta_{i,j}$ is very important in the sense that it must be below some threshold such that image quality of the watermarked image remains good. Concept of weber ratio [5, page 40] is used for determining the value of the threshold. Watermark extraction require original host image and the knowledge of **image key** K. For a particular group, sum of difference between corresponding pixel values of original and watermarked image act as a measure for determination of decoded watermark bit. The algorithm [9] is as follows.

Algorithm 3

1. Read image key K. Let number of subgroup be m.
2. Let the size of attacked watermarked image $I^{W\#}$ and host image I be $2^a \times 2^b$.
3. For $k = 0 \ldots m - 1$ do
  (a) $\sigma_k = 0$.
4. For $i = 0 \ldots 2^a - 1$ do
  (a) For $j = 0 \ldots 2^b - 1$ do
    i. If $|I_{i,j}^{W\#} - I_{i,j}| > |\beta_{i,j}|$ then $I_{i,j}^{W\#} = I_{i,j} + \beta_{i,j}$.
5. For $i = 0 \ldots 2^a - 1$ do
  (a) For $j = 0 \ldots 2^b - 1$ do
    i. If pixel at location (i, j) belongs to group G_k then, $\sigma_k = \sigma_k + (I_{i,j}^{W\#} - I_{i,j})$.
6. For $k = 0 \ldots m - 1$ do
  (a) If $|\sigma_k| \le c_k |\delta_k|$, $q_k = 0$.
  (b) Else $|\sigma_k| > c_k |\delta_k|$, $q_k = 1$.
7. Find q' closest to q and report q' as buyer key B.

Here c_k is known as tolerance factor. It plays an important role in resisting various intentional attacks which may change the intensity of the watermarked image. Due to complex nature of various image transformations it is very difficult to use a particular value of c_k for all the cases. It is better to use a range of values for c_k. The range of c_k is 0.001 to 1. Though use of tolerance factor can limit effect of various image transformations, it may not be able to limit the damage due to non linear geometric bending attack. In non linear attacks, not only the values of individual pixel changes but also their spatial location also changes. A minor variation of watermark insertion and detection algorithm can withstand non-linear geometric bending attack. First host image is spatially divided into number of image blocks having contiguous pixels. Now each subgroup, instead of having number of pixels, consists number of blocks randomly chosen from the available blocks. Exact algorithm [9] is as follows.

Algorithm 4

1. *Let I be the host image of size $2^a \times 2^b$.*
2. *Divide the image into blocks of size $2^\alpha \times 2^\beta$. Let number of such block be $\gamma = 2^{(a+b)-(\alpha+\beta)}$.*
3. *Spatially divide image blocks into several subgroup $G_0, G_1, \ldots, G_{d-2}, G_{d-1}$. Where d is the number of subgroups. Blocks for each subgroup are chosen in random fashion.*
4. *For $j = 0 \ldots \gamma - 1$ do*
    (a) Let image block b_j belongs to subgroup G_k. Then
        i. If $B_k = 0$, then $b_j^w = b_j$.
        ii. If $B_k = 1$, then $b_j^w = b_j + \beta_j$.
5. *Construct the watermarked image I^W using image blocks b^w.*

In algorithm 4 step 4(a)ii means all the pixels of the block is incremented by same amount β_j. Watermark extraction algorithm is also changed accordingly. Due to non-linear attack position of blocks also changes. To correct the block position window based pre-processing algorithm is employed at the time of watermark recovery. We are not presenting the recovery algorithm here, it is similar to algorithm 3.

3.2   Overview of Bruyndonckx et al Scheme

Bruyndonckx et al's scheme [1] like previous scheme is also a block based algorithm which modifies pixel intensities within a block to encode a bit of watermarking information. Encoding a bit is simple and consists of following steps.

- Out of available image blocks a block b_i is chosen in random fashion by the owner. A secret key controls the choice of particular block and provides first level of protection to the owner.
- Pixels of the block is divided into two groups of homogeneous intensities termed as *zone 1* and *zone 2*.
- Each zone is further subdivided into two categories, *category A* and *category B*.
- Thus there is four categories within a block. calculate mean of each categories i.e. m_{1A}, m_{1B}, m_{2A}, m_{2B}. And also calculate mean of each zone i.e. m_1, m_2.
- Intensities of each categories is modified in such a manner that following holds:
    - If watermarking *bit*=1 then
        1. $m_{1A}^* - m_{1B}^* = l$.
        2. $m_{2A}^* - m_{2B}^* = l$.
    - If watermarking *bit*=0 then
        1. $m_{1B}^* - m_{1A}^* = l$.
        2. $m_{2B}^* - m_{2A}^* = l$.

Where $m_{1A}^*, m_{1B}^*, m_{2A}^*, m_{2B}^*$ are the respective mean values of different categories after watermark embedding and l is enforced difference between mean values. To maintain perceptual clarity of the watermarked image, it is necessary that mean values of each zone m_1 and m_2 should remain unchanged even after watermark embedding. Watermark extraction is nothing but calculating $\sum_1 = m_{1A} - m_{1B}$ and $\sum_2 = m_{2A} - m_{2B}$ and the sign of $\sum_1$ and $\sum_2$ determines the decoded bit.

4   Proposed Attack

4.1   Mukherjee et al Scheme

Let us consider a scenario, where one groups n contiguous pixels with similar pixel values from the host image. Here, two pixel are considered similar, if the difference in their intensities is less than a threshold δ. It is expected from watermarking *algorithm 2*, that out of n pixels, $\frac{n}{2}$ pixels will retain their intensities, while another $\frac{n}{4}$ pixel's intensity will be increased. Intensities of remaining $\frac{n}{4}$ pixel expected to decrease. As the amount of change incorporated in each pixel is similar, then $\sum_{i=1}^{n} p_i \approx \sum_{i=1}^{n} p_i'$, where p and p' are the corresponding pixels from host image and watermarked image. If it is possible to identify the similar pixels which form a group in the host image, using the watermarked image only, then one can immediately replace the those pixel's intensity with their average intensity in the watermarked image to remove the watermark. As averaging would remove the inconsistencies in pixel values created by watermark embedding. But when original host image is not available it is difficult, if not impossible to get the group information as existed in the original host image from the watermarked image. Feature based image reconstruction groups the pixel using the concept of connected component and within such connected component, it is expected that there will be little variation of pixel intensity. If there is any variation in pixel intensities, it can be safely assume that it occurred due to watermark embedding and can be removed by replacing pixels with their average intensity. Thus the proposed *algorithm 1* will be successful in removing the watermarks. Simulation result show it is a very strong attack and able to remove the watermark successfully.

Simulation Original host image I is watermarked using algorithm 2 to generate watermarked image I^W. Here also, Reed Muller codes has been used to generate the buyer key as experimented in [9]. Length of such buyer key is 2^n, while the minimum distance between two such buyer key is 2^{n-1}. For simulation we have taken the value of $n = 7$. Thus length of the buyer key B is 128 which can correct 31 bit errors. We are using images of size 128×128 and $\beta_{i,j} = 1$ as used in [9]. Using the recovery *algorithm 3* we recover a key B'. If B and B' matches in at least in $128 - 31 = 97$ places, we can correctly decode the key. Thus bitwise matching below 97 (or 75.78%) will defeat the scheme i.e. buyer key can not be decoded properly.

As mentioned earlier, only the watermarked image I^W is available to the attacker. From the available image I^W, attacker re-constructs an image I^R using the image reconstruction algorithm 1. Re-constructed image is visually indistinguishable from the watermarked image, as can be seen from the figure 1. The image I^R is subjected to watermark recovery. Results are represented in tabular form below. One can see that for all values of tolerance factor c_k bitwise matching value is far less than minimum required value 97 (or 75.78%). Thus our proposed algorithm 1 effectively removes the watermark form the watermarked image.

c_k	Bitwise Matching Value
0.0	49.22%
0.1	49.22%
0.2	50.78%
0.3	49.22%
0.4	51.56%
0.5	50.00%
0.6	52.34%
0.7	52.34%
0.8	53.21%
0.9	53.21%
1.0	51.56%

Image	Bitwise Matching Value
Lena	53.90%
Baboon	49.22%
Peppers	51.56%

Table1. Attack on [9]   **Table 2.** Attack on [1].

Fig. 1. Original, Watermarked (Mukherjee et al Scheme) and Attacked image

If we use algorithm 4 to watermark the host image and corresponding detection algorithm to recover the key, the situation is different. In that case algorithm 1 is unable to remove the watermark and we need to modify algorithm 1 as stated in subsection 4.3.

4.2 Bruyndonckx et al Scheme

Here watermark embedding algorithm divides each block into zones of homogeneous intensities without taking into account image's geometric features. Intensities of constituent pixel of each zone is modified to embed the watermark and amount of change incorporated in each pixel is also similar. From the nature of algorithm it is clear that if an attacker can identify the zones as existed in the original host image, using the watermarked image only, then replacing the constituent pixel's intensity with their average intensity would remove the watermark easily. But the problem of identification of zone, as existed in original host image, can be solved using the image re-construction algorithm 1. Image re-construction algorithm using image's geometric features do identify these zones of homogeneous pixels to remove the watermark. Here we have used 3 different images of size 512×512 to embed 128 bit of watermark. A new image is constructed from the watermarked image and the result of watermark extraction form the reconstructed image is presented in Table 2.

4.3 Improved Cryptanalysis

Analysis of the algorithm 1 makes it clear that it will not be able to discover any anomalies in pixel intensity when watermark embedding algorithm itself block based. Block based watermarking algorithms instead of creating differences in pixel values within a block b_i, normally creates difference in pixel values between adjoining blocks. To overcome this problem, we modify the algorithm 1 in such a manner that it not only takes into account geometric features of the block b_i but also takes into account geometric features of surrounding blocks. Normally we will construct a bigger block b_j^g in such a manner that block b_i remains at the center of b_j^g and block b_j^g instead of b_i will be used for reconstruction. Also step 3(d)viiB of algorithm 1 needs to be modified. Instead of using average of pixel values of a connected component S_j we will use average of those pixel which is not part of b_i but belongs to S_j. In case S_j consists of pixel entirely from b_i then average of intensity is used as before. Proposed algorithm for cryptanalysis is presented below.

Algorithm 5

1. *Read the input image I. Let the size of the image I be $2^a \times 2^b$.*
2. *Divide the image into number of blocks of size $2^\alpha \times 2^\beta$. Let the number of such blocks be $\gamma = 2^{(a+b)-(\alpha+\beta)}$.*
3. *For $i = 0 \ldots \gamma - 1$ do.*
   (a) *Construct a bigger block b_i^g keeping b_i at the center. Size of b_i^g is $(2^\alpha + w) \times (2^\beta + h)$.*
   (b) *Find minimum and maximum pixel values p_{min} & p_{max} from b_i^g.*
   (c) *If $(p_{max} - p_{min} \leq diff)$ then replace all the pixels of block b_i by p_{avg}. Where p_{avg} is the average pixel value of b_i^g and $diff$ is certain pre-defined threshold.*
   (d) *Else*
      i. *$b_{temp} = b_i^g$.*
      ii. *Set threshold $p_t = \frac{p_{min} + p_{max}}{2}$.*
      iii. *Binarize the block b_{temp} using p_t as the threshold i.e. where pixel value is less than p_t mark it as 1(black) otherwise mark it 0(white).*

iv. *Following steps are performed twice. First for black pixels, second for white pixels.*

v. *Find out the number of distinct connected components. Let number of such connected component be n.*

vi. *Label each distinct connected component with different label.*

vii. *Let S_j be the set of pixel locations connected to each other(i.e. they have same label). Where $j = 1 \ldots n$.*

viii. *For $j = 1 \ldots n$ do*

    A. *If S_j consists pixel entirely from b_i then find average P of pixels $b_i^g(k,l)$ where $(k,l) \in S_j$.*

    B. *Else find average P of pixels $b_i^g(k,l)$ where $(k,l) \in S_j$ and (k,l) does not belong to b_i.*

    C. *Set $b_i(k,l) = P$ where $(k,l) \in S_j$ and (k,l) belongs to block b_i.*

4. *Reconstruct the image using image blocks b_i (for $i = 0 \ldots \gamma - 1$).*

Thus **algorithm 5** not only corrects any anomalies in pixel values with in a block b_i but also removes anomalies in pixel value between surrounding blocks of b_i. It is clear that, if a connected component spreads over several blocks, watermark embedding algorithm 4 will creates difference in pixel values among the constituent pixels of that connected component. Algorithm 5 removes this type of anomalies.

Simulation Image of size 256×256 is watermarked using algorithm 4 and the size of individual block is 4×4. Each group G_i consists of 32 such blocks. Thus the number of group is $2^7 = 128$. Other parameters remain as that of sub-section 4.1. This watermarked image is used as input to algorithm 5. Reconstructed image is subjected to watermark extraction and result of that is presented in tabular form below.

c_k	Bitwise Matching Value(%)
0.0	53.21
0.1	53.21
0.2	51.56
0.3	50.78
0.4	50.78
0.5	51.56
0.6	51.56
0.7	53.21
0.8	52.34
0.9	53.21
1	53.21

Table 3. Attack on Mukherjee et al Scheme.

5   Conclusion

In this effort we have presented an image reconstruction method for removal of watermarks and successfully demonstrated it's effectiveness. Both the schemes [9,1] are robust w.r.t. secrecy of watermark's location & content i.e. location of watermark is secret and content of watermark is unknown. But that does not prevent us from removing the watermark. Thus one should not judge the robustness of watermarking schemes on secrecy of location and content alone.

References

1. O. Bruyndonckx, J.-J. Quisquater and B. Macq. Spatial Method for Copyright Labeling of Digital Images. In IEEE workshop on Nonlinear Signal and Image Processing 1995, Greece, pages 456-459.
2. I. J. Cox, J. Kilian, T. Leighton and T. Shamoon. Secure Spread Spectrum Watermaking for Multimedia. *IEEE Transactions on Image Processing*, 6(12):1673–1687, 1997.
3. T. K. Das and S. Maitra. Cryptanalysis of Optimal Differential Energy Watermarking (DEW) and a Modified Robust Scheme. In *INDOCRYPT 2002*, pages 135-148, vol 2551 of Lecture Notes in Computer Science. Springer-Verlag, 2002.
4. F. Ergun, J. Kilian and R. Kumar. A note on the limits of collusion-resistant watermarks. In *Eurocrypt 1999*, pages 140-149, vol 1592 of Lecture Notes in Computer Science. Springer Verlag, 1999.
5. R. C. Gonzalez and R. E. Woods. Digital Image Processing. Addision Wesley, 1992.
6. S. Katzenbeisser, F. A. P. Petitcolas. Information Hiding Techniques for Steganography and Digital Watermarking. Artech House, USA, 2000.
7. D. Kirovski and F. A. P. Petitcolas. Replacement Attack on Arbitrary Watermarking Systems. In ACM workshop on *Digital Rights Management*, 2002.
8. G. C. Langelaar and R. L. Lagendijk. Optimal Differential Energy Watermarking of DCT Encoded Images and Video. *IEEE Transactions on Image Processing*, 10(1):148–158, 2001.
9. D. P. Mukherjee, S. Maitra and S. T. Acton. Spatial Domain Digital Watermarking of Multimedia Objects For Buyer Authentication. To appear in *IEEE Transaction on Multimedia*. (This is an extended version of the following paper:
   S. Maitra and D. P. Mukherjee. Spatial Domain Digital Watermarking with Buyer Authentication. In *INDOCRYPT 2001*, pages 149-161, vol 2247 of Lecture Notes in Computer Science. Springer-Verlag, 2001.)
10. F. A. P. Petitcolas, R. J. Anderson, M. G. Kuhn and D. Aucsmith. Attacks on Copyright Marking Systems. In *2nd Workshop on Information Hiding*, pages 218–238 in volume 1525 of Lecture Notes in Computer Science. Springer Verlag, 1998.
11. F. A. P. Petitcolas and R. J. Anderson. Evaluation of Copyright Marking Systems. In *IEEE Multimedia Systems*, Florence, Italy, June 1999.
12. Y. Wang, J. F. Doherty and R. E. VanDyck. A Wavelet-Based Watermarking Algorithm for Ownership Verification of Digital Images. *IEEE Transactions on Image Processing*, 11(2):77–88, 2002.

More Efficient Password Authenticated Key Exchange Based on RSA

Duncan S. Wong[1], Agnes H. Chan[2], and Feng Zhu[2]

[1] Department of Computer Science
City University of Hong Kong
Hong Kong, China
duncan@cityu.edu.hk

[2] College of Computer Science
Northeastern University
Boston, MA 02115, U.S.A.
{ahchan,zhufeng}@ccs.neu.edu

Abstract. In [17], Zhu, et al. proposed a RSA-based password authenticated key exchange scheme which supports short RSA public exponents. The scheme is the most efficient one among all the RSA-based schemes currently proposed when implemented on low-power asymmetric wireless networks. We observe that its performance can further be improved by proposing two modifications. The first modification shortens the size of the message sent from the server to the client. The second modification dramatically reduces the size of the message sent from the client to the server and therefore can be used to reduce the power consumption of the client for wireless communications in a significant way. We also generalize our modified schemes and formalize the security requirements of all underlying primitives that the generic scheme is constituted. A new primitive called password-keyed permutation family is introduced. We show that the security of our password-keyed permutation family is computationally equivalent to the RSA Problem in the random oracle model.

Keywords: Password Authentication, Key Exchange, Secure Wireless Communications

1 Introduction

We investigate methods of providing efficient password authenticated key exchange (PAKE) for wireless communications between a low-power client and a powerful server. The objective of a password authenticated key exchange scheme is the same as a conventional authenticated key exchange scheme [5]: after two communicating parties successfully executing the scheme, each of them should have certain assurance that it knows each other's true identity (*authentication*), and it shares a new and random session key only with each other and the key is derived from contributions of both parties (*key exchange*). Unlike a

T. Johansson and S. Maitra (Eds.): INDOCRYPT 2003, LNCS 2904, pp. 375–387, 2003.

cryptographic-key authenticated key exchange scheme, the two communicating parties do not have any pre-shared cryptographic symmetric key, certificate or support from a trusted third party. Instead they only share a password. The major difficulty in designing a secure password-based protocol is due to the concern of implicated off-line dictionary attacks against a small password space [3]. A password, a passphrase, or a PIN (Personal Identification Number) generally needs to be easy to remember. Usually it has significantly less randomness than its length suggested or is simply very short in length. In our study, the password space is considered to be so small that an adversary can enumerate it efficiently.

We focus our attention on designing a password authenticated key exchange scheme for wireless communications between a low-power client and a powerful server. A powerful server has comparable computational power and memory capacity to a current desktop machine while a low-power client is as resource constrained as a smart card, a wearable device in a wireless PAN (Personal Area Network), a low-end PDA (Personal Digital Assistant) or a cellular phone. In addition, we consider the client to be mostly battery-powered with inferior communication capability. A typical cellular network in which a mobile unit communicating with a base station, a Bluetooth-based PAN in which a watch-size PDA communicating with a laptop, and a disposable sensor exchanging information with a more capable tandem device in an ad hoc sensor network are some typical examples of our target applications. Therefore, the objectives of our scheme design include optimizing the computational complexity especially at the client side, minimizing memory footprint and reducing the size of the messages exchanged between the two communicating parties. Besides the conventional techniques for pursuing these objectives, such as precomputation and caching, we also explore the fact of receiving radio signal consumes much less power than transmitting in scheme design.

1.1   Related Work

There were many password authenticated key exchange (PAKE) schemes proposed in the last decade, especially recently [3, 9, 16, 4, 13, 11, 6, 8][3]. Most of these schemes are based on Diffie-Hellman key exchange and perform large modular exponentiation operations which may take a long time to compute on a low-power device.

In [3], Bellovin and Merritt investigated the feasibility of using RSA [14] to construct a PAKE. If the RSA public exponent is short, the encryption operation can be done efficiently. However, they also pointed out that an e-residue attack may be feasible if the receiving party has no way to verify whether a RSA public exponent is fraudulent or not, that is, to check if the public exponent is relatively prime to $\phi(n)$ without knowing the factorization of a RSA modulus n. To thwart this attack, the authors considered an interactive protocol for validating the public exponent. However, the protocol was found to be insecure [17].

---

[3] Jablon maintains an updated list at http://www.integritysciences.com/links.html

Two other RSA based schemes were proposed later in [12, 13]. The first one was later found to be insecure, while the second one has to use a large prime for the public exponent. This defeats the purpose of using RSA for low-power clients in our target applications because the computational complexity of performing a RSA encryption is no less than that of carrying out a modular exponentiation in a Diffie-Hellman key exchange based protocol.

In [17], Zhu, et al. modified the interactive protocol in [3] and proposed a scheme which supports short RSA public exponents. The scheme is the most efficient one among all the RSA-based schemes currently proposed when implemented on low-power asymmetric wireless networks. For applications requiring moderate level of security, that is, if 512-bit RSA is used, the scheme takes less than 2.5 seconds of pure computation on a 16MHz Palm V and about 1 second for data transmission if the throughput of a network is only 8kbps. The computation time can be improved to 300 msec and the transmission time can also be reduced to 300 msec if caching is allowed. For applications where 1024-bit RSA is used, the total computation time of the client is estimated to be about 9 seconds according to the performance measurements of various cryptographic operations reported in [15]. It can be reduced dramatically to less than half second if the server's public key is cached at the client side. Since a client is usually communicating with an essentially fixed set of servers in most cases, the first 9-second protocol run can be considered as a one-time setup phase in a system.

1.2 Our Contributions

In [17], the most time consuming part of the scheme is to check the validity of the server's RSA public exponent. The checking mechanism is a two-round interactive protocol which requires 1KB of data sent from the client to the server and another 200B of data sent from the server to the client, if 1024-bit RSA is used with the same configurations of all other parameters specified in [17]. In this paper, we propose two modifications of the interactive protocol for improving its efficiency. The first modification reduces the size of the message sent from the server to the client. It is a straightforward modification but reduces the transmission time, memory footprint and computation complexity all at the same time. The second modification reduces the size of the message sent from the client to the server significantly and hence entails much less power consumption for the low-power client in wireless communications than the original protocol in [17] as receiving radio signal requires much less power than transmitting some. Memory footprint at the client side is also minimized and other optimization techniques such as precomputation and caching are preserved.

On the security analysis, we generalize our modified schemes to a generic one and formalize the security requirements of all underlying primitives that the generic scheme is constituted. In the formalization, we introduce a new primitive called password-keyed permutation family to capture the features of the password related operations in our schemes. We also show that the security of our password-keyed permutation family is computationally equivalent to the RSA Problem in the random oracle model [2].

The rest of the paper is organized as follows. In Sec. 2, the scheme proposed in [17] is reviewed. This is followed by the description of our two modifications in Sec. 3. In Sec. 4, we formalize our modifications by describing a generic scheme and specifying the security requirements of its underlying primitives. We conclude the paper in Sec. 5.

2   A RSA-based PAKE [17] — RSA-PAKE1

In [17], Zhu, et al. proposed a RSA-based password authenticated key exchange (PAKE) scheme which supports short RSA public exponents. It is refined later in [1] to eliminate potential vulnerabilities of using symmetric encryption function. In the following, we review the refined scheme with some modifications so that the final session key is generated from the contributions of both communicating parties. We call the new scheme the RSA-PAKE1.

Define some integer k as a system-wide security parameter. Let two finite-length strings $A, B \in \{0,1\}^*$ denote a powerful server and a low-power client, respectively. Let (n, e) be the RSA public key of A where n is the RSA modulus and e is the public exponent. Suppose A and B share a password $pw \in PW$ where PW denotes a password space in which the password is chosen according to certain probability distribution. Let $G_1, G_2, G_3, G_4, G_5 : \{0,1\}^* \to \{0,1\}^k$ be distinct and independent cryptographically strong hash functions. The protocol proceeds as follows.

1. A selects $r_A \in_R \{0,1\}^k$ and sends $((n, e),\ r_A)$ to B.
2. B checks if (n, e) is a valid public key using an Interactive Protocol (read Sec. 3 for details). It then randomly picks $r_B \in_R \{0,1\}^k$ and $s_B \in_R \mathbb{Z}_n^*$, and computes $\pi = T(pw, A, B, r_A, r_B)$ where $T : \{0,1\}^* \to \mathbb{Z}_n^*$ is a distinct cryptographic hash function. It sends r_B and $z = s_B^e \cdot \pi \bmod n$ to A.
3. A computes π accordingly and obtains s_B from z. It then computes $K = G_1(s_B)$, $c_B = G_3(s_B)$ and $\sigma = G_4(c_A, c_B, A, B)$, randomly picks $c_A \in_R \{0,1\}^k$, and sends $(K \oplus c_A,\ G_2(K, c_A, A, B))$ to B.
4. B computes K and c_B from s_B accordingly. It reveals c_A from the first part of the incoming message and checks if the second part of the incoming message is $G_2(K, c_A, A, B)$. It then computes the session key σ accordingly and sends $G_5(\sigma)$ back to A.
5. A finally checks if the incoming message is $G_5(\sigma)$.

Here we consider the RSA to be a trapdoor permutation over $\mathbb{Z}_n^*$. The operation $s_B^e \cdot \pi \bmod n$ can also be considered as a permutation of $s_B^e \bmod n$ over $\mathbb{Z}_n^*$ where $\pi \in \mathbb{Z}_n^*$. Observing that RSA is a trapdoor permutation also over $\mathbb{Z}_n$, we can use the following two permutation methods to compute z as well. Method 1: Let T be defined as $T : \{0,1\}^* \to \{0,1\}^{|n|-1}$. Compute z as $(s_B^e \bmod n) \oplus \pi$ for all $(s_B^e \bmod n) \in \{0,1\}^{|n|-1}$. Since not all randomly chosen elements in $\mathbb{Z}_n$ after being encrypted are $|n| - 1$ bits long, this method is probabilistic. Method 2: Compute z as $s_B^e + \pi \bmod n$ for all $s_B \in \mathbb{Z}_n$. For simplicity, we skip detail discussions of these two methods and focus our attention on improving the efficiency of the Interactive Protocol for checking the validity of (n, e) in Step 2 above.

3   Improving the Interactive Protocol

In RSA-PAKE1 described above, there is an Interactive Protocol for checking the validity of public keys. To ensure that the RSA cryptosystem works correctly, the public exponent e has to be relatively prime to $\phi(n)$ where n is a RSA modulus. If e is a fraudulent value, an active attacker may be able to launch various e-residue attacks [3, 12, 13].

The idea of using an interactive protocol to detect fraudulent values of e was first discussed in [3]. That is, after a verifier receives (n, e) from a prover, the verifier checks if n and e are odd. Then it picks N (say $N = 10$) integers $m_i \in_R \mathbb{Z}_n^*$, $1 \le i \le N$ and sends $\{c_i \equiv m_i^e \ (\mathrm{mod}\, n)\}_{1 \le i \le N}$ to the prover. The prover computes the e-th root of each c_i as m_i' and sends $\{m_i'\}_{1 \le i \le N}$ back. Correct replies indicate that e has the proper form. In [17], Zhu, et al. found that this preliminary version is insecure. It allows an impersonator of B to test multiple trial passwords in one run of the Interactive Protocol with A. To prevent the attack, RSA-PAKE1 uses the following variant of the Interactive Protocol.

Let $h : \{0,1\}^* \to \{0,1\}^k$ be a collision-free hash function. After the verifier sends out $\{c_i\}_{1 \le i \le N}$, it stores $\{h(m_i)\}_{1 \le i \le N}$ instead of $\{m_i\}_{1 \le i \le N}$ in its memory. Similarly the prover sends back $\{h(m_i')\}_{1 \le i \le N}$. Due to the collision-free property of h, it is negligible to have $h(m) = h(m')$ for $m \ne m'$. The checking mechanism of the interactive protocol is retained and the attack against the preliminary version is prevented. In addition, this also reduces the size of the reply in the Interactive Protocol and reduces the memory footprint of the verifier.

In the following, we describe two modifications. The first one shortens the size of the message sent from the server to the client, while the second one dramatically reduces the size of the message sent from the client to the server and hence saves much more power than the first one for the low-power client in wireless communications as transmitting radio signal requires much more power than receiving some.

3.1   Modification 1

After the verifier sends $\{c_i\}_{1 \le i \le N}$ out, it stores $h(m_1, m_2, \cdots, m_N)$ in its memory. Similarly the prover sends $h(m_1', m_2', \cdots, m_N')$ back. This simple modification reduces the number of hash values to be stored and transferred from N to only one. It also reduces the hash operations at the client side from N times to one single hash operation.

In wireless communications, more power can be saved if the amount of data transmitted is reduced. In Modification 1, we reduce the amount of data needs to be received by the verifier (that is, the low-power client) while the amount of data sent by it remains the same. In Modification 2 below, we reduce the amount of data sent by the low-power client dramatically and therefore significantly reduce the power consumption of it.

3.2 Modification 2

For some odd integers n and $e \geq 3$, let $eRES = \{y = x^e \bmod n \; : \; x \in \mathbb{Z}_n^*\}$ be the e-residue set. If $(e, \phi(n)) \neq 1$, then $|eRES| \leq \phi(n)/3$ since each e-residue has at least 3 e-th roots. When N elements are randomly picked in $\mathbb{Z}_n^*$, the chance of having all of them in $eRES$ is at most 3^{-N}. Base on this fact, we modify the Interactive Protocol as follows.

Define a random function H such that on inputs N, n, e, A, B, and two k-bit binary strings r_1 and r_2, the function generates a sequence of N random elements in $\mathbb{Z}_n^*$. This is denoted as $(c_1, c_2, \cdots, c_N) \leftarrow H(N, n, A, B, r_1, r_2)$. The prover (A) picks $r_1 \in_R \{0,1\}^k$ and sends (n, e, r_1) to the verifier (B). The verifier then replies with $r_2 \in_R \{0,1\}^k$. The prover computes $(c_1, c_2, \cdots, c_N) \leftarrow H(N, n, e, A, B, r_1, r_2)$ and sends $\{m_i = c_i^{1/e} \bmod n\}_{1 \leq i \leq N}$ to the verifier. By using H, the verifier generates $\{c_i\}_{1 \leq i \leq N}$ accordingly and checks if $c_i \equiv m_i^e \pmod n$ for all $1 \leq i \leq N$. The verifier accepts if all the checks are passed. By replacing the original interactive protocol of RSA-PAKE1 with Modification 2, we obtain a scheme shown in Fig. 1. We call it the RSA-PAKE2.

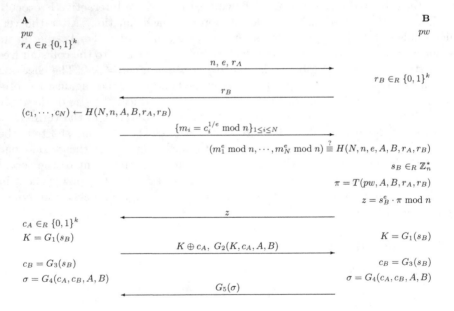

Fig. 1. RSA-PAKE2

In this modification, all the N random numbers are generated dynamically using the random function H. Hence the low-power client B does not need to store any of them in the memory and the memory footprint is further minimized when compared with Modification 1.

On the computation complexity, the number of modular multiplications carried out in RSA-PAKE2 is no more than that in RSA-PAKE1 as well as in Modification 1. Hence the computation time is estimated to be similar to that of the previous variants. That is, it takes 2.5 seconds to compute on a 16MHz Palm V and reduces to 300 msec if the server's public key is cached for subsequent protocol runs after the first protocol run when $N = 10$ and 512-bit RSA is used. For stringent security requirement when 1024-bit RSA is used, the first run of the protocol requires 9 second of pure computation but all the subsequent runs with the same server can be reduced to less than half second.

On the power consumption, the low-power client receives N random numbers rather than sending N numbers as required in RSA-PAKE1 and Modification 1. In addition, only one k-bit long random number is sent by the client in the Interactive Protocol of RSA-PAKE2. This reduces the transmission time of the Interactive Protocol over a 8kbps network by 180 msec for $k = 160$. In addition, since receiving data consumes much less energy than sending data in wireless communications, for 1024-bit RSA, the client of RSA-PAKE2 spends only 20 msec in sending data in the Interactive Protocol, while the client of RSA-PAKE1 spends 1.28 seconds in sending data. RSA-PAKE2's approach of having the low-power client dramatically reduce the message sent with the increase of message received can help the client to save power in a very significant way.

4   Formalization and the Generic Scheme

In this section, we generalize RSA-PAKE2 to a generic scheme and formalize the security requirements of all underlying primitives that the generic scheme is constituted. In the formalization, we introduce a new primitive called password-keyed permutation family to capture the features of the password related operations of RSA-PAKE1 (and the two modifications). We also show that these password related operations satisfy the security requirements of the password-keyed permutation family of the generic scheme if and only if the RSA Problem is hard.

4.1   The Public Key Encryption Function

From the conjectures that the RSA problem is hard and the RSA encryption primitive is a trapdoor one-way permutation over $\mathbb{Z}_n^*$, we formalize the security requirement of the public key encryption function defined by a public key $PK \in PubK(k)$ in the generic scheme to be a trapdoor one-way permutation, given by $\mathcal{E}_{PK} : \mathcal{S}(PK) \to \mathcal{S}(PK)$ where $\mathcal{S}(PK)$ is the set of elements over which the trapdoor permutation is defined. $PubK(k)$ denotes the set of public keys with respect to k. In general, we can relax the requirement to allow any trapdoor one-way function as the encryption function.

4.2   The Password-Keyed Permutation Family

In Fig. 1, z is computed as the encryption of a random element in $\mathbb{Z}_n^*$ followed by a modular multiplication with π where $\pi \in \mathbb{Z}_n^*$ is a function of the password pw with some nonces r_A and r_B and identification information. Similar to the definition of the public key encryption function above, we can also consider the modular multiplication as a permutation over the same set of elements as the public key encryption function. Define $T : \{0,1\}^* \to PwdK(k)$ to be a distinct and independent cryptographic hash function. We assume that the size of $PwdK(k)$ is at least 2^k. Let $\mathcal{P}^{PK} : PwdK(k) \times \mathcal{S}(PK) \to \mathcal{S}(PK)$ be a collection of permutations for every $PK \in PubK(k)$. For simplicity, we usually omit the superscript notation of the public key on $\mathcal{P}$. We call $\mathcal{P}$ a *password-keyed permutation family*. For every $\pi \in PwdK(k)$, we define a permutation $\mathcal{P}_\pi : \mathcal{S}(PK) \to \mathcal{S}(PK)$ by $\mathcal{P}_\pi(x) = \mathcal{P}(\pi, x)$. From these definitions, z is then computed in the generic scheme as $z = \mathcal{P}_\pi(\mathcal{E}_{PK}(s_B))$ where $s_B \in_R \mathcal{S}(PK)$.

We now discuss the security requirements of $\mathcal{P}$. The first requirement of $\mathcal{P}$ is *distinctness*. This means for every pair $(\pi_1, \pi_2) \in PwdK(k) \times PwdK(k)$, $\pi_1 \neq \pi_2$, and for any $y \in \mathcal{S}(PK)$, $\Pr[\mathcal{P}_{\pi_1}(y) = \mathcal{P}_{\pi_2}(y)] \leq \epsilon(k)$ where ϵ is some negligible function. It is not difficult to see that the purpose of having $\mathcal{P}$ be distinct is to prevent disturbing the probability distribution of picking pw from PW, which may provide an adversary with greater advantage in guessing the password.

Besides having $\mathcal{P}$ be distinct, we also require $\mathcal{P}$ to satisfy the following security requirement.

Definition 1. *Given a trapdoor one-way permutation $f : Dom(k) \to Dom(k)$, a password space PW, and a hash function $T : \{0,1\}^* \to PwdK(k)$ behaves like a random oracle, a distinct password-keyed permutation family $\mathcal{P}^f : PwdK(k) \times Dom(k) \to Dom(k)$ is secure if for every probabilistic polynomial-time algorithm E^T and for all sufficiently large k,*

$$\Pr[\, E^T(1^k, A, B, PW, f, r_A) \to (z, r_B, x_1, x_2, \pi_1, \pi_2) \ : \ r_B \in \{0,1\}^k,$$
$$z, x_1, x_2 \in Dom(k), \ \pi_1, \pi_2 \in \Gamma_{A,B,r_A,r_B}, \ z = \mathcal{P}^f(\pi_1, f(x_1)) = \mathcal{P}^f(\pi_2, f(x_2)) \,]$$
$$\leq \epsilon(l)$$

for all $r_A \in \{0,1\}^k$, $A, B \in \{0,1\}^*$ and for some negligible function ϵ where

$$\Gamma_{A,B,r_A,r_B} = \{\, T(pw, A, B, r_A, r_B) \ : pw \in PW \,\}.$$

It means that an attacker should not be able to compute more than one pair of (π, x) such that $\mathcal{P}^f(\pi, f(x))$ produces the same value of z.

To understand the reason of specifying this security requirement, we consider a run of the generic scheme in which an adversary E is impersonating B. Suppose that after receiving PK and r_A from A, E has non-negligible success rate of constructing (z, r_B) and obtaining (at least) two values x_1, x_2 and corresponding $\pi_1, \pi_2 \in \Gamma_{A,B,r_A,r_B}$ such that $z = \mathcal{P}_{\pi_1}^f(f(x_1)) = \mathcal{P}_{\pi_2}^f(f(x_2))$. Then after the third message flow, E can verify if any of x_1 and x_2 is the correct value to generate K. If x_1 (or x_2) is the correct value, then the corresponding password of π_1 (or π_2) must be the password shared between A and B. Otherwise, the two passwords, which generate π_1 and π_2, must not be the correct password. Hence no matter in which case, E can check at least two passwords in each impersonation. On the other hand, if E, impersonating B, constructs a pair (z, r_B) such that it obtains only one value x yielding $z = \mathcal{P}_\pi^f(f(x))$ for some $\pi \in \Gamma_{A,B,r_A,r_B}$, then this is not considered as a successful attack.

We consider this security requirement as a generalization of the 'associativity' problem discovered by Gong et al. in [7] and further exemplified by Jablon in [10]. The idea is also similar to the special characteristic of a password-entangled public-key generation primitive described in [8]. This limits the number of password guesses that the attacker can make to just one guess for each z, and for each run of a PAKE scheme.

Password-Keyed Permutation over $\mathbb{Z}_n^*$. In RSA-PAKE2, both $PwdK(k)$ and $\mathcal{S}(PK)$ are set to $\mathbb{Z}_n^*$. Hence the password-keyed permutation family of RSA-PAKE2 can be written as $\mathcal{P}^{(n,e)} : \mathbb{Z}_n^* \times \mathbb{Z}_n^* \to \mathbb{Z}_n^*$ and is defined as $(\pi, y) \to y \cdot \pi \bmod n$ for all $y, \pi \in \mathbb{Z}_n^*$. It is obvious that $\mathcal{P}^{(n,e)}$ is distinct. The following theorem says that it also satisfies Definition 1.

Theorem 1. *Given a RSA public key (n, e) such that the RSA Problem is hard, a password space PW, and a hash function $T : \{0,1\}^* \to \mathbb{Z}_n^*$ behaves like a random oracle, the password-keyed permutation $(\pi, y) \to y \cdot \pi \bmod n$ is secure for all $\pi, y \in \mathbb{Z}_n^*$.*

A proof is given in Appendix A.

4.3   The Generic PAKE Scheme

We now conclude the generalization and describe the entire generic scheme in the following. It is illustrated in Fig. 2.

The Generic PAKE Scheme

1. A generates a public key pair (PK, SK), picks $r_A \in_R \{0,1\}^k$ and sends PK and r_A to B.
2. B checks if PK is a valid public key. If it is invalid, B rejects the connection. Otherwise, it picks $r_B \in_R \{0,1\}^k$ and $s_B \in_R \mathcal{S}(PK)$, and computes $\pi = T(pw, A, B, r_A, r_B)$. It sends $z = \mathcal{P}_\pi(\mathcal{E}_{PK}(s_B))$ and r_B back to A, and destroys π from its memory.
3. A computes π accordingly and reveals the value of s_B from z. It generates a temporary symmetric key K and B's session key contribution c_B by $G_1(s_B)$ and $G_3(s_B)$, respectively. Then it picks its own session key contribution $c_A \in_R \{0,1\}^k$ and sends $(K \oplus c_A, \ G_2(K, c_A, A, B))$ to B. It later computes the session key σ as $G_4(c_A, c_B, A, B)$ and destroys s_B, π, c_A and c_B from its memory.
4. B computes K and c_B from s_B accordingly and destroys s_B from its memory. It reveals c_A from the first part of the incoming message and checks if the second part of the incoming message is $G_2(K, c_A, A, B)$. If it is false, B

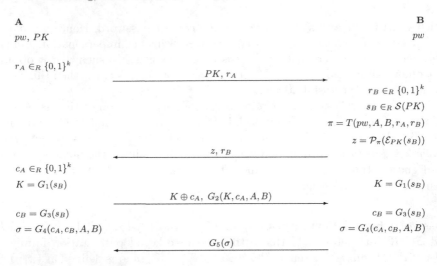

Fig. 2. The Generic PAKE Scheme

rejects the connection. Otherwise, it computes σ accordingly and destroys c_A and c_B from its memory. $G_5(\sigma)$ is then sent back to A and the connection is accepted.

5. A checks if the incoming message is $G_5(\sigma)$. If it is true, A accepts the connection. Otherwise, it rejects the connection.

5   Concluding Remarks

The contributions of the paper can be divided into two parts. In the first part, we propose two modifications on a refined scheme of [17]. The modifications reduce the message size hence improve the network efficiency and memory footprint. More importantly, RSA-PAKE2 dramatically reduces the size of the message sent from the low-power client to the server. This saves a lot of battery power for a portable client from transmitting radio signal in wireless communications.

In the second part, we generalize RSA-PAKE1 and its modifications to a generic scheme and formalize the security requirements of all underlying primitives that the generic scheme is constituted. In the formalization, we introduce a new primitive called password-keyed permutation family to capture the features of the password related operations of RSA-PAKE1 and its modifications. We also show that these password related operations satisfy the security requirements of the password-keyed permutation family of the generic scheme if and only if the RSA problem is hard in the random oracle model. This also implies that RSA-PAKE2 is an instantiation of the generic scheme.

Other instantiations of the generic scheme are also possible. The most challenging problem of designing an instantiation is to devise a secure password-keyed permutation family $\mathcal{P}^{PK}$. For example, an interesting open problem is to devise one which is based on the Quadratic Residuosity Problem.

6   Acknowledgement

We thank Burt Kaliski for his valuable comments and suggestions in the preliminary work of this paper.

References

[1] Feng Bao. Security analaysis of a password authenticated key exchange protocol. to appear in Information Security (ISC 2003), Oct, 2003.

[2] Mihir Bellare and Phillip Rogaway. Random oracles are practical: A paradigm for designing efficient protocols. In *First ACM Conference on Computer and Communications Security*, pages 62–73, Fairfax, 1993. ACM.

[3] S. M. Bellovin and M. Merritt. Encrypted key exchange: Password based protocols secure against dictionary attacks. In *Proceedings 1992 IEEE Symposium on Research in Security and Privacy*, pages 72–84. IEEE Computer Society, 1992.

[4] Victor Boyko, Philip MacKenzie, and Sarvar Patel. Provably secure password-authenticated key exchange using Diffie-Hellman. In *Proc. EUROCRYPT 2000*, pages 156–171, 2000.

[5] Whitfield Diffie, Paul C. Van Oorschot, and Michael J. Wiener. Authentication and authenticated key exchanges. *Designs, Codes, and Cryptography*, 2(2):107–125, June 1992.

[6] R. Gennaro and Y. Lindell. A framework for password-based authenticated key exchange. In *Proc. EUROCRYPT 2003*. Springer-Verlag, 2003. Lecture Notes in Computer Science No. 2656, also in Cryptology ePring Archive: Report 2003/032.

[7] L. Gong, M. A. Lomas, R. M. Needham, and J. H. Saltzer. Protecting poorly chosen secrets from guessing attacks. *IEEE Journal on Selected Areas in Communications*, 11(5):648–656, 1993.

[8] IEEE. *P1363.2 / D10: Standard Specifications for Password-based Public Key Cryptographic Techniques*, Jul 2003.

[9] David P. Jablon. Strong password-only authenticated key exchange. *Computer Communication Review, ACM*, 26(5):5–26, 1996.

[10] David P. Jablon. Extended password key exchange protocols immune to dictionary attack. In *Proceedings of the WETICE'97 Workshop on Enterprise Security*, Cambridge, MA, USA, Jun 1997.

[11] Jonathan Katz, Rafail Ostrovsky, and Moti Yung. Efficient password-authenticated key exchange using human-memorable passwords. In *Proc. EUROCRYPT 2001*. Springer-Verlag, 2001. Lecture Notes in Computer Science No. 2045.

[12] Stefan Lucks. Open key exchange: How to defeat dictionary attacks without encrypting public keys. In *Proc. of the Security Protocols Workshop*, pages 79–90, 1997. LNCS 1361.

[13] Philip MacKenzie, Sarvar Patel, and Ram Swaminathan. Password-authenticated key exchange based on RSA. In *Proc. ASIACRYPT 2000*, pages 599–613, 2000.

[14] Ronald L. Rivest, Adi Shamir, and Leonard M. Adleman. A method for obtaining digital signatures and public-key cryptosystems. *Communications of the ACM*, 21(2):120–126, 1978.

[15] Duncan S. Wong, Hector Ho Fuentes, and Agnes H. Chan. The performance measurement of cryptographic primitives on palm devices. In *Proc. of the 17th Annual Computer Security Applications Conference*, December 2001.

[16] Thomas Wu. The secure remote password protocol. In *1998 Internet Society Symposium on Network and Distributed System Security*, pages 97–111, 1998.

[17] Feng Zhu, Duncan S. Wong, Agnes H. Chan, and Robbie Ye. Password authenticated key exchange based on RSA for imbalanced wireless networks. In *Information Security (ISC 2002)*, pages 150–161. Springer-Verlag, September 2002. Lecture Notes in Computer Science No. 2433.

A   Proof of Theorem 1

Proof. We show that breaking the password-keyed permutation $(\pi, y) \to y \cdot \pi \mod n$ is computationally equivalent to solving the RSA Problem. By breaking the password-keyed permutation, we mean that given a RSA public key (n, e), a password space PW, and an ideal hash function $T : \{0,1\}^* \to \mathbb{Z}_n^*$, there exists a probabilistic polynomial-time algorithm (PPT) E^T such that for some $r_A \in \{0,1\}^k$, $A, B \in \{0,1\}^*$,

$$\Pr[\ E^T(1^k, A, B, PW, (n, e), r_A) \to (z, r_B, x_1, x_2, \pi_1, \pi_2)\ :\ r_B \in \{0,1\}^k,$$
$$z, x_1, x_2 \in \mathbb{Z}_n^*,\ \pi_1, \pi_2 \in \Gamma_{A,B,r_A,r_B},\ z \equiv x_1^e \cdot \pi_1 \equiv x_2^e \cdot \pi_2 \pmod{n}\ ]$$
$$\geq \frac{1}{Q(k)}$$

for some polynomial function Q.

Suppose that **Oracle**$^{\mathbf{RSAP}}$ is a PPT for the RSA Problem with non-negligible probability. That is, given (n, e) and $y \in \mathbb{Z}_n^*$, it is non-negligible to compute $x \leftarrow$ **Oracle**$^{\mathbf{RSAP}}(n, e, y)$ such that $y \equiv x^e \pmod{n}$. We now construct a PPT E^T with access to a random oracle T to break the password-keyed permutation.

For a security parameter k, a server $A \in \{0,1\}^*$, a client $B \in \{0,1\}^*$, a password space PW, a RSA public key (n, e), and a k-bit binary string r_A, E^T proceeds as follows.

$E^T = $ "On inputs 1^k, A, B, PW, (n, e), r_A,

1. randomly generates $r_B \in \{0,1\}^k$,
2. picks $pw_1, pw_2 \in PW$ and computes $\pi_1 = T(pw_1, A, B, r_A, r_B)$ and $\pi_2 = T(pw_2, A, B, r_A, r_B)$ by querying the random oracle of the form T.
3. Constructs two $x_1, x_2 \in \mathbb{Z}_n^*$ such that $x_1^e \cdot \pi_1 \equiv x_2^e \cdot \pi_2 \pmod{n}$. This is done by randomly pick an element $x_1 \in \mathbb{Z}_n^*$, define $z = x_1^e \cdot \pi_1 \mod n$, and compute $x_2 \leftarrow$ **Oracle**$^{\mathbf{RSAP}}(n, e, z \cdot \pi_2^{-1} \mod n)$.
4. Outputs z, r_B, x_1, x_2, π_1 and π_2."

It is easy to see that E breaks the password-keyed permutation with non-negligible probability.

Conversely, suppose that **Oracle**$^{\mathcal{P}}$ is a PPT that breaks the password-keyed permutation $(\pi, y) \to y \cdot \pi \mod n$ with non-negligible probability. We now show that a PPT C can be constructed to solve an instance of the RSA Problem with non-negligible probability, that is given a RSA public key (n, e) and an element $y \in \mathbb{Z}_n^*$, find $x \in \mathbb{Z}_n^*$ such that $y \equiv x^e \pmod{n}$.

Algorithm C uses $\mathbf{Oracle}^{\mathcal{P}}$ as a black box, but has full control over its oracles. That is, C simulates $\mathbf{Oracle}^{\mathcal{P}}$'s view by providing 1^k, $A, B \in \{0, 1\}^*$, a password space PW, (n, e) as A's public key, $r_A \in_R \{0, 1\}^k$ and answers for queries of the form T. In a non-negligible case when $\mathbf{Oracle}^{\mathcal{P}}$ successfully generates $(z, r_B, x_1, x_2, \pi_1, \pi_2)$ such that $z \equiv x_1^e \cdot \pi_1 \equiv x_2^e \cdot \pi_2 \pmod{n}$ for $\pi_1, \pi_2 \in \Gamma_{A,B,r_A,r_B}$, $\mathbf{Oracle}^{\mathcal{P}}$ queries with the form T for at least twice to generate π_1 and π_2 with probability at least $1 - 1/\phi(n)^2$. Suppose C guesses correctly on these two queries. Then C picks $r \in_R \mathbb{Z}_n^*$, and provides $r^e \bmod n$ and y as answers. For all other queries of the form T, C simply picks random elements in $\mathbb{Z}_n^*$ as answers.

Without loss of generality, we assume that $\pi_1 = r^e \bmod n$ and $\pi_2 = y$. After $\mathbf{Oracle}^{\mathcal{P}}$ generates $(z, r_B, x_1, x_2, \pi_1, \pi_2)$, we have

$$z \equiv x_1^e \cdot r^e \equiv x_2^e \cdot y \pmod{n}$$

$$y \equiv (x_1 \cdot x_2^{-1} \cdot r)^e \pmod{n}$$

Hence

$$x \equiv x_1 \cdot x_2^{-1} \cdot r \pmod{n}.$$

For C to make correct guesses of the two queries of the form T that generate π_1 and π_2, we can see that if C makes two random guesses, the probability of guessing correctly is at least $1/Q^2$ where Q is the total number of queries made by $\mathbf{Oracle}^{\mathcal{P}}$. Therefore, C solves the instance of RSA Problem with probability at least $(1 - 1/\phi(n)^2)/Q^2$ of the success probability of $\mathbf{Oracle}^{\mathcal{P}}$. $\square$

A Password-Based Authenticator: Security Proof and Applications[*]

Yvonne Hitchcock[1], Yiu Shing Terry Tin[1], Juan Manuel Gonzalez-Nieto[1], Colin Boyd[1], and Paul Montague[2]

[1] Information Security Research Centre, Queensland University of Technology
GPO Box 2434, Brisbane Q 4001, Australia.
{y.hitchcock, t.tin, j.gonzaleznieto, c.boyd}@qut.edu.au
[2] Motorola Australia Software Centre,
2 Second Ave, Mawson Lakes, SA 5095, Australia.
pmontagu@asc.corp.mot.com

Abstract. A password-based authentication mechanism, first proposed by Halevi and Krawczyk, is used to formally describe a password-based authenticator in the Canetti-Krawczyk proof model. A proof of the security of the authenticator is provided. The possible practical applications of the authenticator are demonstrated by applying it to two key exchange protocols from the ideal world of the Canetti-Krawczyk model to produce two password-based key exchange protocols with provable security in the real world of the model. These two new protocols are almost as efficient as those proposed by Halevi and Krawczyk and have fewer message flows if it is assumed that the client must initiate the protocol. The new authenticator contributes a new component which has been proven secure in the Canetti-Krawczyk model, while the new key exchange protocols are provably secure making them attractive for use in settings where clients must authenticate to a server using a relatively short password.
Keywords: Key management protocols, password authentication, key agreement.

1 Introduction

A major goal of modern cryptography is to enable two or more users on an insecure (adversary controlled) network to communicate in a confidential manner and/or ensure that such communications are authentic. In order to realize this goal, symmetric key cryptographic tools are often used due to their efficiency compared to public key techniques. However, use of such tools requires the creation of a secret key (which is typically at least 100 bits long) known only to the users communicating with each other. Because of the impracticality of each possible pair of users sharing a long term secret key, public key and/or password-based techniques are used to generate such a key when it is required. An advantage of this method of key generation is to keep different sessions independent, which enables the avoidance of replay attacks (since the wrong key

---

[*] Full version of this paper is available at http://sky.fit.qut.edu.au/~boydc/papers/.

T. Johansson and S. Maitra (Eds.): INDOCRYPT 2003, LNCS 2904, pp. 388–401, 2003.

will have been used for the replay) and lessens the impact of key compromise (since only one session will be exposed, not all previous communications).

While public key cryptography can be used to provide a secure method of key agreement, it is not always practical due to the inconvenience and expense of securely storing full length cryptographic keys in some applications. Secure password based key agreement mechanisms (where the only secret information held by one or more of the parties is a short password) are therefore necessary in such environments. One such example is mobile environments, where memory may be scarce and the devices in use may not be tamper resistant. However, because of the short length of the password, special care must be taken when designing protocols to ensure that both the password and the key finally agreed are secret.

1.1   Motivation and Related Work

In the past, a trial and error approach to the security of cryptographic protocols has been taken where cryptographic protocols were proposed together with informal security analyses. However, such protocols were sometimes wrong and usually only partially analysable. Indeed, in some cases, flaws have come to light years after a protocol's proposal and acceptance by the community as being secure.

These problems have led to the development of various formal methods of *proving* the security of a protocol. Bellare and Rogaway [5,6] first proposed a formal model for provable security of protocols. Although their initial model only covered the case where two parties already share a long-term secret, it has been extended, by themselves and others, to cover all the main types of authenticated key exchange (AKE) protocol. The proofs follow the style of most proofs in modern cryptography by reducing the security of the protocol to the security of some underlying primitive. A limitation of these proofs is that they tend to be complex and difficult for practitioners. Even more important from our viewpoint is that they are monolithic, fragile and error prone. A small change in the protocol structure can destroy the proof and leave no indication of how to repair it.

This paper works in the model adopted by Canetti and Krawczyk [8] which we refer to hereafter as the CK-model. Two previous works of Bellare and Rogaway [5] and Bellare, Canetti and Krawczyk [2] form the basis of the CK-model. The former uses the indistinguishability of [9] for defining security while the latter postulates a two-step methodology for a substantial simplification in designing provably secure cryptographic protocols. As a consequence, the CK-model inherits the aforementioned properties of [5,2]. Its modularity is gained by applying a protocol translation tool, called an *authenticator*, to protocols proven secure in a much simplified adversarial setting where authentication of the communication links is not required. The result of such an application is secure protocols in the unsimplified adversarial setting where the full capabilities of the adversary are modelled. Moreover, various basic parts of a protocol can be proven secure independently of each other and then combined to form a single secure protocol. This leads to simpler, less error-prone proofs and the ability to construct a large number of secure protocols from a much smaller number of basic secure components. An overview of the CK-model is given in Section 2.

Although other password-based protocols which have been proven secure do exist, such as those in [12] and [13], they do not use a public key for the server and therefore are not amenable to the CK-model, since it is impossible to separate the key exchange and authentication mechanisms of such protocols [10]. Thus the above advantages of the modular approach used by the CK-model can not be realized by these protocols, making the proof of security of a password-based authenticator and demonstration of its application in the CK-model worthwhile.

1.2   Password-Based Protocols and Their Constraints

Because passwords have in practice very low entropy, it is possible for an attacker to test all possible passwords in a relatively short amount of time, which leads to the requirement that off-line dictionary attacks must be infeasible (that is, an adversary with access to transcripts of one or more sessions must not be able to eliminate a significant number of possible passwords). In addition, on-line dictionary attacks must be infeasible (that is, an active adversary must not be able to abuse the protocol in a way that allows him to eliminate a significant number of possible passwords). Note that in an on-line attack, the adversary can guess a password and attempt to impersonate the user, so at least one password can be eliminated per protocol run. We require that no more than one password can be eliminated per protocol run with non-negligible probability. Other more general security properties required are key authentication (parties participating in the key agreement know the identity of all other parties who could possibly hold a copy of the key [14, p. 490]) and key freshness [14, p. 494] (key freshness is necessary since it is assumed the adversary is able to find the value of old keys). It is also desirable that the protocol be efficient in terms of the number of operations performed and the number of messages transmitted.

1.3   Our Focus

The main focus of this paper is on the security and application of a password based authentication mechanism which was proposed by Halevi and Krawczyk [10] and used in their key agreement protocols. While these key agreement protocols do not have an associated proof of security, the authentication mechanism itself does. However, Boyarsky [7] has criticized both this proof and an earlier version [11] due to an inadequate definition of security. In addition, while Halevi and Krawczyk state that their security formalization and proof provide a basis for a password-based equivalent of an authenticator in the CK-model, the proof they provide can not be used for this purpose for a couple of reasons. Firstly, the proof assumes that there is only one uncorrupted party (in addition to the server) in existence. A valid proof for the CK-model must allow any (polynomially bounded) number of uncorrupted parties. Secondly, the protocol does not actually enable the transmission of an authenticated message, as required by a proper authenticator in the CK-model; the only achievement is that the server knows the client responded to the server's nonce.

In order to overcome these problems, we provide a formal description of a so-called *message transmission* (MT-) authenticator based on the work of Halevi and Krawczyk as well as a proof of its security in the CK-model. This MT-authenticator can be used for constructing authenticators which transform protocols in the simplified adversarial setting to secure protocols in the full adversarial setting, as described in [2].

As a demonstration of the applications of the new MT-authenticator, we apply it to two protocols proposed in the literature for the simplified adversarial setting and thereby obtain two secure key agreement protocols using passwords. While it is possible to obtain other password-based key agreement protocols using this method, the two presented in this paper are the most efficient that can be built from currently available components proven secure in the CK-model. Finally, the new password-based key agreement protocols are compared with those proposed by Halevi and Krawczyk [10] for performance and efficiency.

2 Overview of the Canetti-Krawczyk Approach

Here a brief description of the CK-model is given, but the reader is encouraged to see [2] and [8] for further details. The CK-model defines protocol principals who may simultaneously run multiple local copies of a message driven protocol. Each local copy is called a session and has its own local state. Two sessions are *matching* if they have the same session identifier and the purpose of each session is to establish a key between the particular two parties running the sessions. A powerful adversary attempts to break the protocol by interacting with the principals. In addition to controlling all communications between principals, the adversary is able to *corrupt* any principal, thereby learning all information in the memory of that principal (eg. long-term keys, session states and session keys). The adversary may impersonate a corrupted principal, although the corrupted principal itself is not activated again and produces no further output or messages. The adversary may also *reveal* internal session states or agreed session keys. The adversary must be efficient in the sense of being a probabilistic polynomial time algorithm.

Definition 1 (Informal). *An* AKE *protocol is called* session key (SK-) *secure if the following two conditions are met. Firstly, if two uncorrupted parties complete matching sessions, then they both accept the same key. Secondly, suppose a session key is agreed between two uncorrupted parties and has not been revealed by the adversary. Then the adversary cannot distinguish the key from a random string with probability greater than 1/2 plus a negligible function in the security parameter.*

Two adversarial models are defined: the unauthenticated-links adversarial model (UM) and the authenticated-links adversarial model (AM). The only difference between the two is the amount of control the adversary has over the communications lines between principals. The UM corresponds to the "real world" where the adversary completely controls the network in use, and may modify or create messages from any party to any other party. The AM is a restricted version of the UM where the adversary may choose whether or not to deliver a message, but if a message is delivered, it must have been created by the specified sender and be delivered to the specified recipient without alteration. In addition, any such message may only be delivered once. In this way, authentication mechanisms can be separated from key agreement mechanisms by proving the key agreement secure in the AM, and then applying an authentication mechanism to the key agreement messages so that the overall protocol is secure in the UM. In actual fact, the definition of SK-security in the UM must be relaxed slightly when password-based authentication mechanisms are used, which is addressed in Section 4.

Definition 2 (Informal). *An* authenticator *is a protocol translator that takes an* SK-*secure protocol in the* AM *to an* SK-*secure protocol in the* UM.

Authenticators can be constructed using one or more MT-*authenticators*. An MT-authenticator is a protocol which delivers one message in the UM in an authenticated manner. To translate an SK-secure protocol in the AM to an SK-secure protocol in the UM an MT-authenticator can be applied to each message and the resultant sub-protocols combined to form one overall SK-secure protocol in the UM. However, if the SK-secure protocol in the AM consists of more than one message, the resultant protocol must be optimized, which involves reorder and reuse of message components.

In order to define the security of an MT-authenticator, it is necessary to formally define an MT protocol in the AM as follows. Upon activation within party A on external request (B, m), party A sends the message (A, B, m) to party B and outputs 'A sent m to B.' Upon receipt of a message (A, B, m), B outputs 'B received m from A.' An MT-authenticator is defined to be secure if it *emulates* MT in the UM. Emulation is defined to occur when the global output (which consists of the concatenation of the cumulative output of all parties and the output of the adversary) in the AM is computationally indistinguishable from the global output in the UM. (The output of the adversary is a function of its internal states at the end of the interaction.) Note that in the UM, it is necessary to augment a protocol with an initialization function to allow the required bootstrapping of the cryptographic authentication functions. The CK-approach can now be summarized in the following three steps.

1. Design a basic protocol and prove it SK-secure in the AM.
2. Design an MT-authenticator and prove that it is secure.
3. Apply the MT-authenticator to the AM protocol to produce an automatically secure UM protocol. If necessary, reorder and reuse message components to optimize the resulting protocol.

3   The MT-authenticator

The authentication mechanism of Halevi and Krawczyk [10] is designed to authenticate a client to a server and is based on a randomized encryption of a shared password. It has been modified here to form an MT-authenticator for messages from a client to a server and its security level is discussed in Sect. 4. The authenticator is denoted by $\lambda_{\text{P-ENC}}$ and Fig. 1 gives its specification. The following subsections give a detailed description of various aspects of the authenticator.

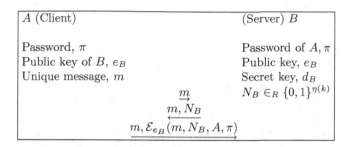

Fig. 1. Password Based Authenticator, $\lambda_{\text{P-ENC}}$

The Parties. Let there be a total of n parties, $P_1, \ldots, P_n$. We split the parties into two disjoint sets, the set of all servers of size s and the set of all clients of size c where $(c + s) = n$. The values n, c and s are polynomial in the security parameter k and may be written as $n(k)$, $c(k)$ and $s(k)$ respectively.

Initialization. The protocol uses an encryption scheme $\Pi = (\mathcal{G}, \mathcal{E}, \mathcal{D})$, which is indistinguishable under adaptive chosen ciphertext attack [4], where $\mathcal{G}$ is the key generation algorithm, and $\mathcal{E}$ and $\mathcal{D}$ are the encryption and decryption algorithms respectively. The initialization function $I_{\text{P-ENC}}$ consists of two parts, I_{Pwd} and I_{PubKey}. I_{Pwd} invokes a password generation algorithm once for each pair of (client, server) parties (C_i, S_j). This algorithm randomly chooses a password, denoted $\pi_{C_i S_j}$ (or simply π when the password is shared between A and B), from a password dictionary, $\mathfrak{D}$, and assigns the password to C_i and S_j. I_{PubKey} invokes, once for each server party S_j, the key generation algorithm $\mathcal{G}$ with security parameter k. Let e_{S_j} and d_{S_j} denote the encryption and decryption keys associated with server party S_j. The public keys, e_{S_j}, are distributed to all client parties C_i. The public information is all public keys: $I_0 = \{e_{S_1}, \ldots, e_{S_s}\}$. The (completely) private information of each server party S_j is the private key d_{S_j}. Shared private information for each S_j is the set of passwords shared by S_j with each C_i, $\{\pi_{C_1 S_j}, \ldots, \pi_{C_c S_j}\}$. Shared private information for each client C_i is the set of passwords shared by C_i with each server S_j, $\{\pi_{C_i S_1}, \ldots, \pi_{C_i S_s}\}$.

Protocol Description. Since the majority of the following discussion is for only one client and server pair, we denote the client by A and the server by B for simplicity. Let γ be the maximum number of unsuccessful attempts per client to complete the protocol with a server. The protocol begins with the initialization function $I_{\text{P-ENC}}$ described above. Each server then sets the number of unsuccessful attempts to complete the protocol with each client to 0. If a client A is activated with an external request to send a unique message m to server B, then A outputs 'A sent message m to B' and sends 'message: m' to B. Upon receipt of 'message: m' from A, B chooses a random value $N_B \in \{0, 1\}^{\eta(k)}$ and sends 'challenge: m, N_B' to A. Upon receipt of 'challenge: m, N_B' from B, A sends 'encryption: $m, \mathcal{E}_{e_B}(m, N_B, A, \pi)$' to B. Upon receipt of 'encryption: $m, \mathcal{E}_{e_B}(m, N_B, A, \pi)$' from A, party B accepts m from A if, when decrypted, m is the same as the cleartext m, A is a valid client, π is the password shared between A and B, B has previously sent 'challenge: m, N_B' (where m and N_B match those in the encryption and the challenge is still outstanding) and the number of unsuccessful attempts to complete the protocol with A is less than or equal to γ. If B accepts m from A then B removes 'challenge: m, N_B' from the list of outstanding challenges and outputs 'B received m from A'. If the value A in the encryption is the identity of a valid client, but B does not accept the message m from A, then B increases the number of unsuccessful attempts to complete the protocol with A by one.

4   Security of MT-authenticator

In what follows we assume that the reader is familiar with the notational conventions and definitions of security for encryption schemes which are commonly used in the literature. In particular, we use the notions of indistinguishability under adaptive chosen ciphertext attack (IND-CCA) [4] and left-or-right indistinguishability under adaptive chosen ciphertext attack (LOR-CCA) [3].

Proposition 1. *Assume that Π is secure in the sense of* IND-CCA *and that passwords are randomly chosen from a dictionary $\mathfrak{D}$ of size $|\mathfrak{D}|$. Then the output of protocol $\lambda_{\text{P-ENC}}$ is the same as that of protocol* MT *in unauthenticated networks with probability $(1 - \epsilon(k))$ where:*

$$\epsilon(k) \leq \left(\nu(k) + \left(1 - \left(1 - \frac{\gamma + 1}{|\mathfrak{D}|} \right)^{s(k)c(k)} \right) \right) \tag{1}$$

and $\nu(k)$ is negligible.

Because the protocol $\lambda_{\text{P-ENC}}$ is password based, this is the best security level we can hope to achieve. In practice $\lambda_{\text{P-ENC}}$ does not achieve 'emulation' in the sense of [2] due to the small size of the dictionary which makes $\epsilon(k)$ non-negligible. An adversary can always guess a password and attempt to use it in $\lambda_{\text{P-ENC}}$. If the attempt to complete the protocol is unsuccessful, the adversary can eliminate that password from the list of possible passwords for the user it attempted to impersonate. However, we show that the probability of the adversary doing better than this is negligible.

Because the probability that the output of $\lambda_{\text{P-ENC}}$ is different from that of MT is not negligible, $\lambda_{\text{P-ENC}}$ can not be used to create SK-secure protocols using the original definition of SK-security. However, the definition can be modified so that the probability of correctly distinguishing the key from a random string is $1/2 + \delta$ where δ is no longer negligible, but no more than a negligible function plus half of the probability of randomly guessing at least one password. In our case, $\delta = \epsilon(k)/2 + \omega(k)$ where $\epsilon(k)$ is defined above and $\omega(k)$ is negligible. (This can be shown by letting D be the event that the adversary guesses the session key correctly and E be the event that the output of the key exchange protocol in the UM is identical to the output of the key exchange protocol in the AM. Then $\Pr(D) = \Pr(D|E)\Pr(E) + \Pr(D|\neg E)\Pr(\neg E) \leq (1/2 + \omega_1(k))(1 - \epsilon(k)) + 1 \cdot \epsilon(k) = 1/2 + \epsilon(k)/2 + \omega(k)$ where $\omega(k)$ and $\omega_1(k)$ are negligible.) We call protocols that satisfy the new definition of security *password-based session key (PBSK-) secure*.

Proof. Due to space limitations we only sketch the proof of Proposition 1 and refer the reader to the full version of the paper. Let $\mathcal{U}$ be a UM-adversary that interacts with $\lambda_{\text{P-ENC}}$. We construct an AM-adversary $\mathcal{A}$ such that the output in the UM and AM is identical with probability $1 - \epsilon(k)$. Adversary $\mathcal{A}$ runs $\mathcal{U}$ on the following simulated interaction with a set of parties running $\lambda_{\text{P-ENC}}$.

1. First $\mathcal{A}$ chooses and distributes keys for the imitated parties, according to function $I_{\text{P-ENC}}$.
2. Next, when $\mathcal{U}$ activates some imitated client party A' for sending a message m to imitated server party B', adversary $\mathcal{A}$ activates client party A in the authenticated network to send m to server B. In addition, $\mathcal{A}$ continues the interaction between $\mathcal{U}$ and the imitated parties running $\lambda_{\text{P-ENC}}$.
3. When some imitated party B' outputs 'B' received $\hat{m}$ from A'', adversary $\mathcal{A}$ activates party B in the authenticated-links model with incoming message $\hat{m}$ from A.
4. When $\mathcal{U}$ corrupts a party, $\mathcal{A}$ corrupts the same party in the authenticated network and hands the corresponding information (from the simulated run) to $\mathcal{U}$.
5. Finally, $\mathcal{A}$ outputs whatever $\mathcal{U}$ outputs.

We first need to show that the above description of the behaviour of $\mathcal{A}$ is a legitimate behaviour of an AM-adversary. The above steps are easy to verify as legal moves for $\mathcal{A}$, except for Step 3. In that case, let $\mathcal{B}$ denote the event that imitated party B' outputs 'B' received $\hat{m}$ from A'' where A' is uncorrupted and the message $(\hat{m}, A, B)$ is not currently in the set U of undelivered messages. In other words, $\mathcal{B}$ is the event where B' outputs 'B' received $\hat{m}$ from A',' and either A was not activated for sending $\hat{m}$ to B or B has already had the same output before. In this event we say that $\mathcal{U}$ broke party A'.

If $\mathcal{B}$ does not occur (that is, Step 3 can always (legally) be carried out), then the above construction is as required. It remains to show that event $\mathcal{B}$ occurs only with low probability. Assume that event $\mathcal{B}$ occurs with probability $\epsilon(k)$.

There are a number of ways in which $\mathcal{B}$ could occur. Firstly, B could output the same nonce twice, coupled with the same message. However, if the length of the nonce, $\eta(k)$, is sufficiently large, the probability of this occurring is negligible, which we denote with $\epsilon_1(k)$.

Obviously $\mathcal{U}$ can attempt to send a message as if from A by guessing the password and including the guess in the final message of the protocol. If a maximum of γ unsuccessful login attempts are allowed for each client, then $\mathcal{U}$ has a probability of at most $\frac{\gamma+1}{|\mathfrak{D}|}$ of succeeding for one particular client and server pair without obtaining any information about the password (apart from the contents of $\mathfrak{D}$). Therefore $\mathcal{U}$ has probability at most $\left(1 - \left(1 - \frac{\gamma+1}{|\mathfrak{D}|}\right)^{s(k)c(k)}\right)$ of succeeding for at least one client and server pair. Then we show that the probability that $\mathcal{B}$ occurs is negligibly higher than this. That is, if $\mathcal{B}$ occurs with probability $\epsilon(k)$ and the function $\epsilon_2(k) \stackrel{\text{def}}{=} \left(\epsilon(k) - \epsilon_1(k) - \left(1 - \left(1 - \frac{\gamma+1}{|\mathfrak{D}|}\right)^{s(k)c(k)}\right)\right)$ is not negligible, then we show that the advantage $\mathbf{Adv}_{\Pi,\mathcal{F}}^{\text{ind}-\text{cca}}(k)$ associated with the encryption scheme for a polynomial time adversary $\mathcal{F}$ is not negligible, which contradicts the assumption that the encryption scheme is secure.

From this point the proof is similar to that of Theorem 5.3 in [1]. However, a few modifications are required for this particular situation. Since the advantage of $\mathcal{F}$ attacking the indistinguishability of the cryptosystem, $\mathbf{Adv}_{\Pi,\mathcal{F}}^{\text{ind}-\text{cca}}(k)$, is negligible, so is the advantage of an adversary $\mathcal{F}$ attacking the left-or-right indistinguishability of the cryptosystem, $\mathbf{Adv}_{\Pi,\mathcal{F}}^{\text{lor}-\text{cca}}(k)$. Let $\mathcal{F}$ be an adversary having polynomial time complexity, and attacking LOR-CCA of Π. Given an encryption key pk, a left-or-right encryption oracle $\mathcal{E}_{pk}(\mathcal{LR}(\cdot,\cdot,b))$ and a decryption oracle $\mathcal{D}_{sk}(\cdot)$, adversary $\mathcal{F}$ runs $\mathcal{U}$ on a simulated interaction with a set of parties running $\lambda_{\text{P-ENC}}$. $\mathcal{F}$ chooses and distributes keys for the imitated parties according to function $I_{\text{P-ENC}}$ with the exception that the public encryption key associated with some server party B^*, chosen at random from the set of servers S, is replaced with the input key pk. A^* is chosen at random from the set of clients C. Note that $\mathcal{F}$ knows the password shared between A^* and B^*, π. The simulation is carried out by $\mathcal{F}$ as per protocol description except for protocol runs between A^* and B^*; in this case, when $\mathcal{U}$ activates A^* to send the third message of the protocol $\lambda_{\text{P-ENC}}$ for the message m and nonce N_B of the server B^* (where A^* has previously output 'A^* sent message m to B^*'), then $\mathcal{F}$ queries the encryption oracle with $\mathcal{E}_{pk}\left(\mathcal{LR}\left((m \parallel N_B \parallel A^* \parallel r), (m \parallel N_B \parallel A^* \parallel \pi), b\right)\right)$ and receives output C, where r is newly chosen for each oracle query and $r \stackrel{R}{\leftarrow} \mathfrak{D}$. $\mathcal{F}$ then sends 'encryption: m, C' to B^*. If $\mathcal{U}$ activates B^* with 'encryption: m, C' where C is not an output of the encryption oracle, $\mathcal{F}$ queries its decryption oracle and finds $p \leftarrow \mathcal{D}_{sk}(C)$. If p is of the form $(m \parallel N_B \parallel A^* \parallel \pi')$ with $\pi' = \pi$, then $\mathcal{F}$ guesses that the bit b associated with the $\mathcal{E}_{pk}(\mathcal{LR}(\cdot,\cdot,b))$ oracle is 1. If the attempt is unsuccessful (i.e $\pi' \neq \pi$ or m, N_B was not previously sent by B^*, or N_B is not an outstanding challenge), $\mathcal{F}$ increments the counter γ of unsuccessful attempts to complete the protocol for A^*. If $\gamma + 1$ attempts have been made for A^* or $\mathcal{U}$ finishes (and there was no successful attempt for A^* which had not used the $\mathcal{E}_{pk}(\mathcal{LR}(\cdot,\cdot,b))$ oracle) then $\mathcal{F}$ guesses that the bit b associated with the $\mathcal{E}_{pk}(\mathcal{LR}(\cdot,\cdot,b))$ oracle is 0.

$\mathcal{U}$'s view of the interaction with $\mathcal{F}$, conditional on the event that $\mathcal{F}$ does not abort the simulation is identically distributed to $\mathcal{U}$'s view of a real interaction with an unauthenticated network if the bit b associated with the $\mathcal{E}_{pk}(\mathcal{LR}(\cdot,\cdot,b))$ oracle is 1. Therefore, the probability of guessing b correctly is the same as that of a successful forgery between A^* and B^*, which is $1 - \left(\left(1 - \frac{\gamma+1}{|\mathfrak{D}|}\right)^{s(k)c(k)} - \epsilon_2(k)\right)^{\frac{1}{s(k)c(k)}}$. On the other hand, if the bit b is 0, since no information given to $\mathcal{U}$ depends in any way on the password, the likelihood of a successful forgery is no more than that of being successful using random guesses, $\frac{\gamma+1}{|\mathfrak{D}|}$. Therefore, the advantage of $\mathcal{F}$ is:

$$\mathbf{Adv}_{\Pi,\mathcal{F}}^{\text{lor-cca}} \geq 1 - \left(\left(1 - \frac{\gamma+1}{|\mathfrak{D}|}\right)^{s(k)c(k)} - \epsilon_2(k)\right)^{\frac{1}{s(k)c(k)}} - \left(\frac{\gamma+1}{\mathfrak{D}}\right) \overset{\text{def}}{=} g(k)$$

$$= 1 - y - \left((1-y)^{p(k)} - \epsilon_2(k)\right)^{\frac{1}{p(k)}}$$

where $y = \frac{\gamma+1}{|\mathfrak{D}|}$ and $p(k) = s(k)c(k)$. Now $g(k)$ is less than or equal to the advantage of the adversary in breaking the encryption scheme. If $g(k)$ is not negligible, we have a non-negligible advantage in breaking the encryption scheme, which contradicts our original assumption. Therefore, assume $g(k)$ is negligible. This implies:

$$g(k) = 1 - y - \left((1-y)^{p(k)} - \epsilon_2(k)\right)^{\frac{1}{p(k)}} \leq \frac{1}{k^c} \text{ for } c \geq 1 \text{ and } k \geq k_c$$

$$(1-y)^{p(k)} - \left((1-y) - \frac{1}{k^c}\right)^{p(k)} \geq \epsilon_2(k) \tag{2}$$

It can be shown that $p(k)a^{p(k)-1}b \geq a^{p(k)} - (a-b)^{p(k)}$ when $b \leq \frac{a}{p(k)}$. Substituting this into (2) gives $\frac{p(k)(1-y)^{p(k)-1}}{k^c} \geq \epsilon_2(k)$ when $k \geq \left(\frac{p(k)}{1-y}\right)^{\frac{1}{c}} = k_d$. Since $(1-y)$ is less than one, $(1-y)^{p(k)-1}$ is also less than one, which implies $\frac{p(k)}{k^c} \geq \epsilon_2(k)$. Now assume that $p(k) \leq k^e$ for $k \geq k_e$. Then $\frac{1}{k^{c-e}} \geq \epsilon_2(k)$ for $k \geq \max(k_c, k_d, k_e)$. Therefore $\epsilon_2(k)$ is negligible, and this completes the proof.   □

The full version of this paper provides details of variations to the authenticator and its proof to allow use of public passwords and/or an encryption scheme indistinguishable under chosen plaintext attack.

5   Application of the Password-Based MT-authenticator

We now show how the password-based MT-authenticator $\lambda_{\text{P-ENC}}$ can be used in practice by applying it to some key exchange protocols which have been proposed in the literature and proven secure in the AM. Specifically, we apply $\lambda_{\text{P-ENC}}$ to protocol 2DH and protocol ENC shown in Figure 2. In the description of the protocols, p and q are two primes such that $q|(p-1)$ and the length of q is k bits.

Protocol 2DH		Protocol ENC	
A	B	A	B
$x \in_R \mathbb{Z}_q$	$y \in_R \mathbb{Z}_q$	$r \in_R \{0,1\}^k$	
		$c = \mathcal{E}_{e_B}(r)$	
$\xrightarrow{\quad A, sid, g^x \quad}$			
$\xleftarrow{\quad B, sid, g^y \quad}$		$\xrightarrow{\quad A, sid, c \quad}$	
$K' = (g^y)^x$	$K = (g^x)^y$		$r' = \mathcal{D}_{d_B}(c)$
		$K = f_r(A,B,sid)$	$K' = f_{r'}(A,B,sid)$

Fig. 2. Two SK-secure AM protocols

A	B
	$\xleftarrow{\quad m \quad}$
$v_A \in_R \{0,1\}^k$	
$\xrightarrow{\quad m, \mathcal{E}_{e_B}(v_A) \quad}$	
$\xleftarrow{\quad m, \mathcal{M}_{v_A}(m,A) \quad}$	

Fig. 3. Encryption-based MT-authenticator, λ_{ENC}

$G = \langle g \rangle$ is a subgroup of $\mathbb{Z}_p^*$ of order q. The parameters (g,p,q) are assumed to be publicly known and all arithmetic is performed in $\mathbb{Z}_p^*$ unless otherwise indicated. Protocol 2DH is the well-known Diffie-Hellman key exchange and has been proven secure in the AM under the decision Diffie-Hellman (DDH) assumption in [8]. Protocol ENC has been proven secure in the AM in [8] under the assumption that the encryption scheme is secure against chosen ciphertext attack and that $\{f_r\}_{r \in \{0,1\}^k}$ is a pseudorandom function family. It is assumed that A has the authentic public key of B, e_B, and that B has the corresponding private key, d_B. Each of the protocols contains a session identifier (sid) whose value is not specified here, but is assumed to be known by protocol participants before the protocol begins. In practice, the session identifier may be determined during protocol execution [8].

Application of λ_{P-ENC} to 2DH. Since 2DH has two message flows, we need to construct a valid authenticator for two-message protocols in the AM. We cannot apply our password-based MT-authenticator to both message flows of 2DH because it would violate the assumption that the client's only secret key is the password. Thus we combine $\lambda_{\text{P-ENC}}$ with another MT-authenticator, λ_{ENC} [2], to form a valid authenticator. λ_{ENC} is shown in Figure 3 and uses an IND-CCA public key encryption scheme and a secure message authentication (MAC) scheme $\mathcal{M}$. $\lambda_{\text{P-ENC}}$ is applied to the first flow of 2DH while λ_{ENC} is applied to the second flow of 2DH. This application yields a new PBSK-secure protocol in the UM, 2DHPE, shown in Figure 4. The length of the nonce from $\lambda_{\text{P-ENC}}$, $\eta(k)$, has been set to k for simplicity.

Application of λ_{P-ENC} to ENC. Since ENC consists of only one message flow, $\lambda_{\text{P-ENC}}$ is the only authenticator which needs to be applied and no optimization

$$A \qquad\qquad\qquad\qquad\qquad\qquad\qquad B$$

$$x \in_R Z_q, v_A \in \{0,1\}^k \qquad\qquad\qquad\qquad\qquad\qquad y \in_R \mathbb{Z}_q$$

$$\xrightarrow{\quad A, sid, g^x, \mathcal{E}_{e_B}(v_A) \quad}$$

$$N_B \in \{0,1\}^k$$

$$\xleftarrow{\quad B, sid, g^y, N_B, \mathcal{M}_{v_A}(B, sid, g^y, A) \quad}$$

$$K' = (g^y)^x$$

$$\xrightarrow{\quad sid, \mathcal{E}_{e_B}(A, sid, g^x, N_B, \pi) \quad}$$

$$K = (g^x)^y$$

Fig. 4. The protocol 2DHPE

or reordering of messages is required. The new UM protocol is shown in Figure 5 and named ENCP. Note that client to server authentication is explicit through use of π, but server to client authentication is implicit through use of the server's public key, e_B.

$$A \qquad\qquad\qquad\qquad\qquad\qquad\qquad B$$

$$r \in_R \{0,1\}^k, c = \mathcal{E}_{e_B}(r)$$

$$\xrightarrow{\quad A, sid, c \quad}$$

$$N_B \in_R \{0,1\}^k$$

$$\xleftarrow{\quad sid, N_B \quad}$$

$$\xrightarrow{\quad sid, \mathcal{E}_{e_B}(A, sid, c, N_B, \pi) \quad}$$

$$r' = \mathcal{D}_{d_B}(c)$$

$$K = f_r(A, B, sid) \qquad\qquad\qquad\qquad K' = f_{r'}(A, B, sid)$$

Fig. 5. The protocol ENCP

Comparison with Halevi and Krawczyk protocols. Since our MT-authenticator is based on the work of Halevi and Krawczyk [10], we compare their results with ours. Their proposal has two key exchanges, one with and one without support for forward secrecy, which we denote HKDH and HK respectively. As previously mentioned, the proof of the authentication mechanism used in these protocols is not adequate for use in the CK-model. Both protocols are claimed to be resistant to off-line dictionary guessing attacks. Two aspects of protocol performance are examined, namely computational requirements and number of message flows. The results are summarized in Table 1 which indicates how many operations a particular protocol needs to perform for a successful protocol run. A number in brackets indicates that that number of operations can be computed off-line. The numbers next to the protocol names indicate the number of message flows in the particular protocol which is initiated by the client. We notice that in terms of on-line computation at the client side our protocols are as efficient and those of Halevi and Krawczyk, but are slightly more expensive in terms of off-line and server-side computations. We stress however that generally on-line client-side computations are more important when considering this type of key-exchange protocols. With respect to message flows, since both HKDH and HK are initiated by the server instead of the client, our protocols 2DHPE and ENCP have the advantage of taking less time to complete the protocol if initiation must be performed by the client. We argue that initiation by clients is a more natural setting as it allows the clients to communicate with the server at will.

Computational Operation	Protocol							
	2DHPE(3)		HKDH(4)		ENCP(3)		HK(4)	
	A	B	A	B	A	B	A	B
Exponentiation	1(1)	1(1)	1(1)	1(1)	0	0	0	0
Asymmetric Encryption	1(1)	0	1	0	1(1)	0	1	0
Asymmetric Decryption	0	2	0	1	0	2	0	1

Table 1. Performance comparison with Halevi and Krawczyk

6   Conclusion

This paper has formally described a password-based authenticator, $\lambda_{\text{P-ENC}}$, in the CK-model and formally evaluated its security. It is impossible for any password-based authenticator to achieve the level of security required for a proper MT-authenticator because the adversary can guess passwords and thereby attempt to impersonate a client on-line with non-negligible probability of succeeding due to the small size of the password dictionary. However, it was proved that $\lambda_{\text{P-ENC}}$ achieves the best possible level of security for a password-based authenticator.

The password-based authenticator was then applied to two protocols that have been proven SK-secure in the AM elsewhere in the literature, resulting in two provably secure key exchange protocols in the UM. Other password-based key exchange protocols can easily be obtained by using $\lambda_{\text{P-ENC}}$ with other components which have been proven secure in the CK-model elsewhere in the literature, but such protocols constructed using presently available components are less efficient than the two presented here. The two new protocols were compared with the two proposed by Halevi and Krawczyk which use the same authentication mechanism, with the result that the new protocols are almost as efficient in terms of number of computations, but better in terms of number of message flows. In addition, the new protocols have associated formal proofs of security, while the Halevi-Krawczyk ones do not. These properties make the new protocols attractive for use in settings where clients must authenticate to a server using a relatively short password.

References

1. J. H. An and M. Bellare. Does encryption with redundancy provide authenticity? In *EUROCRYPT '01*, volume 2045 of *LNCS*, pages 512–528. Springer-Verlag, 2001.

2. M. Bellare, R. Canetti, and H. Krawczyk. A modular approach to the design and analysis of authentication and key exchange protocols. In *STOC '98*, pages 419–428. ACM Press, May 1998.

3. M. Bellare, A. Desai, E. Jokipii, and P. Rogaway. A concrete security treatment of symmetric encryption. In *FOCS*, pages 394–403. IEEE Computer Society Press, 1997.

4. M. Bellare, A. Desai, D. Pointcheval, and P. Rogaway. Relations among notions of security for public-key encryption schemes (extended abstract). In *CRYPTO '98*, volume 1462 of *LNCS*, pages 26–45. Springer-Verlag, 1998.

5. M. Bellare and P. Rogaway. Entity authentication and key distribution (extended abstract). In *CRYPTO '93*, volume 773 of *LNCS*, pages 232–249. Springer-Verlag, 1993.

6. M. Bellare and P. Rogaway. Provably secure session key distribution: The three party case. In *STOC '95*, pages 57–66. ACM Press, 1995.

7. M. Boyarsky. Public-key cryptography and password protocols: The multi-user case. In *Proceedings of the 5th ACM Conference on Computer and Communications Security*, pages 63–72. ACM Press, 1999.

8. R. Canetti and H. Krawczyk. Analysis of key-exchange protocols and their use for building secure channels. In *EUROCRYPT '01*, volume 2045 of *LNCS*, pages 451–472. Springer-Verlag, 2001.

9. S. Goldwasser and S. Micali. Probabilistic encryption. *Journal of Computer and Systems Sciences*, 28(2):270–299, 1984.

10. S. Halevi and H. Krawczyk. Public-key cryptography and password protocols. *ACM Transactions on Information and System Security*, 2(3):230–268, 1999.

11. Shai Halevi and Hugo Krawczyk. Public-key cryptography and password protocols. In *Computer and Communications Security (CCS-98)*, pages 122–131, New York, 1998. ACM Press.

12. J. Katz, R. Ostrovsky, and M. Yung. Efficient password-authenticated key exchange using human-memorable passwords. In *Eurocrypt 2001*, volume 2045 of *LNCS*, pages 475–494. Springer-Verlag, 2001.

13. P. MacKenzie. The PAK suite: Protocols for password-authenticated key exchange. Technical Report 2002-46, DIMACS, 2002.

14. A. Menezes, P. van Oorschot, and S. Vanstone. *Handbook of applied cryptography*. CRC Press, 1997. ISBN 0-8493-8523-7.

Stronger Security Bounds for OMAC, TMAC, and XCBC[*]

Tetsu Iwata and Kaoru Kurosawa

Department of Computer and Information Sciences,
Ibaraki University
4–12–1 Nakanarusawa, Hitachi, Ibaraki 316-8511, Japan
{iwata, kurosawa}@cis.ibaraki.ac.jp

Abstract. OMAC, TMAC and XCBC are CBC-type MAC schemes which are provably secure for arbitrary message length. In this paper, we present a more tight upper bound on $\mathsf{Adv}^{\mathsf{mac}}$ for each scheme, where $\mathsf{Adv}^{\mathsf{mac}}$ denotes the maximum success (forgery) probability of adversaries. Our bounds are expressed in terms of the *total length* of all queries of an adversary to the MAC generation oracle while the previous bounds are expressed in terms of the *maximum length* of each query. In particular, a significant improvement occurs if the lengths of queries are heavily unbalanced.

Category. Message authentication codes (MACs), Secret key cryptography, Block ciphers.

Key words. OMAC, TMAC, XCBC, modes of operation, block cipher, provable security.

1 Introduction

1.1 Background

The CBC MAC [5,7] is a well-known method to generate a message authentication code (MAC) based on a block cipher E. We denote the CBC MAC value of a message M by $\mathrm{CBC}_K(M)$, where K is the key of E. While Bellare, Kilian, and Rogaway proved that the CBC MAC is secure for fixed length messages [1], it is *not* secure for variable length messages.

Therefore, several variants of CBC MAC have been proposed which are provably secure for variable length messages. They include EMAC, XCBC, TMAC and then OMAC.

EMAC (Encrypted MAC) is obtained by encrypting $\mathrm{CBC}_{K_1}(M)$ by E again with a new key K_2 [2]. That is, $\mathrm{EMAC}_{K_1,K_2}(M) = E_{K_2}(\mathrm{CBC}_{K_1}(M))$. Petrank and Rackoff proved that EMAC is secure if the message length is a multiple of n, where n is the block length of E [14].

---

[*] A long version of this paper is available as [9].

T. Johansson and S. Maitra (Eds.): INDOCRYPT 2003, LNCS 2904, pp. 402–415, 2003.

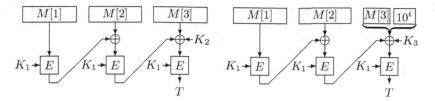

Fig. 1. Illustration of XCBC.

For arbitrary length messages, we can simply append the minimal 10^i to a message M so that the length is a multiple of n. In this method, however, we must append an entire extra block 10^{n-1} if the size of the message is already a multiple of n. This is a "wasting" of one block cipher invocation.

Black and Rogaway next proposed XCBC to solve the above problem [3]. XCBC takes *three* keys: K_1 for E, and K_2 and K_3. In XCBC, we do not append 10^{n-1} if the size of the message is already a multiple of n. Only if this is not the case, we append the minimal 10^i. In order to distinguish them, K_2 or K_3 is XORed before encrypting the last block. XCBC is now described as follows (see Fig. 1).

- If $|M| = mn$ for some $m > 0$, then XCBC computes exactly the same as the CBC MAC, except for XORing an n-bit key K_2 before encrypting the last block.
- Otherwise, 10^i padding ($i = n - |M| - 1 \bmod n$) is appended to M and XCBC computes exactly the same as the CBC MAC for the padded message, except for XORing another n-bit key K_3 before encrypting the last block.

Kurosawa and Iwata then proposed TMAC which requires *two* keys, K_1 and K_2 [10]. TMAC is obtained from XCBC by replacing (K_2, K_3) with $(K_2 \cdot \mathsf{u}, K_2)$, where u is some non-zero constant and "$\cdot$" denotes multiplication in $\mathrm{GF}(2^n)$.

Finally, Iwata and Kurosawa proposed OMAC which requires only *one* key K of the block cipher E [8]. OMAC is a generic name for OMAC1 and OMAC2. Let $L = E_K(0^n)$. Then OMAC1 is obtained by replacing (K_2, K_3) with $(L \cdot \mathsf{u}, L \cdot \mathsf{u}^2)$ in XCBC. Similarly, OMAC2 is obtained from XCBC by replacing (K_2, K_3) with $(L \cdot \mathsf{u}, L \cdot \mathsf{u}^{-1})$.

Remark. XCBC, TMAC and OMAC were proposed to NIST (National Institute of Standards and Technology) and they are currently under the consideration for recommendation. See [13] for details.

1.2 Our Contribution

XCBC, TMAC and OMAC are all provably secure against chosen message attack. Indeed, the authors showed an upper bound on $\mathsf{Adv}^{\mathsf{mac}}$ for each scheme, where $\mathsf{Adv}^{\mathsf{mac}}$ denotes the maximum success (forgery) probability of adversaries.

In this paper, we present a more tight upper bound on $\mathtt{Adv}^{\mathsf{mac}}$ for each scheme by using a more specific parameter. Consider adversaries who run in time at most t and query at most q messages to the MAC generation oracle.

1. The previous bounds are expressed in terms of the maximum length of each query.
2. Our bounds are expressed in terms of the total length of all queries.

More precisely,

1. Table 1 shows the previous bounds on $\mathtt{Adv}_F^{\mathsf{mac}}(t, q, m)$ which is defined as the maximum forgery probability of adversaries such that each query is at most m blocks, where 1 block is n bits, and
2. Table 2 shows our bounds on $\mathtt{Adv}_F^{\mathsf{mac}}(t, q, \sigma)$ which is defined as the maximum forgery probability of adversaries such that the total length of all queries are at most σ blocks,

where F is XCBC, TMAC or OMAC and n is the block length of the underlying block cipher E. In these tables, $\mathtt{Adv}_E^{\mathsf{prp}}(t', q')$ is the the maximum distinguishing probability between the block cipher E and a randomly chosen permutation, where the maximum is over all adversaries who run in time at most t' and make at most q' queries.

In general, $\sigma \leq mq$, where σ is the total block length of all queries, q is the number of queries, and m is the maximum block length among all queries.

A significant improvement occurs if all queries are very short (say, 1 block) except for one very long query (m blocks). For example, suppose that $n = 64$ (for example, Triple DES [4]), $m = 2^{16}$ and $q = 2^{16} + 1$. It is easy to see that $\sigma = 2^{16} + 2^{16} = 2^{17}$. In this case, our bounds shown in Table 2 are still meaningful while the previous bounds shown in Table 1 are useless because they become larger than one.

1.3 Our Collision Bound

To show our security bounds, we derive upper bounds on some collision probabilities. For q distinct messages $M^{(1)}, \ldots, M^{(q)}$ such that each $|M^{(i)}|$ is a multiple of n, let $\sigma = (|M^{(1)}| + \cdots + |M^{(q)}|)/n$.

For XCBC and TMAC, we consider a collision such that

$$\mathrm{CBC}_P(M^{(i)}) = \mathrm{CBC}_P(M^{(j)})$$

for some $i \neq j$, where CBC_P denotes the CBC MAC with a randomly chosen permutation P as the underlying block cipher E. We then prove that

$$\Pr(1 \leq {}^{\exists}i < {}^{\exists}j \leq q, \mathrm{CBC}_P(M^{(i)}) = \mathrm{CBC}_P(M^{(j)})) \leq \frac{\sigma^2}{2^n}$$

for any $M^{(1)}, \ldots, M^{(q)}$. It is formally stated in Lemma 5.1 and proved in [9].

Table 1. Previous security bounds of XCBC, TMAC and OMAC.

Name	Security Bound
XCBC [3, Corollary 2]	$\mathrm{Adv}^{\mathrm{mac}}_{\mathrm{XCBC}}(t, q, m) \leq \dfrac{(4m^2 + 1)q^2 + 1}{2^n} + 3 \cdot \mathrm{Adv}^{\mathrm{prp}}_{E}(t', q'),$ where $t' = t + O(mq)$ and $q' = mq$.
TMAC [10, Theorem 5.1]	$\mathrm{Adv}^{\mathrm{mac}}_{\mathrm{TMAC}}(t, q, m) \leq \dfrac{(3m^2 + 1)q^2 + 1}{2^n} + \mathrm{Adv}^{\mathrm{prp}}_{E}(t', q'),$ where $t' = t + O(mq)$ and $q' = mq$.
OMAC [8, Theorem 5.1]	$\mathrm{Adv}^{\mathrm{mac}}_{\mathrm{OMAC}}(t, q, m) \leq \dfrac{(5m^2 + 1)q^2 + 1}{2^n} + \mathrm{Adv}^{\mathrm{prp}}_{E}(t', q'),$ where $t' = t + O(mq)$ and $q' = mq + 1$.

Table 2. Security bounds of XCBC, TMAC and OMAC obtained in this paper.

Name	Security Bound
XCBC	$\mathrm{Adv}^{\mathrm{mac}}_{\mathrm{XCBC}}(t, q, \sigma) \leq \dfrac{3\sigma^2 + 1}{2^n} + \mathrm{Adv}^{\mathrm{prp}}_{E}(t', q'),$ where $t' = t + O(\sigma)$ and $q' = \sigma$.
TMAC	$\mathrm{Adv}^{\mathrm{mac}}_{\mathrm{TMAC}}(t, q, \sigma) \leq \dfrac{3\sigma^2 + 1}{2^n} + \mathrm{Adv}^{\mathrm{prp}}_{E}(t', q'),$ where $t' = t + O(\sigma)$ and $q' = \sigma$.
OMAC	$\mathrm{Adv}^{\mathrm{mac}}_{\mathrm{OMAC}}(t, q, \sigma) \leq \dfrac{4\sigma^2 + 1}{2^n} + \mathrm{Adv}^{\mathrm{prp}}_{E}(t', q'),$ where $t' = t + O(\sigma)$ and $q' = \sigma + 1$.

For OMAC, we consider MOMAC-E, a variant of the CBC MAC, as follows. Let a message be $M = M[1] \circ M[2] \circ \cdots \circ M[m]$, where $|M[1]| = |M[2]| = \cdots = |M[m]| = n$ and $m \geq 2$. Let P_1 and P_2 be two independent randomly chosen permutations. Then (1) Let $Y[1] = P_1(M[1])$, (2) For $i = 2, \ldots, m - 1$, compute $Y[i] = P_2(M[i] \oplus Y[i - 1])$. (3) Finally define MOMAC-$\mathrm{E}_{P_1, P_2}(M) = M[m] \oplus Y[m - 1]$. We show that

$$\Pr(1 \leq {}^{\exists}i < {}^{\exists}j \leq q, \text{MOMAC-}\mathrm{E}_{P_1, P_2}(M^{(i)}) = \text{MOMAC-}\mathrm{E}_{P_1, P_2}(M^{(j)}))$$
$$\leq \frac{(\sigma - q)^2}{2^n} \ .$$

It is formally stated in Lemma 4.1 and proved in [9].

2 Preliminaries

2.1 Notation

For a set A, $x \xleftarrow{R} A$ means that x is chosen from A uniformly at random. If $a, b \in \{0, 1\}^*$ are equal-length strings then $a \oplus b$ is their bitwise XOR. If

$a, b \in \{0,1\}^*$ are strings then $a \circ b$ denote their concatenation. For simplicity, we sometimes write ab for $a \circ b$ if there is no confusion.

For an n-bit string $a = a_{n-1} \cdots a_1 a_0 \in \{0,1\}^n$, let $a \ll 1 = a_{n-2} \cdots a_1 a_0 0$ denote the n-bit string which is a left shift of a by 1 bit, while $a \gg 1 = 0 a_{n-1} \cdots a_2 a_1$ denote the n-bit string which is a right shift of a by 1 bit.

If $a \in \{0,1\}^*$ is a string then $|a|$ denotes its length in bits. For any bit string $a \in \{0,1\}^*$ such that $|a| \le n$, we let

$$\mathrm{pad}_n(a) = \begin{cases} a 10^{n-|a|-1} & \text{if } |a| < n, \\ a & \text{if } |a| = n. \end{cases} \tag{1}$$

Define $\|a\|_n = \max\{1, \lceil |a|/n \rceil\}$, where the empty string counts as one block. In pseudocode, we write "Partition M into $M[1] \cdots M[m]$" as shorthand for "Let $m = \|M\|_n$, and let $M[1], \ldots, M[m]$ be bit strings such that $M[1] \cdots M[m] = M$ and $|M[i]| = n$ for $1 \le i < m$."

2.2   CBC MAC

A block cipher E is a function $E : \mathcal{K}_E \times \{0,1\}^n \to \{0,1\}^n$, where $\mathcal{K}_E$ is the set of keys and $E(K, \cdot) = E_K(\cdot)$ is a permutation on $\{0,1\}^n$. n is called the block length of E.

The CBC MAC [5,7] is the simplest and most well-known MAC scheme based on block ciphers E. For a message $M = M[1] \circ M[2] \circ \cdots \circ M[m]$ such that $|M[1]| = |M[2]| = \cdots = |M[m]| = n$, let $Y[0] = 0^n$ and $Y[i] = E_K(M[i] \oplus Y[i-1])$ for $i = 1, \ldots, m$. Then the CBC MAC of M under key K is defined as $\mathrm{CBC}_K(M) = Y[m]$.

Bellare, Kilian, and Rogaway proved that the CBC MAC is secure for fixed length messages [1]. However, it is well known that CBC MAC is *not* secure for variable length messages.

2.3   XCBC, TMAC, and OMAC

XCBC, TMAC and OMAC are CBC-type MAC schemes which are provably secure for arbitrary message length.

- Each scheme takes a message $M \in \{0,1\}^*$ and produces a tag in $\{0,1\}^n$.
- Each scheme is defined by using a block cipher $E : \mathcal{K}_E \times \{0,1\}^n \to \{0,1\}^n$.

XCBC.   XCBC takes three keys $(K_1, K_2, K_3) \in \mathcal{K}_E \times \{0,1\}^n \times \{0,1\}^n$. The algorithm of XCBC is described in Fig. 2 and illustrated in Fig. 1, where $\mathrm{pad}_n(\cdot)$ is defined in (1).

$$
\begin{array}{|l|}
\hline
\textbf{Algorithm } \mathrm{XCBC}_{K_1,K_2,K_3}(M) \\
Y[0] \leftarrow 0^n \\
\text{Partition } M \text{ into } M[1]\cdots M[m] \\
\textbf{for } i \leftarrow 1 \textbf{ to } m-1 \textbf{ do} \\
\qquad X[i] \leftarrow M[i] \oplus Y[i-1] \\
\qquad Y[i] \leftarrow E_{K_1}(X[i]) \\
X[m] \leftarrow \mathtt{pad}_n(M[m]) \oplus Y[m-1] \\
\textbf{if } |M[m]| = n \textbf{ then } X[m] \leftarrow X[m] \oplus K_2 \\
\qquad\qquad\qquad \textbf{else } X[m] \leftarrow X[m] \oplus K_3 \\
T \leftarrow E_{K_1}(X[m]) \\
\textbf{return } T \\
\hline
\end{array}
$$

Fig. 2. Definition of XCBC.

TMAC-family and TMAC. TMAC takes two keys $(K_1, K_2) \in \mathcal{K}_E \times \{0,1\}^n$. In general, TMAC-family is defined by not only a block cipher E but also (1) a universal hash function $H : \mathcal{K}_H \times X \to \{0,1\}^n$ where $\mathcal{K}_H$ is the set of keys and X is the domain and (2) two distinct constants $\mathtt{Cst}_1, \mathtt{Cst}_2 \in X$. They must satisfy the following three conditions for sufficiently small $\epsilon_1, \epsilon_2, \epsilon_3$. (We write $H_K(\cdot)$ for $H(K, \cdot)$.)

1. $\forall y \in \{0,1\}^n, \#\{K \in \mathcal{K}_H \mid H_K(\mathtt{Cst}_1) = y\} \le \epsilon_1 \cdot \#\mathcal{K}_H$
2. $\forall y \in \{0,1\}^n, \#\{K \in \mathcal{K}_H \mid H_K(\mathtt{Cst}_2) = y\} \le \epsilon_2 \cdot \#\mathcal{K}_H$
3. $\forall y \in \{0,1\}^n, \#\{K \in \mathcal{K}_H \mid H_K(\mathtt{Cst}_1) \oplus H_K(\mathtt{Cst}_2) = y\} \le \epsilon_3 \cdot \#\mathcal{K}_H$

The algorithm of TMAC-family is described in Fig. 3 and illustrated in Fig. 4.

TMAC is obtained by letting $\mathcal{K}_H = \{0,1\}^n$, $H_K(x) = K \cdot x$, $\mathtt{Cst}_1 = \mathtt{u}$ and $\mathtt{Cst}_2 = 1$, where "$\cdot$" denotes multiplication over $\mathrm{GF}(2^n)$ (See Appendix A for details). Equivalently, TMAC is obtained by letting $H_{K_2}(\mathtt{Cst}_1) = K_2 \cdot \mathtt{u}$ and $H_{K_2}(\mathtt{Cst}_2) = K_2$. The above three conditions are satisfied with $\epsilon_1 = \epsilon_2 = \epsilon_3 = 2^{-n}$.

OMAC-family, OMAC1, and OMAC2. OMAC is a generic name for OMAC1 and OMAC2, where OMAC1 and OMAC2 take just one key $K \in \mathcal{K}_E$. In general, OMAC-family is defined by not only a block cipher E but also (1) a universal hash function $H : \{0,1\}^n \times X \to \{0,1\}^n$ where X is the domain, (2) two distinct constants $\mathtt{Cst}_1, \mathtt{Cst}_2 \in X$ and (3) an arbitrary n-bit constant $\mathtt{Cst} \in \{0,1\}^n$. (The set of keys of H is $\{0,1\}^n$.) They must satisfy the following six conditions for sufficiently small $\epsilon_1, \epsilon_2, \ldots, \epsilon_6$.

1. $\forall y \in \{0,1\}^n, \#\{L \in \{0,1\}^n \mid H_L(\mathtt{Cst}_1) = y\} \le \epsilon_1 \cdot 2^n$
2. $\forall y \in \{0,1\}^n, \#\{L \in \{0,1\}^n \mid H_L(\mathtt{Cst}_2) = y\} \le \epsilon_2 \cdot 2^n$
3. $\forall y \in \{0,1\}^n, \#\{L \in \{0,1\}^n \mid H_L(\mathtt{Cst}_1) \oplus H_L(\mathtt{Cst}_2) = y\} \le \epsilon_3 \cdot 2^n$
4. $\forall y \in \{0,1\}^n, \#\{L \in \{0,1\}^n \mid H_L(\mathtt{Cst}_1) \oplus L = y\} \le \epsilon_4 \cdot 2^n$
5. $\forall y \in \{0,1\}^n, \#\{L \in \{0,1\}^n \mid H_L(\mathtt{Cst}_2) \oplus L = y\} \le \epsilon_5 \cdot 2^n$
6. $\forall y \in \{0,1\}^n, \#\{L \in \{0,1\}^n \mid H_L(\mathtt{Cst}_1) \oplus H_L(\mathtt{Cst}_2) \oplus L = y\} \le \epsilon_6 \cdot 2^n$

The algorithm of OMAC-family is described in Fig. 5 and illustrated in Fig. 6.

```
Algorithm TMAC-family_{K_1,K_2}(M)
Y[0] ← 0^n
Partition M into M[1] ··· M[m]
for i ← 1 to m − 1 do
        X[i] ← M[i] ⊕ Y[i − 1]
        Y[i] ← E_{K_1}(X[i])
X[m] ← pad_n(M[m]) ⊕ Y[m − 1]
if |M[m]| = n then X[m] ← X[m] ⊕ H_{K_2}(Cst_1)
             else X[m] ← X[m] ⊕ H_{K_2}(Cst_2)
T ← E_{K_1}(X[m])
return T
```

Fig. 3. Definition of TMAC-family.

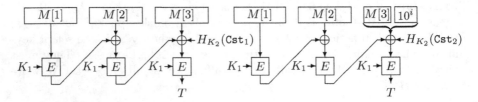

Fig. 4. Illustration of TMAC-family.

OMAC1 is obtained by letting $\mathtt{Cst} = 0^n$, $H_L(x) = L \cdot x$, $\mathtt{Cst}_1 = \mathtt{u}$ and $\mathtt{Cst}_2 = \mathtt{u}^2$, where "·" denotes multiplication over $\mathrm{GF}(2^n)$. Equivalently, OMAC1 is obtained by letting $L = E_K(0^n)$, $H_L(\mathtt{Cst}_1) = L \cdot \mathtt{u}$ and $H_L(\mathtt{Cst}_2) = L \cdot \mathtt{u}^2$. OMAC2 is the same as OMAC1 except for $\mathtt{Cst}_2 = \mathtt{u}^{-1}$. Equivalently, OMAC2 is obtained by letting $L = E_K(0^n)$, $H_L(\mathtt{Cst}_1) = L \cdot \mathtt{u}$ and $H_L(\mathtt{Cst}_2) = L \cdot \mathtt{u}^{-1}$. The above six conditions are satisfied with $\epsilon_1 = \cdots = \epsilon_6 = 2^{-n}$ for both OMAC1 and OMAC2.

3   Stronger Security Bounds

3.1   Definitions of Security

Our definitions follow from [1,6,12]. Let $\mathrm{Perm}(n)$ denote the set of all permutations on $\{0,1\}^n$. We say that P is a random permutation if P is randomly chosen from $\mathrm{Perm}(n)$.

The security of a block cipher E can be quantified as $\mathsf{Adv}_E^{\mathrm{prp}}(t, q)$, the maximum advantage that an adversary $\mathcal{A}$ can obtain when trying to distinguish $E_K(\cdot)$ (with a randomly chosen key K) from a random permutation $P(\cdot)$, where the maximum is over all adversaries who run in time at most t, and make at most q queries to an oracle (which is either $E_K(\cdot)$ or $P(\cdot)$). This advantage is defined as follows.

```
Algorithm OMAC-family_K(M)
L ← E_K(Cst)
Y[0] ← 0^n
Partition M into M[1]···M[m]
for i ← 1 to m − 1 do
        X[i] ← M[i] ⊕ Y[i − 1]
        Y[i] ← E_K(X[i])
X[m] ← pad_n(M[m]) ⊕ Y[m − 1]
if |M[m]| = n then X[m] ← X[m] ⊕ H_L(Cst_1)
            else X[m] ← X[m] ⊕ H_L(Cst_2)
T ← E_K(X[m])
return T
```

Fig. 5. Definition of OMAC-family.

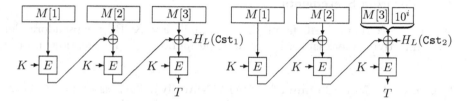

Fig. 6. Illustration of OMAC-family.

$$\begin{cases} \mathsf{Adv}_E^{\mathsf{prp}}(\mathcal{A}) \stackrel{\text{def}}{=} \left| \Pr(K \stackrel{R}{\leftarrow} \mathcal{K}_E : \mathcal{A}^{E_K(\cdot)} = 1) - \Pr(P \stackrel{R}{\leftarrow} \mathrm{Perm}(n) : \mathcal{A}^{P(\cdot)} = 1) \right| \\ \mathsf{Adv}_E^{\mathsf{prp}}(t, q) \stackrel{\text{def}}{=} \max_{\mathcal{A}} \{ \mathsf{Adv}_E^{\mathsf{prp}}(\mathcal{A}) \} \end{cases}$$

We say that a block cipher E is secure if $\mathsf{Adv}_E^{\mathsf{prp}}(t, q)$ is sufficiently small (prp stands for PseudoRandom Permutation).

Similarly, a MAC algorithm is a map $F : \mathcal{K}_F \times \{0,1\}^* \to \{0,1\}^n$, where $\mathcal{K}_F$ is a set of keys and we write $F_K(\cdot)$ for $F(K, \cdot)$. We say that an adversary $\mathcal{A}^{F_K(\cdot)}$ *forges* if $\mathcal{A}$ outputs $(M, F_K(M))$ where $\mathcal{A}$ never queried M to its oracle $F_K(\cdot)$. Then we define the advantage as

$$\begin{cases} \mathsf{Adv}_F^{\mathsf{mac}}(\mathcal{A}) \stackrel{\text{def}}{=} \Pr(K \stackrel{R}{\leftarrow} \mathcal{K}_F : \mathcal{A}^{F_K(\cdot)} \text{ forges}) \\ \mathsf{Adv}_F^{\mathsf{mac}}(t, q, \sigma) \stackrel{\text{def}}{=} \max_{\mathcal{A}} \{ \mathsf{Adv}_F^{\mathsf{mac}}(\mathcal{A}) \} \end{cases}$$

where the maximum is over all adversaries who run in time at most t, and make at most q queries, having aggregate length of at most σ blocks, where the aggregate length of q queries $M^{(1)}, \ldots, M^{(q)}$ is $\sigma = \sum_{1 \le i \le q} \|M^{(i)}\|_n$. We say that a MAC algorithm is secure if $\mathsf{Adv}_F^{\mathsf{mac}}(t, q, \sigma)$ is sufficiently small.

Let $\mathrm{Rand}(*, n)$ denote the set of all functions from $\{0,1\}^*$ to $\{0,1\}^n$. This set is given a probability measure by asserting that a random element R of

Rand$(*, n)$ associates to each string $M \in \{0,1\}^*$ a random string $R(M) \in \{0,1\}^n$. Then we define the advantage as

$$
\begin{cases}
\mathrm{Adv}_F^{\mathrm{viprf}}(\mathcal{A}) \stackrel{\mathrm{def}}{=} \left| \Pr(K \stackrel{R}{\leftarrow} \mathcal{K}_F : \mathcal{A}^{F_K(\cdot)} = 1) - \Pr(R \stackrel{R}{\leftarrow} \mathrm{Rand}(*, n) : \mathcal{A}^{R(\cdot)} = 1) \right| \\
\mathrm{Adv}_F^{\mathrm{viprf}}(t, q, \sigma) \stackrel{\mathrm{def}}{=} \max_{\mathcal{A}} \left\{ \mathrm{Adv}_F^{\mathrm{viprf}}(\mathcal{A}) \right\}
\end{cases}
$$

where the maximum is over all adversaries who run in time at most t, make at most q queries, having aggregate length of at most σ blocks. We say that a MAC algorithm is pseudorandom if $\mathrm{Adv}_F^{\mathrm{viprf}}(t, q, \sigma)$ is sufficiently small (viprf stands for Variable-length Input PseudoRandom Function).

Without loss of generality, adversaries are assumed to never ask a query outside the domain of the oracle, and to never repeat a query.

3.2   Theorem Statements

We first prove that OMAC-family, TMAC-family and XCBC are pseudorandom if the underlying block cipher is a random permutation P (information-theoretic result).

Lemma 3.1 (Main Lemma for OMAC-family). *Suppose that H, Cst_1 and Cst_2 satisfy the conditions in Sec. 2.3 for some sufficiently small $\epsilon_1, \ldots, \epsilon_6$, and let Cst be an arbitrarily n-bit constant. Suppose that a random permutation $P \in \mathrm{Perm}(n)$ is used in OMAC-family as the underlying block cipher. Let $\mathcal{A}$ be an adversary which asks at most q queries, having aggregate length of at most σ blocks. Assume $\sigma \leq 2^n/2$. Then*

$$
\begin{aligned}
\Big| \Pr(P \stackrel{R}{\leftarrow} \mathrm{Perm}(n) &: \mathcal{A}^{\mathrm{OMAC\text{-}family}_P(\cdot)} = 1) \\
&- \Pr(R \stackrel{R}{\leftarrow} \mathrm{Rand}(*, n) : \mathcal{A}^{R(\cdot)} = 1) \Big| \leq \frac{\sigma^2}{2} \cdot \left(\frac{5}{2^n} + 3\epsilon \right),
\end{aligned} \tag{2}
$$

where $\epsilon = \max\{\epsilon_1, \ldots, \epsilon_6\}$.

Lemma 3.2 (Main Lemma for TMAC-family). *Suppose that H, Cst_1 and Cst_2 satisfy the conditions in Sec. 2.3 for some sufficiently small $\epsilon_1, \epsilon_2, \epsilon_3$. Suppose that a random permutation $P \in \mathrm{Perm}(n)$ is used in TMAC-family as the underlying block cipher. Let $\mathcal{A}$ be an adversary which asks at most q queries, having aggregate length of at most σ blocks. Assume $\sigma \leq 2^n/2$. Then*

$$
\begin{aligned}
\Big| \Pr(P \stackrel{R}{\leftarrow} \mathrm{Perm}(n), K_2 \stackrel{R}{\leftarrow} \mathcal{K}_H &: \mathcal{A}^{\mathrm{TMAC\text{-}family}_{P,K_2}(\cdot)} = 1) \\
&- \Pr(R \stackrel{R}{\leftarrow} \mathrm{Rand}(*, n) : \mathcal{A}^{R(\cdot)} = 1) \Big| \leq \frac{\sigma^2}{2} \cdot \left(\frac{5}{2^n} + \epsilon \right),
\end{aligned} \tag{3}
$$

where $\epsilon = \max\{\epsilon_1, \epsilon_2, \epsilon_3\}$.

Lemma 3.3 (Main Lemma for XCBC). *Suppose that a random permutation* $P \in Perm(n)$ *is used in XCBC as the underlying block cipher. Let $\mathcal{A}$ be an adversary which asks at most q queries, having aggregate length of at most σ blocks. Assume $\sigma \leq 2^n/2$. Then*

$$\left| \Pr(P \xleftarrow{R} Perm(n), K_2, K_3 \xleftarrow{R} \{0,1\}^n : \mathcal{A}^{\mathrm{XCBC}_{P,K_2,K_3}(\cdot)} = 1) \right.$$
$$\left. - \Pr(R \xleftarrow{R} Rand(*, n) : \mathcal{A}^{R(\cdot)} = 1) \right| \leq \frac{3\sigma^2}{2^n} . \tag{4}$$

Main part of the proofs are given in Sec. 4 (for OMAC), and Sec. 5 (for TMAC and XCBC), respectively.

Given the above lemmas, it is standard to pass to the following complexity-theoretic result (For example, see [1, Section 3.2]). It shows that OMAC, TMAC and XCBC are pseudorandom if the underlying block cipher is secure.

Corollary 3.1. *Let $E : \mathcal{K}_E \times \{0,1\}^n \to \{0,1\}^n$ be the underlying block cipher used in OMAC, TMAC and XCBC. Then*

- $\mathrm{Adv}_{\mathrm{OMAC}}^{\mathrm{viprf}}(t, q, \sigma) \leq \dfrac{4\sigma^2}{2^n} + \mathrm{Adv}_E^{\mathrm{prp}}(t', q')$, *where $t' = t + O(\sigma)$ and $q' = \sigma + 1$,*
- $\mathrm{Adv}_{\mathrm{TMAC}}^{\mathrm{viprf}}(t, q, \sigma) \leq \dfrac{3\sigma^2}{2^n} + \mathrm{Adv}_E^{\mathrm{prp}}(t', q')$, *where $t' = t + O(\sigma)$ and $q' = \sigma$, and*
- $\mathrm{Adv}_{\mathrm{XCBC}}^{\mathrm{viprf}}(t, q, \sigma) \leq \dfrac{3\sigma^2}{2^n} + \mathrm{Adv}_E^{\mathrm{prp}}(t', q')$, *where $t' = t + O(\sigma)$ and $q' = \sigma$.*

Finally, we obtain the following theorem in the usual way (For example, see [1, Proposition 2.7]). It shows that OMAC, TMAC and XCBC are secure as MACs if the underlying block cipher is secure.

Theorem 3.1. *Let $E : \mathcal{K}_E \times \{0,1\}^n \to \{0,1\}^n$ be the underlying block cipher used in OMAC, TMAC and XCBC. Then*

- $\mathrm{Adv}_{\mathrm{OMAC}}^{\mathrm{mac}}(t, q, \sigma) \leq \dfrac{4\sigma^2 + 1}{2^n} + \mathrm{Adv}_E^{\mathrm{prp}}(t', q')$, *where $t' = t + O(\sigma)$ and $q' = \sigma + 1$,*
- $\mathrm{Adv}_{\mathrm{TMAC}}^{\mathrm{mac}}(t, q, \sigma) \leq \dfrac{3\sigma^2 + 1}{2^n} + \mathrm{Adv}_E^{\mathrm{prp}}(t', q')$, *where $t' = t + O(\sigma)$ and $q' = \sigma$, and*
- $\mathrm{Adv}_{\mathrm{XCBC}}^{\mathrm{mac}}(t, q, \sigma) \leq \dfrac{3\sigma^2 + 1}{2^n} + \mathrm{Adv}_E^{\mathrm{prp}}(t', q')$, *where $t' = t + O(\sigma)$ and $q' = \sigma$.*

4   Proof for OMAC-family

Intuitively, proof of Lemma 3.1 works as follows: We first construct MOMAC (Modified OMAC) and show that MOMAC and OMAC are indistinguishable. We then show that MOMAC is a secure MAC. Since MOMAC and OMAC are indistinguishable, OMAC is a secure MAC.

The improvement of the security bound for OMAC comes from the improvement of the security bound for MOMAC. To show the improved security

```
Algorithm MOMAC-E_{P_1,P_2}(M)
Partition M into M[1] ··· M[m]
X[1] ← M[1]
Y[1] ← P_1(X[1])
for i ← 2 to m − 1 do
        X[i] ← M[i] ⊕ Y[i − 1]
        Y[i] ← P_2(X[i])
X[m] ← M[m] ⊕ Y[m − 1]
return X[m]
```

Fig. 7. Definition of MOMAC-E. Note that $|M| = mn$ for some $m \geq 2$.

bound for MOMAC, we consider MOMAC-E (Modified OMAC without final encryption). It takes a message M such that $|M| = mn$ for some $m \geq 2$. It uses two independent random permutations $P_1, P_2 \in \mathrm{Perm}(n)$. The algorithm MOMAC-E$_{P_1,P_2}(\cdot)$ is described in Fig. 7.

The improved security bound for MOMAC (and Lemma 3.1) can be proved by using the following simple lemma.

Lemma 4.1 (MOMAC-E Collision Bound). *Let* $q, m_1, \ldots, m_q$ *and* σ *be integers such that* $m_i \geq 2$, $\sigma = m_1 + \cdots + m_q$, *and* $\sigma \leq 2^n/2$. *Let* $M^{(1)}, \ldots, M^{(q)}$ *be fixed and distinct bit strings such that* $|M^{(i)}| = m_i n$. *Then the probability of collision,*

$$\Pr(P_1, P_2 \xleftarrow{R} \mathrm{Perm}(n) :$$
$$1 \leq {}^\exists i < {}^\exists j \leq q, \mathit{MOMAC\text{-}E}_{P_1,P_2}(M^{(i)}) = \mathit{MOMAC\text{-}E}_{P_1,P_2}(M^{(j)}))$$

is at most $\frac{(\sigma-q)^2}{2^n}$.

A proof of this lemma is given in [9]. The rest of proof of Lemma 3.1 is similar to the proof of Lemma 5.1 in [8]. A full proof is in [9].

5   Proofs for TMAC-family and XCBC

Similarly to the proof of Lemma 3.1, proof of Lemma 3.2 (resp. Lemma 3.3) works as follows: We first consider FCBC defined by Black and Rogaway [3] and show that FCBC and TMAC (resp. XCBC) are indistinguishable. Since FCBC is a secure MAC, TMAC (resp. XCBC) is a secure MAC.

The improvement of the security bound for TMAC (resp. XCBC) comes from the improvement of the security bound for FCBC. To show the improved security bound for FCBC, we consider CBC-E (CBC MAC without final encryption). It takes a message M such that $|M| = mn$ for some $m \geq 1$. It is obtained from the CBC MAC by removing the final encryption. More precisely, the algorithm CBC-E$_P(\cdot)$ is described in Fig. 8, where $P \in \mathrm{Perm}(n)$ is a random permutation.

> **Algorithm** CBC-E$_P(M)$
> $Y[0] \leftarrow 0^n$
> Partition M into $M[1] \cdots M[m]$
> **for** $i \leftarrow 1$ **to** $m - 1$ **do**
> $\qquad X[i] \leftarrow M[i] \oplus Y[i - 1]$
> $\qquad Y[i] \leftarrow P(X[i])$
> $X[m] \leftarrow M[m] \oplus Y[m - 1]$
> **return** $X[m]$

Fig. 8. Definition of CBC-E.

The improved security bound for FCBC (and Lemma 3.2 and Lemma 3.3) can be proved by using the following simple lemma.

Lemma 5.1 (CBC-E Collision Bound). *Let q, $m_1, \ldots, m_q$ and σ be integers such that $m_i \geq 1$, $\sigma = m_1 + \cdots + m_q$, and $\sigma \leq 2^n/2$. Let $M^{(1)}, \ldots, M^{(q)}$ be fixed and distinct bit strings such that $|M^{(i)}| = m_i n$. Then*

$$\Pr(P \stackrel{R}{\leftarrow} \mathrm{Perm}(n) : 1 \leq {}^{\exists}i < {}^{\exists}j \leq q, \mathrm{CBC\text{-}E}_P(M^{(i)}) = \mathrm{CBC\text{-}E}_P(M^{(j)})) \leq \frac{\sigma^2}{2^n} .$$

A proof of this lemma is given in [9]. The rest of proofs of Lemma 3.2 and 3.3 are similar to the proof Lemma 5.1 in [10] and Theorem 4 in [3], respectively. Full proofs are in [9].

References

1. M. Bellare, J. Kilian, and P. Rogaway. The security of the cipher block chaining message authentication code. *JCSS*, Vol. 61, No. 3, pp. 362–399, 2000. Earlier version in *Advances in Cryptology — CRYPTO '94, LNCS 839*, pp. 341–358, Springer-Verlag, 1994.
2. A. Berendschot, B. den Boer, J. P. Boly, A. Bosselaers, J. Brandt, D. Chaum, I. Damgård, M. Dichtl, W. Fumy, M. van der Ham, C. J. A. Jansen, P. Landrock, B. Preneel, G. Roelofsen, P. de Rooij, and J. Vandewalle. Final Report of RACE Integrity Primitives. *LNCS 1007*, Springer-Verlag, 1995.
3. J. Black and P. Rogaway. CBC MACs for arbitrary-length messages: The three key constructions. *Advances in Cryptology — CRYPTO 2000, LNCS 1880*, pp. 197–215, Springer-Verlag, 2000.
4. FIPS Publication 46-3. Data Encryption Standard (DES). U. S. Department of Commerce / National Institute of Standards and Technology, October 25, 1999.
5. FIPS 113. Computer data authentication. Federal Information Processing Standards Publication 113, U. S. Department of Commerce / National Bureau of Standards, National Technical Information Service, Springfield, Virginia, 1994.
6. O. Goldreigh, S. Goldwasser and S. Micali. How to construct random functions. *J. ACM*, Vol. 33, No. 4, pp. 792–807, October 1986.
7. ISO/IEC 9797-1. Information technology — security techniques — data integrity mechanism using a cryptographic check function employing a block cipher algorithm. International Organization for Standards, Geneva, Switzerland, 1999. Second edition.

8. T. Iwata and K. Kurosawa. OMAC: One-Key CBC MAC. Pre-proceedings of *Fast Software Encryption, FSE 2003*, pp. 137–161, 2003. To appear in *LNCS*, Springer-Verlag. See http://crypt.cis.ibaraki.ac.jp/omac/omac.html.
9. T. Iwata and K. Kurosawa. Stronger security bounds for OMAC, TMAC and XCBC. Long version of this paper. Available at Cryptology ePrint Archive, Report 2003/082, http://eprint.iacr.org/. See also http://crypt.cis.ibaraki.ac.jp/omac/omac.html.
10. K. Kurosawa and T. Iwata. TMAC: Two-Key CBC MAC. *Topics in Cryptology — CT-RSA 2003, LNCS 2612*, pp. 33–49, Springer-Verlag, 2003.
11. R. Lidl and H. Niederreiter. Introduction to finite fields and their applications, revised edition. Cambridge University Press, 1994.
12. M. Luby and C. Rackoff. How to construct pseudorandom permutations from pseudorandom functions. *SIAM J. Comput.*, Vol. 17, No. 2, pp. 373–386, April 1988.
13. National Institute of Standards and Technology (NIST). Modes of operation for symmetric key block ciphers. http://csrc.nist.gov/CryptoToolkit/modes/.
14. E. Petrank and C. Rackoff. CBC MAC for real-time data sources. *J.Cryptology*, Vol. 13, No. 3, pp. 315–338, Springer-Verlag, 2000.

A   The Field with 2^n Points

We interchangeably think of a point a in $GF(2^n)$ in any of the following ways:

1. as an abstract point in a field;
2. as an n-bit string $a_{n-1} \cdots a_1 a_0 \in \{0,1\}^n$;
3. as a formal polynomial $a(u) = a_{n-1}u^{n-1} + \cdots + a_1u + a_0$ with binary coefficients.

To add two points in $GF(2^n)$, take their bitwise XOR. We denote this operation by $a \oplus b$.

Multiplication. To multiply two points, fix some irreducible polynomial $f(u)$ having binary coefficients and degree n. To be concrete, choose the lexicographically first polynomial among the irreducible degree n polynomials having a minimum number of coefficients. We list some indicated polynomials (See [11, Chapter 10] for other polynomials).

$$\begin{cases} f(u) = u^{64} + u^4 + u^3 + u + 1 & \text{for } n = 64, \\ f(u) = u^{128} + u^7 + u^2 + u + 1 & \text{for } n = 128, \text{ and} \\ f(u) = u^{256} + u^{10} + u^5 + u^2 + 1 & \text{for } n = 256. \end{cases}$$

To multiply two points $a \in GF(2^n)$ and $b \in GF(2^n)$, regard a and b as polynomials $a(u) = a_{n-1}u^{n-1} + \cdots + a_1u + a_0$ and $b(u) = b_{n-1}u^{n-1} + \cdots + b_1u + b_0$, form their product $c(u)$ where one adds and multiplies coefficients in $GF(2)$, and take the remainder when dividing $c(u)$ by $f(u)$.

Note that it is particularly easy to multiply a point $a \in \{0,1\}^n$ by u. We show a method for $n = 128$, where $f(u) = u^{128} + u^7 + u^2 + u + 1$. Then multiplying $a = a_{127} \cdots a_1 a_0$ by u yields a product $a_{127}u^{128} + a_{126}u^{127} + \cdots + a_1u^2 + a_0u$.

Thus, if $a_{127} = 0$, then $a \cdot \mathbf{u} = a \ll 1$. If $a_{127} = 1$, then we must add $\mathbf{u}^{128}$ to $a \ll 1$. Since $\mathbf{u}^{128} + \mathbf{u}^7 + \mathbf{u}^2 + \mathbf{u} + 1 = 0$ we have $\mathbf{u}^{128} = \mathbf{u}^7 + \mathbf{u}^2 + \mathbf{u} + 1$, so adding $\mathbf{u}^{128}$ means to xor by $0^{120}10000111$. In summary, when $n = 128$,

$$a \cdot \mathbf{u} = \begin{cases} a \ll 1 & \text{if } a_{127} = 0, \\ (a \ll 1) \oplus 0^{120}10000111 & \text{otherwise.} \end{cases}$$

Division. Also, note that it is easy to divide a point $a \in \{0,1\}^n$ by $\mathbf{u}$, meaning that one multiplies a by the multiplicative inverse of $\mathbf{u}$ in the field: $a \cdot \mathbf{u}^{-1}$. We show a method for $n = 128$. Then multiplying $a = a_{127} \cdots a_1 a_0$ by $\mathbf{u}^{-1}$ yields a product $a_{127}\mathbf{u}^{126} + a_{126}\mathbf{u}^{125} + \cdots + a_2\mathbf{u} + a_1 + a_0\mathbf{u}^{-1}$. Thus, if $a_0 = 0$, then $a \cdot \mathbf{u}^{-1} = a \gg 1$. If $a_0 = 1$, then we must add $\mathbf{u}^{-1}$ to $a \gg 1$. Since $\mathbf{u}^{128} + \mathbf{u}^7 + \mathbf{u}^2 + \mathbf{u} + 1 = 0$ we have $\mathbf{u}^{127} = \mathbf{u}^6 + \mathbf{u} + 1 + \mathbf{u}^{-1}$, so adding $\mathbf{u}^{-1} = \mathbf{u}^{127} + \mathbf{u}^6 + \mathbf{u} + 1$ means to xor by $10^{120}1000011$. In summary, when $n = 128$,

$$a \cdot \mathbf{u}^{-1} = \begin{cases} a \gg 1 & \text{if } a_0 = 0, \\ (a \gg 1) \oplus 10^{120}1000011 & \text{otherwise.} \end{cases}$$

Progressive Verification:
The Case of Message Authentication
(Extended Abstract)

Marc Fischlin[*]

Department of Computer Science & Engineering,
University of California, San Diego, USA
mfischlin@cs.ucsd.edu
http://www-cse.ucsd.edu/~mfischlin/

Abstract. We introduce the concept of progressive verification for cryptographic primitives like message authentication codes, signatures and identification. This principle overcomes the traditional property that the verifier remains oblivious about the validity of the verified instance until the full verification procedure is completed. Progressive verification basically says that the more work the verifier invests, the better can the verifier approximate the decision.

In this work we focus on message authentication. We present a comprehensive formal framework and describe several constructions of such message authentication codes, called pv-MACs (for progressively verifiable MACs). We briefly discuss implications to other areas like signatures and identification but leave it as an open problem to find satisfactory solutions for these primitives.

1 Introduction

Cryptographic primitives like signatures, message authentication codes or identification involve verification procedures that assure the verifier of the validity of the input. This means that the verifier performs a certain number of verification steps and finally outputs a reliable decision; the error probability of this decision is usually negligible.

Consider the following experiment. We start the verification algorithm on some instance. After some t steps, e.g., after half of the full verification, we stop the algorithm and ask for a decision about the correctness of the given instance. In this case, most verification procedures cannot predict the result better than before the start when $t = 0$. We call this all-or-nothing verification: in order to give a reliable decision one must either run the full verification procedure or need not start at all. The situation is given in the left part of Figure 1.

The idea of progressive verification, as displayed in the right part of Figure 1, is to relate the error probability of the decision to the running time of the verifier.

---

[*] This work was supported by the Emmy Noether Programme Fi 940/1-1 of the German Research Foundation (DFG).

T. Johansson and S. Maitra (Eds.): INDOCRYPT 2003, LNCS 2904, pp. 416–429, 2003.

Namely, progressive verification ensures that the confidence grows with the work the verifier invests. Put differently,

> The error probability of the verifier's decision decreases nontrivially with the number of performed steps of the verification procedure.

Note that *the aim of progressive verification in general is to save time for both valid and invalid inputs*. Specifically, the verifier can choose a confidence level at the outset, prematurely terminate the verification according to this level, and can finally decide about validity of the given input. Also, *the concept* is not limited to message authentication (on which we focus here) but rather *applies to verification procedures in general*.

Assuming that the verifier never rejects correct inputs we can reconceive progressive verification as a method to spot flawed inputs prematurely. Specifically, to turn such an error detection procedure into a progressive verification algorithm let the verifier, if asked at some point during the verification, simply predict authenticity of the input if no error has been found yet. Viewed this way the paradigm then says: the likeliness of rejecting fallacious inputs grows nontrivially with the performed work. Indeed we will usually adopt this more convenient viewpoint.

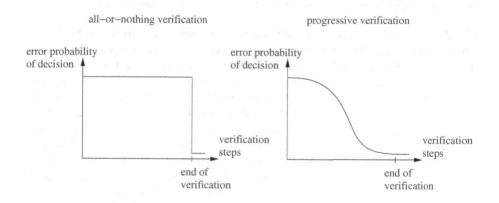

Fig. 1. Idea of Progressive Verification

In this work we introduce the concept of progressive verification by virtue of message authentication codes. Nowadays, all popular constructions of MACs first divide the message into blocks, then apply a pseudorandom or universal hash function to each block (possibly inserting the result from one evaluation into the input of another), and finally combine the result to a short MAC. Examples include XOR MAC [3], UMAC [5], XCBC MAC [6] and PMAC [7].

In case of such block-oriented MACs, verification progress can refer to the number of message blocks that have been inspected and for which the block function has been applied. Progressive verification then says that it suffices to look at, say, 75% of the blocks, and still be able to detect incorrect inputs quite often. Here, however, the limitations of progressive verification show. If the adversary changes only a single block then it is likely that the verifier will not be able to notice errors quite early. This is particularly true if the adversary controls the order of blocks in which the verifier receives them (for example, if the adversary delays the corresponding IP package until all other blocks have been received).

Still, the aim of progressive verification of MACs is not to spot errors with overwhelming probabilty instantaneously. It suffice to reduce the number of processed blocks for an accurate or fallacious input *on the average* —as long as the additional effort for the progressive verifiability does not outweigh this. In particular, the size of MACs should not increase too much and the extra work to compute and verify the progressively verifiable MAC should not grow significantly compared to an ordinary MAC. For example, doubling the number of block cipher invocations in order to reduce the workload for incorrect inputs by 50% is arguable.

Obviously, the overhead for progressive verification does not pay off for short messages of a few blocks only. Therefore progressive verification aims at applications where large messages are involved (like for authentication of large files); in such cases a few additional block ciphers calls merely add neglectable overhead, and doubling or tripling the short MAC of a few bits is usually not critical.

Our Results. We provide a formal framework for progressively verifiable MACs. This includes stringent definitions and discussions. Then we present two constructions of progressively verifiable MACs; a third one has been omitted for space reasons. One of these solutions will serve as an example to get acquainted to this field. The other one provides a reasonably good solution, allowing to spot errors after about 50% − 55% of the message while increasing the MAC size by a factor of roughly three. Note that 50% is a lower bound for such algorithms if the adversary inserts a single incorrect block only in a valid message-MAC pair: the verifier will access this block and be able to detect the error only after 50% on the average. However, if the adversary is supposed to tamper more blocks then one can go below this bound. We will touch this issue and implications to other areas like signatures at the end.

Related Work. Partially checking the validity of signatures or MACs has already been used in the context of incremental cryptography [1,2]. Although the primary interest there is the fast *computation* of signatures and MACs of related messages, local checks turned out to be useful countermeasure against so-called substitution attacks. However, the results in [9,10] indicate that checking the local validity of signatures or MACs by accessing only a few blocks yields impractically large checksums. Indeed, the idea here is similar to the incremental case: try to detect errors by inspecting only a part of the message blocks. Luckily, our aim is to detect messages as soon as possible with some (possibly small, yet) noticeable probability. By this, we can somewhat bypass the results in [9,10] which do not give useful lower bounds for such cases.

Interestingly, some identification systems already have the property of being progressively verifiable. Namely, if the protocol proceeds in sequential rounds and in each round a cheating prover will be caught with some small probability, say, 1/2. Then there is some chance that the verifier will not have to run the whole protocol when communicating with an adversary. For instance, some identification protocols like [8] use this technique of repeating atomic protocols in order to reduce the soundness error. However, such protocols are most times superseded by identification schemes running in a single round and providing the same soundness level, like [12,14]. Although the latter schemes are not known to support progressive verification their superior running time characteristics certainly beat the benefits of the progressive verifiability of the former schemes.

Progressive verification also raises the issue of timing attacks [13]. That is, the adversary may submit messages with fake MACs or signatures and deduce useful information from the verifier's respond time (e.g., how many blocks have been processed). Such attacks must be treated with care. In fact, we can easily extend our model such that the adversary actually learns the number of accessed blocks before rejection. Alternatively, one can try to avoid such attacks completely, i.e., by replying to the sender only after a predetermined amount of time, independent whether an error has been detected early or not.

Finally, in a related paper [11] we show how to improve the verification time for hash chains via progressive verification. In that example, the verifier determines a variable security bound at the beginning of the verification and then merely performs a corresponding fraction of the hash function evaluations, for both valid and input inputs.

Organization. In Section 2 we define progressive verifiable (pv) MACs. In Section 3 we present two constructions of such pv-MACs. In Section 4 we discuss extensions of our model as well as applications of the concept to other areas like signature schemes.

2   Definition

The definition of progressive verification mimics the one of adaptive chosen-message attacks on MAC schemes. The basic difference is how the verification proceeds and what the adversary's success goal is. Before dipping into the technical details we give a brief overview and then fill in the details.

Overview. For the definition we will modify the well-known adaptive chosen-message attacks scenario on message authentication codes. Let $\mathcal{A}$ be an adversary mounting such a chosen-message attack, i.e., $\mathcal{A}$ has access to oracles MAC and VF producing, respectively, verifying MACs with some secret key. Here we substitute the oracle VF by a so-called progressive verification step which we specify below.

Upon termination, in classical forgeability attacks the adversary's task is to come up with a new message and a valid MAC for this message. As for progressive verification, the adversary $\mathcal{A}$ now aborts the experimental phase and engages in another final progressive verification for a new message. We call this the adversary's *attempt*.

For the attempt we measure the probability that the verifier accesses at most m of n blocks of the adversarial message before detecting an error, i.e., after a fraction $p = \frac{m}{n}$ of all blocks. We will denote this probability by $\Delta_{\mathcal{A}}^{\mathsf{order}}(p)$. The superscript order indicates the order of accesses, i.e., if determined by the verifier, at random or by the adversary. Note that, in this sense, $1 - \Delta_{\mathcal{A}}^{\mathsf{order}}(1)$ denotes the probability of $\mathcal{A}$ forging a MAC for a new message.

For a progressively verifiable MAC we will demand that the detection probability is positive for some $p < 1$. This should hold for any adversary $\mathcal{A}$ bounded by a certain running time and query complexity. This means that for any such bounded adversary the verifer sometimes spots incorrect inputs before reading the whole message. Note that this definition is quite liberal: it suffices for example to identify incorrect inputs at the second to last block with some very small probability. However, the higher the probability for small fractions, the better of course.

Progressive Verification Protocol. We now specify the progressive verification procedure. We define it as an interactive protocol between $\mathcal{A}$ and a verifier $\mathcal{V}$ involving a third party $\mathcal{T}$. This party $\mathcal{T}$ is considered to be trustworthy, i.e., follows its prescribed program and does not cooperate maliciously with the adversary. *We note that $\mathcal{T}$ is a virtual party introduced for definitional reasons only; the actual progressive verification is of course carried out locally at the verifier's site.*

Instructively, one may think of $\mathcal{T}$ as the communication channel from the adversary's output chain to the verifier's memory. In particular, the order in which the verifier is able to process the message blocks can depend on:

- the verifier's choice, e.g., if the verifier loads the whole message into the memory and then decides on the order. Here, the verifier's choice may be fixed in advance (nonadaptive) or depend on intermediate results (adaptive). We denote these orders by V.nonad and V.adapt, respectively.
- on random delays, say, depending on the transportation over the Internet. We denote this order by T.rnd where we presume that the distribution is clear from the context. If not stated differently, then we assume the uniform distribution on the set of all orders (although we do not claim that this is the appropriate description of delays of Internet packages).
- the adversary's decision, e.g., if the adversary delays the packages maliciously according to a nonadaptive behavior (A.nonad). Concerning adaptive choices of the adversary we refer for the discussion following the formal description of progressive verification steps.

The fully specified progressive verification protocol is given in Figure 2. At the beginning of the procedure $\mathcal{A}$ commits to a message $M^* = m_1^* || \ldots || m_n^*$ of n blocks (each block, possibly except for the final one, consisting of B bits). $\mathcal{A}$ also sends a putative MAC τ^* to $\mathcal{T}$ who forwards τ^* and the length $|M^*|$ of the message to $\mathcal{V}$. Then, in each of the n rounds, $\mathcal{T}$ delivers one of the message blocks to the verifier. The order of the blocks is determined accoding to the parameter order, i.e., chosen by the verifier, the adversary or at random by $\mathcal{T}$. At the end of each round $\mathcal{V}$ can return a REJECT notification to $\mathcal{T}$. Party $\mathcal{T}$ then hands the final decision to the adversary.

Progressive Verification for order $\in$ {V.nonad, V.adapt, T.rnd, A.nonad}

- $\mathcal{A}$ submits message $M^* = m_1^* || \ldots || m_n^*$ of n blocks and putative MAC τ^* to $\mathcal{T}$
- $\mathcal{T}$ forwards $|M^*|$ and τ^* to $\mathcal{V}$
- $\mathcal{V}$ may send REJECT to $\mathcal{T}$ and stop; $\mathcal{T}$ then sends REJECT to $\mathcal{A}$

- // DETERMINE ORDER FOR NONADAPTIVE AND RANDOM ORDERS:
        if order = V.nonad then $\mathcal{V}$ sends a permutation π to $\mathcal{T}$
        if order = A.nonad then $\mathcal{A}$ sends a permutation π to $\mathcal{T}$
        if order = T.rnd   then $\mathcal{T}$ generates a random permutation π

- for $j = 1$ to n do   // LET $\Pi_{j-1}^c = \{1, \ldots, n\} - \{\pi(1), \ldots, \pi(j-1)\}$

    • // DETERMINE NEXT BLOCK FOR ADAPTIVE ORDER:
        if order = V.adapt then $\mathcal{V}$ submits $\pi(j) \in \Pi_{j-1}^c$ to $\mathcal{T}$

    • $\mathcal{T}$ forwards $(\pi(j), m_{\pi(j)}^*)$ to $\mathcal{V}$
    • $\mathcal{V}$ may send REJECT to $\mathcal{T}$ and stop; $\mathcal{T}$ then forwards REJECT to $\mathcal{A}$

- if $\mathcal{V}$ has not stopped yet then $\mathcal{V}$ sends ACCEPT to $\mathcal{T}$ who sends ACCEPT to $\mathcal{A}$

Fig. 2. Progressive Verification Protocol

There are numerous variations to the progressive verification protocol (which we do not investigate in detail in this extended abstract here). For example, one could let $\mathcal{T}$ not pass the length $|M^*|$ of the message to $\mathcal{V}$ at the outset. Also note that one could demand that $\mathcal{V}$ determines the order of accesses before even seeing the MAC τ^*; indeed, in two of our solutions $\mathcal{V}$ actually bases the order on τ^*. Also, to capture timing attacks, one may let $\mathcal{T}$ forward the decision together with the round number j to the adversary upon rejection.

Furthermore, in each loop $\mathcal{A}$ could be informed by $\mathcal{T}$ about the block number sent to $\mathcal{V}$. In this case, one could for example let the adversary at the end of each round decide to alter message blocks that have not been delivered yet. Depending on whether the message length is given to $\mathcal{V}$ in advance one could then also let the adaptive adversary decide to change the length of the message during the rounds (but such that the length never drops below a previously delivered block number). Or, instead of letting the adversary decide maliciously where to put incorrect message blocks, those message blocks can be disturbed accidently.

Restrictions on the Attempt. We restrict the input that the adversary can use in her final attempt after the experimental phase. Namely, we demand that the adversary's message M^* has not been submitted to MAC earlier. This is the classical restriction from unforgeability definitions of MACs. Here, this allows to measure how well the progressive verifcation works for tampered inputs.

Detection Probabilities. Let $\mathbb{Q}_{[0,1]} := \mathbb{Q} \cap [0,1]$. For $p \in \mathbb{Q}_{[0,1]}$ denote by $\Delta_{\mathcal{A}}^{\text{order}}(p)$ the probability that $\mathcal{V}$ outputs REJECT after having accessed $m = \lfloor pn \rfloor$ blocks in the attempt with adversary $\mathcal{A}$. Obviously $\Delta_{\mathcal{A}}^{\text{order}}(0) = 0$. Moreover, $\Delta_{\mathcal{A}}^{\text{order}}(1)$ equals the probability that the adversary will be caught at all. We call $\Delta_{\mathcal{A}}^{\text{order}}(p)$ the *cumulative detection probability* as it describes the probability of finding errors with a fraction of p *or even less* blocks.

The cumulative detection probability $\Delta_{\mathcal{A}}^{\text{order}}$ is defined with respect to a specific adversary $\mathcal{A}$. To extend this defintion to sets of adversaries we parameterize adversaries by a vector (t, q, b) describing the running time t of the adversary (in some fixed computational model), the number of queries $q = (q_{\text{MAC}}, q_{\mathcal{V}})$ to MAC and $\mathcal{V}$, and the maximum number of blocks $b = (b_{\text{MAC}}, b_{\mathcal{V}})$ in each submission. Such an adversary is called (t, q, b)-*bounded*.

We let $\Delta_{(t,q,b)}^{\text{order}}(p)$ be a function such that for any (t, q, b)-bounded adversary $\mathcal{A}$ we have

$$\Delta_{(t,q,b)}^{\text{order}}(p) \leq \Delta_{\mathcal{A}}^{\text{order}}(p) \qquad \text{for all } p \in \mathbb{Q}_{[0,1]}.$$

Note that with $\Delta_{(t,q,b)}^{\text{order}}(p)$ the function $\Delta_{(t,q,b)}^{\text{order}}(p)/2$ for example is also an appropriate function. Of course, we usually seek the best security bound. It is also straightforward to derive a classical asymptotic definition relating the adversary's parameters polynomially to a security parameter.

We remark that there are some subtle points in the definition here. For example, if we consider an adversary $\mathcal{A}$ that always outputs a message of two blocks in her attempt, then $\Delta_{\mathcal{A}}^{\text{order}}(p) = 0$ for $p < 0.5$ because $\lfloor np \rfloor = 0$ for such values $n = 2$, $p < 0.5$. Therefore this bound carries over to the set of adversaries and $\Delta_{(t,q,b)}^{\text{order}}(p) = 0$ for $p < 0.5$. Thus, usually some restriction on the length of the message blocks applies, say, the progressive verification procedure is only run on messages of $n \geq 100$ blocks to provide a sufficient granularity of 1%. Since our solutions build on top of an ordinary MAC we can simply run the basic verification procedure if $n < 100$ and invoke the progressive verification for $n \geq 100$.

Formally, we can include a lower bound $b_{\mathcal{V}}^L$ of the number of blocks for progressive verification runs in the bound (t, q, b) on the adversary. Then we can "adjust" the small error caused by the granularity by subtracting $1/b_{\mathcal{V}}^L$ from $\Delta_{(t,q,b)}^{\text{order}}(p)$.

Defining Progressively Verifiable MACs. Each "ordinary" MAC can be considered as a pv-MAC with $\Delta_{(t,q,b)}^{\text{order}}(p) = 0$ for $p < 1$. We thus rule out such trivial solutions in the following definition; yet, we merely demand that premature error detection may happen sometimes:

Definition 1. *Let* order $\in \{$V.nonad, V.adapt, T.rnd, A.nonad$\}$. *Then a message authentication code is* $(t, \boldsymbol{q}, \boldsymbol{b})$-*progressively verifiable with respect to* order *if there is a function* $\Delta^{\text{order}}_{(t,\boldsymbol{q},\boldsymbol{b})}$ *with*

$$\Delta^{\text{order}}_{(t,\boldsymbol{q},\boldsymbol{b})}(p) > 0$$

for some $p < 1$.

It is again easy to infer an asymptotic definition.

We sometimes call a MAC which is $(t, \boldsymbol{q}, \boldsymbol{b})$-progressively verifiable with respect to some order simply a pv-MAC, prescinding the bound on the running time as well as the type of order. Accordingly, we sometimes write $\Delta^{\text{order}}(p)$ or even $\Delta(p)$ etc. if the details are irrelevant or clear from the context.

From the cumulative detection probability $\Delta(p)$ one can deduce the corresponding density function $\delta(p)$ describing the detection probability after reading exactly $m = pn$ of n blocks. This also allows to define the average number of blocks the verifier accesses. We provide the formal definitions in the full version.

3   Constructions of pv-MACs

We start with a straightforward approach to warm up. We then move on to a more sophisticated approach in Section 3.2. Interestingly, our solutions below do not exclude each other: in fact, all solutions can be combined easily to provide improved detection performance. All solutions work for a nonadaptively chosen order by the verifier.

3.1   Divide-and-Conquer Construction

The divide-and-conquer method works as follows. Assume that we are given some secure MAC scheme. For a message of n blocks (where we assume for simplicity that n is always even), instead of MACing all n blocks together, we individually compute a MAC for the first $n/2$ blocks and then for the remaining $n/2$ blocks where we prepend the fixed-length MAC of the first half to the message (possibly padded to block length). This is done with independent keys for each part. The complete MAC is given by the concatenation of both individual MACs.

We assume that the verifier determines the order of block accesses. That is, the verifier tosses a coin and, depending on the outcome, first starts to read all the blocks of the left or right half of the message. The verifier then checks the MAC of this part (by prepending the putative MAC of the left half if the right half is checked). If this MAC is valid then the verifier continues with the other half and verifies the other MAC. $\mathcal{V}$ accepts if and only if both tests succeed.

It is not hard to see that the verifier will detect errors after reading half of the input with probability 0.5, except if the adversary $\mathcal{A}$ manages to forge a MAC of the underlying scheme which happens with some small probability $\epsilon_{\mathcal{A}}$ only. We omit a formal proof in this version.

Altogether, the cumulative detection function $\Delta_{\mathcal{A}}^{\mathsf{V,nonad}}(p)$ for this progressive verification with nonadaptively chosen order is given by:

$$\Delta_{\mathcal{A}}^{\mathsf{V,nonad}}(p) = \begin{cases} 0 & \text{if } 0 \leq p < 0.5 \\ 0.5 - 2\epsilon_{\mathcal{A}} - \frac{1}{b_{\mathcal{V}}^L} & \text{if } 0.5 \leq p < 1 \\ 1 - 2\epsilon_{\mathcal{A}} - \frac{1}{b_{\mathcal{V}}^L} & \text{if } p = 1 \end{cases}$$

Neglecting $\epsilon_{\mathcal{A}}$ and the lower bound $\frac{1}{b_{\mathcal{V}}^L}$ on the block number the verifier therefore reads on the average 75% of the blocks.

There are several drawbacks nested in the divide-and-conquer approach. The main drawback is that the expected number of inspected blocks for fallacious inputs is quite large: about 75% of the input have to be processed on the average. Also, the cumulative detection function is constant in the intervals $[0, 0.5)$ and $[0.5, 1)$. This is even true if the adversary tampers more than a single block in a given message-MAC pair. Then the verifier still needs to process at least 50% of the input —even if the adversary submits a random message with a random value as MAC. It is preferrable of course to be able to detect such completely faulty inputs earlier.

There are some advantages to the divide-and-conquer method, though. First, it works with any underlying MAC scheme. Second, the probabilistic verifier decides upon the order at random for each new progressive verification run. Hence, even if the adversary, in a timing attack, gets to know the number of processed blocks in previous runs, the error in the attempt will be found with probability approximately 0.5 after half of the blocks.

Certain variations to the basic scheme apply, of course. For instance one can divide the message into three equally large parts and output three MACs allowing to spot incorrect inputs after 66.66% of the blocks and so on. However, we are still left with the problem of a mainly constant cumulative detection probability.

3.2   Partial-Intermediate-Result Construction

In this section we address a solution where the cumulative detection probability is roughly proportional to the work. For this we first recall some basics about CBC-MACs before describing our partial-intermediate-result construction.

Preliminaries. To build a MAC we essentially compute an XCBC-MAC [6] of the message, using a block cipher $E_K : \{0,1\}^B \to \{0,1\}^B$ and keys $K \in \{0,1\}^k$, $K_2, K_3 \in \{0,1\}^B$ for input/output length B. Specifically, the XCBC-MAC of a message $M = m_1 || \ldots || m_n$ of B-bit blocks $m_1, \ldots, m_n$ (with $|m_n| \leq B$) is given by:

- if $|m_n| = B$ then let $K^* = K_2$
  else if $|m_n| < B$ then let $K^* = K_3$ and pad M with 10^j for $j = B - 1 - |m_n|$

- set $C_0 = 0^B$ and for $j = 1$ to $n - 1$ compute $C_j = E_K(C_{j-1} \oplus m_i)$
- return $C = E_K(K^* \oplus C_{n-1} \oplus m_n)$ where m_n has possibly been padded

That is, one pads the message only if necessary and, depending on this padding, one inserts the key K_2 or K_3 in the final step of the CBC-MAC computations.[1]

Outline. Instead of starting the XCBC-computation with the first message block here we start at a random bit position start $\in \{0, 1, \ldots, |M| - 1\}$ of the (unpadded) message M. That is, we run the XCBC-MAC computation for the rotated message with offset start, and we also prepend start (padded to B bits) to this rotated message. The value start will later be appended in encrypted form to the original MAC as well.

During the MAC computation we write down the least significant bits of the intermediate results $C_{\lfloor j \cdot \text{width} \rfloor + 1}$ for width $= \frac{n}{1+B/2}$. We call the corresponding message blocks *check blocks*.[2] By the choice of the value width we get $B/2$ extra bits lsb which can then be encrypted as a single block together with the $B/2$ bits representing the value start. We assume a lower bound of $b_\mathcal{V}^L \geq 1 + B/2$ on the block length of progressively verified messages to ensure that all check blocks are distinct. Figure 3 shows the computation of the XCBC-MAC in our case.

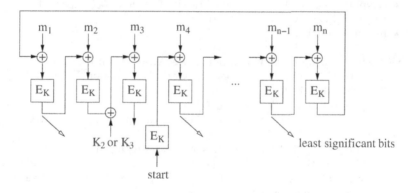

Fig. 3. Progressively Verifiable PIR-XCBC-MAC computation

In addition we compute another (CBC-)MAC σ for the XCBC-MAC C, the encryption e of lsb$||$start and the bit length of the message. In this case, a CBC-MAC suffices as we apply this MAC computation only to fixed length inputs of three blocks. Our complete MAC is given by (C, e, σ).

---

[1] We remark that any (reasonable) kind of CBC-MAC works for our construction here; we use the XCBC-MAC instead of the CBC-MAC in order to deal with variable input length right away.

[2] The reason for taking the blocks with number $\lfloor j \cdot \text{width} \rfloor + 1$ instead of $\lfloor j \cdot \text{width} \rfloor$ is that we have prepended the value start as the new first block.

To verify the MAC (C, e, σ) progressively, first check the CBC-MAC σ. If it is valid then decrypt lsb||start and start re-computing the XCBC-MAC beginning at the random position start. Each time after processing a further check block compare the least significant bit of this check block with the decrypted value; if they match then continue, else stop with output REJECT. The formal description of our construction PIR-XCBC-MAC (for *partial intermediate results*) appears in Figure 4.

PIR-XCBC-MAC for V.nonad order

− Key Generation: pick k-bit keys K and B-bit keys $K_2, K_3, K_{\text{ENC}}, K_{\text{CBC}}$

− Compute MAC for message M of n blocks:
  - pick random value start $\in \{0, 1, \ldots, |M| - 1\}$, encode with $B/2$ bits
  - rotate M by offset start and prepend block $0^{B/2}||$start to get M_{start} of $n + 1$ blocks
  - compute XCBC-MAC C of M_{start} with keys K, K_2, K_3
  - let lsb_j be least significant bit of intermediate value $C_{\lfloor j \cdot \text{width}\rfloor + 1}$ of XCBC-MAC computation, $j = 1, 2, \ldots, B/2$; let $\text{lsb} = \text{lsb}_1||\ldots||\text{lsb}_{B/2}$
  - compute $e := E_K(K_{\text{ENC}} \oplus C) \oplus \text{lsb}||$start
  - compute CBC-MAC σ for $C||e||\,|M|$ with function $E_K(K_{\text{CBC}} \oplus \cdot)$, where $|M|$ is encoded as B bit string
  - return (C, e, σ)

− Progressive Verification of M and putative MAC (C, e, σ):
  - verify CBC-MAC σ for message $C||e||\,|M|$ with function $E_K(K_{\text{CBC}} \oplus \cdot)$; if verification fails then return REJECT
  - compute lsb||start $:= E_K(K_{\text{ENC}} \oplus C) \oplus e$
  - compute M_{start} from M as for MAC computation, using offset start
  - verify MAC:
    * compute XCBC-MAC of M_{start} with keys K, K_2, K_3
    * if during this computation the least significant bit of some intermediate value $C_{\lfloor j \cdot \text{width}\rfloor + 1}$ does not match the previously decrypted value lsb_j then stop and output REJECT
    * finally verify that the computed XCBC-MAC matches the given one C; if not then output REJECT

Fig. 4. Progressively verifiable XCBC-MAC

Design Principle. The idea of the construction is as follows. Using a CBC-type of MAC for M ensures that, if the adversary tampers a single block for a valid massage-MAC pair, then after this incorrect block is processed, *all* subsequent intermediate results will be essentially independent of the previously computed values. This is also referred to as the error propagation property of CBC-like

MACs. In particular, this means that the least significant bits of these intermediate results are likely to be independent from the encoded ones. Hence, soon after we process a fallacious block we will find the error with probability 50%, and with probability 25% after an additional check block and so on.

Starting at a random position with the computation and verification ensures that the adversary remains oblivious about the order of the blocks and thus cannot deliberately tamper the block which will be processed last. MACing the length of the message together with C and e basically prevents attacks in which the adversary appends an extra block to messages, e.g., if the adversary appends a block $11\ldots1$ to the message $m = 11\ldots1\|\ldots\|11\ldots1$ such that the verifier will not access a truly tampered block during computations. Similarly, prepending the value start to the rotated message prevents that two cyclically shifted messages (say, $00\ldots0\|11\ldots1$ and $11\ldots1\|00\ldots0$) of the same length are accidently rotated to the same message (e.g., to $00\ldots0\|11\ldots1$).

The following theorem says that the cumulative detection probability grows linear with the number of accessed blocks. Hence, the detection probability is roughly uniformly distributed. Furthermore, the proof of the theorem also shows that the MAC is unforgeable in the classical sense.

Theorem 1. *PIR-XCBC-MAC is a $(t, \boldsymbol{q}, \boldsymbol{b})$-progressively verifiable MAC with respect to order V.nonad. The cumulative detection probability is given by*

$$\Delta^{\text{V.nonad}}_{(t,\boldsymbol{q},\boldsymbol{b})}(p) = p - \frac{4}{B} - bad$$

where

$$bad \le \boldsymbol{Adv}^{prp}_{(t',q')} + \frac{18.5(q_{\text{MAC}} + q_{\mathcal{V}})^2 + 6(q_{\text{MAC}} + q_{\mathcal{V}})^2(b_{\text{MAC}} + b_{\mathcal{V}} + 5)^2}{2^B} + \frac{1}{b_{\mathcal{V}}^L}$$

for the advantage $\boldsymbol{Adv}^{prp}_{(t',q')}$ of any adversary running in time $t' = t + \mathcal{O}(BS)$ and making at most $q' = q_{\text{MAC}}(b_{\text{MAC}} + 5) + q_{\mathcal{V}}(b_{\mathcal{V}} + 5)$ queries to distinguish the block cipher E from a truly random permutation.

The bound $\Delta^{\text{V.nonad}}_{(t,\boldsymbol{q},\boldsymbol{b})}(p)$ essentially stems from the fact that the adversary is oblivious about our random start position. Hence, the verification algorithm usually approaches an incorrect block after processing a fraction of $\frac{m}{n}$ blocks. Then it takes a few more check blocks to detect the error, namely, 2 on the average which are processed after another fraction of $\frac{2\text{width}}{n} \approx \frac{4}{B}$ blocks. The amount bad originates from the (in)security of the deployed cryptographic schemes. Conclusively, the average number of blocks is roughly $50\% + 4/B$. The verifier thus reads on the average about 53% of an invalid submission for $B \ge 128$.

The proof as well as an alternative construction is omitted for space reasons.

4   Discussion

Our solutions show that progressive verification can be achieved. We have focused on message authentication and presented reasonable constructions of pv-MACs. Yet, some open problems remain, both in the field of MACs and other cryptographic primitives. In this section we give some possible future directions.

Improved Solutions for pv-MACs. We have already observed that our model of progressive verification can be varied, say, by allowing the adversary to adaptively decide upon the message based on the verifier's order. Finding good solutions for such cases is still open. Similarly, all our constructions rely on (non-adaptively) chosen orders by the verifier. Coming up with nontrivial solutions for random or adversarial orders is a challenging task.

Another interesting extension is to investigate the relationship between the number of incorrect and accessed blocks. For example, in our model the quality of the progressive verification procedure is measured against adversaries that need to tamper only a single block in a given message. This immmediately yields a lower bound of 50% for the average number of block accesses. It would be interesting to see

- how our constructions perform in the case that there are more incorrect blocks, possibly distributed over the message according to some specific distribution, and
- if better solutions can be found if the number of fallacious blocks must exceed a certain lower bound; this may be especially interesting for parallelizable computations of MACs where the "avalanche" effect of CBC-MACs disappears.

Progressive Verification for other Primitives. Concerning other cryptographic primitives we remark that one can transfer the solutions to the hash-and-sign principle. Namely, assume that an iterated hash functions like SHA or RIPEMD is applied to the message. These hash functions process the message similarly to CBC-MACs, and we can output the least significant bits of check blocks in addition the hash value. Then we sign the actual hash value as well as the extra output, e.g., with a number-theoretic signature function. The complete signature is given by the output of the signature function and the additional vector.

What is the advantage of the constructions in this case? Assume that the message is large such that the signing time is dominated by the hash evaluation. Say that the adversary is supposed to change one of the first blocks. Then the adversary can in fact bias the intermediate values by choosing the message blocks adaptively. In fact, she will be able to find a message such that the progressive verification procedure has to re-compute at least most parts of the hash computation. Nevertheless, in order to fool the verifier drastically the adversary has to invest some work first. This leads to a more balanced workload between the adversary and the verifier. For example, this may have applications for protection against denial-of-service attacks.

Finally, we remark that it would also be interesting to find nontrivial signature schemes in which the underlying number-theoretic function is somewhat progressively verifiable.

Acknowledgment

We thank the anonymous reviewers for their comments.

References

1. M. BELLARE, O. GOLDREICH, S. GOLDWASSER: Incremental Cryptography: The Case of Hashing and Signing, *Advances in Cryptology — Proceedings of Crypto '94, Lecture Notes in Computer Science, Vol. 839, pp. 216–233, Springer-Verlag*, 1994.
2. M. BELLARE, O. GOLDREICH, S. GOLDWASSER: Incremental Cryptography and Application to Virus Protection, *Proceedings of the 27th ACM Symposium on Theory of Computing (STOC), pp. 45–56*, 1995.
3. M. BELLARE, R. GUERIN, P. ROGAWAY: XOR MACs: New Methods for Message Authentication Using Finite Pseudorandom Funtions, *Advances in Cryptology — Proceedings of Crypto '95, Lecture Notes in Computer Science, Vol. 963, pp. 15–29, Springer-Verlag*, 1995.
4. M. BELLARE, J. KILIAN, P. ROGAWAY: The Security of Cipher Block Chaining Message Authentication Code, *Advances in Cryptology — Proceedings of Crypto '94, Lecture Notes in Computer Science, Vol. 839, pp. 341–358, Springer-Verlag*, 1994.
5. J. BLACK, S. HALEVI, H. KRAWCZYK, T. KROVETZ, P. ROGAWAY: UMAC: Fast and Secure Message Authentication, *Advances in Cryptology — Proceedings Crypto '99, Lecture Notes in Computer Science, vol. 1666, pp. 216–233, Springer-Verlag*, 1999.
6. J. BLACK, P. ROGAWAY: CBC MACs for Arbitrary-Length Messages: The Three-Key Construction, *Advances in Cryptology — Proceedings of Crypto 2000, Lecture Notes in Computer Science, Vol. 1880, pp. 197–215, Springer-Verlag*, 2000.
7. J. BLACK, P. ROGAWAY: A Block-Cipher Mode of Operation for Parallelizable Message Authentication, *Advances in Cryptology — Proceedings of Eurocrypt 2002, Lecture Notes in Computer Science, Vol. 2332, pp. 384–397, Springer-Verlag*, 2002.
8. U. FEIGE, A. FIAT, A. SHAMIR: Zero Knowledge Proofs of Identity, *Journal of Cryptology, Vol. 1, pp. 77–94, Springer-Verlag*, 1988.
9. M. FISCHLIN: Incremental Cryptography and Memory Checkers, *Advances in Cryptology — Proceedings of Eurocrypt '97, Lecture Notes in Computer Science, Vol. 1233, pp. 393–408, Springer-Verlag*, 1997.
10. M. FISCHLIN: Lower Bounds for the Signature Size of Incremental Schemes, *Proceedings of the 38th IEEE Symposium on Foundations of Computer Science (FOCS), pp. 438–447, IEEE Computer Society Press*, 1997.
11. M. FISCHLIN: Fast Verification of Hash Chains, *to appear in RSA Security 2004 Cryptographer's Track, Lecture Notes in Computer Science, Springer-Verlag*, 2004.
12. L.C. GUILLOU, J.-J. QUISQUATER: A Practical Zero-Knowledge Protocol Fitted to Security Microprocessors Minimizing Both Transmission and Memory, *Advances in Cryptology — Proceedings of Eurocrypt '88, Lecture Notes in Computer Science, Vol. 330, pp. 123–128, Springer-Verlag*, 1988.
13. P.C. KOCHER: Timing Attacks on Implementations of Diffie-Hellman, RSA, DSS, and Other Systems, *Advances in Cryptology — Proceedings of Crypto '96, Lecture Notes in Computer Science, Vol. 1109, pp. 104–113, Springer-Verlag*, 1996.
14. C.P. SCHNORR: Efficient Signature Generation by Smart Cards, *Journal of Cryptology, Vol. 4, pp. 161–174*, 1991.

Author Index

Lecture Notes in Computer Science

For information about Vols. 1–2821
please contact your bookseller or Springer-Verlag

Vol. 2859: B. Apolloni, M. Marinaro, R. Tagliaferri (Eds.), Neural Nets. Proceedings, 2003. X, 376 pages. 2003.

Vol. 2860: D. Geist, E. Tronci (Eds.), Correct Hardware Design and Verification Methods. Proceedings, 2003. XII, 426 pages. 2003.

Vol. 2861: C. Bliek, C. Jermann, A. Neumaier (Eds.), Global Optimization and Constraint Satisfaction. Proceedings, 2002. XII, 239 pages. 2003.

Vol. 2862: D. Feitelson, L. Rudolph, U. Schwiegelshohn (Eds.), Job Scheduling Strategies for Parallel Processing. Proceedings, 2003. VII, 269 pages. 2003.

Vol. 2863: P. Stevens, J. Whittle, G. Booch (Eds.), «UML» 2003 – The Unified Modeling Language. Proceedings, 2003. XIV, 415 pages. 2003.

Vol. 2864: A.K. Dey, A. Schmidt, J.F. McCarthy (Eds.), UbiComp 2003: Ubiquitous Computing. Proceedings, 2003. XVII, 368 pages. 2003.

Vol. 2865: S. Pierre, M. Barbeau, E. Kranakis (Eds.), Ad-Hoc, Mobile, and Wireless Networks. Proceedings, 2003. X, 293 pages. 2003.

Vol. 2867: M. Brunner, A. Keller (Eds.), Self-Managing Distributed Systems. Proceedings, 2003. XIII, 274 pages. 2003.

Vol. 2868: P. Perner, R. Brause, H.-G. Holzhütter (Eds.), Medical Data Analysis. Proceedings, 2003. VIII, 127 pages. 2003.

Vol. 2869: A. Yazici, C. Şener (Eds.), Computer and Information Sciences – ISCIS 2003. Proceedings, 2003. XIX, 1110 pages. 2003.

Vol. 2870: D. Fensel, K. Sycara, J. Mylopoulos (Eds.), The Semantic Web - ISWC 2003. Proceedings, 2003. XV, 931 pages. 2003.

Vol. 2871: N. Zhong, Z.W. Raś, S. Tsumoto, E. Suzuki (Eds.), Foundations of Intelligent Systems. Proceedings, 2003. XV, 697 pages. 2003. (Subseries LNAI)

Vol. 2873: J. Lawry, J. Shanahan, A. Ralescu (Eds.), Modelling with Words. XIII, 229 pages. 2003. (Subseries LNAI)

Vol. 2874: C. Priami (Ed.), Global Computing. Proceedings, 2003. XIX, 255 pages. 2003.

Vol. 2875: E. Aarts, R. Collier, E. van Loenen, B. de Ruyter (Eds.), Ambient Intelligence. Proceedings, 2003. XI, 432 pages. 2003.

Vol. 2876: M. Schroeder, G. Wagner (Eds.), Rules and Rule Markup Languages for the Semantic Web. Proceedings, 2003. VII, 173 pages. 2003.

Vol. 2877: T. Böhme, G. Heyer, H. Unger (Eds.), Innovative Internet Community Systems. Proceedings, 2003. VIII, 263 pages. 2003.

Vol. 2878: R.E. Ellis, T.M. Peters (Eds.), Medical Image Computing and Computer-Assisted Intervention - MICCAI 2003. Part I. Proceedings, 2003. XXXIII, 819 pages. 2003.

Vol. 2879: R.E. Ellis, T.M. Peters (Eds.), Medical Image Computing and Computer-Assisted Intervention - MICCAI 2003. Part II. Proceedings, 2003. XXXIV, 1003 pages. 2003.

Vol. 2880: H.L. Bodlaender (Ed.), Graph-Theoretic Concepts in Computer Science. Proceedings, 2003. XI, 386 pages. 2003.

Vol. 2881: E. Horlait, T. Magedanz, R.H. Glitho (Eds.), Mobile Agents for Telecommunication Applications. Proceedings, 2003. IX, 297 pages. 2003.

Vol. 2883: J. Schaeffer, M. Müller, Y. Björnsson (Eds.), Computers and Games. Proceedings, 2002. XI, 431 pages. 2003.

Vol. 2884: E. Najm, U. Nestmann, P. Stevens (Eds.), Formal Methods for Open Object-Based Distributed Systems. Proceedings, 2003. X, 293 pages. 2003.

Vol. 2885: J.S. Dong, J. Woodcock (Eds.), Formal Methods and Software Engineering. Proceedings, 2003. XI, 683 pages. 2003.

Vol. 2886: I. Nyström, G. Sanniti di Baja, S. Svensson (Eds.), Discrete Geometry for Computer Imagery. Proceedings, 2003. XII, 556 pages. 2003.

Vol. 2887: T. Johansson (Ed.), Fast Software Encryption. Proceedings, 2003. IX, 397 pages. 2003.

Vol. 2888: R. Meersman, Zahir Tari, D.C. Schmidt et al. (Eds.), On The Move to Meaningful Internet Systems 2003: CoopIS, DOA, and ODBASE. Proceedings, 2003. XXI, 1546 pages. 2003.

Vol. 2889: Robert Meersman, Zahir Tari et al. (Eds.), On The Move to Meaningful Internet Systems 2003: OTM 2003 Workshops. Proceedings, 2003. XXI, 1096 pages. 2003.

Vol. 2891: J. Lee, M. Barley (Eds.), Intelligent Agents and Multi-Agent Systems. Proceedings, 2003. X, 215 pages. 2003. (Subseries LNAI)

Vol. 2893: J.-B. Stefani, I. Demeure, D. Hagimont (Eds.), Distributed Applications and Interoperable Systems. Proceedings, 2003. XIII, 311 pages. 2003.

Vol. 2894: C.S. Laih (Ed.), Advances in Cryptology - ASIACRYPT 2003. Proceedings, 2003. XIII, 543 pages. 2003.

Vol. 2895: A. Ohori (Ed.), Programming Languages and Systems. Proceedings, 2003. XIII, 427 pages. 2003.

Vol. 2897: O. Balet, G. Subsol, P. Torguet (Eds.), Virtual Storytelling. Proceedings, 2003. XI, 240 pages. 2003.

Vol. 2898: K.G. Paterson (Ed.), Cryptography and Coding. Proceedings, 2003. IX, 385 pages. 2003.

Vol. 2899: G. Ventre, R. Canonico (Eds.), Interactive Multimedia on Next Generation Networks. Proceedings, 2003. XIV, 420 pages. 2003.

Vol. 2901: F. Bry, N. Henze, J. Maluszyński (Eds.), Principles and Practice of Semantic Web Reasoning. Proceedings, 2003. X, 209 pages. 2003.

Vol. 2902: F. Moura Pires, S. Abreu (Eds.), Progress in Artificial Intelligence. Proceedings, 2003. XV, 504 pages. 2003. (Subseries LNAI).

Vol. 2903: T.D. Gedeon, L.C.C. Fung (Eds.), AI 2003: Advances in Artificial Intelligence. Proceedings, 2003. XVI, 1075 pages. 2003. (Subseries LNAI).

Vol. 2904: T. Johansson, S. Maitra (Eds.), Progress in Cryptology – INDOCRYPT 2003. Proceedings, 2003. XI, 431 pages. 2003.

Vol. 2905: A. Sanfeliu, J. Ruiz-Shulcloper (Eds.), Progress in Pattern Recognition, Speech and Image Analysis. Proceedings, 2003. XVII, 693 pages. 2003.

Vol. 2916: C. Palamidessi (Ed.), Logic Programming. Proceedings, 2003. XII, 520 pages. 2003.